Development
of
Aircraft Certification
Requirements

Richard L Newman

and

Madeleine Kolb

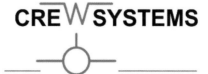

Published by
Crew Systems
Post Office Box 25054
, Seattle, Washington, USA.

ISBN: 978-1-66786-180-7 (paperback)
ISBN: 978-1-66786-791-5 (hard cover)
ISBN: 978-1-66786-181-4 (e-book)

Table of Contents

PART III EQUIPMENT APPROVAL AND CERTIFICATION PROCEDURES

PART IV: CURRENT STATUS OF AIRCRAFT CERTIFICATION

List of Figures

List of Tables

Acknowledgements

It is difficult to identify the many colleagues and mentors who have influenced the ideas presented in this book. We have indeed been fortunate in the sheer number of individuals who have helped us along the way, To those whom we have forgotten, please accept our apologies and know that we appreciate your efforts.

There are, however, a few who deserved mention by name: Dale Dunford, Loran Haworth, Joe Jacobsen, Robert C. Jones, and Gerry Lakin. The many interesting, occasionally heated, discussions with these colleagues at the FAA's Transport Airplane Directorate -- helped focus our thoughts on the goals of certification.

We'd like to acknowledge other colleagues: Mark Anderson, Dr Chris Belcastro, Mike Borfitz, Steve Boyd, Tom Carr, Shawn Coyle, Dennis Cryder, Dave Ellis, John Foster, Capt Tom Foxworth, Dave Gollings, Chris Grable, Kevin Greeley, Dr Loren Groff, Fred Hoerner, Dave Klyde, Dr Ron Kruk, Dr Brian Lee, Dennis Newton, Dr Angus Rupert, Dr Jerry Snyder, and Capt Robert Sumwalt.

We also had mentors in our early aviation career, chronologically: Capt Jerry Fredrickson, Al Wilson, Mark Baldwin, and Dr Gerry Gregorek.

These colleagues and mentors are primarily those of only one of us. Unfortunately, Madeleine Kolb passed away before this book was completed. Other than our colleagues at the FAA mentioned above, I can only mention those that I can remember those that she mentioned. These include Susan Boyland and Annete Kovite of the FAA. She was also always citing Professor Bradley of the University of Washington.

One former colleague comes to mind for a special vote of thanks. Don Bateman was the driving force behind the widespread incorporation of Ground Proximity Warning System (GPWS), which reduced then rate of Controlled Flight Into Terrain accidents by 99.8%. The discussion of GPWS begins on page 143. This book is respectfully dedicated to Don Bateman, who has saved more lives than anyone in the history of aviation.

C. D. "Don" Bateman*

About The Authors

Madeleine Kolb

Ms Madeleine Kolb received her bachelor's degree in Zoology and her master's degree in Genetics from the University of Washington. After serving in the Peace Corps, she was employed at the Massachusetts Executive Office of Environmental Affairs (EOEA). Working in various programs within the EOEA, Madeleine prepared technical reports, status reports, draft and final regulations, findings, master plans, etc.

Later, as Project Manager at GZA, an engineering consulting firm near Boston, she prepared over 300 technical reports for a variety of clients, including Teradyne, General Dynamics, Biogen, Boston University, and the Roman Catholic Archdiocese of Boston.

After her father died, Madeleine returned to Seattle and to the UW, earning a Certificate in Technical Writing and Editing. As a technical writer and editor at the FAA, she worked with aeronautical engineers and others to create a variety of technical, regulatory documents.

Madeleine was active in Toastmasters and volunteered to give presentations at work at every opportunity. One was on the occasion of the 100th Anniversary of the Wright Brothers Flight in 1903 or the retirement lunch of a fellow worker whom she roasted—in a light-hearted, teasing sort of way.

After retiring from the FAA, Madeleine began writing blog posts, completed the requirements for her Distinguished Toastmasters (DTM) award, competed in writing and speaking contests, taking flying lessons, and, of course, began writing this book with her husband.

Richard L Newman

Dr Richard L. (Dick) Newman received his bachelor's degree in chemistry from Rensselaer Polytechnic Institute. Following graduation, he began graduate studies at Penn State University. After completing all-but-dissertation for his doctorate, Dick left school to fly for Northwest Airlines. He left airline flying because of its adverse affect on domestic life and helped design wind tunnels. At the same time, Dick completed the requirements for a master's degree in engineering from Purdue.

Dick was then employed as a contract flight test engineer, first on an US Air Force program and then on a Canadian STOL project in Ottawa. Canada. When the latter program ended, he returned to the States and began working as an independent contractor on both FAA and Air Force contracts. He was able to use an Air Force Contract Report on head-up display design requirements as his Ph D dissertation.

Dick continued as a flight test consultant working both on and off-site for a variety of domestic and foreign organizations.

In 2001, Dick joined the faculty of Embry-Riddle Aeronautical University as an associate professor. In 2003, he was recruited to join the FAA in the Transport Standards Staff.

Following retirement from the FAA, Dick spent three years at the Naval Air Systems Command developing protocols for approving experimental military aircraft to fly in the National Airspace System (NAS)

His book *Head-Up Displays: Designing the Way Ahead* is the basic reference for head-up display design, certification, and training. A later book *Cockpit Displays: Test and Evaluation* provides the methodology for simulator and flight testing of modern flight displays.

Madeleine Kolb and Richard L Newman

Following their retirements from the FAA, Dick and Madeleine were married at the top of the Space Needle. Dick began giving Madeleine flying lessons and Madeleine introduced Dick to Toastmasters. Together they began to write this book.

Unfortunately, Madeleine passed away in 2019 and is buried at the Tahoma National Cemetery, in Kent, Washington.

Prologue

Aviation has a tradition of always having a second check or third check. Doug (the PF) calls for the landing gear to be extended, John (the PNF) lowers the landing gear and check its status. In maintenance, Joe (the mechanic) tightens the nut and Pete (the inspector) checks the torque. Does that mean one is better than the other? No, it's just that we always want a second pair of eyes on the task.

During design, Peter does the stress analysis and Dennis checks the numbers. Following the fabrication of the prototype structure, Frank will test the design to construction. Similarly, during certification, the certification authority (or its designee) will check the applicant's design, not because they're better or because they don't trust the applicant. It's because that's the way things work in aviation.

The accident investigation boards around the globe act as inspectors of the certification authorities around the world. They keep us (the certification engineers) honest. They are supporting members of the certification team.

We hope that our readers will see this teamwork between the certification authorities and the accident investigators in following pages.

PART I:
INTRODUCTION AND BACKGROUND

Chapter 1: Airworthiness Philosophy

Scope of Book

This is a book on civil aircraft airworthiness. As such we shall be primarily concerned with civil aircraft. A civil aircraft is defined by what it isn't. The International Civil Aviation Organization (ICAO) defines a civil aircraft as an aircraft other than a State Aircraft.

ICAO describes *State Aircraft* as those belonging to the military, police, or customs services of a particular country. The United States uses the term *Public Aircraft*, which is essentially any aircraft used by a federal agency or state and local agencies when operated for a government purpose. *Public Aircraft* is defined in Federal legislation. These definitions are included in Appendix I: Abbreviations and Definitions.

What is "Airworthy"?

Airworthy is defined in the regulations: *Airworthy* means that the aircraft conforms to its type design and is in a condition for safe operation.[1]* While the regulation cited applies to aircraft records and not to the aircraft itself, it is, nevertheless, the basis for the common definition of *airworthy*:†

> **Airworthy Aircraft**: An aircraft that conforms to its type design and is in a condition for safe flight.‡

By this definition, an aircraft without a Type Certificate cannot be airworthy. This is important since the whole purpose of aircraft certification is to approve a given aircraft type as meeting the appropriate airworthiness standards and to approve a particular model of that type as airworthy. Without a Type Certificate, an inspector can only state that the aircraft is "in a safe condition for flight".

Safety Standards

The next definition that is needed is one for *safety*. The military standard for system safety[2] defines *mishap*, *risk*, and *safety* as

> **Mishap**: An unplanned event or series of events resulting in death, injury, occupational illness, damage to or loss of equipment or property, or damage to the environment.

> **Risk**: An expression of the effect and possibility of a mishap in terms of potential mishap severity and probability of occurrence.

> **Safety**: Freedom from those conditions than can cause death, injury, occupational illness, damage to or loss of equipment or property, or damage to the environment.

Safety can be thought of in terms of risk. The lower the risk the greater the safety.

---

* Superscript numbers denote references listed at the end of each chapter.
† Definitions used in this book are listed in Appendix 1.
‡ We will include definitions as needed in this format where appropriate to show important definitions.

Freedom from Risk

This is the risk of injury to the general public (people on the ground), to the flying public (people in other aircraft), or to passengers and crew on our own aircraft.

There is the risk of damage to property on the ground, to other aircraft, or to our own aircraft. Normally, aircraft certification analysis does not consider damage to property on the ground. Nor is damage to property on the ground a determining factor in determining if a civilian mishap is an accident or not.[3] *

This will be of interest as we ponder the airworthiness certification of unmanned aircraft (UA). The consensus today is that ownship damage is not a concern in UA safety.

Level of Safety (Aircraft Operations)

In terms of civil aircraft operations, the following operations are listed in order of levels of safety:

✈ Airline Operations   ↑   Arrows show
✈ Air Taxi Operations ↑   increasing
✈ Business Jets   ↑   level of
✈ General Aviation   ↑   safety

There are special categories of operations which are allowed to operate with some operational restrictions, such as unpopulated areas, restrictions to daylight hours, or restrictions to visual flight conditions:

✈ Agricultural Operations
✈ Flight Test
✈ Ferry Flights

Level of Safety (Aircraft Categories)

The following are the aircraft categories:

✈ Transport Category   Airline operation assumed
✈ Commuter Category   Airline operation assumed
✈ Normal Category   Varies with engine type and weight[5]
                                        Varies with number of passengers†
✈ Restricted Category   Implied assumption of risk

Measure of Safety

There is only one valid measure of safety: the accident record. There is an axiom of aviation that we always need a second opinion. A mechanic tightens the wrench; the inspector checks the torque and the safety wire. A single accident report provides a second opinion about an individual aircraft, pilot, or mechanic. Accident statistics or trends provide second opinions on aircraft standards, pilot standards, or mechanic standards. In a sense, the accident investigation boards of the world act as the inspectors of aircraft certification.

Throughout the book, we have included references to significant accidents which influenced changes in certification rules. Appendix 4 contains descriptions of these accidents.

Basic Airworthiness

There are several aspects of airworthiness approval: Type Design Approval, Production Approval, Individual Aircraft Airworthiness Approval, Continued Airworthiness Approval, and

---

* Military mishaps do consider property on the ground in determining if the mishap is an accident.[4]
† Subsequent to amendment 23-64.[5]

Equipment Approval. The following section is based on FAA approval. Other nations will have similar procedures.

Type Design Approval

The proven record of manned aircraft shows that the industry and regulators know how to design aircraft in terms of structures, systems, propulsion, etc. They know how to tailor the requirements for different type of aircraft: small light airplanes, helicopters, large jet transports, etc. The requirements grew from problems in the past. The set of regulations grew from nothing prior to the 1920s. In that decade, the first requirements were written based on the accidents happening at the time.

Since 1965, most civil aircraft airworthiness requirements have similar organizations. Table I shows this organization.

These requirements are for approval for the design to receive a Type Certificate (TC). The TC allows individual aircraft to be approved as airworthy by the definition on page 227. The applicant (The person or organization seeking the TC) will present an application package to show compliance with the requirements. [6]

Table 1 Standard Organization of Aircraft Airworthiness Requirements

Subpart	Description of Content
General	This Subpart describes the type of aircraft being treated, such as Small Airplanes, Transport (or Large) Airplanes, Small Rotorcraft, or Transport Rotorcraft. It often includes other documents by reference, such as technical reports of other standards. It my also include special requirements.
Flight	This Subpart requires proof of compliance with the flight requirements, such as stability & control or performance. These requirements will be evaluated over the appropriate combinations of load (weight and center of gravity) and aircraft configuration by flight test on actual aircraft supplemented by calculations. The amount of testing will vary from a little range for light aircraft to an extensive range for transports.
Structure	This Subpart demonstrating that the aircraft structure be demonstrated to withstand the required loads. This will include static testing.
Design and Construction	This Subpart covers materials and workmanship. It also includes cockpit controls and pilot view. It also includes various items dealing with occupant crashworthiness protection, fire protection, and emergency evacuation.
Powerplant	This Subpart covers the powerplant installation and is usually predicated on using certified engines and propellers. It also covers the fuel system, the engine induction system and engine cooling. Engine and propeller are also included.
Equipment	This Subpart covers all installed equipment: flight instruments, automatic pilots, interior and exterior lighting, airframe ice protection, etc.
Operating Limitations and Information	This Subpart requires the applicant to establish all necessary operating limitations. It also requires the establishment of the kinds of operations for which the aircraft is approved (such day/visual, night/visual, day/instruments, etc. The applicant must also prepare an Instructions for Continued Airworthiness document.
Electrical Wiring Interconnection Systems	This Subpart organizes transport airplane design requirements for aircraft wiring into a single section of the airworthiness regulations.
Engines	This Subpart outlines airworthiness approval for the installation of non-certified engines in certain light aircraft, such as motorgliders.
Propellers	This Subpart outlines airworthiness approval for the installation of non-certified propellers in certain light aircraft, such as motorgliders.
Appendices	Typical Appendices have included • Airworthiness Criteria for Helicopter Instrument Flight, • Criteria for Helicopter Category A, • Extended Operations (ETOPS), • HIRF Environment and Equipment HIRF Testing, • Icing Certification, • Instructions for Continued Airworthiness, and • Simplified Design Load Criteria for Conventional Single Engine Airplanes of 2722 kg or Less Maximum Weight

The FAA finds compliance with the requirements and may use designated engineering representatives (DERs). Compliance finding is discretionary on the part of the FAA. In other words, the FAA need not check every paragraph or review every engineering report or test report.[7] If the application package is compliant with the regulations, the FAA must issue the TC unless there are unsafe conditions not covered be the requirements.

Production Approval
Production approval allows the TC holder (or licensee) to mass produce the aircraft model. The Production Certificate (PC) is limited to a specific model. It is primarily concerned with quality control:

→ Who are the inspectors?
→ What are the inspection procedures?
→ What happens to rejected parts
    Can they be reworked?
    How will any non-compliant parts be disposed?;
→ How are production test flights accomplished? and
→ Is there a proper audit trail?

Individual Aircraft Approval
Most aircraft will have a standard Airworthiness Certificate (AWC). Normally this will be based on having been manufactured under a Production Certificate. There is an alternative method. An approved inspector can do a conformity inspection to show the aircraft is in compliance with its TC. This is the approach taken if the owner constructs an aircraft out of spare parts.[8]
    Other FAA flight approvals include Restricted, or Experimental Airworthiness Certificates which can be issued for prescribed purposes

Continued Airworthiness
Aircraft must be maintained in an airworthy state. This will involve inspections, maintenance, and repairs as necessary. The Type Design must include Instructions for Continued Airworthiness (ICA). General maintenance standards are found in FAR 43 or EASA Part M

Airworthiness Directives (AD's)
An Airworthiness Directive (AD) is issued when an *unsafe condition exists* and it condition is likely to be found in other aircraft. Procedures are found in FAR 39. Once an unsafe condition is determined to exist, the FAA must issue an AD. Directives are often taken from manufacturer's service bulletins. The operator may propose alternate means of compliance.
    The aircraft need not have a Type Certificate to be subject to an AD. Although unusual, an AD can apply to an aircraft with an Experimental Airworthiness Certificate, if there is an unsafe condition which can occur in these aircraft.

Aircraft Certification Categories
Aircraft Categories are groupings of aircraft based on intended use.

Acrobatic Category
These are small airplanes intended for use without restrictions to types of maneuvers. The only restrictions are those indicated by flight test. They were certified under the following regulations: CAR 4, CAR 4a, CAR 3, and FAR 23. New Acrobatic Category certifications were discontinued with the adoption of amendment 23-64.[5]

Commuter Category

These are multi-engine airplanes intended for regional airline service. Commuter Category airplanes are limited to nineteen passenger seats and maximum takeoff weights (MTOW) of 19,000 lbs. They were certified under Part 23 between February 1987 and August 2017. New Commuter Category certifications were discontinued with the adoption of amendment 23-64.[5]

Light Category

Originally, Light Category airplanes were limited to MTOW of 1000 lb and wing loading of 6 psf certified under Bulletin 7[9] or CAR 4.[10] Light Category was no longer available for new certifications with the cancellation of CAR 4.

Normal Category (Airplane)

Until 1941, the Normal Category applied to all airplanes other than those certified under the Light Category. In 1941, the Acrobatic and Transport Categories were added to CAR 4.

Originally, Normal Category airplanes were airplanes intended for nonacrobatic, nonscheduled passenger and nonscheduled cargo operations, according to the definition in CAR 3.02.[11] They were certified under the following regulations: CAR 4, CAR 4a, CAR 3, and Part 23. Subsequent to amendment 23-62, *all* airplanes certified under Part 23 will be certified in Normal Category.[12]

Normal Category (Rotorcraft)

Normal Category Rotorcraft are rotorcraft with MTOW of 6,000 lb or less and nine or fewer passenger seats. After October 18, 1999, the weight limit for Normal Category Rotorcraft was raised to 7,000 lb.[13]

Transport Category Airplanes

The term Transport Category is not defined in the regulations. The FAA provides a circular definition: "Transport airplanes are airplanes for which a type certificate is applied for in the Transport Category and that meet the Transport Category airworthiness requirements".* In other words, a Transport Category airplane is one certified as a Transport Category airplane under Part 25. This is not very helpful. We propose a more useful definition:

> **Transport Category Aircraft**: Aircraft intended for commercial airline operations carrying passengers or cargo for hire.

Transport Category Rotorcraft

We have found no definition in either Part 1 or Part 29. It would seem that the above definition would be applicable.

Utility Category

These are small airplanes intended for normal operations and limited acrobatic maneuvers. These airplanes are not suited for snap or inverted maneuvers. They were certified under Part 23 (or predecessor regulations). New Utility Category certifications were discontinued with the adoption of amendment 23-64.[12]

---

* https://www.faa.gov/aircraft/air_cert/design_approvals/transport/ accessed July 31, 2017

Requirements for Different Categories

Small vs. Transport Airplanes

<u>Transport Airplane</u>: Transport airplane requirements* are very detailed.[10] [14] For example, there are detailed requirements for

- ✈ Defined takeoff path considering that, in the event of an engine failure, *at any point*, during the takeoff that the airplane can either stop safely or continue the takeoff safely;
- ✈ Detailed requirements for V_1, V_R, V_2, V_{EF}, takeoff run (TOR), takeoff distance (TOD), accelerate-stop distance (ASD);
- ✈ Climb gradients, and airplane configurations;
- ✈ Performance corrections for weight, altitude, and temperature (WAT) and runway slope; and
- ✈ Detailed list of stall speed requirements scheduled by aircraft weight and configuration.

Figure 1 shows a sketch of transport airplane takeoff.

During performance testing, we can see that the choice of the engine failure speed, V_{EF} has a strong influence on the ASD. A V_{EF} of zero would result in an ASD of zero. The ASD would increase as V_{EF} increases.

On the other hand, a V_{EF} of zero would lead to a engine-out takeoff and a long TOR and TOD. These distances would decrease as V_{EF} increases. Flight testing will provide data to find the optimum VEF for a given runway, surface wind, and air temperature.

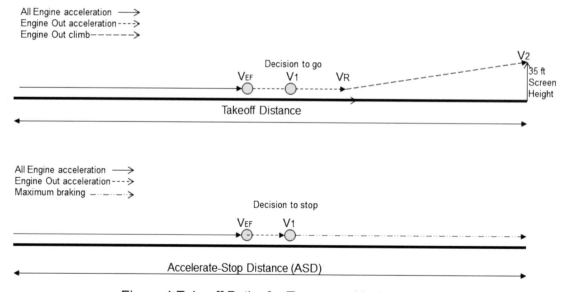

Figure 1 Takeoff Paths for Transport Airplanes

Transport airplane test procedures are objective and extremely detailed, except for workload which can be quite subjective.

Structural requirements include fatigue test and analysis, for the projected lifetime of the aircraft.

Systems analysis based on "continued safe flight and landing" following any single failure unless that failure is "extremely improbable" ($p < 10^{-9}$).

---

* Such as CAR 4b in the 1950s[10] or FAR 25 since the 1960s.[14]

<u>Small Airplanes</u>: The small airplane requirements are somewhat less detailed for airplanes less than 6,000 lb.* The performance requirements are generally less rigorous than for transport airplanes. For example, takeoff data need only be shown at sea level and runway slope correction is not required. Figure 2 shows a sketch of small airplane takeoff considerations.

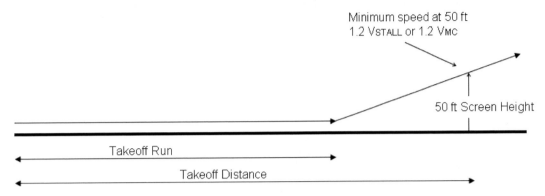

Figure 2 Takeoff Path for Small Airplanes

There is no requirement for small multi-engine airplanes to demonstrate a continued takeoff following an engine failure. In fact, there was no requirement that multi-engine airplanes have a guaranteed "stay-up" performance following an engine failure.

Single-engine airplanes and multi-engine airplanes without sufficient engine-out performance were required to have a stall-speed 61 kts or less.

Handling requirements are generally more rigorous:

+ Stalls required to have relatively benign characteristics.
+ Spins must be demonstrated for single-engine airplanes at aft CG, maximum weight, and controls limits set 1° beyond normal limits in the adverse direction

    Spins not approved: 1-turn spin recover in 1 turn

    Spins approved: 6 turn spin recover in one and a half turns.

Systems certification is based of a probability of a catastrophic accident of $p < 10^{-6}$ (for single-engine piston-powered airplanes under 6,000 lb).[16] We will be discussed this later in Chapter 7.

Small Airplanes vs. Small Helicopters

Small helicopters have some requirements based on rotary-wing characteristics. Rotor speed limits must be established to allow for safe autorotation following an engine failure. Rotor blade pitch limits must be established and a low rotor rpm warning is required to warn the pilot.

Performance data or helicopters must be provided throughout the desired envelope. For reciprocating powered helicopters, the data must be based on 80% relative humidity (RH). For turbine powered helicopters, the RH will be 80% for temperature at or below ISA decreasing linearly to 34% RH at ISA+50°F. Above ISA+50°F, the RH will be constant at 34%.

The hovering in ground effect (HIGE) ceiling must be determined over the desired WAT range. The HIGE ceiling must be at least

+ Reciprocating power: at least 4000 ft (ISA).
+ Turbine power: 2500 ft (ISA+22°F).

Takeoff performance must be determined up to the maximum altitude for which certification is desired.

---

* Such as CAR 3 in the 1950s[11] or FAR 23 since the 1960s.[15]

The minimum rate intended of descent airspeed and the best angle of glide airspeed must be determine in autorotation at maximum weight and rotor speed selected by the applicant.

At slow speeds near the ground, a helicopter will have to trade altitude for airspeed to obtain the rotor rpm to complete a safe autorotation. If there is any combination of low height and slow speed (including hover) from which a safe autorotation and landing cannot be made following an engine failure, a limiting height/speed (H/V) diagram must be established. Figure 3 shows such a diagram for a small helicopter.[17] The H/V diagram must be established from sea level to the maximum altitude capability of the aircraft or 7000 ft density altitude (DA), whichever is less.

The cross-hatched area on the left shows the region to be avoided at sea level and the single hatched area shows the additional conditions to be avoided at high altitude. Figure 3 also shows a high speed low-altitude region where the pilot has limited reaction time to avoid ground impact. The H/V diagram also shows a narrow takeoff profile.

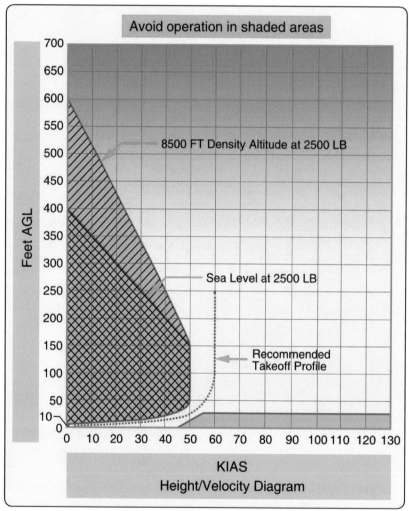

Figure 3 Exemplar Height-Velocity Diagram
From FAA's Helicopter Flying Handbook[17]

Wind velocities from all azimuths must be established without loss of control (including crosswind takeoffs, sideward flight and rearward flight. This is to be shown from sea level to the maximum takeoff and landing altitude capability of the rotorcraft or 7000 ft DA, whichever is less.

The rotorcraft must be free from ground resonance.

The design maneuvering load factor must be at least +3.5/-1.0. The rotorcraft must be able to withstand the loads resulting from a vertical gust of ±30 ft/sec.

Transport Helicopters

Category A transport helicopters have the same takeoff requirements as transport airplanes. At any point in the takeoff path, following an engine failure, the pilot can either stop within the confines of the runway or heliport or continue the takeoff with acceptable performance. Operating from an airport runway, the helicopter can do a normal takeoff down the runway until reaching a safe speed. It's much the same as in Figure 1, except that the helicopter is not rolling down the runway but is flying just above it.

The difficulty lies in operation from a helipad. In this case, the helicopter may have to hover and "back up" keeping the helipad in sight. At the decision point, the helicopter can either land back on the pad or descend and accelerate until gaining enough airspeed to climb away.[18] This is a very daunting task.

Category B transport helicopters do not have a guaranteed "stay-up capability." Like small multi-engine airplanes they must be controllable after an engine failure to a safe forced landing.

Engines and Propeller Requirements

Engine requirements list some generalizations and some details for the design and construction. Material suitability must be based on experience or test. The materials must conform to approved specifications. The engine must be designed to minimize the occurrence and spread of fire. Engine design and construction must minimize the development of unsafe conditions between overhaul periods. It must also allow for necessary cooling. Each spark ignition system must have a dual ignition system with two spark plugs for each cylinder with two separate electrical systems.

The engine must undergo a vibration survey covering from idle speed to the greater of 110% maximum cruising rpm or 103% maximum takeoff rpm. Any vibration found must not lead to stresses above the endurance limit stress of the material.

The major requirement is a 150 hour dynamometer endurance test. Operating conditions vary with the specific installation. These tests follow speed schedules which vary with engine types. The speed schedules vary between high-power and low-power settings. Following completion of the test, a teardown inspection examines parts for excessive wear or impending failure.

Propeller requirements list some generalizations and some details for the design and construction. Material suitability must be based on experience or test. The materials must conform to specifications acceptable to the Administrator.* The propeller must be designed so that no single failure can cause propeller pitch below the low pitch stop. The propeller controls must be designed, constructed, and validated to show no single failure can result in a hazardous effect. The maximum stress must be acceptable to the Administrator.

The propeller hub, blade, and counterweights must engine must be tested for 1 hour at rotational speed providing twice the centrifugal load that would be present at maximum operational rotation speed. Components attached to the propeller must be subjected to a load equivalent to 159%† of the centrifugal load that would be present at the maximum rotational speed. The major requirement is a 110 hour dynamometer endurance test.

---

*   The use of the capitalized term Administrator refers the FAA Admistrator or a person to whom he has delegated his authority in the matter concerned.[19]

†   This is the correct value in §35.35, as published in Amendment 35-8.

Comparison of International Requirements

Airworthiness requirements have been created by many agencies. This section will provide a brief summary of differences in approach among various agencies. The following discussions are not detailed review, but rather an indication of the general approach to certification of various types of aircraft.

International Civil Aviation Organization (ICAO)

The International Civil Aviation Organization (ICAO) is an agency of the United Nations. ICAO Annex 8[20] is a list of high level requirements, such as

- ✈ Takeoff performance is scheduled for pressure altitude
  (and, optionally, temperature) in no-wind conditions.
- ✈ The structure must be free from flutter within the operational
  envelope.

      Annex 8 can be thought of as a specification for National Airworthiness Requirements and is not binding on member states. Member states can deviate if they notify ICAO of differences. Such differences are published by ICAO.

Organization Scientifique et Technique Internationale du Vol a Voile

Organization Scientifique et Technique Internationale du Vol a Voile * (OSTIV) was formed in 1949 as an international nonprofit organization. Papers published in the OSTIV journal, *Technical Soaring* resulted in recommendations for sailplane airworthiness.[21] These recommendations were ultimately incorporated into JAR 22 and EASA CS 22 for sailplanes and powered sailplanes.

US Airworthiness Requirements

The major milestone in aircraft certification was the redrafting of the major airworthiness rules into FAR 21 through FAR 39. At the same time, the rules were written into parallel paragraphs using a standard style:

- ✈ FAR 21, Certification Procedures for Products and Articles
- ✈ FAR 23, Normal, Utility, Acrobatic, and Commuter Category
  Airplanes (formerly CAR 3);
- ✈ FAR 25, Transport Category Airplanes (formerly CAR 4);
- ✈ FAR 27, Normal Category Rotorcraft (formerly CAR 6);
- ✈ FAR 29, Transport Category Rotorcraft (formerly CAR 7);
- ✈ FAR 31, Manned Free Balloons;
- ✈ FAR 33, Aircraft Engines (formerly CAR 13); and
- ✈ FAR 35 , Propellers (formerly CAR 14).

      The new regulations were edited with no changes to the meaning, but with common wording and common sections to the extent possible. The individual paragraphs are numbered identically, which makes comparisons much easier. For example, the paragraph on Longitudinal Stability use paragraphs §25.173, §27.173, and §29.173. This led to the abbreviation §2x.173 for these paragraphs together. Appendix 2 lists the individual paragraphs for the major aircraft types.

      That this was a significant accomplishment is shown by the adoption of the same format and for the most part, the same wording by the Joint Airworthiness Requirements (JAR) and by the European Aviation Safety Agency (EASA). Table 2 shows the common airworthiness numbering.

---

*   English translation: *International Scientific and Technical Soaring Organization.*

Table 2 Common Airworthiness Numbering

Part	FAR Title (EASA Title)[a]	FAR	JAR	CS
Part 21	Certification Procedures for Products and Articles (Design, Certification and Production)	✓	✓	✓
Part 22	No FAR Title (Sailplanes and Powered Sailplanes	-	✓	✓
Part 23[b]	Normal, Utility, Aerobatic, and Commuter Category Airplanes (Normal, Utility, Aerobatic, and Commuter Category Aeroplanes)	✓	✓	✓
Part 23[c]	Normal Category Airplanes (Normal Category Aeroplanes)	✓	✓	✓
Part 25	Transport Category Airplanes (Large Aeroplanes)	✓	✓	✓
Part 26	Continued Airworthiness and Safety Improvements for Transport Category Airplanes (Additional Airworthiness Specifications for Operations)	✓	-	✓
Part 27	Normal Category Rotorcraft (Small Rotorcraft)	✓	✓	✓
Part 29	Transport Category Rotorcraft (Large Rotorcraft)	✓	✓	✓
Part 31	Manned Free Balloons (Gas Balloons or Hot Air Balloons)	✓	-	✓
Part 33 (Part E)	Aircraft Engines	✓	✓	✓
Part 34	Fuel Venting and Exhaust Emission Requirements for Turbine Engine Powered Airplanes (Aircraft Engine Emissions and Fuel Venting)	✓	-	✓
Part 35 (Part P)	Propellers	✓	✓	✓
Part LSA	No FAR Title (Light Sport Aeroplanes)	-	-	✓
Part VLA	No FAR Title (Very Light Aeroplanes)	-	✓	✓
Part VLR	No FAR Title (Very Light Rotorcraft)	-	✓	✓
Notes: (a) JAR or CS title in parentheses (b) Amendment 23-63 to FAR 23 (Amendment 4 to CS-23) (c) Amendment 23-64 to FAR 23 (Amendment 5 to CS-23)				

There are currently no US airworthiness standards for gliders, airships, or very light airplanes. There is a provision in F

FAR 21, §21.17,[22] that allows the FAA to use appropriate requirements from other sources. Three Advisory Circulars (AC's) provide guidance for certifying these aircraft using European requirements.[23][24][25]

British Civil Airworthiness Regulations (BCAR)

The British Civil Airworthiness Regulations (BCAR) were probabilistic in nature with prescribed maximum accident rate targets with flexibility to achieve targets. Compliance demonstration required more analysis than with the FARs. Manufactures complained that compliance finding was much more subjective than in the US.

There were fairly rigorous stall requirements in the BCAR. As a result, many foreign manufactured transport airplanes required a stick pusher installation to obtain British certification.

The BCAR were superseded by the Joint Airworthiness Requirements (JAR).

Concorde Airworthiness

The *Concorde* development was a joint effort of France and the United Kingdom.. At the outset of the program in the late 1960s, the UK airworthiness was based on BCAR, while France used what was substantially that of the US CAR 4b.[26]. The US did validate the *Concordes*' airworthiness and issued a US Type Certificate.[27]

The *Concordes* were fitted with an early analog Fly-by-Wire (FBW) system. Unlike later digital systems, the primary reason was not to tailor the flight control algorithms, but rather to compensate for the expansion of the aircraft from thermal heating during cruise.[28]

Fourteen *Concordes* flew for three airlines. British Airways and Air France flew the majority, Braniff flew an interchange service taking over from the European carriers, reregistering them in the US, and flying them on to Dallas, TX, using Braniff crews. The British or French registrations were covered with tape and a temporary "N" registration displayed. On the return flights, this procedure was replayed. The flights within the US operated at subsonic speeds.[29]

Ultimately, *Concorde* operations ended following the crash of an Air France aircraft after it ran over some debris on the runway at Paris. The wing fuel tank was damaged and caught fire and crashed. All 109 souls on board and four on the ground were killed.

Figure 4 *Concorde*, G-BOAG, taken at JFK Airport, New York, March 28, 1995*

While the *Concorde* captured the imagination of most (including your authors), it was neither a commercial nor a safety success. While it only had one fatal accident, had had a catastrophic accident rate of 4.2×10^{-6}, far below the threshold of 10^{-9}.† But it certainly was beautiful.

Joint Airworthiness Requirements (JAR)

In 1970, several European countries formed the Joint Aviation Authorities (JAA) with the purpose of developing a common certification standard for transport airplanes.‡ This common set of Joint Airworthiness Requirements (JAR) is organized with the same part numbers with identical paragraph numbering as the FAA, as shown in Table 2.

The certification process itself was run by each national Civil Aviation Authority (CAA), not by JAA. At this point in time, JAA only exists as a training organization.

European Aviation Safety Agency (EASA)

The European Aviation Safety Agency (EASA) is the European Union agency for aviation safety. Unlike JAA in which each member state performed its own certification to a common set of standards, EASA is the sole agency for aircraft certification in Europe. The EASA standards for certi-

---

*   Concorde, G-BOAG, taken at JFK Airport, New York, March 28, 1995. Downloaded from www.google.com/search?q=Concorde++Pictures+Public+Domain&ei=QbijYpCsOIrCuvQP9a6FyAI& ved=0ahUKEwjQ0K_S6qP4AhUKoY4IHXVXASkQ4dUDCA0&uact=5&oq=Concorde++Pictures+Pu blic+Domain&gs_lcp=Cgdnd3Mtd2l6EANKBAhBGAFKBAhGGABQrAxYoFZgul1oBHAAeACAA YgBiAHTFZIBBDAuMjOYAQCgAQHAAQE&sclient=gws-wiz, accessed June 10, 2022

†   See discussion in Chapter 7.

‡   https://www.skybrary.aero/index.php/JAA, accessed July 14, 2017

fication are called Certification Specifications (CS). These standards use identical organization and paragraph numbering, also shown in Table 2.

Individual Aircraft Airworthiness Certificates (AWC's)

Standard Airworthiness Certificates

Standard Airworthiness Certificates may be issued to Acrobatic, Commuter, Manned Balloons, Normal, Restricted, Transport, and Utility Category aircraft by showing conformity with their TC. They may also be issued to other types of aircraft, such as gliders, or light sport aircraft.

FAR 21, Subpart B[30] governs the issuing of TCs and Subpart H[31] governs the issuing of an individual AWC. Table 3, copied from FAA Order 8139.2J,[32] shows the categories and the types of an Airworthiness Certificate for each category.

Table 3 Summary of Airworthiness Regulations, Policies, and Procedures (From FAA Order 8139.2J[32])

Airworthiness Certificate Aircraft Category/ Aircraft Category	14 CFR Part				Policy
	Design Standard	Design Approval	Production Approval	Airworthiness Certificate	
Standard Airworthiness Certificates for:		TC per 21.21/29	PC per 21/G	21.183(a): new under 21/G	Chapter 3
Normal	23, 17			21.183(b): new under 21/F	
Utility	23			21.183(c): new import, foreign SOM	
Acrobatic	23				
Commuter	23				
Transport	25, 29			21.183(d): used & surplus military	
Manned Free Balloons	31				
Special Classes (includes UAS	21.17(b)				
Airplanes with turbine engines	34				
Special Airworthiness Certificates for:					Chapter 4
Primary	21.17(f)	21.24/29 TC	PC per 21/G	21.184	Chapter 5
Restricted	Varies	21.25/29	Varies	21.185	Chapter 6
Multiple (Restricted & __)	Varies	Varies	Varies	21.187	Chapter 7
Limited	DOD	TC	N/A	21.189	Chapter 8
SLSA	ASTM	N/A	N/A	21.190	Chapter 9
Experimental, purpose of:	N/A	N/A	N/A	21.191/3	
R&D				21.191(a)	Chapter 10
Showing Compliance				21.191(b)	Chapter 10
Crew Training				21.191(c)	Chapter 11
Exhibition				21.191(d)	Chapter 12
Air Racing				21.191(e)	Chapter 13
Market surveys				21.191(f)/195	Chapter 14
Amateur –built				21.191(g)	Chapter 15
Primary kit-built				21.191(h)	Chapter 16
ELSA				21.191(i)	Chapter 17
SFP (US-registered)	N/A	N/A	N/A	21.197/199	Chapter 18
Provisional	Varies	TC per 21/C	PC per 21/G	Part 2/l	Chapter 19
SFA (Foreign-registered)	N/A	N/A	N/A	Part 375, 91.715	Chapter 22

Special Airworthiness Certificates

Experimental Category: Aircraft can be issued an Experimental AWC for one of several specific purposes, spelled out in §21.191. These purposes are

> ✈ Research and Development;
> ✈ Showing Compliance with Airworthiness Regulations;
> ✈ Crew Training (Training of the applicant's Flight Crews);
> ✈ Exhibition at Air Shows;*
> ✈ Marketing Surveys;
> ✈ Operating Amateur-Built Aircraft; and
> ✈ Operating Kit-Built Primary Category Aircraft.[33]

Experimental aircraft are required to carry a two to four inches high placard saying "EX-PERIMENTAL" at each entrance to the cabin, cockpit, or pilot station.

Generally, the application for issuance of an experimental AWC will require a *program letter* outlining the purpose for which the aircraft will be used; the estimated time or number of flights required, and the proposed geographic area. Normally, strict geographical constraints will be imposed in operational limits.

Most Experimental AWCs will normally have a time limit (up to a year) and have strict geographical limitations. They will also have a prohibition against carrying occupants not required for the purpose of the flight.

Experimental Certificates for Amateur Built and Kit Built Primary aircraft will be restricted to a flight test area for the first 50 hours of operation. After this, they will normally have the geographical limits relaxed. Their certificates will not expire every year. They also may carry passengers.

Limited Category: These aircraft are surplus military aircraft. They may not be used to carry passengers or cargo for hire. Limited Category TCs were approved under CAR 9. New Limited TCs are no longer being approved. These aircraft are required to carry placard saying "LIMITED" ", two to four inches high, at each entrance to the cabin, cockpit, or pilot station.[34]

Light Sport Aircraft: The LSA Category covers light airplanes, gliders, powered parachutes, weight-shift control (WSC) and lighter-than air aircraft (balloons or airships). LSA must weight 1320 lb or less† and there are performance restrictions as well1. An LSA operating under a special AWC must display a two to four inches high placard," LIGHT SPORT" at each entrance to the cabin, cockpit, or pilot station

There are two restrictions that an applicant must be aware of. The aircraft must have met the LSA requirements since its original certification. This would prevent an applicant from attempting to recertify a 1350 lb aircraft by amending the TC to a 1320 lb gross weight. Also, while a gyroplane meets the definition of an LSA, Current FAA policy does not allow issuing a Special AWC for a LSA gyroplane.[32]

Primary Category: An airplane or rotorcraft may be issued a Special AWC as a Primary Category aircraft‡ under §21.24 and weighs 2700 lb or less and has a maximum of four seats. There are other restrictions for stall speed or rotor disk loading.[35] The applicant must show that the aircraft meets the applicable worthiness requirements of FAR 23, FAR 27, FAR 31, FAR 33, and FAR 35, or other criteria the FAA may find as appropriate.

The Primary Category has not been particularly successful in that there are few TCs issued. An early FAA evaluation of a Primary Category airplane has been published in the *Journal of Aircraft*.[36]

Restricted Category: These aircraft are limited to an approved special purpose operations, such as agricultural application, aerial surveying, etc. These aircraft are required to carry a two to four

---

*   Including training, practice and flying to and from events.

†   1,430 lb for water aircraft.

‡   This section does not apply to Primary Category aircraft assembled from kits which must be licensed in the Experimental Category,

inch high placard saying "RESTRICTED" at each entrance to the cabin, cockpit, or pilot station. The special purpose operations are

+ Agricultural (spraying, dusting, seeding, and animal control;
+ Forest and wildlife conservation;
+ Aerial surveying;
+ Patrolling (pipelines, powerlines, and canals);
+ Weather control (cloud seeding);
+ Aerial advertising (skywriting, banner towing, etc.); and
+ Any other operation specified by the FAA. [37]

Restricted Category aircraft can be certified under §21.25, CAR 8, or §21.29 or previously certified in another category and modified for the particular purpose. It is possible for a given aircraft to be certified in another Category and in the Restricted Category. It may be possible to switch between the two Categories if the TCDS so provides. This may be desirable, for example, to carry a ground crew member from to and from a working site. Restricted Category Aircraft will have operating limitations on carriage of passengers.

Special Flight Permits (SFP)

A special airworthiness certificate issued to a US registered aircraft that may not meet applicable airworthiness requirements but is capable of safe flight. SFPs can be issued for

+ Production test flights;
+ Customer demonstration flights;
+ Flight to a location where maintenance is to be performed;
+ Ferrying a non-air-carrier four-engine aircraft with one inoperative engine to a base where maintenance is to be performed;
+ Ferrying an airplane to the base of a purchaser or to a storage location; and
+ Operating an overweight aircraft for flight beyond the normal range over water or land areas without adequate refueling stops.

International Operations

FAA-issued special AWCs or SFPs do not authorize flight in foreign nations without that nation's concurrence. This can be as simple as a phone call and a fax transmission in each direction.

References

1   "Definitions," *General Requirements*, 14 CFR Part 3, §3.5(a), amendment 3-1, effective October 17, 2005

2   *Standard Practice for System Safety*, MIL-STD-882D, February 10, 2000

3   "Definitions," *Notification and Reporting of Aircraft Accidents or Incidents and Overdue Aircraft and Preservation of Aircraft Wreckage, Mail, Cargo, and Records*, 49CFRPart 830, §830.2

4   *DOD Accident/Mishap/Incident Classification Guide and CSSO List*, Reference DOD Instruction 6055.07, June 2011

5   *Revision of Airworthiness Standards for Normal, Utility, Acrobatic, and Commuter Category Airplanes, Final Rule*, Docket No. FAA-2015-1621; 14 CFR Part 23, amendment 23-64, effective August 30, 2017

6   *The FAA and Industry Guide to Product Certification*, Published by Aerospace Industries Association, Aircraft Electronics Association, General Aviation Manufacturers Association, and Federal Aviation Administration, Third Edition, May 2017

7   M E F Plave,, "United States v. Varig Airlines: The Supreme Court Narrows the Scope of Government Liability Under the Federal Tort Claims Act," *Journal of Air Law and Commerce*: 51 1985, 197-239

8   *Standard Airworthiness Certification of Surplus Military Aircraft and Aircraft Built from Spare and Surplus Parts*, FAA Advisory Circular AC-21-13, April 5, 1973

9   *Airworthiness Requirements for Aircraft*, US Department of Commerce, Aeronautics Bulletin 7A, effective June 1, 1928

10   *Airplane Airworthiness Transport Categories*, Civil Aeronautics Board CAR 4b, As amended to December 31, 1953

11   *Airplane Airworthiness: Normal, Utility, Acrobatic, and Restricted Purpose Categories*, Civil Aeronautics Board CAR 3, effective November 13, 1945

12   "Certification of Normal Category Airplanes," *Airworthiness Standards: Normal Category Airplanes*, 14 CFR Part 23, §23.2005, amendment 23-64, effective August 30, 2017

13   *Airworthiness Standards, Normal Category Rotorcraft*, 14 CFR Part 27, amendment 27-37 effective 10/18/1999

14   *Airworthiness Standards: Transport Category Airplanes*, 14 CFR Part 25,

15   *Airworthiness Standards; Normal, Utility, and Acrobatic Category Airplanes*, 14 CFR Part 23, Generic Reference for FAR 23 prior to amendment 23-63, March 21, 2017

16   *System Safety Analysis and Assessment for Part 23 Airplanes*, FAA Advisory Circular 23.1309-E, 11/17/2011

17   *Helicopter Flying Handbook*, FAA-H-8083-21B, 2019

18   S Coyle, *Cyclic and Collective: More Art and Science of Flying Helicopters*, Lebanon, OH: Eagle Eye Solutions, 2009, pp. 402-404

19   *Definitions and Abbreviations*, 14 CFR Part 1, §1.1, amendment 1-73, June 27, 2018

20   *Airworthiness of Aircraft*, ICAO Annex 8, 10th edition April 2008

21   *OSTIV Airworthiness Standards for Sailplanes*, Delft, Netherlands, OSTIV 1966

22   "Designation of applicable regulations", *Certification Procedures for Products and Articles*, 14 CFR Part 21, §21.17, amendment 21-100, effective August 30, 2017

23   *Type Certification—Airships*, FAA Advisory Circular AC-21.17-1A, September 25, 1992

24   *Type Certification—Fixed-Wing Gliders (Sailplanes), Including Powered Gliders*, FAA Advisory Circular AC-21.17-2A, February 10, 1993

25   *Type Certification of Very Light Airplanes Under FAR §21.17(b)*, FAA Advisory Circular AC-21.17-3, December 21, 1992

26   N F Harpur, "The Airworthiness Certification of Concorde," *SAE Transactions*, 1971, 2507-2513 (SAE Paper 710756)

27   "Bae/SNIAS, *Concorde* Type 1," *Type Certificate Data Sheet*, TCDS A45EU, Revision 3, November 24, 2009

28   Hallion, R, *NASA's Contributions to Aeronautics: Aerodynamics, Structures, Propulsion,Ccontrols*, Washington National Aeronautics and Space Administration, 2010

29   http://www.heritageconcorde.com/braniff-airways-concorde-operations   , accessed June 14, 2022

30   *Type Certificates*, 14 CFRPart 21, Subpart B, amendment 21-92, effective April 16, 2011

31   *Airworthiness Certificates*, 14 CFR Part 21, Subpart H, amendment 21-92, effective April 16, 2011

32   *Airworthiness Certification of Aircraft*, FAA Order 8130.2J, July 21, 2017

33   "Issue of Type Certificate: Restricted Category Aircraft," *Certification Procedures for Products and Articles*, 14 CFR §21.25 (b), amendment 21-92, effective April 16, 2011

34   "Display of Marks; General," *Identification and Registration Marking*, 14 CFR §45.23 (b), amendment 45-24, effective September 1, 2004

35   "Issuance of Type Certificate: Primary Category Aircraft," *Certification Procedures for Products and Articles,*14 CFR §21.24, amendment 21-92, effective April 16, 2011

36   M W Anderson, "Flight Test Certification of Primary Category Aircraft Using TP101-41E Sportplane Design Standard," *Journal of Aircraft*: 32 [3], 1995, 631-635

37   "Issue of Type Certificate: Restricted Category Aircraft," *Certification Procedures for Products and Articles*, 14 CFR §21.25 (b), amendment 21-92, effective April 16, 2011

Chapter 2:
A Brief Word on Regulations

**Glory, flying regulations
Have them read at every station
Crucify the man who breaks one
The Air Force is shot to hell***

In this book we will generally be discussing US regulations since we are more familiar with them. However, we will discuss differences with other regulations, such as the International Civil Aviation Organization (ICAO) or the British Civil Airworthiness Regulations (BCAR).

In referring to US regulations, we will use "CAR" to denote Civil Air Regulations and "FAR" to denote Federal Aviation Regulations. Where we are referring to multiple national and international regulations, such as JAR or EASA CSs, we may use Part to denote common references, such as Part 25 to refer to FAR 25, JAR 25, or CS 25.

Why Regulations and Not Laws

The major reason for establishing regulations over laws is the special knowledge required. Most legislators do not possess the special technical knowledge to draft legislation that is required for complex technical industries, such as aviation. It is logical that such an agency would have the expertise necessary to create appropriate regulations. Therefore, rather than passing a law directly enumerating a set of rules, Congress passes a law creating an agency with a mandate to write regulations.

The Code of Federal Regulations (CFR)

14 CFR is an abbreviation for Title 14 of the Code of Federal Regulations, and FAR is one for Federal Aviation Regulation. All FARs are part of 14 CFR; but not all regulations in 14 CFR are FARs. Title 14 covers aeronautics and space.

- ✈ 14 CFR Parts 1 through 199 are the Federal Aviation Regulations (FARs);
- ✈ 14 CFR Parts 200 through 399 are the Air Carrier Economic Regulations;
- ✈ 14 CFR Parts 400 through 1299 are the Commercial Space Transportation Regulations;
- ✈ 14 CFR Parts 1300 through 1399 are the National Aeronautics and Space Regulations; and
- ✈ 14 CFR Parts 1300-1399 are the Air Transportation System Stabilization Regulations;

It may come as a surprise to the reader, but the Federal Aviation Regulations, the Air Carrier Economic Regulations, and the Air Transportation System Stabilization Regulations are not in the category of transportation regulations, but rather in aeronautics and space. Table 4 lists sections of the Code of Federal Regulations of interest in aviation.

---

* Chorus from Oscar Brand, *Glory, Flying Regulations*, © 1959 Elektra Entertainment. It is sung to the tune of *Battle Hymn of the Republic*.

Table 4 Government Regulations of Interest in Aviation

Title	Area or Agency	Regulations
14 CFR	Aeronautics and Space	Federal Aviation Regulations Air Carrier Economic Regulations Commercial Space Transportation NASA Regulations Air Transportation System Stabilization
19 CFR	Homeland Security	International Landing Rights Airports
40 CFR	Environmental Protection Agency	Aircraft and Engine Emissions
47 CFR	Federal Communications Commission	Frequency Allocation
49 CFR	Transportation	Drug and Alcohol Testing Motor Vehicle Safety Standards NTSB Regulations TSA Regulations

Why 14 CFR and not FAR?

Why do Government officials avoid using "FAR" to mean Federal Aviation Regulation? The answer is possible confusion between the Federal Acquisition Regulations* (in 48 CFR) and the Federal Aviation Regulations (in 14 CFR). This didn't seem to be a problem until 2002. In May 2002, FAA Order 1329.46C stated:

Do *not* use the acronym "FAR" to refer to FAA's regulation. Neither the Department of Transportation or the Office of the Federal Register allow us [the FAA] to use "FAR" for our regulations.[1]   (emphasis in the original text)

This hadn't been seen as a problem before. The previous version of the order, published in September 2000 did not mention any difficulty with the term "FAR."[2] In any event, the prohibition against "FAR" applies to government offices only.

It is interesting to observe that SFAR is still allowed for Special Federal Aviation Regulation. The FAA website still refers to current and historical Federal Aviation Regulations. There doesn't seem to be a problem with the term, only with the acronym.

Since this book will never again refer to the Federal Acquisition Regulations, we can use FAR in the text without confusion.† We will, however, use the formal abbreviation, 14 CFR, for reference citations.

We want to emphasize: All Federal Aviations Regulations are part of 14 CFR; but not all 14 CFR regulations are Federal Aviation Regulations.

Rules of Construction

We shall use the following rules of construction:

Part x: A particular section of the US or other rules, FAR, JAR, etc. using the common paragraph designation.

FAR x: A particular section of the US Federal Aviation Regulations.

JAR x: A particular section of the Joint Airworthiness Requirements.

CS x: A particular section of the EASA Certification Standards.

---

* Interesting trivia: The FAA is exempt from complying with the Federal Acquisition Regulations.
† Besides, "14 CFR" doesn't exactly roll off the tongue.

Administrative Procedures Act of 1946

To create or amend a regulation, the agency must follow the procedures in the *Administrative Procedures Act of 1946* (APA).[3] The agency must explain its rationale for the rule, discuss the comments, and include the rationale and discussion as a preamble to the final rule. When writing or changing a regulation, that agency must:

> ✈ Draft a proposed regulation or a change to a regulation;
> ✈ Publish the proposal as a Notice of Proposed Rulemaking (NPRM);
> ✈ Allow for a public comment period (usually 60 days); and
> ✈ Address each comment.

The agency must address the comments received. If there are many substantially similar comments, they can be addressed as a group. We have spent many hours addressing each comment during our tenure at the FAA. Following the comment period and disposing of each comment, the agency has the following options:

> ✈ Publish a final rule identical to the NPRM;
> ✈ Publish a less restrictive rule; or
> ✈ Withdraw the rule.

The responses to the comments will be published in the final rule preamble.

Furthermore, the APA allows private citizens to petition an agency to change an existing regulation.

NPRMs and final rules are published in the Federal Register, the official daily publication for rules, proposed rules, proposed rules, notices, and executive orders. The FAA has published an Advisory Circular, AC-11-2A,[4] to describe agency procedures for rule making as well as instructions to add one's name to the mailing distribution list. FAR 11*describes FAA procedures for making, amending, or repealing FARs.[5]

On occasion, a government agency may issue an Advance Notice of Proposed Rulemaking (ANPRM). The ANPRM is used as a means of obtaining public input at an early stage before the agency has done significant development on its own. ANPRMs are not mentioned in the APA. Therefore an agency is not required to issue one.

The FAA's Regulatory and Guidance Library (RGL) provides a convenient path to the aviation regulations†. The Section marked Code of Federal Regulations (Title 14) contains all of the regulations controlled by the FAA and many of those controlled by the Department of Transportation. The RGL also contains many previous versions of regulations and their preambles. This is important when researching an aircraft model's certification basis which might have been years in the past.

If you are ever confused about an FAA regulation,‡ read the preamble. It will explain the rationale and process of developing the rule. Read the applicability – it may not apply in a particular case. If you still don't understand, write a letter to the Agency's General Counsel. They will usually reply in a couple of months.

---

*   Part 11 is one of those regulations that a written in "question and answer" format. Both of your authors find this format noisome and hope the practice will have an early demise.

†   http://rgl.faa.gov/, accessed July 14, 2017

‡   Or any government regulation, for that matter.

Relief from Regulations

Exemptions

Section 11.15 describes exemptions as "relief from the requirements of a current regulation"[6] An individual or entity may petition for an exemption. The FAA will follow the normal APA steps as in any rulemaking.

Waivers

The FAA may issue waivers for temporary relief for some provisions of the Operating Rules:

FAR   91: General Operating and Flight Rules,[7]
FAR 105: Parachute Jumping,[8] and,
FAR 107: Small Unmanned Aircraft Systems.[9] [10]

Regulations, Orders, or Advisory Material

Regulatory material has several classifications: Regulations, Orders, Notices, and Advisory Materials. Under the aegis of the legislation governing its operation, the FAA issues regulation as describes. It also issues orders and advisory materials.

FAA Orders

An FAA Order describes procedures for use by FAA personnel in their duties. For example, FAA Order 8110.4C describes the procedures and practices the FAA uses in issuing Type Certificates.[11] Another Order of interest in certification is FAA Order 8110.37F, dealing with Designated Engineering Representatives.[12]

Notices are temporary orders. They are issued with an expiration date, normally one year. They use the same numbering scheme as orders but the number is preceded by an "N".

FAA Advisory Circulars

Advisory Circulars (AC's) are published by the FAA and contain guidance for compliance with various regulations. From an aircraft certification point of view, they provide acceptable means of compliance. They are only binding on the FAA, not on applicants. That is, if an applicant wishes to use, for example, AC-25-11B[13] as a means of compliance, the FAA must accept it. The applicant is free to choose another means.

FAA Policy Letters

FAA Policy Letters are similar in effect to Advisory Circulars – binding on the FAA, but not required for the applicant.

Operating Rules versus Certification Rules

Certification rules (e. g. FAR 23 or FAR 25) apply to the issuance of a TC. As such they apply to the applicant and to the FAA. They do not change for a particular aircraft model and can only be changed by an Airworthiness Directive.

Table 5 lists the current set of Federal Aviation Regulations. Note that only Subchapter C: Aircraft are certification rules (Parts 21 through 49). Subchapters F and G are operating rules.

Operating Rules (e. g. FAR 91 or FAR 121) apply to the operation of the aircraft. As such they apply to the pilot or operator of that aircraft. In addition, they change on the effective date of the amendment. By comparison, changes to the certification rules normal take effect for applications received after the effective date of the rule.

Table 5 Federal Aviation Regulations

Part	Title
Subchapter A: Definitions and General Requirements	
1	Definitions and Abbreviations
3	General Requirements
5	Safety Management Systems
Subchapter B: Procedural Rules	
11	General Rulemaking Procedures
13	Investigative and Enforcement Procedures
14	Rules Implementing the Equal Access to Justice Act of 1980
15	Administrative Claims Under the Federal Tort Claims Act
16	Rules of Practice for Federally Assisted Airport Enforcement Procedures
17	Procedures for Protests and Contracts Disputes
Subchapter C: Aircraft	
21	Certification Procedures for Products and Parts
23	Airworthiness Standards: Normal, Utility, Acrobatic, and Commuter Category Airplanes (pre-amdt. 23-64) Airworthiness Standards: Normal Category Airplanes (post-amdt. 23-64)
25	Airworthiness Standards: Transport Category Airplanes
26	Continued Airworthiness and Safety Improvements for Transport Category Airplanes
27	Airworthiness Standards: Normal Category Rotorcraft
29	Airworthiness Standards: Transport Category Rotorcraft
31	Airworthiness Standards: Manned Free Balloons
33	Airworthiness Standards: Aircraft Engines
34	Fuel Venting and Exhaust Emission Requirements for Turbine Engine Powered Airplanes
35	Airworthiness Standards: Propellers
36	Noise Standards: Aircraft Type and Airworthiness Certification
39	Airworthiness Directives
43	Maintenance, Preventive Maintenance, Rebuilding, and Alteration
45	Identification and Registration Marking
47	Aircraft Registration
49	Recording of Aircraft Titles and Security Documents
Subchapter D: Airmen	
60	Flight Simulation Training Device Initial and Continuing Qualification and Use
61	Certification: Pilots, Flight Instructors, and Ground Instructors
63	Certification: Flight Crewmembers Other Than Pilots
65	Certifications: Airmen Other Than Flight Crewmembers
Subchapter E: Airspace	
71	Designation of Class A, B, C, D, and E Airspace Areas; Air Traffic Service Routes; and Reporting Points
73	Special Use Airspace
75	Objects Affecting Navigable Airspace
Subchapter F: Air Traffic and General Operating Rules	
91	General Operating and Flight Rules
93	Special Air Traffic Rules
95	IFR Altitudes
97	Standard Instrument Procedures
99	Security Control of Air Traffic
101	Moored Balloons, Kites, Amateur Rockets, and Unmanned Free Balloons
103	Ultralight Vehicles
105	Parachute Jumping
107	Small Unmanned Aircraft Systems
75	Objects Affecting Navigable Airspace
Subchapter G: Air Carriers and Operators for Compensation or Hire: Certification and Operations	
110	General Requirements
117	Flight and Duty Limitations and Rest Requirements: Flightcrew Members
119	Certification: Air Carriers and Commercial Operators
120	Drug and Alcohol Testing Program
121	Operating Requirements: Domestic, Flag, and Supplemental Operations
125	Certification and Operations: Airplanes Having a Seating Capacity of 20 or More Passengers or a Maximum Payload Capacity of 6,000 Pounds of More; and Rules Governing Persons on Board Such Aircraft

Part	Title
Subchapter G: Continued	
129	Operations: Foreign Air Carriers and Foreign Operators of U.S.-Registered Aircraft Engaged in Common Carriage
133	Rotorcraft External Load Operations
135	Operating Requirements: Commuter and On Demand Operations and Rules Governing Persons on Board Such Aircraft
137	Agricultural Aircraft Operations
139	Certification of Airports
Subchapter H: Schools and Other Certified Agencies	
141	Pilot Schools
142	Training Centers
145	Repair Stations
147	Aviation Maintenance Technician Schools
Subchapter I: Airports	
150	Airport Noise Compatibility Planning
151	Federal Aid to Airports
152	Airport Aid Program
153	Airport Operations
155	Release of Airport Property from Surplus Property Disposal Restrictions
156	State Block Grant Pilot Programs
157	Notice of Construction, Alteration, Activation, and Deactivation of Airports
161	Notice and Approval of Airport Noise and Access Restrictions
169	Expenditure of Federal Funds for Nonmilitary Airports or Air Navigation Facilities Thereon
Subchapter J: Navigational Facilities	
170	Establishment and Discontinuance Criteria for Air Traffic Control Services and Navigational Facilities
171	Non-Federal Navigation Facilities
Subchapter K: Administrative Regulations	
183	Representatives of the Administrator
185	Testimony by Employees and Production of Records in Legal Proceedings, and Service of Legal Process and Pleadings
187	Fees
189	Use of Federal Aviation Administration Communications System
193	Protection of Voluntarily Submitted Information
Subchapter N: War Risk Insurance	
198	Aviation Insurance

As an example, during a recent flight instructor renewal course, the instructor stated that a Normal Category airplane could not be flown intentionally at bank angles beyond 60°. The rationale for such a prohibition was given as §23.3. However §23.1 says that Part 23 applies to "each person who applies for such a [Type] certificate". So Part 23 does not apply to a pilot.

The question was then raised, "What rule, applying to a pilot, would ensure operating within the limitations of the TC? The answer is: §91.9 which require pilots to comply with operating limitations in the flight manual and with posted markings and placards.

Application to Foreign Agencies

The previous discussion was based on US practice. Foreign agencies have similar arrangements. However some of the requirements and limitations may not apply. For example, EASA may decline to use one of their published Acceptable Means of Compliance (AMC) publications on a particular project.

References

1   *Advisory Circular System*, FAA Order 1320.46C, May 31, 2002

2   *Advisory Circular System*, FAA Order 1320.46B, September 25, 2000

3   *Administrative Procedures Act*, Public Law 79-404, enacted June 11, 1946

4   *Notice of Proposed Rulemaking Distribution System*, FAA Advisory Circular AC-11-2A, July 26, 1984

5   *General Rulemaking Procedures*, 14 CFR Part 11, effective September 20, 2000

6   "What is a Petition For Exemption," *General Rulemaking Procedures*, 14 CFR Part 11, §11.15, effective September 20, 200026

7   "Policy and Procedures," *General Operating and Flight Rules*, 14 CFR Part 91, §91.903, amendment 91-0, effective September 30, 1963

8   'Parachute Operations Over or Into a Congested Area or an Open-air Assembly of Persons," *Parachute Operations*, 14 CFR Part 105, §105.17, effective July 9, 2001

9   "Waiver Policy and Requirements," *Small Unmanned Aircraft Systems*, 14 CFR Part 107, §107.200, effective August 29, 2016

10 "List of Regulations Subject to Waivers," *Small Unmanned Aircraft Systems*, 14 CFR Part 107, §107.205, effective March 16, 2021

11 *Type Certification*, FAA Order 8110.4C, March 28, 2007, Change 6, effective March 6, 2017

12 *Designated Engineering Representative (DER) Handbook*, FAA Order 8110.37F, August 31, 2017

13 *Electronic Flight Displays*, FAA Advisory Circular AC-25-11B, October 7, 2014

Chapter 3: The Early Days

"Sacrifices must be made".*

In this chapter, we will trace the development of the US airworthiness standards from their inception to the current day. Why is this history important? There are several reasons. First it will give us some understanding of the rationale for the development of new rules and changes to existing rules. This is important to airplane designers since gives insight into the development. It also provides the basis for alternatives in the event the particular wording of a requirement causes difficulty in finding compliance with that particular requirement.

A second reason is that airworthiness requirements never expire. Once a particular design is certified, its certification basis is permanent and will not change with subsequent amendments. In the US, with some exceptions,† modifications to an existing design will use the same certification rules as were found in the original certification.

Aircraft Certification began in the 1920s with British, French, and US standards. What prompted these standards? To find out we must go back to the early days or aviation

Safety in the Early Days of Aviation

We shall look at aviation safety from the gliders of the 1890s up to World War I and then during the early days of the airmail.‡

The Age of Gliders

The first of the glider founding fathers was Sir George Cayley. He conceived of a fixed-wing aircraft and constructed one from a carriage. He persuaded his coachman, one Alan Appleby, to ride the vehicle down a hill. It became airborne, but crashed after several seconds. Appleby is reported to have given notice saying he had been hired to drive, not fly.

The premier name from "The Age of Gliders" is, of course, Otto Lilienthal who performed approximately 2,000 glider flights from 1891 to his death in 1896.[1] Lilienthal was a German engineer who flew what we would call *hangliders*. These were controlled by shifting the pilot's weight. Lilienthal perished in August 1896 when his glider pitched up, probably from a gust and he was unable to recover. At the time of his death, it is estimated that he had accumulated about five hours of flight time. In addition to being one of the first serious students of aviation, Lilienthal had four patents for aviation devices and, with his brother, wrote a book, *Birdflight as the Basis of Aviation: A Contribution Towards a System of Aviation.*[2]

One of Lilienthal's students, Percy Pilcher flew about 1,000 flights between 1895 and 1899. Pilcher, a Scottish engineer, was killed when his glider's structure collapsed in flight.

During the late 1890s, Octave Chanute, a retired engineer, designed and developed gliders and sponsored several young pilots to fly them along the coast of Lake Michigan. He published their results in a book, *Progress in Flying Machines.*[3]

---

*   Reportedly the last words of Otto Lilienthal (1848-1896) after his fatal crash.
†   An Airworthiness Directive (AD) can require changes to correct an unsafe condition.
‡   We will skip 1914-1917 (World War I) because of the paucity of civilian mishap data.

Figure 5 Lilenthal in Flight*

The final team of glider flyers was the Wright Brothers. Wilbur and Orville Wright began their flying in the fall of 1900 along the sand dunes near Kitty Hawk, North Carolina. They drew heavily on the work of Lilienthal and Chanute. They departed from their predecessors and substituted aerodynamic control for weight shift control.[4] They began flying aircraft to learn the techniques of such control, first as kites in 1900, then as man-carrying gliders. During the period from 1900 to 1902, the Wrights made about 1,000 glider flights.[1]

The glider pioneers, flew approximately 5,000 flights during this period. Holanda[1] estimates the total flight time at about 20 hours (~14 sec per flight). What were the causes of the fatal crashes? Lilienthal's mishap was a loss of control likely a result of the shortcomings of the flight controls.† Pilcher's death was likely the result of airframe structural failure.

Early Airplanes (1903 – 1908)

During this period, there were approximately 1,000 flights. About 500 were flown by the Wright Brothers, mostly solo.. They had acquired about 40 hours of flying time. Their longest flight was about 40 minutes.

In September 1908, the Wrights were demonstrating their *Flyer* to the US Army. During the flight, a propeller blade separated resulting in loss of control. Army Lieutenant Thomas Selfridge was flying as an observer and died in the crash.[5]

During the 1903-1908 time period, the rest of the world made about 500 flights, the longest of which was 44 minutes.[1] The safety rate for this period was 1,000 flights per fatality. The likely cause for the Wright *Flyer* accident was structural failure.

---

*   Library of Congress Picture
†   Although with an estimated 5 hours of flight experience, we cannot rule out pilot error.

Figure 6 Scene Following the 1908 Wright *Flyer* Fatal Accident*

The "Wild West"

"The First 348" was a list created by Herman Hoernes and published in the French Magazine *Aero Manuel* between 1911 and 1914. This list began with Thomas Selfridge and included only powered airplane casualties. Chronologically, the list shows the fatalities for each year:

1908	1
1909	3
1910	29
1911	79
1912	143
1913	92

Aero Manuel ceased publication in August 1913 and the list was not continued. Nationalities of the decedents included were 92 Frenchmen, 77 Germans, 52 Americans, 26 Russians, and 25 Italians.

Hollanda estimates the flying hours for the period covered by "The First 348" to be one million hours.[1] This is equivalent to a rate of one fatality per 2800 flights. The causes were similar to the previous eras: poor handling exacerbated by lack of training, inadequate airframe structure, and a new cause: engine failure.

---

*   US Army Photograph

The Airmail (1918 to 1921)

Following World War I, the first real use for civil aviation was to deliver airmail. The first flight took place May 15, 1918. The US Army operated the airmail system then turned it over to the US Post Office. The official in charge was Otto Praeger. He operated the airmail service based on the postal motto:

Neither snow nor rain nor heat nor gloom of night shall stay these couriers from the swift completion of their appointed rounds.

The results should have been predictable. The fledgling industry was simply not ready for scheduled operations in all weather. Pilots were not permitted to cancel flights for weather conditions. The instruments and navigation systems in the aircraft of that time would only allow visual flight. The stretch of mountains between New York and Cleveland became known as "Hell's stretch".

There were an estimated 11,250 airmail flights during this period with 27 fatalities.[1] This works out to one fatality per 417 flights. The causes were much the same as before with the addition of accidents caused by weather.

The Airmail, Part Two

In 1921, a new administration came to Washington. Under the new Postmaster General, safety became the primary consideration. Pilots were now allowed to cancel flights if the conditions were unsafe. Fatalities per flight dropped to 0.2 per thousand flights – the lowest ever. Of course, engineering improvements didn't hurt.

Safety in Aviation's Early Years

Safety in the years leading up to the 1920 can be summarized in Table 6, which is based on Holanda's book.[1] Even as dramatic the improvement in the 1930s was, recent scheduled airline accident data is even more remarkable. In 2117, there were 41.8 million airline flights worldwide with 19 fatalities.[6] This is a rate of 0.0005. Clearly, the industry has been doing something right.

Table 6 Safety Before the Advent of Aircraft Certification

Period	Years	Flights	Fatalities	Rate[a]
Age of Gliders	1891-1902	5,000	2	0.4
Early Airplanes	1903-1908	1,000	1	1.0
The Wild West	1908-1913	1,000,000	347	0.3
Airmail, Part 1	1918-1921	11,250	27	2.4
Airmail, Part 2	1921-1927	91,875	16	0.2
US Airlines[b]	1930-1939	8,950,000	434	0.05
Notes (a) Rate = fatalities per 1,000 flights.				
(b) Airline rates in the 1930s shown for comparison				

Similar causes, plus engine failure, continued through the early airplanes up to 1913. Again, pilot error can be considered as causing many of the loss of control accidents. As Nevil Shute* said in his autobiography:

In those days, a pilot was expected to be brave and resolute, a daredevil who was not afraid to take risks.[7]

---

*   Nevil Shute Norway (1899-1960) was an early aeronautical engineer, who as a design engineer on the British R100 airship, flew from London to Canada and back. He flew the Atlantic twice in 1930. He later became a novelist, using his first and middle names, Nevil Shute, as his *nom de plume*.

In 1918, with the airmail, aircraft were now expected to operate routinely in all kinds of weather. Thus a new cause for accidents was added to the list: weather related accidents. These included the hazards of low visibility, thunderstorms, and airframe icing. These mishap causes were the impetus behind aviation research and aviation regulation, as we shall see.

It can be argued that the improvements in aviation safety came about because of aviation research over the years, and not because of certification requirements. The counter argument to this was the popularity of unregulated ultralight aircraft in the 1980s. As stated during the presentation by Gregorek, Bragg, and Newman:

> **These aircraft have terrible handling qualities, abysmal structural integrity, and poor engine reliabilities – we make up for this by requiring no pilot training.[8]**

The First Airworthiness Requirements

The initial airworthiness requirements were fairly simplistic by today's standards. In general, they required a set of drawings with vague statements about the strength of the structure. As with early military specifications,* they listed specific performance requirements. One of these required a one hour endurance test during which the aircraft "must be steered in all directions without difficulty and at all times under perfect control and equilibrium.

British Requirements (1921)

The requirements for issuing an airworthiness certificate were issued in 1921 and reprinted in a NACA report.[10] The strength of the structure was to be checked along the general lines in use at the end of the war. Engines approved for use in the war were approved for civil use. Similarly designed and tested engines may receive approval. Materials and workmanship must inspected by the Air Ministry. No details were given.

The requirements for stability and control are unusual:

> **"Normally the stability and effectiveness of controls can be determined within reasonable limits from the drawings of a new type aircraft".†**

An evaluation flight by an official pilot would take place following the constructor's flight trials. Further flights would only be flown if the official pilot found the behavior of the airplane presented new problems. No criteria for such action were discussed.

French Requirements (1923)

The requirements for issuing an airworthiness certificate were published in 1923 and, like the British requirements, were reprinted in a NACA report.[11] The actual requirements are quite vague. The material to be submitted included drawings, a note giving the strength computations, an inspection guide, and a signed declaration of conformity.

Airplanes were divided into three classes: Touring, Racing/Experimental, and Commercial. There were a number of specific requirements for commercial airplanes. A partial list follows:

Landing: The pilot must have a clear field of vision downwards and when "near the ground," have a field of vision from ten meters to the horizon.

---

* Such as the Army's specification for a heavier-than-air flying machine, which called for a demonstrated speed of 40 miles per hour averaged in both directions.[9]

† Could this be the origin of the test pilots' expression "If it looks good, it will fly good"?

<u>Controls</u>: The airplane must maintain straight flight stick-free. The control forces must allow a pilot of average strength to fly the required demonstrations. If the airplane has flight duration of more than four hours, it must be equipped with dual controls.

<u>Radio</u>: If more than ten persons can be carried, the airplane must be fitted with a radio telegraph. A radio telephone may be substituted, if accepted by the government. Two years after adoption of these regulations, all airplanes, regardless of seating capacity, must be equipped with radio apparatus as described.

<u>Engines</u>: Dual ignition was required.

Flight Demonstrations:

- ✈ Takeoff run less than 984 ft (300 m); *
- ✈ Landing run less than 820 ft (250 m);
- ✈ Climb of at least 1 hour 15 minutes
  Must reach at altitude of at least 11,483 ft (3500 m) in 60 minutes; and.
- ✈ Five successive figure eights at an altitude between 1640 and 3280 ft (500 and 1,000 m). This test must be completed in 15 minutes.
  (Airplanes with a total load over 5512 lb (2500 kg) may complete the test in 20 minutes.

These tests seem simplistic in today's environment. In particular, the figure eight demonstration (essentially ten complete circles in alternating directions in fifteen or twenty minutes) seems strange at this late date.

References

1   R Hollanda, *A History of Aviation Safety*, Bloomington, IN: AuthorHouse, 2009, pp. 2-3
2   O Lilienthal, *Birdflight as the Basis of Aviation*, 1889, republished 2001, Hummelstown, PA: Markowski International Publishers; Translated by A W Isenthal, 1911; Introduction and addendum by G Lilienthal
3   O Chanute, *Progress in Flying Machines*, Courier Corporation, 1894, Reprinted Long Beach, CA: Lorenz & Herweg, 1976
4   O Wright, *The Early History of the Airplane*, Prabhat Prakashan, 1922
5   J D Martin, *The First United States Army Aircraft Accident Report*, Department of the Army, 1999; reprinted copy of Accident Report on the Wright5 Flyer, ca. 1908
6   *Annual Review 2018*, International Air Transport Association, 2019
7   Shute, N, *Slide Rule: The Autobiography of an Engineer*, Cresskill, NJ: Paper Tiger, 1954
8   G M Gregorek, M Bragg, and R L Newman, "Practical Considerations of Ultralight Flight Testing," *Proceedings Second Flight Testing Conference and Technical Display*, Las Vegas, November 1983
9   *Advertisement and Specification for a Heavier-than-Air Flying Machine*, Signal Corps Specification No. 486, December 23, 1907
10   *British Certificates of Airworthiness*, NACA Technical Memorandum 23, May 1921; reprinted from Aeronautics, April 1921
11   *Regulations Governing the Issuance of Certificates of Airworthiness in France*, NACA Technical Note 155, August 1923; reprinted from *Bulletin de la Navigation Aérienne*, March 1923

---

*   With a head wind of less than 6 m/sec (11.7 knots)

PART II:
HISTORICAL DEVELOPMENT

Chapter 4.
History of US Aviation Agencies

"We trained hard—but it seemed that every time we were beginning to form up into teams we were reorganized. I was to learn later in life that we tend to meet any new situation by reorganizing, and what a wonderful method it can be for creating the illusion of progress while actually producing confusion, inefficiency, and demoralization." Petronius Abiter, ca. 61 CE*

We will begin with the regulatory agencies, Bureau of Air Commerce through to the Federal Aviation Administration. Because accident investigation is the tool by which we measure our performance, this will include these agencies. Finally, aeronautical research helps progress and often provides the data needed,. The final section of this chapter will cover the aeronautical research organizations.

Organization of Regulatory Agencies

Overall Development
The Federal Aviation Administration and its aircraft certification organization did not leap into being fully formed. Indeed they began small and grew over the years. Each major change was triggered by a defining event – usually a prominent aircraft accident. The cure had a profound effect on the entire organization.

Figure 7 traces the development of these organizations from the Air Commerce Act of 1926 through to the Federal Aviation Act of 1958 and beyond.

Aviation accident investigation plays a major part in changes in aviation regulation. It is a basic tenet of the aviation industry that every action should be followed by an inspector or a monitor verifying that the act was performed properly. To a large extent, then, the accident investigation organizations serve as the quality checks for aircraft certification. We have traced the development of the regulatory agencies and the accident investigating agencies.

Creation of the Aeronautics Branch
In the mid-1920s, there was a great deal of discussion about the new aviation industry. Should the military have control over US aviation? In 1925, President Coolidge appointed a presidential aircraft board to study US national aviation. This board was also called the Morrow Board after its chairman, Dwight Morrow,† completed its work in 1925. The Morrow Board recommended a civilian aviation bureau separate from the military.[2]

The Air Commerce Act of 1926 was signed into law on May 29, 1926 by President Coolidge. To carry out the functions under the Air Commerce Act, the secretary of commerce created the Aeronautics Branch within the Commerce Department. The Post Office's airmail navigation services were transferred to this Aeronautics Branch in July 1927.[3]

---

* Widely attributed to Petronius Arbiter (ca. 27-66 CE), It should be attributed to Charles Ogburn;[1] http://www.quotationspage.com/quote/25618.html, accessed April 29, 2020.

† Dwight Morrow was the father-in-law of Charles Lindbergh.

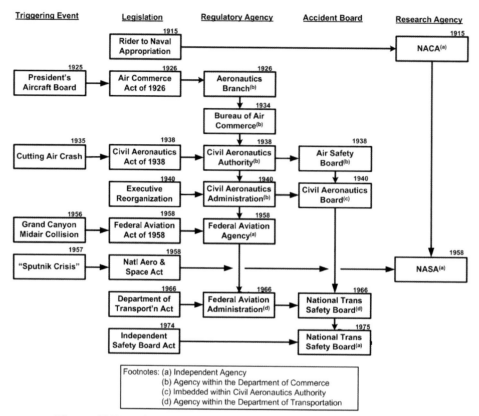

Figure 7 Development of US Aviation Regulatory Agencies

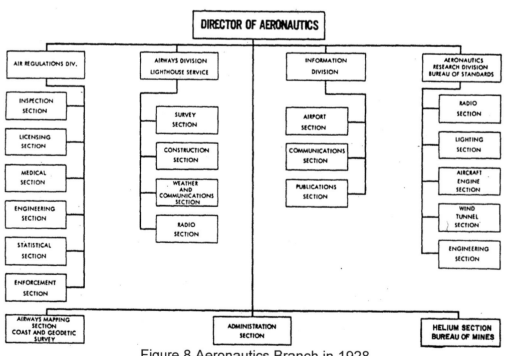

Figure 8 Aeronautics Branch in 1928
From *FAA Historical Chronology*[3]

Initially, the Aeronautics Branch was relatively small and drew on other branches within the Department of Commerce. Figure 8 shows the organization of the Aeronautics Branch. It is interesting to note the support from other agencies, such as engineering support from the bureau of standards and airways development support from the lighthouse service. The bureau of mines even got into the act with their helium section.*

The aviation industry was fairly small in these early days, so it was reasonable to have the small bureau in Washington. However, the need to have certification facilities distributed throughout the country soon became clear. In 1933, manufacturers were allowed to conduct aircraft engine endurance tests locally rather than of shipping the engines to Washington for testing.

Creation of the Bureau of Aeronautics

In 1934, the name of the Aeronautics Branch was changed to Bureau of Air Commerce to more accurately reflect its duties. By this time, there were no longer any major functions that were structurally part of other government bureaus. This is reflected in the new organization chart shown in Figure 9.

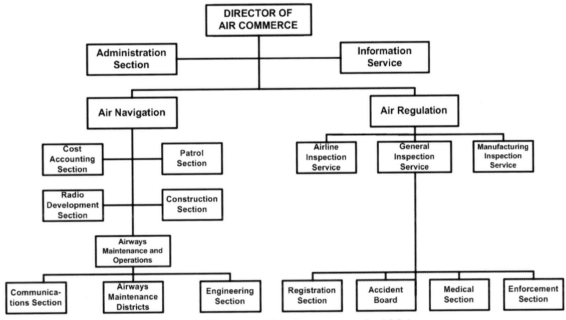

Figure 9 Bureau of Air Commerce in 1934
From *FAA Historical Chronology*[3]

Creation of the Civil Aeronautics Authority

Following the crash of TWA Flight 2 in May 1935[4] and the resulting death of Senator Bronson Cutting, the aviation industry was criticized for deficiencies in weather forecasting, poor judgment by the pilot, and dispatching the flight with known radio problems and below minimum weather.† This led to the Civil Aeronautics Act of 1938 and the creation of the Civil Aeronautics Authority (CAA). This placed all civil aviation activities in a single agency.[3] Figure 9 shows the organization chart of the new CAA.

---

*   At the time, the future of long-range air transport was thought to be helium-filled airships.
†   This accident is described in Appendix 4 on page 262.

The new Air Safety Board had the responsibility to investigate accidents, determine their probable cause, and make recommendations to prevent future accidents.

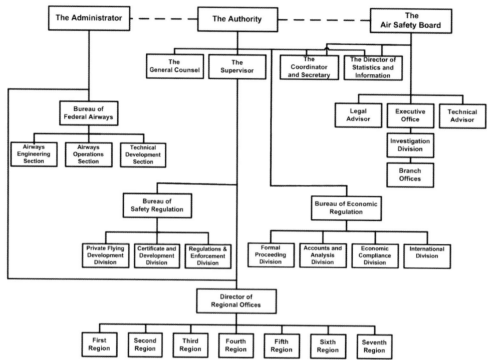

Figure 10 Civil Aeronautics Authority in 1939
From *FAA Historical Chronology*[3]

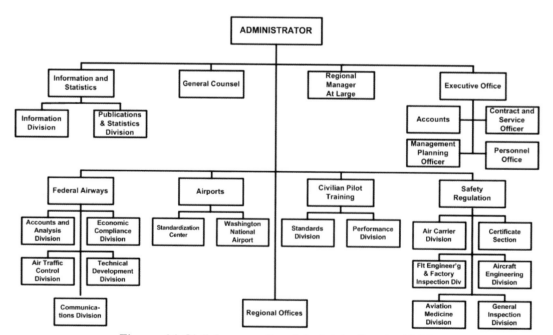

Figure 11 Civil Aeronautics Administration in 1940
From *FAA Historical Chronology*[3]

Reorganization and Creation of the Civil Aeronautics Administration

The reorganization of the CAA, under President Roosevelt's Reorganization Plan went into effect June 30, 1940. This reorganization was intended to clarify the relations between the CAA administrator and the five-member board. This board was now designated the Civil Aeronautics Board (CAB). The CAB absorbed the accident investigation duties formerly held by the Air Safety Board which was abolished. The CAA was renamed the Civil Aeronautics Administration. Figure 11 shows the CAB organization chart in 1940.

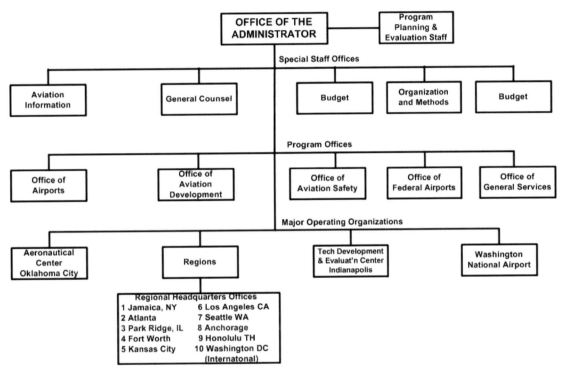

Figure 12 Civil Aeronautics Administration in 1949
From *FAA Historical Chronology*[3]

Reorganization of CAA

In 1945, the CAA administrator reorganized the CAA into regions. The result amounted to a significant decentralization in operations, as shown in Figure 12 Headquarters would now concentrate on policy. One effect was the increase in the use of designees. The unintended consequences of this new policy would be a lack of standardization between certification offices. There were only minor changes in the organization through the rest of the CAA's tenure. Figure 13 shows the CAA organization just prior to being replaced by the FAA.

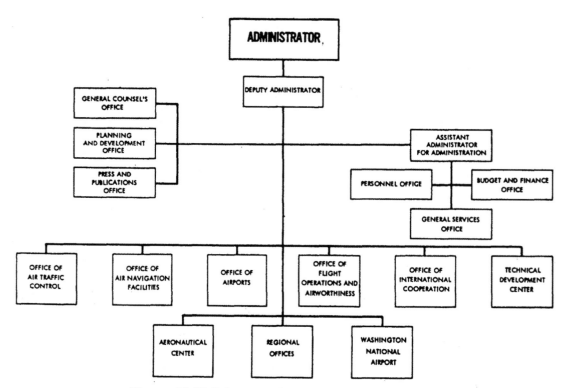

Figure 13 Civil Aeronautics Administration in 1957
From *FAA Historical Chronology*[3]

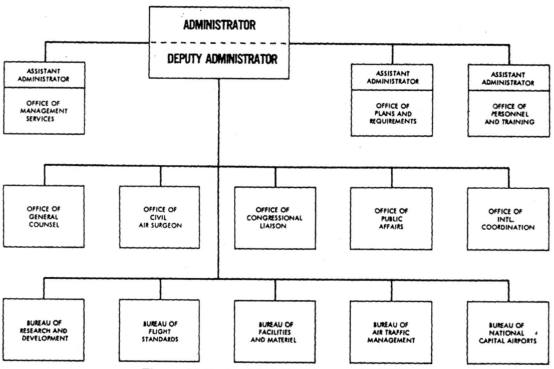

Figure 14 Federal Aviation Agency in 1959
From *FAA Historical Chronology*[3]

Creation of Federal Aviation Agency

On June 30, 1956, TWA Flight 2, a Lockheed *Constellation* and United Flight 718, a DC-7 collided over the Grand Canyon.[5]* Both airplanes were flying under visual flight rules (VFR) in uncongested airspace. In the aftermath, Congress held hearings to review the problems of airspace and air traffic management. These hearings would lead to the creation of the Federal Aviation Agency.[3] Initially, there was little change from the CAA. Figure 14 shows the early organization chart.

Centralized FAA Organization

In January 1960, FAA Administrator Quesada reorganized the agency to a centralized concept. The headquarters were given authority to exercise direct supervision over all field offices, except in Alaska and Hawaii. The position of regional administrator wise abolished and regional managers were appointed to carry out administrative and support functions required by field offices. Figure 15shows the FAA organization following this reorganization.[3]

In 1961, the new administrator reorganized the agency to delegate broad operational responsibilities to regional offices.† This was intended to allow the headquarters to concentrate on policy issues. Figure 16 shows the new, decentralized organization.[3]

A further move to decentralize the agency took place over the next 2 years. Under this program, an area manager would head each of the 18 area offices, and would have line responsibility over four basic operating programs: air traffic, flight standards, airway facilities, and airports, programs that had previously been in the hands of the regional directors and the regional program division chiefs.

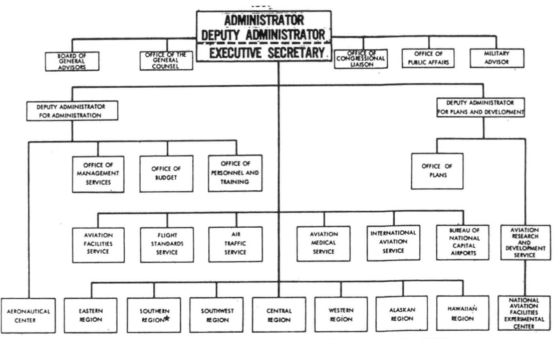

Figure 15 Centralized Federal Aviation Agency in 1961
From *FAA Historical Chronology*[3]

---

*   This accident is described in Appendix 4 on page 350.

†   In researching this history, it seemed to us that each new administrator was trying to make the agency fit his own goals. This led us to include the quote at the top of page 33.

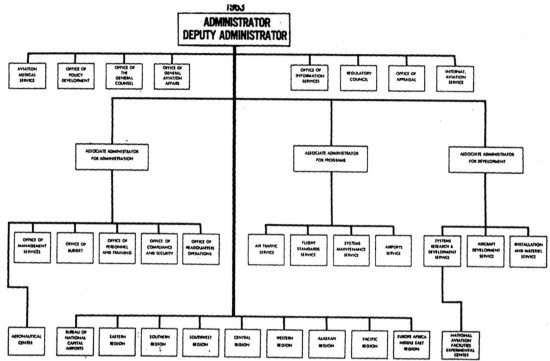

Figure 16 Decentralized Federal Aviation Agency in 1963
From *FAA Historical Chronology*[3]

Creation of Department of Transportation

In 1966, President Johnson signed the Department of Transportation Act, bringing 31 previously scattered Federal elements, including FAA, into a single cabinet-level Department of Transportation (DOT). DOT absorbed the following agencies:

✈ Bureau of Public Roads;
✈ Federal Aviation Agency;
✈ US Coast Guard;
✈ Maritime Administration
✈ Interstate Commerce Commission;
✈ Great Lakes Pilotage Administration;
✈ St. Lawrence Seaway Development Corporation;
✈ Alaska Railroad; and
✈ Minor transportation-related activities of other agencies.[3]

DOT began operations on April 1, 1967. At the same time, FAA ceased to be the independent Federal Aviation Agency and became the Federal Aviation Administration within the new department.[3]

Figure 17 shows an early FAA organization under the new DOT.

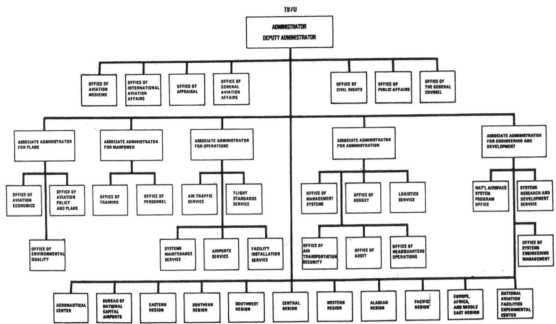

Figure 17 Federal Aviation Administration in 1970
From *FAA Historical Chronology*[3]

Effect on Aircraft Certification

We have seen a growth in the size of the variously named US aviation agencies from the relatively small Aeronautics Branch to the much larger Federal Aviation Administration. We have seen a small group in the nation's capital grow to an organization spread throughout the country, indeed throughout the world. To this point, the discussion of agency organization is been background information for the real issue – "What was the effect on aircraft certification?"

The FAA is not just aircraft certification. Aircraft certification is but a small part of the overall FAA. Aircraft certification accounts for about 1,300 positions within the FAA (out of about 48,000). By comparison, air traffic control ahs about 15,000 positions.

In the late 1970s the FAA was spread over the country. There were ACO's everywhere from Atlanta to Anchorage. The major effect was two-fold: Manufacturers had better access to ACOs, and there was a lack of standardization between ACO's.

It is no secret that there were differences between ACO's. There was a perception within the certification communities that some offices were "easier" than others. Some applicants sought excuses to apply to an ACO far from their place of business. This became known as "ACO shopping.

In our experience, dealing with many different ACO's, both as a DER and as an FAA engineer, the biggest problem wasn't "easy" versus "hard" ACO's. It was just that each office had its own procedures and preferences. As an FAA engineer working on a specific project, it was cumbersome to deal with the different ACO's, particularly if the project involved a Supplemental Type Certificate (STC) in one ACO on an airplane originally certified in different ACO.

Establishment of Lead Regions

In 1980, the FAA established the lead region concept under which designated FAA regions assumed certain responsibilities on a nationwide basis. FAA assigned lead regions to perform national headquarters staff functions relative to various aspects of aircraft certification. The initial lead regions were:

- ✈ Central Region: small aircraft;
- ✈ Great Lakes Region: propellers;
- ✈ New England Region: engines; and
- ✈ Southwest Region: rotorcraft.[3]

Later, certificating regions were designated as the final certification authority for certain categories of aircraft, parts, or materials. The certificating regions were;

- ✈ New England Region: engines (effective January 1980);
- ✈ Great Lakes Region: propellers (effective July 1980);
- ✈ Northwest Region: transport airplanes (effective November 1980); and
- ✈ Central Region: small aircraft (effective January 1981).[3]

In 1981, this confusing arrangement with lead regions and certificating regions was modified to form four aircraft certification directorates. These directorates assumed the certification responsibilities previously assigned to regions under the lead region concept. However, they also received additional responsibilities to improve the certification process. The directorates were managed by the following regions:

- ✈ Central Region: small aircraft;
- ✈ New England Region: engines and propellers
- ✈ Northwest Mountain Region: transport aircraft; and
- ✈ Southwest Region: rotorcraft.[3]

The authority of the directorates extended beyond regional boundaries. For example, ACOs in the Central, Southern, and Great Lakes regions reported directly to the Small Airplane Certification Directorate at the Central Region headquarters.[6]

Office of Aviation Safety to Report to Administrator

In a move intended to sharpen FAA's focus on safety, the Office of Aviation Safety would now reports directly to the Administrator instead of to the Associate Administrator for Aviation Standards. This change was effective in August 1984.[3]

FAA Office of Airworthiness

"The [FAA] Office of Airworthiness has two prime functions: to establish minimum standards for the design and manufacture of all US [civil] aircraft and to certify that all aircraft meet these standards prior to introduction into service. Airworthiness standards prescribe explicit flight, structural, design and construction, powerplant, and equipment requirements.

"The office issues 'type' certificates to prototype aircraft built in conformance to airworthiness standards after successful testing. Manufacturers try to ensure that individual aircraft conform to the type [design] to obtain airworthiness certification. If major changes are made in an aircraft design, a new type certificate is required. However, if less extensive changes are made, FAA amends a type certificate and issues a supplemental one.* As pilots must have additional and expensive training to operate a new type of aircraft, manufacturers and airlines prefer continuous supplemental certificates and pilot type ratings.

---

*   Where the original text uses *supplemental*, actual term should be *amended*

"Four FAA regional offices have certification authority for aircraft and certain systems:
- ✈ Central Region (Kansas City) certificates general aviation aircraft,
- ✈ New England Region (Boston) certificates engine and propulsion systems,
- ✈ Northwest Mountain Region (Seattle) certificates large commercial aircraft; and
- ✈ Southwest Region (Fort Worth) certificates rotorcraft.

"This decentralized management lends itself to internal FAA disagreements over regulatory actions and sometimes to outright contradictions".*

Reorganization of Aircraft Certification Service

In June 2017, the Aircraft Certification Service was reorganized to consolidate the flow of policies and avoid the sometimes cumbersome arrangement of ACO's.[8] The organization is shown in Figures 17 through 19.

This change has not been in place long enough for a fair assessment. However, we are convinced that it can only help.

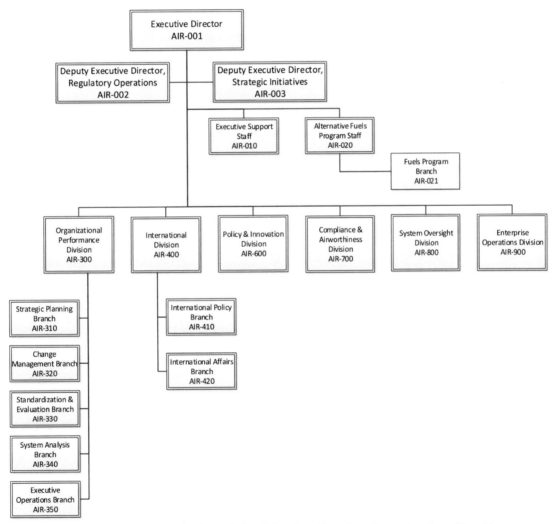

Figure 18 Realigned Aircraft Certification Service Organization Chart†

---

*   *Safe Skies for Tomorrow*, p. 56.[7]
†   From *FAA Historical Chronology*[3]

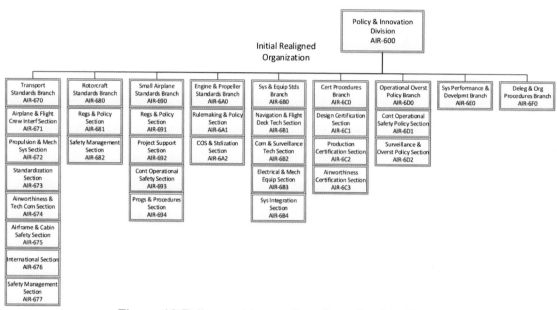

Figure 19 Policy and Innovation Organization Chart
From *FAA Historical Chronology*[3]

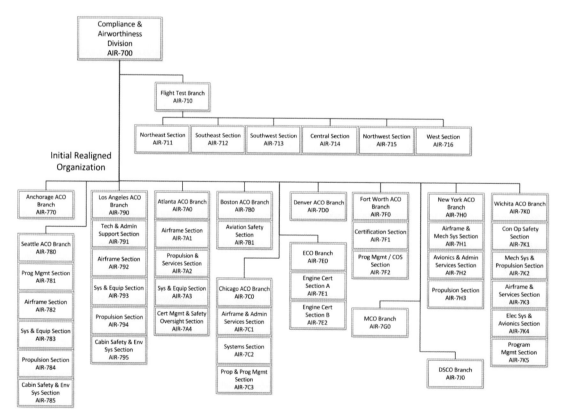

Figure 20 Revised ACO Organization
From *FAA Historical Chronology*[3]

National Transportation Safety Board (NTSB)

In 1967, the National Transportation Safety Board (NTSB) was created within the DOT; The NTSB absorbed the safety functions from CAB and Interstate Commerce Commission.[3]

In 1974, in passing the Transportation Safety Act, Congress declared that the NTSB could not properly conduct its responsibility of determining the probable cause of transportation without total separation and independence. The result was an independent agency. Board members are nominated by the president and confirmed by the senate to serve five-year terms. Figure 21 shows the NTSB organization chart. The NTSB has no authority to regulate, fund, or be directly involved in any mode of transportation. Its functions are to conduct investigation and make recommendations from an objective viewpoint.

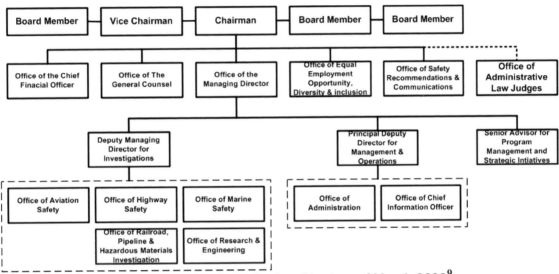

Figure 21 NTSB Organization Chart as of March 2020[9]

In addition to its investigative role, the NTSB also serves as an appeals board for airmen, mechanics, or mariners if the FAA or Coast Guard takes a certificate action. Its administrative law judges (ALJs) hear, consider, and issue decisions for appeals filed with the board. Either the certificate holder or the FAA may appeal the decision of the ALJ to the full board.[10]

Since its inception, the NTSB has investigated more than 149,000 aviation accidents and thousands of surface transportation accidents. NTSB investigators travel throughout the country and to every corner of the world to investigate significant accidents and develop factual records and safety recommendations with one aim—to ensure that such accidents never happen again.

Including the NTSB's role and organization is appropriate when writing on aircraft certification since the NTSB, in a sense, is the inspector for airworthiness certificates. The accident record is the final measure of the success or failure of granting of airworthiness certificates.

Research Agencies and Organizations

National Advisory Committee for Aeronautics

The National Advisory Committee for Aeronautics was founded in 1915 (See Figure 7 on page 36). The enabling legislation slipped through Congress as a rider to the Naval Appropriations Bill in March 1915.[11] In the 1920s and 1930s, NACA-sponsored research helped pave the way for better airplanes. It had a profound effect on the development of airplane requirements as well.

We must remember that, at the time, very little was known about airplane handling, such as how to recover from a spin or design an airplane that would resist the spin. Some of the first research sponsored by the NACA was measuring the structural loads in flight by James Doolittle.[12][13] Other work provided a recommended recovery for accidental spins, the NACA spin recovery technique.[14] In a 1936 report, the NACA correlated measured stability of a wide variety of airplanes with pilots' subjective observations.[15] All of these studies led to quantitative data to improve airplane safety.

In the 1940s, NACA flew research flights into various weather patterns and determined the envelope of icing clouds that could threaten airplanes.[16][17] This research led directly to the icing envelopes in Part 25, Appendix C.

National Aeronautics and Space Administration

In 1958, in response to the "Sputnik Crisis," Congress passed the Aeronautics and Space Act reorganizing and renaming the NACA into National Aeronautics and Space Administration. NASA was established as an independent agency.

In 2000, NASA flew the Shuttle Radar Topography Mission (SRTM) aboard the Space Shuttle *Endeavour*. This mission was a joint program of NASA, the National Geospatial-Intelligence Agency,* and the German and Italian Space Agencies. It used dual radar antennas to acquire interferometric data processed to digital topographic data at 1 arc-sec resolution.[18] This mission led directly to the development of Terrain Awareness and Warning Systems.

Guggenheim Foundation

The Guggenheim Foundation also played a significant role in developing academic research and teaching organizations and in sponsoring aviation research.[19]

References

1   C Ogburn, Jr, *The Marauders*, New York: Harper & Brothers, 1959, p. 60
2   *Report of President's Aircraft Board*, Washington: US Government Printing Office, November 1925
3   E Preston *et al. FAA Historical Chronology: Civil and the Federal Aviation Government*, Washington: Federal Aviation Administration, 1998
4   *Accident Report to TWA Flight 6, near Atlanta, Mo., May 6, 1935*, Bureau of Air Commerce, June 1935
5   *Accident Investigation Report. Trans World Airlines, Inc., Lockheed 1049A, N 6202C, and United Air Lines, Inc., Douglas DC-7, N 6324C, Grand Canyon, Arizona, June 30, 1956*, CAB Report, File No. 1-0090, April 1957
6   *Aircraft Certification Directorate Procedures*, FAA Order 8100.5, October 1, 1982, Superseded by FAA Order 8100.5A, dated September 30, 2003
7   *Safe Skies for Tomorrow: Aviation Safety in a Competitive Environment*, Washington: Office of Technology Assessment, 1980, reprinted by University of Michigan Library
8   *Aircraft Certification Service Organizational Realignment References*, FAA Order 8100.18, 06/27/2017
9   From http:// www.ntsb.gov/about/organization/Pages/default.aspx, downloaded October 31, 2020
10   From http:// www.ntsb.gov/about/organization/ALJ/Pages/office_alj.aspx, downloaded October 31, 2020
11   R E Bilstein, *Orders of Magnitude. A History of the NACA and NASA, 1915-1990*, Washington: National Aeronautics and Space Administration, 1989. The NASA History Series.

---

*   Formerly known as National Imagery and Mapping Agency (NIMA).

12   J D Doolittle, *Wing Loads as Determined in Flight*, Thesis: Massachusetts Institute of Technology, 1924

13   J D Doolittle, *Accelerations in Flight*, NACA Report 203, 1925

14   W H McAvoy, *Piloting Technique for Recovery from Spins*, NACA TN-555, February 1936

15   H A Soulé, *Flight Measurements of the Dynamic Longitudinal Stability of Several Airplanes and a Correlation of the Measurements with Pilots' Observations of Handling Characteristics*, NACA Report 578, July 1936

16   W Lewis, *A Flight Investigation of the Meteorological Conditions Conducive to the Formation of Ice on Airplanes*, NACA TN-1393, 1947

17   A Jones and W Lewis, *Recommended Values of Meteorological Factors to be Considered in the Design of Aircraft Ice Protection Equipment*, NACA TN-1855, 1949

18   T G Farr *et al.*, "The Shuttle Radar Topography Mission", *Reviews of Geophysics*: 45, June 2007, RG2004,   doi:10.1029/2005RG000183;   http://agupubs.onlinelibrary.wiley.com/doi/epdf /10.1029/2005RG000183, accessed November 5, 2020

19   R P Hallion, *Legacy of* Flight. The Guggenheim Contribution of American Aviation, Seattle: University of Washington Press, 1977

Chapter 5: US Airworthiness Development: 1926-1965

In this chapter, we shall follow the development of US airworthiness requirements from the Air Commerce Regulations through the Civil Air Regulations (CAR) up to their replacement by the Federal Aviation Regulations (FAR).

In this chapter, we will look at the various sets of airworthiness rules promulgated before 1965. Each subsection will present a brief summary of the inception of the rules and then discuss amendments giving a highlight of the significant changes. It is important to remember that there are airplanes certified under these rules still flying today (and likely to have future modifications which will need to be certified under these rules.

At this late date, it is difficult to determine some of the specific dates of these changes. These rules were all written before the advent of electronic records. Many documents are missing; most are scanned images from poor paper copies.

Figure 22 shows a timeline for the growth of aircraft certification requirements.

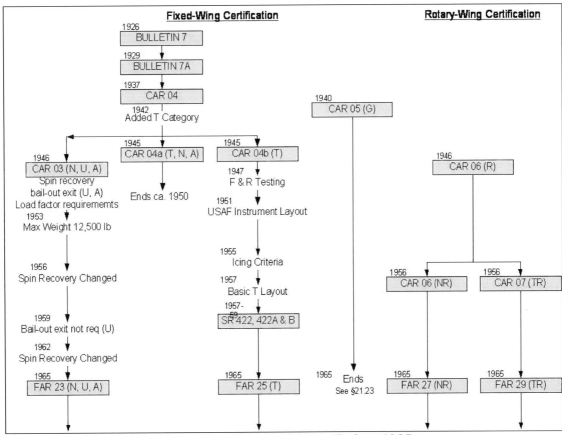

Figure 22 US Airworthiness Before 1965

We shall review the development of the aircraft standards in chronological order:

Rule	Aircraft Type	Year	Page
Bulletin 7A	Airplanes	1926	p. 52
CAR 4	Airplanes	1937	p. 54
CAR 5	Gliders	1940	p. 60
CAR 4a	Airplanes	1945	p. 57
CAR 4b	Airplanes	1945	p. 57
CAR 3	Airplanes	1946	p. 61
CAR 6	Helicopters	1946	p. 66
CAR 7	Helicopters	1956	p. 68

We will also examine two special aircraft categories, Restricted Category, CAR 8 on page 70 and Limited Category (for military surplus aircraft), CAR 9 on page 70.

Finally, we will look at the development of Aircraft Engine Requirements (Bulletin 7G and CAR 13) on page 72 and Propeller Requirements (Bulletin 7G and CAR 15) on page 74.

Aeronautics Bulletin 7A (1926)

On May 20, 1926, the Air Commerce Act of 1926 was signed into law. In August, William P MacCracken took office as the head of the aeronautical branch of the Department of Commerce. On December 31st, the first Air Commerce Regulations became effective.[1] Subsequently, regulations covering the airworthiness approval of airplanes were promulgated, Bulletin 7A,[2] Bulletin 14,[3] and Bulletin 26.[4]

Figure 23 Buhl CA-3: First Airplane Certified in US*

*   https://images.search.yahoo.com/yhs/search?p=Picture+of+Buhl+CA-3+Airplane&fr=yhs-symantec-ext_onb&hspart=symantec&hsimp=yhs-ext_onb&imgurl=https%3A%2F%2Fupload.wikimedia.org%2Fwikipedia%2Fcommons%2Fthumb%2F5%2F54%2FJ4_buhlverville.JPG%2F300px-J4_buhlverville.
JPG#id=35&iurl=https%3A%2F%2Fupload.wikimedia.org%2Fwikipedia%2Fcommons%2Fthumb%2F5%2F54%2FJ4_buhlverville.JPG%2F300px-J4_buhlverville.JPG&action=click, accessed November 27, 2020

Because of a lack of data, the initial airworthiness requirements did not include lighter-than-air aircraft. Heavier-than-air aircraft were divided into three classifications*:

 I. Normal airplanes (all aircraft other than light airplanes or gliders;

 II. Light airplanes (aircraft less than 1,000 lb gross weight with a wing loading of not more than 6 lb/ft^3); and

 III. Gliders (aircraft having no power unit.†[5]

No specific glider requirements were published. The necessary requirements were to be determined on a case-by-case basis.

The applicant for a Type Certificate (TC) was to provide a set of drawings and a drawing list, a weight and balance diagram, and a stress analysis. The stress analysis requirements follow the pioneering work of Doolittle[6][7] and are found in Bulletin 26.[4] For the first time, the airworthiness approval follows the rational approach of later regulations.

Flight Tests

Flight test, too, changed from the demonstrations (such as flying figure eight patterns). We now have requirements like "shall have adequate control," and "shall be longitudinally, laterally, and directionally stable".

There was a requirement for all airplanes to be able to recover from a six turn spin with neutral controls and power off. Airplanes of more than 4,000 lb gross weight will not be subject to the spin test unless the Secretary of Commerce believed it to be advisable.

The maximum limited [air]speed was not to be exceeded in any flight test. This differs philosophically from modern regulations. Today, an adequate margin between demonstrated airspeed and placarded maximum airspeed must be demonstrated.

Figure 24 Douglas DC-3: Certified Under Bulletin 7A‡

---

* At the time, helicopters and autogyros had not been developed

† With the adoption of CAR 5, the definition of a glider was changed to "A glider is a heavier-than-air aircraft the free flight of which does not depend principally upon a power generating unit".† This allowed for so-called *self-launching* gliders.

‡ Copyright © Akira Uekawa, used by permission.

Seats and Safety Belts

In their review of crashworthiness requirements, the NTSB cited a safety belt requirement that "Seatbelts and their attachments shall be capable of withstanding a load of 1,000 pounds applied in the same manner as would be applied in a crash.[8] The earliest regulatory reference we could identify was Bulletin 7A in 1934.[9]

Representative Aircraft Certified

The first airplane to receive a Type Certificate was the Buhl CA-3 (TC: ATC-1). Other well-known models certified under Bulletin 7A include Beech 18, Beech 17, Douglas DC-3(TC: A-618), Lockheed L-10, and Waco UPF-7.

Civil Air Regulation 4 (1937)

On May 6, 1935, a Transcontinental and Western Air (TWA) DC-2 crashed near Atlanta, MO, killing five of the eight persons aboard.* Senator Bronson M. Cutting of New Mexico was among the fatalities. A Bureau of Air Commerce report cited the accident's causes as the US Weather Bureau's failure to predict hazardous weather and misjudgments by the pilot and TWA ground personnel.[10] In June 1936, however, a committee chaired by Sen. Royal S. Copeland issued a report alleging that the tragedy was caused by malfunctioning navigational aides and voicing other criticisms of the Bureau of Air Commerce.[11] The controversy gave impetus to a number of reforms that eventually resulted in the Civil Aeronautics Act of 1938.

CAR 4, *Airplane Airworthiness*, was issued in November 1937[12] For the purpose of these regulations, airplanes were dividend into Normal and Light Airplanes. Light airplanes were defined as airplanes less than 1,000 lb and wing loading no more than 6 lb/sq ft. This definition is identical to that in Bulletin 7A.

Procedurally, the following data was required for the issuance of a Type Certificate (TC):

- ✈ Set of Drawings
- ✈ Drawing List
- ✈ Structural Analysis
- ✈ Test Reports
- ✈ Material Specifications
- ✈ Equipment List
- ✈ Weight and Balance Report
    - Balance Diagram
    - Weight Table
- ✈ Structural Report
- ✈ Structural Analysis

The initial version of CAR 4 required that passenger seats be securely fastened in place.[13] [14] Loads were to be based on a passenger weighing 170 lb.[15] Seat belts were to be approved for either one or two persons, based on the strength of the belts. Belts for one person were to be capable of supporting a load of 1,000 lb; belts for two persons, load of 2,000 lb.[16]†

There was a provision for certifying a single airplane without a Type Certificate. In this case, the secretary of commerce would require such ground and flight tests deemed necessary.

---

*   This accident is described in Appendix 4 on page 262.
†   Belts for glider use only needed to be capable of supporting 850 lb.

Performance demonstrations include:

> ✈ Landing speed: <65 MPH up to 20,000 lb
>   <70 MPH 30,000 or more
> (airplanes not approved for passengers could add 5 MPH to these limits))
> ✈ Takeoff in 1,000 ft (SLS, calm winds)
> ✈ Initial climb rate will be 8 ×stall speed (or 300 ft/min)

In the equipment requirements, CAR 4 differentiated between airline carrier-passengers (ACP), airline carrier-goods (ACG), and non-airline carrier (NAC).

Maneuvering load factors range from 4.33 at 1,000 lb GW to 1.5 at 45,000 lb GW and is shown in the figure. The specific values "are derived largely from experience with conventional type of airplanes and shall be considered as minimum values.[17] No further discussion of the required load factors was presented.

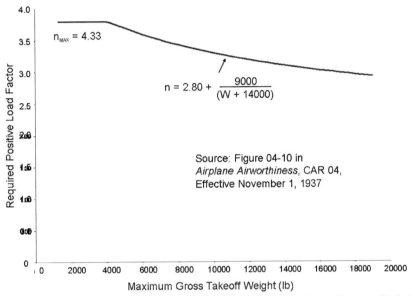

$$n = 2.80 + \frac{9000}{(W + 14000)}$$

Source: Figure 04-10 in
Airplane Airworthiness, CAR 04,
Effective November 1, 1937

Figure 25 CAR 4 Load Factor Variation with Airplane Gross Weight

As with Bulletin 7A, there was no flight test performed beyond the certified flight envelope.

1938: Creation of the CAA

On August 22, 1938, the Bureau of Air Commerce and the Bureau of Air Mail were transferred the Civil Aeronautics Authority (CAA), an independent federal agency.

Also beginning in 1938, Civil Aeronautics Manuals (CAM's) were published containing both the airworthiness regulations and explanatory and supplemental material supporting the regulations. The notation followed the rules, so that, for example, the manual material CAM 04.030 corresponded to the rule CAR 4.030.

1941: Aircraft Categories

In 1941, CAR 4 was changed to include three airplane categories:[18]

> ✈ Acrobatic Category (A)
> ✈ Normal Category (N)
> ✈ Transport Category (T)

There was no stated definition of these categories. There were also no specific requirements for Normal or Acrobatic Categories.

T-category airplanes now had to provide takeoff and climb data over a range of weights and altitudes. Takeoff data now had to consider failure of the critical engine at a point that would allow either continued takeoff and climb or safely rejecting the takeoff. Takeoff data would now include the distance from liftoff to clearing a fifty-foot "screen height". No temperature account-ability was required. All the performance data was to be provided in an "Airplane Operating Manual".

1944 Publication of CAM 04T

In 1944, CAM 04T was published.[19] This document identified all of the requirements of CAR 4 that applied specifically to Transport Category airplanes. It also interpreted these regulations and indicated means of compliance.

The initial published version required protection for emergency landings

Category	A
Upward	2 g,
Forward	6 g
Sideward	1.5 g,
Downward	4.5 g.[20]

The standard weight for passengers and crew was set at 170 lbs per person.[21]

Figure 26 Piper J-3 *Cub*: Certified Under CAR 4*

Airplanes Certified Under CAR 4

Airplanes certified under CAR 4 include the Boeing 377 *Stratocruiser* (TC: ATC-812), the Piper J-3 *Cub* (TC: ATC-660), and the Taylorcraft BC-45.

Termination of CAR 4

No new certification projects under CAR 4 were accepted after December 1945. However CAR 4a and CAR 4b replaced CAR 4

---

* http:// flyawaysimulation.com/media/images2/images/piper-cub-J-3-D-EOJM-fsx1.jpg, accessed November 27, 2020

Civil Air Regulation 4a (1945)

In 1945, the CAA split CAR 4 into CAR 4a and CAR 4b.. The earliest version of CAR 4a found was an amendment effective November 22, 1947.[22] Based on inspection of this amendment and the earliest complete version of CAR 4a (dated, April 7, 1950),[23] there were no significant changes from CAR 4.

Like CAR 4, CAR 4a was divided into three Categories: Normal Category , Transport Category, and Acrobatic Categories. There was no stated definition of these categories.

Airplanes Certified Under CAR 4a

Airplanes certified under CAR 4a include Cessna 140, Ercoupe 415, Aeronca 7AC *Champion*, and Boeing 75 *Kaydet*.

Termination of CAR 4a

CAR 4a was terminated as a certification basis in approximately 1950. We were unable to find any records stating the exact date.

Civil Air Regulation 4b (1945)

In 1945, the CAA split CAR 4 into CAR 4a and CAR 4b. The earliest complete version of CAR 4b found in government archives was the version dated December 31, 1953.[24] Based on the individual amendments that preceded this document, it appears that CAR 4b was a reorganization of previous requirements dealing solely with transport airplanes.

The performance calculations in CAR 4b were based on aircraft weight, wind, and air density corrected for temperature and humidity. However, while 100 % of the humidity effect was considered only 50% of the temperature effect was considered.

Figure 27 Lockheed L-1649 *Starliner* Certified Under CAR 4b (NASA Photo)*

---

*   A Trans World Airlines Lockheed L-1649 *Starliner* (civil registration N7301C) in flight in the late 1950s or the early 1960s, during a test flight. Note the flight test pitot boom attached to the right wing-tip. This aircraft was written off after an accident at Bogotá-Eldorado Airport, Colombia, on 18 December 1966, while in service with *Aerocondor de Colombia*. 17 of 59 passengers and crew were killed. NASA Photo AILS A83-0499-18

1947: Function and Reliability Tests

In March 1947, CAR 4b was revised to require service tests specifically designed to ascertain reliability and proper functioning of the airplane and its equipment.[25]

1951: Change to Instrument Layout

In 1951, CAR 4b was revised to place greater emphasis on cockpit standardization.[26] All newly certified transport airplanes would have to have cockpit instruments arranged according to the US Air Force standard.[27][28] This arrangement is shown in Figure 28.

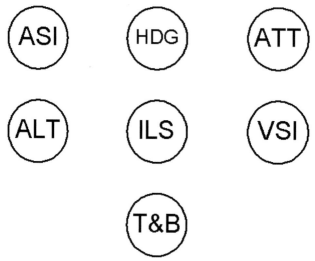

Figure 28 Basic Flight Instrument Panel Arrangement (CAR 4b, 1951)

The rationale for this change was

> **"[I]t is important from a national defense standpoint that there be a substantial similarity not only between civil aircraft operated by different air carriers but between civil and military aircraft, since it may be expected that civil aircraft will be used by the military..."[26]**

1953: Revised Maneuvering Loads in Part 4b

The positive maneuvering load factor was to be selected by the applicant, but must be at least 2.5. The negative maneuvering load factor was to -1.0 up to the design cruise speed, V_C and then decrease linearly to zero at the design dive speed, V_D. Note that the load factor requirements no longer were a function of the aircraft weight.[29]

1955: Incorporation of Icing Protection

CAR 4b (§4b.640) was revised to incorporate a comprehensive and detailed set of standards intended to provide protection in type of icing conditions which might be reasonably expected during operations.[30] The changes were based on research reports published by the NACA.[31][32] These reports were based on a number of flight encounters with icing conditions within clouds during the late 1940s. These data were developed into two icing envelopes shown in Figures 9 and 10. These envelopes have been incorporated into all US certification rules and were the only icing envelopes in use until 2015.

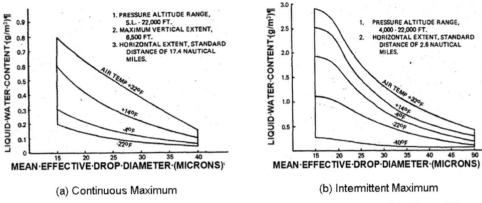

(a) Continuous Maximum   (b) Intermittent Maximum

Figure 29 Continuous and Intermittent Maximum Atmospheric Icing Conditions

It is important to realize that these envelopes did not consider flight in freezing precipitation, a fact that appeared to surprise many following the accident to EHF 4184 in October 1994.*

1957: Change to Instrument Layout – Version II

In 1957, in response to a petition from the Air Line Pilots Association (ALPA), CAR 4b was again revised to require a different standard cockpit arrangement.[33] All newly certified transport airplanes would have to have cockpit instruments arranged according to the "Basic T" in Figure 10. This "Basic T" has since become the standard for all airplanes.

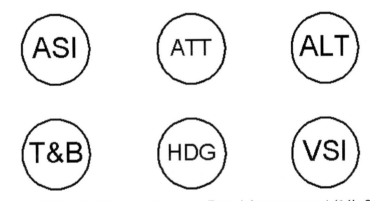

Figure 30 Basic Flight Instrument Panel Arrangement (§4b.611)

1957 through 1959: Turbine-Powered Airplanes

In 1957, there were a number of turboprop and turbojet certification projects within the FAA. The FAA became concerned that there was one area in which CAR 4b would be inadequate for certifying turbine-powered airplanes' performance. Since many of these projects were in place, changing the text of CAR 4b would not apply to active projects. Accordingly, a Special Civil Air Regulation (SR) was adopted that would apply whether or not there was an active project. The result was SR-422,[34] which applied to all turbine-powered airplanes not yet certificated.

SR-422 introduced full temperature accountability in all stages of performance, except for landing distance.† Because of the introduction of full temperature accountability, the specific climb requirements were also reevaluated. One of the more significant was the change to minimum climb gradients rather then climb rates.

---

*   Thie accident, known by the location "Roselawn," is described in Appendix 4 on page 334.

†   Landing distance is strongly influenced by pilot technique and is normally analyzed empirically.

In 1958, SR-422A was issued to modify the takeoff requirements and introduced a new requirement for the all-engine-takeoff in addition to the engine failure at V_1 case. The regulation now permitted "unbalanced field lengths" and allowed for the use of clearways and stopways.

In 1959, SR-422B was issued to include the concept of a certified rotation speed, V_R. It also had some minor changes.

Airplanes Certified Under CAR 4b

Airplanes certified under CAR 4b include Boeing 707, Boeing 727, Lockheed L-188 *Electra*, and Douglas DC-7.

Termination

In 1965, CAR 4b, SR-422, SR-422A, and SR-422B were cancelled and replaced by FAR 25.

Figure 31 Boeing 727 – Certified Under CAR 4b*

Civil Air Regulation 5 (1940)

In May 1940, the CAA published a new set of airworthiness requirements, CAR 5, for gliders.[35] This new rule required that gliders would have to be registered with the CAA. Gliders were divided into three classifications as shown in Table 7.

As with CAR 4, there was a provision to allow for a single aircraft to be certified without obtaining a TC.

7

Table 7 Glider Classifications in CAR 5

Class	Description	Airplane Tow	Blind Flying
Class I	High Performance Intermediate Utility	Permitted	Permitted
Class II	High Performance Intermediate Utility	Permitted	Not Permitted
Class III	Intermediate Utility	Not Permitted	Not Permitted

As glider pilots ourselves, we were surprised at the restriction of Class III to ground tows only.

1962: Addition of Rotary-Wing Gliders

In 1962, CAR 5 was amended to allow for rotary-wing gliders. To the best of our knowledge, there have been none certified.

Airplanes Certified Under CAR 5

Gliders certified under CAR 5 include the Schweizer SGS 2-32 (shown in Figure 32 as the Navy -26). The last glider certified under CAR 5 was the Aer-Pegaso M-100S which was approved in February 1965.

Figure 32 Schweizer SGS 2-32 – Certified Under CAR 5*

Termination of CAR 5

In 1965, CAR 5 was cancelled without replacement by the adoption of the Federal Aviation Regulations.

Civil Air Regulation 3 (1945)

In 1945, The Civil Aeronautics Board (CAB) published CAR 3 covering Normal, Utility, Acrobatic, and Restricted Purpose Categories.[36] The rule applied to non-transport airplanes, however

---

*   US Navy Photo. The aircraft shown was designated an X-26 and used by the US Navy Test Pilot School at Patuxent River NAS. It is not clear if this means one of your authors (RLN) can claim to have flown an "X plane".

current certification projects could continue under CAR 4 through 1946. Four Categories were described, for the first time with definitions:

Normal Category

Normal Category airplanes are intended for non-acrobatic, non-scheduled passenger and cargo operations. The required design maneuvering load factor for Normal Category airplanes was a function of weight.

$$n+ \quad = \quad 2.1 + [24000 / (W + 10000)] \tag{1}$$

$$n- \quad = \quad -0.4 \times n+ \tag{2}$$

The positive design load factor (n+) need not be greater than 3.8 for airplanes with a MTOW less than 4117.6 lb. For airplanes with a MTOW greater than 50000 lb, the positive design load factor must not be less than 2.5. This variation with weight is shown in the Figure 33.

In the range of airplane design weights between 3000-12000 lb, these load factors were 3-5 percent less than the original CAR 4 values, shown in Figure 25.

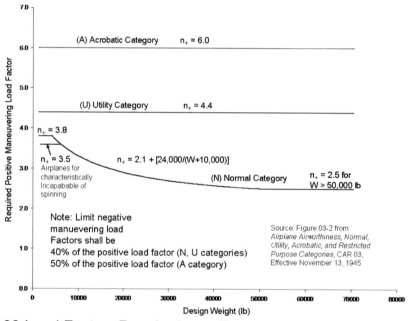

Figure 33 Load Factors Required as a Function of Gross Weight (CAR 3)

All airplanes certified in Normal Category were to be placarded against spins unless demonstrated to be "characteristically incapable of spinning".* All airplanes of 4,000 lb or less gross weight were required to recover from a one-turn spin with the controls assisted to overcome friction in not more than one and one-half additional turns and without exceeding either the maximum load factor or the maximum allowed airspeed. These tests were to be conducted at any permissible weight and CG.

It must not be possible to obtain an uncontrollable spin by means of any possible use of the controls.

All seats, berths, and supporting structure in airplanes certified in the Normal Category must be designed for a passenger weight of 170 lb.

---

* In which case they were to be certified "Characteristically Incapable of Spinning,"

The initial published version of CAR 3 required protection for emergency landings

Category	N or U	A
Upward	3 g	4.5 g,
Forward	9 g	9 g,
Sideward	1.5 g	1.5 g.[37]

Remarkably, provision was required for safety belts or harness in all seats or berths, but actual installation was not required. Nott until 1950 were approved safety belts actually required for all airplanes manufactured after January 1, 1951.[38]

For airplanes with retractable landing gear, the structure must be designed to protect the occupants and a wheels-up landing with moderate descent velocity. The structure must be designed to protect the occupants in a complete turnover unless the "configuration renders the possibility of turnover remote". The assumed occupant weights were 170 lb for Normal Category and 190 lb for Utility and Acrobatic Categories.[39] This increase was that these occupants were assumed to be wearing 20 lb parachutes

Airplanes to be certified as "Characteristically Incapable of Spinning" must be demonstrated in flight that it was would not spin with:

➤ Gross weight 5% greater than the weight for which approval is desired;

➤ CG 3% aft of the rearmost CG for which approval is desired;

➤ Upward elevator travel 4° beyond the limit for which approval is desired; and

➤ Rudder travel 7° beyond the limit for which approval is desired.

These airplanes need not be designed with a design maneuvering factor greater then 3.5 when using Equation (1).*

Figure 34 *Ercoupe*: Certified as "Characteristically Incapable of Spinning"†

Utility Category

Utility Category airplanes are intended for normal operations and limited acrobatic maneuvers. Such airplanes are not suited for snap or inverted maneuvers. Their positive design maneuvering

---

*   Equation (1) is found on page 62.

†   https://i.pinimg.com/originals/13/ad/5b/13ad5bb700780659309368b339d06065.jpg, accessed November 27, 2020

load factor was to be at least +4.4. As before, the negative load factor was to be at least 40% of the whatever value was chosen for the positive load factor.

Airplanes to be certified in the Utility Category were required to satisfy all spin requirements for either the Normal Category (including placard requirements) or the Acrobatic Category.

Airplanes certified in the Utility Category were to be provided with exits to permit all occupants to quickly bail out with parachutes.

All seats, berths, and supporting structure in airplanes certified in the Utility Category must be designed for a passenger weight of 190 lb (which includes a 20 lb parachute).

Acrobatic Category

Acrobatic Category airplanes have no specific maneuver restrictions except those required, based on flight tests. The positive design maneuvering load factor was to be at least +6.0. The negative design maneuvering load factor was to be 50% of whatever value was chosen for the positive load factor.

All Acrobatic Category airplanes must be capable of spinning and must be able to recover from a six-turn spin with controls free in not more than four additional turns. It also must be able to recover at any point from a spin up to six turns with normal recovery technique in not more than one and one-half additional turns and without exceeding either the maximum load factor or the maximum allowed airspeed. These tests were to be conducted at any permissible weight and CG.

It must not be possible to obtain an uncontrollable spin by means of any possible use of the controls.

Airplanes certified in the Acrobatic Category were to be provided with exits to permit all occupants to quickly bail out with parachutes.

All seats, berths, and supporting structure in airplanes certified in the Acrobatic Category must be designed for a passenger weight of 190 lb (which includes a 20 lb parachute).

Restricted Purpose

There airplanes were intended to be operated for restricted purposes not logically encompassed by the foregoing Categories. The requirements for Restricted Purpose Category are those for any one of the foregoing Categories which are not incompatible with the nature of the special purpose, plus any suitable operating limitations which the Administrator finds will provide an appropriate level of safety.

All Categories

While CAR 3 is often thought of as applying only to small airplanes, at its inception, there was no maximum weight specified for any of the categories. The Curtiss C-46 (*Commando*) with a maximum takeoff weight of 46,800 lb was certified under CAR 3 (Normal Category).

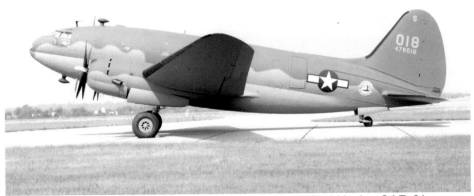

Figure 35 Curtiss C-46 *Commando*: Certified Under CAR 3*

Airplanes could be certified under multiple categories, provided requirements for each category were met.

In addition to maneuvering loads, airplanes had to be designed to withstand a vertical 30 ft/sec gust at V_C decreasing linearly with speed to a 15 ft/sec gust at V_D.

The never-exceed speed, V_{NE} was to be no greater than the lesser of 90 % of V_D, the design dive speed or 90 % of V_{DF}, the demonstrated dive speed.

1951: Stall and Spin Changes

In January 1951, the spin requirements were changed for Normal Category airplanes to recover from a one-turn spin with normal recovery control inputs in not more than one additional turn.

In addition, a requirement for a clear and distinct stall warning with the flaps and landing gear in any position was added.

Additional performance was required for airplanes with MTOW of more than 6,000 lb.

1952 Spin Changes – Version II

The acrobatic spin requirements were changed to require recovery from a six-turn spin or at any point from a six-turn spin in not more than one and one-half additional turns with the controls in the normal recovery position.

1953: Maximum Weight Now 12,500 lb

CAR 3 now stated a maximum gross weight of 12,500 lb. In addition Airplane Flight Manuals were no longer required for airplanes of 6,000 lb or less.

1957: Spin Changes – Version III

Airplanes "characteristically incapable of spinning" now had to use Equation (1)† except that n+ need not be greater than 3.8 up from 3.5.

1962: Spin Changes – Version IV

Spin requirements were changed (again). Normal Category airplanes must recover from a 1-turn spin in 1 additional turn with the controls as required

---

*   National Museum of the Air Force Photograph
†   Equation (1) is found on page 62.

Acrobatic Category airplanes must recover from at any point in a 6-turn spin in 1-1/2 additional turns the controls as required (no change). They must also recover from a 1-turn flaps-down spin in 1 additional turn. They may retract the flaps during the recovery provided the airplane has a placard against intentional flaps-down spins.

All multi-engine airplanes must have "no undo spinning tendency."

Termination of CAR 3
In 1965, CAR 3 was cancelled and replaced by FAR 23.

Airplanes Certified Under CAR 3
Airplanes certified under CAR 3 include Beech 35 *Bonanza*, Beech 90 *King Air*, Cessna 150, Cessna 172, Curtiss C-46 *Commando*, Ercoupe, Lear 23, and Taylorcraft BC-65.

Figure 36 Taylorcraft BC-65: Certified Under CAR 3*

Civil Air Regulation 6 (1945)

In 1945, the CAB published CAR 6 which provided airworthiness requirements for rotorcraft.[40] The first helicopter certified was the Bell 47 (Army H-13 *Sioux*)

There were no weight limits. Initially, the flight requirements were quite vague: "All rotorcraft shall have such general performance and flight characteristics as to provide reasonable safety during any [appropriate] maneuvers."

In 1956, rotorcraft certification was split into two with the creation of a new CAR 7 for aircraft over 6,000 lb. CAR 6 was now limited to 6,000 lb.

Normal Category Rotorcraft
In 1956, rotorcraft certification was split into two with the creation of a new CAR 7. CAR 6 was now limited to 6,000 lb aircraft and now labeled Normal Category.[41]

---

*   www.airliners.net, accessed November 27, 2020

Small helicopters have some requirements based on rotary-wing characteristics. Rotor speed limits must be established to allow for safe autorotation following an engine failure. Rotor blade pitch limits must be established and a low rotor rpm warning is required to warn the pilot.

Helicopter best rate-of-climb speed was to be determined at sea level standard conditions. For multi-engine helicopters, the best rate-of-climb (or minimum rate-of-descent) performance was to be determined with one engine inoperative (OEI).

The hovering in ground effect (HIGE) ceilings for helicopters was to be determined over the range of weights, altitudes, and temperatures (WAT) over which the helicopter was to be certified. This ceiling was to be at least 4,000 ft under standard conditions.

Takeoff performance was to be determined up to the maximum altitude for which certification is desired.

The minimum rate intended of descent airspeed and the best angle of glide airspeed must be determined in autorotation at maximum weight and rotor speed selected by the applicant.

Flight load requirements were stated in 06.212. The structure was to be designed for flight maneuvering load factors of +3.5 and -1.0 throughout the flight envelope.[42] These load factors could be reduced to +2.5 and -0.5 if proven by analysis and flight demonstrations. The structure was to be substantiated for vertical gusts up to ±30 ft/sec throughout the airspeed range, including hovering.[43]

The first requirements for emergency landing conditions were found in 1956. The assumed occupant weight was 170 lb.[44][45] The design crash loads for occupant protection were set at

Direction	Forces
Upward	1.5 g,
Forward	4 g,
Sideward	2 g,
Downward	4 g.[46][47]

We were unable to find any discussion regarding data supporting the choice of crash loads.*

Figure 37 Bell 47 (*Sioux*): Certified Under CAR 6†

---

* No copies of the NPRMs were located.

† http://images.search.yahoo.com/yhs/search;_ylt=AwrWp2fJKcFfPy8ASQ8PxQt.;_ylu=Y29sbwNncTEEcG9 ak0.pinimg.com%2Foriginals%2F32%2Fb1%2F8a%2F32b18a0974cf3cb1b56259da48666187.jpg&acti on=click, accessed November 27, 2020

Aircraft Certified Under CAR 6

The first helicopter certified was the Bell 47 (Army H-13 *Sioux*). Other helicopters certified under CAR 6 include Bell 206 *Jet Ranger* and Hughes 369 *Cayuse*. Gyroplanes certified include the Umbaugh 18A

Termination of CAR 6

In 1965, CAR 6 was cancelled and replaced by FAR 27.

Civil Air Regulation 7 (1956)

In 1956, the CAB published a new CAR 7 for Transport Category rotorcraft. Two such categories were created: Transport Category A for multi-engine rotorcraft and Transport Category B for single or multi-engine rotorcraft of 20,000 lb or less weight.[48]

Helicopter performance was to cover the range of atmospheric conditions – including sea level standard (SLS) conditions. Engine power was to be based on 80% relative humidity (RH) or 0.7 in Hg water vapor pressure, whichever is lower.

Rotor speed limits must be established. A range of power-on rotor speeds was to be established to provide an adequate margin consistent with appropriate maneuvers. A means to prevent rotor speeds substantially below these limits was to be provided. A range of power-off rotor speed limits shall be established to allow for safe autorotation.

Flight load requirements were the same as Part 06. The structure was to be designed for flight maneuvering load factors of +3.5 and -1.0 throughout the flight envelope.[49] These load factors could be reduced to +2.5 and -0.5 if shown to be extremely improbable by analysis and flight demonstrations. The structure was to be substantiated for vertical gusts up to ±30 ft/sec throughout the airspeed range, including hovering.[50]

Category A Performance Requirements

If a range of heights exists at any speed (including zero) within which it is not possible to make a safe landing following failure of the critical engine with TO power on the remaining engine(s), a height/velocity (H/V) diagram must be established.

The takeoff path must be defined with all engines operating up to a defined engine-failure point at which the critical engine is assumed to fail suddenly. From that point on it shall be possible to continue the takeoff with the engine failed to a point where an engine-out climb is established, or discontinue the takeoff and land safely in the takeoff area. The climb speed, known as the takeoff safety speed (V_{TOSS}), must be established.

This is similar in intent to transport airplane takeoff certification. The major distinction is that the helicopter may be airborne at the decision point and that the decision point must be at or above V_{TOSS}. Takeoff performance must be established at all weights, altitudes, and temperatures (WAT) selected for certification.

Takeoff climb at V_{TOSS} must be shall be at least 100 ft/min with the landing gear extended, the CG in the most unfavorable position, the critical engine inoperative, and the remaining engine(s) at takeoff power.

Enroute climb performance shall be determined over the range of aircraft weights, altitudes and temperatures selected for certification (WAT), with the landing gear retracted, the CG in the most unfavorable position, the critical engine inoperative, and the remaining engine(s) at maximum continuous power.

Hovering performance shall be determined over the WAT (for takeoff) in ground effect with the landing gear extended.

Landing performance shall be determined over the WAT with the operating engines not exceeding their approved limitations. The approach and landing must not require an exceptional degree of skill on the part of the pilot or exceptionally favorable conditions.

Category B Performance Requirements

If a range of heights exists at any speed (including zero) within which it is not possible to make a safe landing following complete engine failure, an H/V diagram must be established. Optionally, multi-engine helicopters may comply with the Category A requirements.

Takeoff performance shall be established with the most unfavorable CG location. The takeoff shall allow a safe landing at any point along the flight path.

Enroute climb performance shall be determined over the WAT, with the landing gear retracted, the CG in the most unfavorable position, and engine(s) at maximum continuous power.

Hovering performance shall be determined over the WAT (for takeoff) in ground effect with the landing gear extended. At maximum weight, the hovering ceiling shall be at least 4,000 ft under standard atmospheric conditions.

Structure

Proof of compliance was to be established by tests unless the structure is of a type where experience shows that analysis is reliable. Tests included:

- ✈ A 200 hour dynamic and endurance test of the rotors and rotor drives including controls. This test was to be conducted on the aircraft and the power was to be absorbed by the actual rotors to be installed. There was a very specific schedule for this test.
- ✈ For Category A: A 150-hour endurance test for each gearbox in the rotor drive system.
- ✈ A static loading test on the control system.
- ✈ An operation test on each control system. This test was to operate each control from the pilot compartment with the entire system loaded. There was to be no jamming, excessive friction, or excessive deflection.
- ✈ Flight stress measurements on main and auxiliary rotors and on pylon structures. The structure must be designed to withstand these loads. The service life of these parts was to be based on fatigue tests or other methods acceptable to the Administrator.
- ✈ A ground vibration test to determine natural frequencies of the blades and other structural components.

Crashworthiness requirements were identical to those in CAR 6.[51]

1958: Category A Takeoff Change

The takeoff provisions of §7.113 were amended to all the critical decision points to be at an airspeed less than V_{TOSS}.[52]

1959 Deletion of Ground Vibration Test

The ground vibration test to determine natural frequencies was deleted. It was based on the conclusion that any such natural frequency would become apparent during other ground and flight tests.[52]

1962 Deletion of Gearbox Test

The 150-hour gearbox test was eliminated as it was concluded that any problems in the gearbox would become evident during other ground and flight tests.[53]

Helicopters Certified Under CAR 7

Helicopters certified under CAR 7 include Vertol 107* and Sikorsky S-61.

Figure 38 Sikorsky S-61: Certified Under CAR 7†

Termination of CAR 7

In 1965, CAR 7 was cancelled and replaced by FAR 29.

Civil Air Regulation 8 (1950)

In 1950‡, the CAB published a new CAR 8 for Restricted Category aircraft.[54] These are aircraft intended for a specific purpose for which strict compliance with existing TC requirements would be impractical. The following aircraft types are eligible:

↛ An aircraft type which has not been previously type certificated but which is shown by the applicant to comply with all of the airworthiness requirements of any other aircraft category, except those requirements that the Administrator finds inappropriate for the special purpose for which the aircraft is to be used;

↛ An aircraft type which has been manufactured in accordance with the requirements of, and accepted for use by, a US military service and subsequently modified for a special purpose (whether or not such aircraft has been issued a TC under the provisions of CAR 9; or

---

*   Type Certificate now owned by Columbia Helicopters, Aurora, OR.

†   https://images.search.yahoo.com/yhs/search;_ylt=AwrVk.PQKsFf4nkA7SgPxQt.;_ylu=Y29sbwNncTE
     EcG9zAzEEdnRpZAMEc2VjA3Nj?p=Photographs+of+Sikorsky+S-61+helicopter&fr=yhs-symantec-
     ext_onb&hspart=symantec&hsimp=yhs-ext_onb#id=8&iurl=https%3A%2F%2Fupload.wikimedia.org
     %2Fwikipedia%2Fcommons%2F4%2F47%2FBray_Air_Display_2012_-_Sikorsky_S-61N_Helicopte
     r_(7625461204).jpg&action=click, accessed November 27, 2020

‡   We were unable to locate any copies of CAR 8 published prior to 1950.

✈ A modification of a TC issued for an aircraft which has subsequently been modified for a special purpose when, upon inspection, the Administrator finds that the modification conform to good aeronautical practice and than no feature or characteristic of the aircraft renders it unsafe when operated un limitations prescribed for its use.

Appropriate operating limitations will be issued as are necessary for the safe operation of the aircraft and for the protection of the public. No passenger, not required for the special purpose, will be carried during the special purpose operations. If multiple airworthiness operations are issued, separate operating limitations shall be issued.

Aircraft Certified Under CAR 8

Aircraft certified under CAR 8 include Grumman G-164 *AgCat*.

Termination

In 1965, CAR 8 was cancelled and replaced by FAR §21.25.

Figure 39 Grumman G-164 *AgCat*: Certified Under CAR 8*

---

*   https://images.search.yahoo.com/yhs/search;_ylt=AwrT4R.kMMFf.IwArO42nIlQ;_ylu=c2VjA3NlYXJ jaARzbGsDYnV0dG9u;_ylc=X1MDMTM1MTE5NTcwMgRfcgMyBGFjdG4DY2xrBGNzcmNwdmlk AzdkWUVZREV3TGpLeTBmdjhYMzgyT2dWZk1qWXdNUUFBQUFDDWWRKU-jAEZnIDeWhzLXN5bWFudGVjLWV4dF9vbmIEZnIyA3NhLWdwbGBdwcmlkA0dIdFh0M244Um42 SWVrVklnYW-ZnaEEEbl9zdWdnAzAEb3JpZ2luA2ltYWdlcy5zZWFyY2gueWFob28uY29tBHBvcwMwBHBxc3Ry AwRwcXN0cmwDBHFzdHJsAzM0BHF1ZXJ5A3Bob3RvZ3JhcGhzJTIwb2YlMjBHcnVtbWFuJTIw QUdDQVQEdF9zdG1wAzE2MDY0OTY1MjA-?p=photographs+of+Grumman+AGCAT&fr=yhs-symantec-ext_onb&fr2=sb-top-images.search&ei=UTF-8&n=60&x=wrt&hsimp=yhs-ext_onb&hspart=symantec, accessed November 27, 2020

Civil Air Regulation 9 (1946)

In 1946, the CAB published a new CAR 9 for Limited Category aircraft — surplus military aircraft.[55]

CAR 9 stated four requirements:
1. The aircraft was a make and model originally designed, manufactured, and accepted for service by a military service of the US for combat or other specialized service;
2. There was no civilian aircraft of essentially the same basic model for which a TC has been issued;
3. The service record of this make and model in military service does not disclose any unsafe characteristics; and
4. Application was made prior to December 31, 1947 (later extended to June 30, 1965)

Aircraft certified as Limited Category were not allowed to carry passengers or cargo for hire.

Aircraft Certified Under CAR 9

Aircraft certified under CAR 9 include Lockheed L-382* *Hercules*, North American B-25 *Mitchell*, and Stinson L-1 *Sentinel*.

Termination

In 1965, CAR 9 was cancelled and replaced by FAR §21.27.

Engine Requirements (Bulletin 7G and CAR 13)

Aeronautics Bulletin 7G (1934)

Aeronautics Bulletin 7G contained the requirements for the design, construction, and testing of engines for installation in licensed aircraft.[56] These requirements did not apply to Light Category airplanes. This bulletin dealt with conventional internal combustion engines using a gasoline fuel.†

Design requirements included
> ✈ Dual ignition systems with at least two spark plugs per cylinder (not required for engines under 100 horsepower);
> ✈ All engines and engine accessories to be designed and constructed to reduce to a minimum the chances of in-flight failure and so far as possible to prevent fires in the air or in the event of a crash;
> ✈ Provision for installation of a means for preventing ice forming in the carburetor;
> ✈ No undue vibration at speeds which are to be used continuously; and
> ✈ Fuel system design shall provide satisfactory operation at all normal flight attitudes.

There was a requirement for an endurance test, but the hours were not given. A tear-down inspection was required following the test.

---

*   Except L-328J, which was certified under FAR 25.
†   Special rulings for other types of engines would be made as the occasion develops.

Civil Air Regulation 13 (1937)

Aeronautics Bulletin 7G was replaced by Civil Air Regulation 13 in 1937. The earliest documented version of CAR 13 we could find from the Department of Transportation archives was the version effective November 1, 1937.[57]

Six design requirements were listed

+ Dual ignition systems with at least two spark plugs per cylinder (not required for engines under 100 horsepower);
+ All engines and engine accessories to be designed and constructed to reduce the chances of in-flight failure and of fires;
+ Operate smoothly with no undue vibration;
+ Provide a means for preventing carburetor ice; and
+ Fuel system design should provide satisfactory operation in normal flight.

Engine testing for a sea level engine included a block test to be run in periods of at least 5 hours on consecutive workdays. Not more than three forced stops were allowed. Each forced stop was to add five hours to the test. The test duration was at least 100 hours which to be

+ 50 hours at full throttle at rated speed (plus or minds 3%) and
+ 50 hours at 75% of proposed rated power at propeller load speed.

A calibration power versus speed calibration shall be made before or after the block test.

A 10-hour flight test was to include a full throttle climb to 15,000 ft or to the service ceiling of the airplane.

If a takeoff rating in excess of the block test rating was desired, then the engine was to be run 10 hours at TO power to be run in 5-hour periods. This test may be run as the last 10 hours of the block test or as a separate test. An alternate TO power test of alternate 5 minutes at TO power and 5 minutes at idle. If the alternate schedule is chosen, the test duration was to be twenty hours to be on consecutive workdays.

Following completion of the block test and calibration test, a complete tear-down and detailed inspection of engine parts shall be made with particular attention to excessive wear or signs of failure.

Altitude engines will use the preceding tests except that the ground tests will be conducted at the proposed rated power instead of full power. The flight test will be conducted at rated power below the rated altitude.

Engines intended for installation in light aircraft (as defined in CAR 4) may be tested to show the maximum power and speed. In this case, the engine model will be issued a Power Rating, and not a Type Certificate.

1941 Dual Ignition

In 1941, CAR 13 was amended to require dual ignition on all engines regardless of power.[58]

1956 Inclusion of Turbine Engines

In 1956, CAR 13 was amended to include turbine engines.[59]

Turbine engines now required to be free of detrimental surge throughout the operating range in the minimum ambient air temperature in which it is to be operated. The induction system was to be designed to minimize the danger of ice accretion.

For turbine engines, a vibration survey shall be conducted to investigate the vibration characteristics. If critical vibration is found, the engine design shall be made to eliminate it or the endurance test must include operation under the most adverse vibration condition for a period sufficient to operate without fatigue failure.

Prior to the endurance test, the power control(s) shall be adjusted to produce the maximum allowable gas temperatures and rotor speeds at takeoff condition. This setting should not be changed during the calibration and endurance testing.

The 150-hour endurance test shall consist of 30 periods of 5 hours each. Each 5 hours period shall be

> ✈ 60 minutes with alternating periods of
>> 5 minutes at TO power(thrust) and
>> 5 minutes at idle;
> ✈ 30 minutes at the greater of
>> 91% TO power(thrust) or
>> MC power(thrust);
> ✈ 90 minutes at MC power(thrust);
> ✈ 60 minutes at 90% MC power(thrust);
> ✈ 30 minutes at 75% MC power(thrust); and
> ✈ 30 minutes with alternating periods of
>> 30 seconds of TO power (thrust) and
>> ~5 minutes at idle.

The acceleration/deceleration steps between idle and TO should be accomplished in ~1 sec.

During the endurance tests, 75 engine starts should be made. Thirty of these should follow a 2-hour shutdown. It is acceptable to have the remaining starts after the completion of the 150-hour endurance testing.

A complete engine teardown inspection is required after completion of the above testing. A detailed inspection of the engine parts is to check for fatigue and wear.

1959 Changes

In 1959, CAM 13 was published to include further amendments since the previous section.[60]

Significant changes for reciprocating engines included requirements for induction systems to be self-draining to prevent liquid locks when the airplane is parked and for the crankcase to be vented to the atmosphere. There were also four-plus pages of testing requirements for approving new sparkplugs!

Turbine engines were now required to have dual ignition systems.

There were also additional requirements for thrust reversers for turbojet engines and negative-torque systems for turboprop engines.

Termination of CAR 13

In 1965, CAR 13 was cancelled and replaced by FAR 33.

Propeller Requirements (Bulletin 7G and CAR 14)

Aeronautical Bulletin 7G (1934)

Aeronautics Bulletin 7G also contains the requirements for the design, construction, and testing of propellers for licensed aircraft.[56] These requirements did not apply to Light Category airplanes.

Propeller design requirements included:

> ✈ Operation without excessive vibration or flutter;
> ✈ Fabricated with suitable materials;
> ✈ Smooth surfaces, faired with no abrupt curvature or irregularities;

➤ Metal surfaces should be machine smooth without tool marks;

➤ All changes in cross-section should be faired with as large fillet as possible; and

➤ Wood propellers to be finished so that the wood grain is visible.

Propeller testing requirements for metal or variable pitch propellers were:

➤ Fifty-hour endurance block test rigidly mounted on an internal combustion engine of the same general characteristics to be used in service;

➤ Tests may be run without stopping or broken into 5 hours (or more) at a time; and

➤ Variable pitch propellers shall operated over the pitch range at least once per hour. The engine power may be reduced to prevent overspeed when changing pitch.

A 10-hour block test was required for fixed-pitch wood propellers.

Vibration surveys were recommended for metal propellers.

A detailed inspection for failures or suspected failures was to follow the propeller testing.

Civil Air Regulation 14 (1937)

CAR 14 replaced Aeronautics Bulletin 7G in November 1937 with no significant changes.[61]

CAM 14 (1938)

The first Civil Aeronautics Manual (CAM) for propellers was issued in 1938.[62] There were no significant regulatory changes. One recommendation was interesting:

**Wood used should be purchased
from a reputable lumber company.**

Termination of CAR 14

In 1965, CAR 14 was cancelled and replaced by FAR 35.

References

1   *Air Commerce Regulations*, US Department of Commerce Aeronautics Bulletin 7, effective December 31, 1926

2   *Airworthiness Requirements for Aircraft*, Civil Aeronautics Board, Aeronautics Bulletin 7A, effective October 1, 1934

3   *Requirements for Approved Type Certificates: Airplane Structures, Airplane Engines, Airplane Propellers*, US Department of Commerce Aeronautics Bulletin 14, effective June 1, 1928

4   *Design Information for Aircraft*, US Department of Commerce Aeronautics Bulletin 26, October 1934

5   "Definitions", *Glider Airworthiness*, US Department of Commerce CAR 5, §5.1(b), effective March 5, 1952

6   J D Doolittle, *Wing Loads as Determined in Flight*, Thesis: Massachusetts Institute of Technology, 1924

7   J D Doolittle, *Accelerations in Flight*, NACA Report 203, 1925

8   *The Status of General Aviation Crashworthiness*, NTSB Safety Report NTSB-SR-80-2, December 17, 1980 Appendix A

9   *Airworthiness Requirements for Aircraft*, Civil Aeronautics Board, Aeronautics Bulletin 7A, effective October 1, 1934

10   *Accident Report to TWA Flight 6, near Atlanta, Mo., May 6, 1935*, Bureau of Air Commerce, June 1935

11   Komons, N A, *The Cutting Air Crash: Case Study In Early Federal Aviation Policy*, Washington: Government Printing Office, 1984

12   *Airplane Airworthiness*, US Department of Commerce, CAR 4, effective November 1, 1937

13   "Seats," *Airplane Airworthiness*, CAR 04, §04.5890, effective November 1, 1937

14   "Passenger Chairs," *Airplane Airworthiness*, CAR 04, §§04.463, effective November 1, 1937

15   "Passenger Loads," *Airplane Airworthiness*, CAR 04, §§04.264-04.2640, effective November 1, 1937

16   "Safety Belts," *Aircraft Equipment Airworthiness*, CAR 15, §§15.30-ff, , effective November 1, 1937

17   "Maneuvering Load Factors," *Airplane Airworthiness*, CAR 04, §04.2120, effective November 1, 1937

18   *Airplane Airworthiness*, US Department of Commerce, CAR 4, effective August 15, 1942

19   *Transport Category Requirements*, US Department of Commerce CAM 04T, November 1, 1944

20   "Emergency Landing Conditions. General", *Airplane Airworthiness: Transport Categories*, CAR 04, §04.260, amendment 04-0, effective November 9, 1945

21   "Unit Weights for Design Purposes," *Airplane Airworthiness: Transport Categories*, CAR 4b, §04.071, amendments 04-0, effective November 9, 1945

22   *Differentiation of Numbers of Parts 04*, Civil Aeronautics Board, amendment 04-5, effective November 9, 1944

23   *Part 4a-Airplane Airworthiness*, Civil Aeronautics Board CAR 4a, amended to April 7, 1960

24   *Part 4b -- Airplane Airworthiness. Transport Categories*, Civil Aeronautics Board CAR 4b, As mended to December 31, 1953

25   "Service Tests for Aircraft," *Airplane Airworthiness. Transport Categories*, Civil Aeronautics Board CAR 4b, amendment 4b-2, effective January 17, 19517

26   "Cockpit Standardization," *Airplane Airworthiness. Transport Categories*, Civil Aeronautics Board CAR 4b, amendment 4b-2, effective January 17, 1951

27   E T Long et al., *At the Controls: The Smithsonian National Air and Space Museum Book of Cockpits*, Niagara Falls, NY: Boston Mills Press, 2001

28   R L Newman and W R Ercoline, "The Basic T: The Development of the Instrument Panel," *Presented to Aerospace Medical Association Annual Meeting*, Orlando, May 2006

29   "Loads," *Airplane Airworthiness. Transport Categories*, Civil Aeronautics Board CAR 4b, §4b.200, as amended to December 31, 1953

30   "Ice Protection," *Airplane Airworthiness. Transport Categories*, Civil Aeronautics Board CAR 4b, §4b.640, amendment 4b-2, effective August, 25, 1955

31   W Lewis, *A Flight Investigation of the Meteorological Conditions Conducive to the Formation of Ice on Airplanes*, NACA TN-1393, 1947

32   A Jones and W Lewis, *Recommended Values of Meteorological Factors to be Considered in the Design of Aircraft Ice Protection Equipment*, NACA TN-1855, 1949

33   "Arrangement and Visibility of Instrument Installations," *Airplane Airworthiness*. Transport Categories, CAM 4b, §4b.611, revised January 1958

34   *Turbine-Powered Transport Category Airplanes of Current Design*, Special Civil Air Regulation SR-422, effective August 27, 1957

35   *Glider Airworthiness*, Civil Aeronautics Authority CAR 5, June 1, 1940

36   *Airplane Airworthiness: Normal, Utility, Acrobatic, and Restricted Purpose Categories*, Civil Aeronautics Board CAR 3, effective November 13, 1945

37   "Emergency Provisions: Protection', *Airplane Airworthiness: Normal, Utility, Acrobatic, and Restricted Purpose Categories*, CAR 03, §03.3811, amendment 03-0, effective November 13, 1945

38   "Safety Belts," *Airplane Airworthiness: Normal, Utility, Acrobatic, and Restricted Purpose Categories*, CAR 03, §3. 715, amendment 3-2, effective February 6, 1950

39   "Weights",, *Airplane Airworthiness: Normal, Utility, Acrobatic, and Restricted Purpose Categories*, CAR 03, §3. 071, amendment 03-0, effective November 13, 1945

40   *Rotorcraft Airworthiness*, Civil Aeronautics Board CAR 6, effective May 24, 1946

41 *Rotorcraft Airworthiness: Normal Category*, Civil Aeronautics Board CAR 6, Amended December 20, 1956

42 "Maneuvering Flight Conditions," *Rotorcraft Airworthiness*, CAR 06, §06.2120, Effective May 24, 1946

43 "Gust Conditions," *Rotorcraft Airworthiness*, CAR 06, §06.2121, Effective May 24, 1946

44 "Weights, Crew and Passengers," *Rotorcraft Airworthiness. Normal Category*, CAR 6, §6.1(d)(5)(iii), As amended to December 20, 1956.

45 "Weights, Crew and Passengers," *Rotorcraft Airworthiness. Transport Categories*, CAR 7, §7.1(d)(5)(iii), effective August 1, 1956

46 "Emergency Landing Conditions. General", *Rotorcraft Airworthiness. Normal Category*, CAR 6, §6.260(a), As amended to December 20, 1956

47 "Emergency Landing Conditions. General", *Rotorcraft Airworthiness. Transport Categories*, CAR 7, §7.260(a), effective August 1, 1956

48 *Rotorcraft Airworthiness: Transport Categories*, Civil Aeronautics Board CAR 7, effective August 1, 1956

49 "Maneuvering Flight Conditions," *Rotorcraft Airworthiness; Transport Categories*, CAR 07, §07.212, Effective August 1, 1956

50 "Gust Conditions," *Rotorcraft Airworthiness; Transport Categories*, CAR 07, §07.213, Effective August 1, 1956

51 "Emergency Landing Conditions. General", *Rotorcraft Airworthiness. Transport Categories*, CAR 7, §7.260(a), effective August 1, 1956

52 *Rotorcraft Airworthiness: Transport Categories. Miscellaneous Amendments Resulting from the 1957 Airworthiness Review*, Civil Aeronautical Board CAR 7, amendment 7-2, effective May 17, 1958

53 *Rotorcraft Airworthiness: Transport Categories. Miscellaneous Amendments Resulting from first Airworthiness Review*, Federal Aviation Agency CAR 7, amendment 7-5, effective May 3, 1962

54 *Aircraft Airworthiness Restricted Category*, Civil Aeronautics Board CAR 8, effective October 11, 1950

55 *Aircraft Airworthiness Limited Category*, Civil Aeronautics Board CAR 9, effective November 21, 1946

56 *Airworthiness Requirements for Engines and Propellers*, Department of Commerce Aeronautics Bulletin 7G, As amended July 1, 1334

57 *Aircraft Engine Airworthiness*, US Department of Commerce CAR 13 effective November 1, 1937

58 *Aircraft Engine Airworthiness*, US Department of Commerce CAR 13, effective August 1, 1941

59 *Aircraft Engine Airworthiness*, US Department of Commerce CAR 13, effective June 15, 1956

60 *Aircraft Engine Airworthiness*, Federal Aviation Agency CAM 13, October 1, 1959

61 *Aircraft Propeller Airworthiness*, US Department of Commerce CAR 14, effective November 1, 1937

62 *Aircraft Propeller Airworthiness*, US Department of Commerce CAM 14, December 1, 1938

Chapter 6: US Airworthiness Development 1965-2017

On June 30, 1956, two airline transports collided over the Grand Canyon in good weather. All 128 people aboard the United DC-7 and the TWA *Constellation* perished in the crash. Like the accident which killed Senator Cutting (see page 54) 21 years earlier, the results was a government wide review of aviation regulations. The chief result was the creation of the Federal Aviation Agency (FAA) on December 31, 1958.

In the years that followed, there was a complete review of all of the existing US aviation regulations, leading to a complete rewrite of many regulations. In 1965, the Civil Air Regulations were eliminated in a sweeping change, and replaced with Federal Aviation Regulations (FARs). The new FARs didn't have new material, but rather they were a reorganization to make the rules easier to use.

The following outlines the re-organization of the airworthiness regulations:

CAR 1 became FAR 21*
CAR 3 became FAR 23
CAR 4b† became FAR 25
CAR 5 now covered by FAR21, §21.17(b)
CAR 6 became FAR 27
CAR 7 became FAR 29
CAR 8 now covered by FAR 21, §21.25
CAR 9 now covered by FAR 21, §21.27
CAR 10 now covered by FAR 21, §21.29
CAR 13 became FAR 33
CAR 14 became FAR 35

The new airworthiness regulations used the same organization in each of the separate parts. There was now a standard arrangement:

Subpart A	General
Subpart B	Flight
Subpart C	Structure
Subpart D	Design and Construction
Subpart E	Powerplant
Subpart F	Equipment
Subpart G	Operating Limitations

Paragraphs were also numbered identically, for example:

Paragraph 173: Static longitudinal stability

§23.173	formerly CAR 3.114 and parts of 3.115
§25.173	formerly CAR 4b.151
§27.173	formerly CAR 6.123(b) (less (1)-(4)
§29.173	formerly CAR 7.123(b) (less (1)-(4)

---

* Some paragraphs transferred to Parts 39, 45, and 91
† Including SR-422, -422A, and -422B

Paragraph 1303: Flight and navigation instruments
 §23.1303   formerly CAR 3.655 (b)
 §25.1303   formerly CAR 4b.603
 §27.1303   formerly CAR 6.603
 §29.1303   formerly CAR 7.603

Initially, all of the paragraph numbers had odd numbers, such as §25.1301, §25.1303, §25.1305, etc. This allowed for the insertion of new paragraphs should the need arise. This standardization of paragraph numbering has been a great help to certification engineers who deal with different aircraft categories. It was adopted in the Joint Airworthiness Requirements (JAR) and in EASA's Certification Specifications. Initially, only odd-numbered paragraphs were used to all for insertion of new requirements. The common listing of paragraphs can be seen in Appendix 2.

This similarity led to using the expression Part 2x to denote similar paragraphs in the new airworthiness regulations. That is not to say that the rules have been static. On the contrary, many of the rules have been changed to deal with developing problems. Figure 37 shows the major changes in these rules.

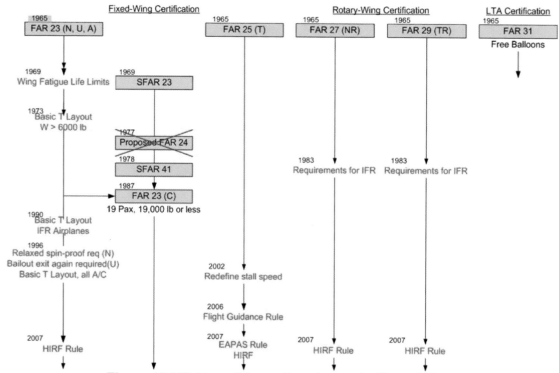

Figure 40 US Airworthiness Requirements Since 1965

In this chapter, we shall review these in chronological order for each category of aircraft
We placed Part 25 first because many of the significant changes, such as HIRF, and complex system issues were led by the Transport Directorate. We have also separated the Commuter

Category airplanes from the other categories in Part 23. For rotorcraft, we have also placed the Transport Rotorcraft ahead of Small Rotorcraft.

Part 25: Transport Airplanes

The initial issue of Part 25 requirements was substantially identical to the final version of CAR 4b including SR-422, -422A, and -422B.[1]

1967: Side-Facing Seats

Transport airplanes now required each occupant seated in a sideward facing seat protected from head injury by safety belt and shoulder harness; safety belt and elimination of injurious object; or safety belt and energy absorbing rest to support arms, shoulders, head, and spine.[2]

In 1969, this was changed to apply for each occupant seated in a seat oriented more than 18 degrees to the longitudinal plane also be protected from head injury by safety belt and shoulder harness; safety belt and elimination of injurious object; or safety belt and energy absorbing rest to support arms, shoulders, head, and spine.[3] This was proposed in response to an industry request for passengers seated in lounge seats at relatively small angles to the fuselage axis.

1970: Equipment Risk Model

The Equipment, Systems, and Installations (§25.1309) now included a risk model which would require catastrophic outcomes to be extremely improbable.[4]

Figure 41 Boeing 737: Certified Under Part 25*

1970: Revised Maneuvering Loads

The previously constant limit maneuvering load factors of +2.5/-1.0 were changed to use Equations (1) and (2) on page 62.[5] The rational recognized the factor that designers of smaller business jet were choosing maneuver load factors higher than the current minimum specified values. In Part 25 The new variation was consistent with the Part 23 requirements for Normal Category airplanes.

---

* http://www.varig-airlines.com/en/b737200.htm, accessed July 30, 2017

1988: Increased Crash Loads and Requirement for Dynamic Seat Testing

§25.561 increased the downward crash load for Transport Category airplanes from 4.5 to 6.0g and the rearward load was established at 1.5 g. The sideward crash load was increased from 1.5 to 4 g on the seats and 3 g on the airframe. Items of mass in the cabin were now required to be restrained with same loads as occupant seats if the items of mass were capable of injuring an occupant.[6]

Figure 42 A320 Certified Under Part 25 with Special Conditions.*

1994: Electrical and Electronic Systems Lightning Protection

With the advent of new transport designs with more electronic systems, the FAA became concerned about the vulnerability of these systems to the effects of lightning. The NPRM commented that the existing rules had become fragmented and incomplete. The outcome of the rulemaking process was the promulgation of §25.1316 in 1994.[7]

1997: Increased Crash Loads

The sideward crash loads were increased from 1.5 to 3 g on the airframe and 4 g on the seats and attachments. Items of mass in passenger compartments must now be restrained for the following loads:

Upward Crash Load	3.0 g
Forward Crash Load	9.0 g
Sideward Crash Load	4.5 g
Downward Crash Load	1.5 g.[8]

2002: New Reference Stall Speed

This revision required a demonstrated stall speed to be corrected to a 1-g load factor. Historically, the stall speed, V_S, was the lowest speed recorded during the stall demonstration. The new *reference stall speed*, V_{SR}, may not be loss than the 1-g stall speed. To keep operational speeds consistent, related speeds use slightly different multiplying factors. V_{REF} was $1.3V_{S0}$ and will now be $1.23V_{SR0}$. This made stall speed consistent with JAR.[9]†

---

* British Midland A320-200 (G-MIDR) shown on approach to Heathrow Airport, Photographed by Adrian Pingstone in September 2003 and released to the public domain. Downloaded from http://commons.wikimedia.org/wiki/File:Bmi.a320.750pix.jpg, accessed June 21, 2022

† The Final Rule was not available on RGL. The reference shown is for the NPRM.

2006: Flight Guidance Rule

The main impetus was an accident in Nagoya, Japan, on April 26, 1994, involving an Airbus A300-600 operated by China Airlines.[10]* Contributing to that accident were conflicting actions taken by the flight crew and the airplane's autopilot. The flight crew tried to correct the autopilot's directions. The combination of out-of-trim conditions, high engine thrust, and flaps that were retracted too far led to a stall, which resulted in an accident involving 264 fatalities. Although this particular accident involved an A300-600, other accidents, incidents, and safety indicators demonstrate that this problem is not confined to any one airplane type, manufacturer, operator, or geographic region.

On July 13, 1996, a McDonnell Douglas MD-11 operated by American Airlines experienced an in-flight upset near Westerly, Rhode Island.[11]† When the airplane was cleared to descend to 24,000 feet, the first officer initiated a descent by means of the autopilot. With approximately 1,000 feet left in the descent, the captain became concerned that the airplane might not level off at the assigned altitude and instructed the first officer to slow the rate of descent. The first officer adjusted the pitch thumbwheel on the autopilot control panel. This maneuver proved ineffective. The captain then took manual control of the airplane, began applying back pressure to the control column, and disconnected the autopilot. Flight data recorder data show the airplane experienced an immediate 2.3 g pitch upset followed by more oscillations, resulting in four injuries.

The change revised, reorganized, and added additional material to Sec. §25.1329.[12] It addressed the autopilot, autothrust, and flight director in a single section and changed the name of §25.1329 from "Automatic pilot system" to "Flight guidance system" to reflect the inclusion of autothrust and flight director. Essentially, the revision ensured that the flight guidance would not place the airplane in a hazardous situation without a warning, so called "bark before bite" rule.

2007: HIRF Rule

The electromagnetic HIRF environment results from the transmission of electromagnetic energy from radar, radio, television, and other ground-based, shipborne, or airborne radio frequency (RF) transmitters. This environment has the capability of adversely affecting the operation of aircraft electrical and electronic systems. There were reports of incidents to military aircraft in the 1970s.

Although the HIRF environment did not pose a significant threat to earlier generations of aircraft, in the late 1970s designs for civil aircraft were first proposed that included flight-critical electronic controls, electronic displays, and electronic engine controls, such as those used in military aircraft. These systems are more susceptible to the adverse effects of operation in the HIRF environment. Accidents and incidents involving civil aircraft with flight-critical electrical and electronic systems have also brought attention to the need to protect these critical systems from high-intensity radiated fields.‡

In the early 1990s, the FAA began issuing special conditions to apply the HIRF standards to all aircraft. Finally, in 2007, these were added directly as §25.1317.[14]

The HIRF rule was widely accepted by the industry. The only controversy seemed to be whether to call it High Energy Radiated Fields (HERF) or High Intensity Radiated Fields (HIRF). Yet it took 27 years to make the changes to the regulations, fifteen years of repeated special conditions (See Figure 43). How long would a controversial rule take?

---

*   This accident is described in Appendix 4 on page 357.

†   This accident is described in Appendix 4 on page 358.

‡   While we applaud the forward thinking actions in developing the HIRF rule, a search of the NTSB database shows only one accident caused by HIRF. That accident was a double engine failure in an airship.[13] This accident is described in Appendix 4 on page 330.

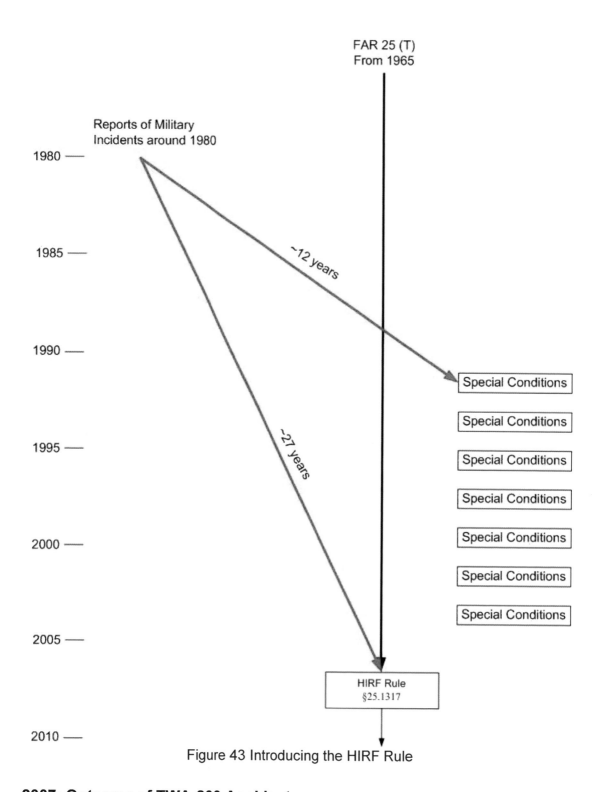

Figure 43 Introducing the HIRF Rule

2007: Outcome of TWA-800 Accident

Fuel tank explosions have been a constant threat with serious aviation safety implications for many years. Since 1960, 18 airplanes have been damaged or destroyed as the result of a fuel tank explosion. One of these was a Boeing 747 (TWA 800) off Long Island, New York in 1996.*[15]

---

*   This accident is described in Appendix 4 on page 396.

The FAA instituted a program after the TWA 800 accident to ensure the airworthiness of aging aircraft fuel systems and wiring.[16] This was the Enhanced Airworthiness Program for Airplane Systems (EAPAS). In May 2001, the FAA issued SFAR 88 to detect and correct potential system failures that can cause ignition.[17]

In 2007, the FAA amended regulations for certification and operations of Transport Category airplanes. These changes were deemed necessary to help ensure continued safety of commercial airplanes and improve the design, installation, and maintenance of airplane electrical wiring systems and align those requirements as closely as possible with the requirements for fuel tank system safety. This final rule organized and clarified design requirements for wire systems by moving existing regulatory references to wiring into a single section of the regulations specifically for wiring and by adding new certification rules. It also required holders of type certificates for certain Transport Category airplanes to conduct analyses of their airplanes and make necessary changes to existing Instructions for Continued Airworthiness (ICA) to improve maintenance procedures for wire systems.[18]

Another outcome of the EAPAS program was Part 26, a continued airworthiness program for transport airplanes.[19] This rule included EAPAS/Fuel Tank Safety, Widespread Fatigue Damage, Fuel Tank Flammability, and Damage Tolerance. This rule should probably have been placed with maintenance regulations, not with Type Certification.

2013: Human Factors Rule

This amendment was intended to minimize the occurrence of design-related flight crew errors. This will enable a flight crew member to detect and manage his or her errors when the errors occur.[20] Paragraph §25.1302, the "human factors" rule, uses 323 words to expand on the nature of equipment requiring pilot intervention and could be inferred §25.1301 and §25.1309. In our opinion, it is too general to be useful and is, as a result, unnecessary.

The FAA "Lessons Learned" website, states that this rule is a direct outcome of the Airbus accident* in Nagoya, Japan in 1994.[10] In view of the nineteen year interval, this does not seem reasonable.

The NPRM is no longer available, but the final rule preamble does not support any such linkage. The final rule also does not mention any accidents. It does state that the adopting the rule eliminated some regulatory differences between the US and EASA without affecting current industry design practices, which may have been the underlying reason.

2015: Ice Protection

On October 31, 1994, an ATR-72 was in a holding pattern and was descending to a newly assigned altitude of 8,000 feet when it experienced an uncommanded roll excursion and crashed during a rapid descent.† The loss of control was attributed to a sudden and unexpected aileron hinge moment reversal that occurred after a ridge of ice accreted beyond the deice boots.[21] It is worth noting that the accident occurred while holding with flaps extended at relatively high speed to reduce the deck angle for the flight attendants. The roll excursion happened as the flaps were retracted.

There were, of course, several factors in this mishap. The airplane was flying in freezing precipitation, which was not considered during the NACA research flights leading up to the original icing certification criteria in 1955.(See discussion on page 58).

As result, the FAA amended the ice protection regulations in Part 25 by adding a new §15.1420 and a new Appendix O, "Supercooled Large Drop Icing Conditions." The sections ap-

* This accident is described in Appendix 4 on page 357.
† This accident is described in Appendix 4 on page 334.

ply only to airplanes with a MTOW less than 60,000 lb or with reversible flight controls.[22] This was the first major change to icing certification in 60 years.

Fly-by-Wire

An astute reader will notice that there has been no mention of certification requirements for fly-by-wire (FBW) airplanes. All FBW transport airplanes certified to date have relied on special conditions as part of their certification basis. This includes A320, A330/340, A350, B-777, B-787, ERJ-170, ERJ-190, and SAAB-2000.*

Aircraft Certified Under Part 25

Aircraft certified under Part 25 include Airbus A320, Airbus A330/340, Boeing 737, Boeing 747, Boeing 777and McDonnell-Douglas MD-80. The Airbus A320 and A330/340 and the Boeing 777 were all certified with special conditions for the fly-by-wire flight control systems.

Figure 44 Boeing 777 Certified Under Part 25 with Special Conditions†

---

* The FBW requirements for the *Concorde*, the only other civil FBW transport airplane are not available. We have not been able to locate any copies of the *Concorde* requirements.
† Boeing 777-200ER at Seattle-Tacoma Airport, Phone taken by Greg Goebel and released into the public domain. Downloaded from http://www.public-domain-image.com/public-domain-images-pictures-free-stock-photos/transportation-vehicles-public-domain-images-pictures/aeroplanes-aircrafts-public-domain-images-pictures/boeing-777-200er-at-seattle-tacom.jpg, accessed June 10, 2022.

Part 23: Small Airplanes (N, U, and A Categories)

The initial issue of Part 23 requirements was substantially id265entical to the final version of CAR 3.[23]

1969: Fatigue Life Limits

The first major change came in 1969, when the wing structure now required fatigue life limits.[24] The summary of the comments received indicated this change wasn't necessary since existing design practices at the time were conservative and that most fatigue instances were caused by prior damage or corrosion. Nevertheless, the FAA did not agree and the rule was adopted. One of the first airplanes certified following amendment 23-7 was the Grumman American AA-5 which had a wing spar life limit of 12,500 hr.[25]

1969: Protection from Head Injury

Part 23 airplanes now required that each occupant be protected from head injury with safety belt and shoulder harness; safety belt and elimination of injurious object; or safety belt and energy absorbing rest to support arms, shoulders, head, and spine.[26]

1973: Instrument Arrangement

In 1973, the Basic T arrangement (see Figure 30 on page 59) was now required on all Part 23 airplanes over 6,000 lb.[27] Hitherto; this arrangement had only been required on Transport Category airplanes

1977: Front Seat Shoulder Harness

Part 23 airplanes now required that *front seats* have shoulder harnesses installed.[28] No design requirements were published. A *front seat* was defined as a located at a flight crew member station or any seat located along-side such a seat.*

1985: Shoulder Harness for Each Occupant

Part 23 airplanes must now protect each occupant from serious head injury by a safety belt and shoulder harness that is designed to prevent the head from contacting any injurious object for each forward and aft-facing seat. For other seat orientations, the seat and restraint means must be designed to provide a level of occupant protection equivalent to that provided for forward and aft-facing seats with safety belts and shoulder harnesses installed.[29]

In addition, a Special Retroactive Requirement applied this change to all newly manufactured airplanes, even if they were previously certified.[30] †

---

*   This awkward definition leads to inconsistencies, such as a Piper *Cub* which has two tandem pilot seats. The airplane is customarily flown from the seat in the rear. The seat in back is obviously a *front seat*. Many operators remove the flight controls from the forward station. So, in this case, the seat in front is no longer a *front seat*"

†   Since §23.2 is no longer current, it would follow that this "special retroactive requirement" is no longer applicable for newly manufactured airplanes.

1988: Additional Crash Safety Requirements

Items of mass, within the cabin, that could injure an occupant must be restrained for the following loads

Direction of Load	Load
Upward Crash Load	3.0 g,
Forward Crash Load	18.0 g,
Sideward Crash Load	4.5 g.[31]

A new paragraph, §23.562, now requires dynamic tests or rational analysis supported by dynamic tests to show that Head Injury Criteria (HIC) will not exceed 1,000.[32] These tests will require the use of an approved Anthropomorphic Test Dummy (ATD) with a nominal weight of 170 lb and seated in the normal upright position.[33]

1990: Instrument Arrangement – Version II

In 1990, the Basic T arrangement was required on all Part 23 airplanes which were not limited to Visual Flight Rules (VFR).

Also in 1990, a new section, §23.1311, "Electronic Display Instrument Systems," was introduced.[34] Among other requirements, was the requirement for display of information essential for the continued safe flight and landing remain immediately available to pilots following any single failure* without the need for immediate pilot action.

1991: Spins – Version V†

Normal Category airplanes could now be tested to show they were "spin-resistant".[35] There was still the option to certify them as "characteristically incapable of spinning".

Figure 45 Grumman American AA-5 *Traveler.* Certified Under Part 23‡

---

*   or combination of failures
†   Continuing the trend in CAR 3
‡   AA-5 (*Traveler*) registered G-BEFF, photographed by Adrian Pingstone in July 2005 released to the public domain. Accessed July 30, 2017 from https://en.wikipedia.org/wiki/Grumman_American_AA-5#/media/File:Grumman.aa-5.traveller.g-bezf.arp.jpg

1993: Stall Speed for Single-Engine Airplanes

One of the carry-overs from CAR 3 was a maximum stall speed for single-engine airplanes of 70 MPH in the landing configuration (V_{S0}). In 1969, the units were changed to 61 knots. In 1993 a new paragraph, §23.562(d) was added allowing an applicant to trade crashworthiness for stall speed.[36] This paragraph increased the crashworthiness requirements by the factor $(V_{S0}/61)^2$.

1996: Spins – Version VI

Several changes were made to Part 23.[37] A bailout exit was again required for Utility Category airplanes approved for intentional spins. Airplanes could now be tested to show they were "spin-resistant."

1996: Additional Crash Loads

A 6.0 g downward vertical crash load was added to §23.561.[38]

1996: Instrument Arrangement – Version III

The Basic T was now required for all Part 23 airplanes.[39]

1996: Miscellaneous Amendments

Several changes were made to harmonize FAA requirements with the European requirements.[40] These included requiring the determination of glide distance and speed for single-engine airplanes. The amendment also deleted the "characteristically incapable of spinning" option.

2007: HIRF Rule

In 2007, the HIRF rule was added to Part 23.[41] The discussion can be found with amendment 25-122 on page 83. This was added as §23.1308. This number differs from the numbering of §§ 25.1317, 27.1317, and 29.1317. No rationale was offered for this difference.

2011: Lightning Rule

Existing regulations for the lightning protection of electrical and electronic systems installed on aircraft certificated under Part 23, 27 and 29 require the applicant only to "consider" the effects of lightning. Unlike system lightning protection regulations for Part 25 airplanes, these regulations had not been significantly amended since they were first adopted and did not reflect current advances in technology. Adopted in the 1960s, these regulations require that the aircraft be protected against catastrophic effects of lightning, but do not have specific requirements for electrical and electronic system lightning protection. At the time, most aircraft contained mechanical systems, or simple electrical and electronic systems. Airframe components were made from aluminum materials, with high electrical conductivity, and offered good protection against lightning.

The early 1980s ushered in Part 25 designs that routinely included more complex electrical and electronic systems. In addition, there has been a trend for increased use of composite aircraft materials with less inherent lightning protection than aluminum. As electrical and electronic systems became more common on Part 23 airplanes, the FAA issued §25.1316 in 1994 specifically requiring protection for electrical and electronic systems on Part 25 Transport Category airplanes.

§23.1306,* §27.1316, and §29.1316 now establish consistent performance standards for lightning protection of aircraft electrical and electronic systems against the catastrophic, hazard-

---

* Again, no rational was offered for the different paragraph numbering between the different FAR parts.

ous or major failures of the functions these systems provide. The standards for protection against catastrophic failure would require an applicant to show that the function that the system performs would not be adversely affected during or after the time the aircraft is exposed to lightning, and that the system that was affected would automatically recover normal operation of that function in a timely manner after the aircraft is exposed to lightning. The standards for protection against hazardous or major failure would require the applicant to show that the affected function would recover normal operation in a timely manner after the aircraft is exposed to lightning.[42]

The performance standards would also be imposed according to the aircraft' potential for exposure to lightning. The standards for all aircraft operated under instrument flight rules would meet more stringent requirements than aircraft certificated to Part 23 and Part 27 standards approved solely for operations under visual flight rules. This proposal ensured that protection would be applied to aircraft according to their potential for exposure to lightning.

2012: Light Jet Certification

This amendment made several changes related to light jet certification. The majority of this rule allows manufacturers of jets to achieve product certification without the numerous special conditions, equivalent level of safety (ELOS) findings, and exemptions previously required to certificate these products. Therefore, this final rule reduces the certification burden on the applicant and allows the FAA to focus resources on other safety-critical items. In addition, this final rule enhances safety by requiring additional battery endurance requirements and increasing the climb gradient performance for certain Part 23 airplanes.[43] The change added requirements for airplanes with aft, fuselage-mounted engines

2017: Complete Rewrite of Part 23

The most noticeable change in amendment 23-64 is the abandoning of the common paragraph number scheme that served the certification team so well since 1965.[44] The *New Part 23* also eliminated Acrobatic, Utility, and Commuter Categories. All Part 23 airplanes will now be Normal Category with limits of 19,000 lb MTOW and 19 passengers. The major change in The *New Part 23* is a lack of objective standards. The implications of this amendment will be discussed later in Chapter 10.

2019: Application for Extra EA-300

In July 2019, Extra Flugzeugbau applied for an Acrobatic Category for their Extra EA-300 in the Acrobatic Category which, at the time was no longer available under the New Part 23. The certification basis was agreed to as Part 23 as amended by Amendments 23-1 through 23-62. While the required certification limits for maneuvering load factors in the certification basis were +6/-3; the actual limits approved for the airplane were $\pm$ 10 (solo) and $\pm$ 8 (dual).

Aircraft Certified Under Part 23

Aircraft certified under Part 25 include Grumman American AA-5 *Traveler*, Cessna CE-525, Commander 112, Extra EA-300, and Piper PA-46.

Figure 46 Extra EA-300: Certified Under Part 23 in Acrobatic Category*

Commuter Category

Initial Part 23 rules limited weight, not occupants. This was not an issue until lighter engines and structure permitted airplanes, such as the *Twin Otter* to carry significant payloads. These were adopted by the "commuter airlines" to serve feeder routes for the major airlines. These commuter airlines often operated scheduled service under the air taxi operating rules (Part 135).

1969: SFAR 23

In 1969, SFAR 23 was issued to provide additional standards for operations under Part 135.[45] SFAR 23 applied to Part 23 airplanes certified to carry more than 10 occupants. The term *occupant* was chosen and includes flight crew. The rule included provisions for emergency exits and some fire protection measures.†

1978: SFAR 41

In 1978, SFAR 41[46] was issued to provide additional standards for operations under Part 135. This rule was for airplanes "with a passenger seating configuration, excluding pilot seats, of 10 seats or more (but not more than 19 seats". SFAR 41 imposed some additional performance requirements as well as passenger exits.

1983: Proposed Part 24

In 1983, the FAA proposed Part 24 to provide standards for "Commuter" airplanes. The proposal was not well-received, partly because it proposed probabilistic design criteria.

---

* Extra EA-300 from http://flightaware.com, accessed July 22. 2021. The airplane was designed by Walter Extra and manufactured by Extra Flugzeugbau.

† The text of the rule is not available on RGL. The discussion is base on the rule preamble.

Figure 47 DHC-6 *Twin Otter.* Certified Under CAR 3 and SFAR 23*

1987: Commuter Category

In 1987, the Commuter Category was codified with Amendment 23-24. It was limited to multi-engine propeller airplanes up to 19,000 lb and up to 19 passengers. The performance requirements were substantially those for propeller transport airplanes.

2007: HIRF Rule

In 2007, the HIRF rule was added to Part 23.[41] The discussion can be found with amendment 25-122 on page 83.

2011: Lightning Rule

Amendment 23-61 established §23.1306 covering lightning protection for Part 23 aircraft.[42] This amendment also covered Part 23 Commuter airplanes. See discussion on page 89.

Systems requirements are essentially identical to transport airplanes, except Commuter Category airplanes require a standby attitude indicator as part of the airworthiness certification. Transport Category airplanes do not, unless operating as an air carrier.

2012: Jet-Powered Commuter Airplanes

Amendment 23-62, discussed previously on page 90, allowed for the certification of jet-powered Commuter Category airplanes.[47]

Elimination of Commuter Category

The revision of Part 23 eliminated the Commuter Category which was merged into the Normal Category.[44]

---

*   NASA Photo by Joe Copalman. This particular aircraft, N-607NA, was used in the NASA icing research program. One of your authors (RLN) flew this airplane as part of The Ohio State University's participation in that program.

Figure 48 Fairchild SA-227 Certified under CAR 3 and SFAR-41*

Part 29: Transport Rotorcraft

The initial issue of Part 29 requirements was substantially identical to CAR 7.[48] There were two performance categories in the initial version:

> Category A provides the most rigid rules, requiring independent engines, fuel systems and electrical systems. Category A requires that no single failure in these areas can cause simultaneous loss of more than one engine. In Category A, occupants are assured of a safe flight capability with an engine failure at any point during flight. and

> Category B rotorcraft are not required to have the capability for continued flight in the event of an engine failure and must only demonstrate a safe landing to a defined surface after drift-down or autorotation.

1983: Criteria for Transport Certification

In 1983, the regulations were amended to add limitations on both maximum weight and passenger seats for Category B. Category B was restricted to rotorcraft with a maximum weight of 20,000 pounds or less and nine or fewer passenger seats.

Rotorcraft with a maximum weight of 20,000 pounds or less but with 10 or more passenger seats could be certified as Category B if compliance with §§29.67(a)(2), 29.87, 29.1517, and subparts C, D, E, and F of Part 29 were met.[49]

1983: Appendix B. Criteria for Instrument Flight

As stated in the NPRM,[50] these changes "proposed revisions to the applicability sections of Parts 27 and 29 which allow for increased utilization for rotorcraft engaged primarily in utility or cargo operations and provides additional protection for rotorcraft carrying 10 or more passengers. The rotorcraft industry is the fastest growing segment in aviation today. The increasing use and the advent of IFR flight in rotorcraft demonstrates the need to review these parts."

Amendment 29-21 added Appendix B to Part 29, *Airworthiness Criteria for Helicopter Instrument Flight*.[51] Appendix B imposed additional stability requirements with different requirements for single or two-pilot operations. It also mandated a standby attitude indicator. Instrument systems and other systems essential for IFR flight must be shown to be protected from both Continuous Maximum and Intermittent Maximum Icing Envelopes from Part-25, Appendix C. (See page 59).

---

*   Wikipedia photo, accessed July 30, 2017

Figure 49 Sikorsky S-76 certified under FAR 29*

1984-1996: Additional Crashworthiness Requirements

There were several changes to the crashworthiness requirements during this period. These revisions to Part 29 were made in lockstep with those to Part 27. Since the changes are identical, they are listed here.

Shoulder Harness Requirements

In 1984, §§27.785 and 29.785 now required a shoulder harness at each crew member seat and each seat beside a crew member seat. For all other seats, each occupant must be protected from head injury by safety belt and shoulder harness; safety belt; elimination of injurious object; or safety belt and energy absorbing rest to support arms, shoulders, head, and spine.[52][53]

Increased Crash Load and Dynamic Test Requirements

In 1989, §§27.561 and 29.561 were updated to specify the following design ultimate inertia forces for the occupants as well as items of mass inside the cabin:

Upward Crash Load	4.0 g
Forward Crash Load	16.0 g
Sideward Crash Load	8.0 g
Downward Crash Load	20.0 g.

These changes to the helicopter design requirements were based on published research findings: the: US Army's *Crash Survival Design Guide*[54] and the FAA's *Analysis of Rotorcraft Crash Dynamics for Development of Improved Crashworthiness Design Criteria*[55] which were cited in the NPRM.[56]

The supporting structure for items of mass above or behind the crew and passenger compartments (including rotors, transmissions, and engines) must resist the following ultimate inertial loads

Upward Crash Load	1.5 g
Forward Crash Load	8.0 g
Sideward Crash Load	2.0 g
Downward Crash Load	4.0 g (after intended displacement of seat).[57][58]

---

*   Source: http://en.wikipedia.org/wiki/Sikorsky_S-76#/media/File:Pesca_1_(3737941060).jpg

New §§27.562 and 29.562 now require dynamic tests or rational analysis supported by dynamic tests to show that HIC will not exceed 1,000.[59][60] These tests were to use an approved Anthropomorphic Test Device (ATD).[33]

Additional Protection for Fuselage Fuel Tanks
In 1994, §§27.561 and 29.561 were updated to require additional structure to protect fuselage fuel tanks below the passenger floor from rupture in a crash.1996.[61][62]

Additional Rearward Crash Load
In 1996, §§27.561 and 29.561 were updated to add a 1.5 g rearward crash load.[63][64]

2007: HIRF Rule

In 2007, the HIRF rule was added as §29.1317.[65] The discussion can be found with amendment 25-122 on page 83.

2011: Lightning Rule

The lightning rule was added as §29.1316 for rotorcraft.[72] See the discussion on page 89.

Aircraft Certified Under Part 29

Aircraft certified under Part 29 include Airbus Helicopters BK-117 and Sikorsky S-76.

Part 27: Small Rotorcraft

The initial issue of Part 27 requirements was substantially identical to the final version of CAR 6.[66] The original weight limit for these rotorcraft was 6,000 lb, the final weight limit in CAR 6.

1983: Appendix B. Criteria for Instrument Flight

These rule changes were prompted by changing helicopter use in instrument flight. They were a result of the *Rotorcraft Regulatory Review Program*.[67]

Amendment 27-19 added Appendix B to Part 27, *Airworthiness Criteria for Helicopter Instrument Flight*.[68] Appendix B imposed additional stability requirements with different requirements for single or two-pilot operations. It also mandated a standby attitude indicator. Instrument systems and other systems essential for IFR flight must be shown to be protected from both Continuous Maximum and Intermittent Maximum Icing Envelopes from Part-25, Appendix C. (See Figure 29.)

1984-1996: Additional Crashworthiness Requirements

There were several changes to the crashworthiness requirements during this period. These revisions to Part 27 were made in lockstep with those to Part 29 and can be found on page 94.

1996: Provision for Performance Category A

New §27.1(c) added an optional basis for Normal Category multi-engine rotorcraft to be certificated to Category A performance requirements by meeting certain design and performance requirements of part 29.[69] These Part 29 sections were listed in the new Appendix C to Part 27.

Figure 50 Robinson R-22, Certified Under FAR 27*

2007: HIRF Rule

The HIRF rule was added as §§27.1317 and 29.1317.[70] The discussion can be found with amendment 25-122 on page 83.

2011: Lightning Rule

The lightning rule was added as §§27.1316 and 29.1316 for rotorcraft.[71] [72] See the discussion on page 89.

Aircraft Certified Under Part 27

Aircraft certified under Part 27 include Airbus Helicopter BO-105, Robinson R-22, and Robinson R-66.

Part 33 : Aircraft Engines

The initial issue of Part 33 requirements was substantially identical to the final version of CAR 13.[73]

1993: Electrical and Electronic Controls

This amendment added a new §33.28 with new requirements for electrical and electronic engine controls.[74] This required these control systems to be:

↟ Adequately described in the instruction manual;
↟ Designed so that power failure or data interruption not result in an unacceptable power or thrust change or prevent continued safe operation;
↟ Designed so that no single failure will result in an unsafe condition; and
↟ Use software designed and implemented to prevent unsafe conditions.

---

*   http://www.dreamstime.com/photos-images/robinson-r44.html

1996: Windmilling Control and Contingency Ratings

In 1996, there were two separate changes.

Amendment 33-17. In propeller-driven airplanes, it was possible to feather the propeller to keep the airflow from driving the propeller and engine. In a turbojet engine, the engine continued to rotate potentially causing further damage. This amendment provides for preventing unsafe conditions resulting from uncontrolled engine windmilling.[75] This change was harmonized with foreign authorities.

Amendment 33-18. This amendment established new contingency ratings for a 30 second and a 2 minute one-engine inoperative (OEI) contingency rating for helicopter engines to supplement the existing 2-1/2 minute OEI rating. These contingency ratings all increased power for multi-engine helicopters following an engine failure.[76]

1998: Revised Rain and Hail Ingestion Standards

Following a number of turbine engine power losses, the FAA and other authorities concluded that current requirements for high bypass turbine engines were lacking in protection from compressor surging and engine flameout. This amendment[77] added additional requirements; including a new Appendix B.[78] This amendment was coordinated with Transport Canada and the Joint Airworthiness Authorities (JAA).

1998: Bird Ingestion

Following a DC-10 takeoff accident,[79]* the National Transportation Safety Board (NTSB) issued a Safety Recommendation A-76-64,[80] recommending that the FAA, "amend 14 CFR 33.77 to increase the maximum number of birds ... required to be ingested into turbine engines with large inlets." In response to the recommendation, the FAA sponsored an industry wide study of the types, sizes, and quantities of birds that had been ingested into aircraft turbine engines of all sizes, and the resulting affects on engine performance.

Based on the study results, the FAA concluded that §33.77 should be modified to increase the severity of the bird ingestion testing requirements regarding large, high bypass ratio engines. Amendment 33-20 changes the size distribution of birds to be considered during the design and testing of new engines.[81]

2007: Early ETOPS Approval

Amendment 33-21 provided new requirements for Early ETOPS where the aircraft was approved for ETOPS without the extensive service history normally used.

2008: Electrical and Electronic Controls

This change updates standards for engine control systems to reflect current industry practices and harmonizes these standards with EASA.[82] The rule includes HIRF as an item to be considered.

2015: Engine Induction Systems

This amendment adds requirements for turbine engines to operate in conditions described in new flight conditions containing supercooled large drop, mixed phase, and ice crystal icing conditions.[83] These conditions are contained in the new Appendix D, Mixed Phase and Ice Crystal Icing Envelope (Deep Convective Clouds).

---

*   This accident is described in Appendix 4 on page 259.

Part 35: Propellers

The initial issue of Part 35 requirements was substantially identical to the final version of CAR 14.[84]

2008: Changes to Address New Technology

Amendment 35-8 addressed several areas of new technology or changed risk: Safety Analysis, Bird Impact, Propeller Fatigue, and Lightning.[85] This amendment was first proposed in 1974, taking thirty four years![86]

Safety Analysis. Amendment 35-8 revised the text and title of §35.15 to require applicants to conduct a safety analysis to ensure the collective risk from all propeller failure conditions is acceptably low.

Bird Impact: The amendment required the applicant to demonstrate, by tests or analysis based on tests or experience on similar designs, that the propeller can withstand the impact of a 4-pound bird at the critical location(s) and critical flight condition(s) of a typical installation without causing a major or hazardous propeller effect. This does not apply to fixed-pitch wood propellers of conventional design.

Propeller Fatigue: Fatigue of composite propellers should be tested. §35.37 now expands the requirement to all materials and components (including controls system components, if applicable) whose failure would cause a hazardous propeller effect.

Lightning: New §35.18 addressed lightning strikes. The applicant must demonstrate, by tests, analysis based on tests, or experience on similar designs, that the propeller can withstand a lightning strike without causing a major or hazardous propeller effect. The limit to which the propeller has been qualified must be documented in the appropriate manuals. This section does not apply to fixed-pitch wood propellers of conventional design.

2008: Changes to Address New Technology

Amendment 35-9 addressed several areas of new technology or changed risk: Safety Analysis, Bird Impact, Propeller Fatigue, and Lightning.[87][88]

This safety analysis should:

+ Assume a typical installation;
+ Consider likely consequences of failure;
+ Consider latent failures and secondary failures; and
+ Consider multiple failures.

References

1   *Airworthiness Standards; Transport Category Airplanes*, 14 CFR Part 25, amendment 25-0, effective February 1, 1965

2   "Seats, Berths, Safety Belts, and Harnesses", *Airworthiness Standards: Transport Category Airplanes*, 14 CFR Part25, §25.785, amendment 25-15, effective October 24, 1967

3   "Seats, Berths, Safety Belts, and Harnesses", *Airworthiness Standards: Transport Category Airplanes*, 14 CFR Part25, §25.785, amendment 25-20, effective April 23, 1969

4   "Equipment systems and installations," *Airworthiness Standards; Transport Category Airplanes*, 14 CFR Part 25, §25.1316, amendment 25-23, effective May 8, 1970

5   "Limit Maneuvering Load Factors", *Airworthiness Standards: Transport Category Airplanes*, 14 CFR Part25, §25.561, amendment 25-23, effective April 8, 1970

6   "Structure. General", *Airworthiness Standards: Transport Category Airplanes*, 14 CFR Part25, §25.561, amendment 25-64, effective June 16, 1988

7   "System Lightning Protection," *Airworthiness Standards; Transport Category Airplanes*, 14 CFR Part 25, §25.1316, amendment 25-80, effective May 3, 1994

8   "Structure. General", *Airworthiness Standards: Transport Category Airplanes*, 14 CFR Part25, §25.561, amendment 25-91 effective July 29, 1997

9   "Notice of Proposed Rulemaking," *Federal Register*: 61 [12], January 18, 1996, pp. 1259-1268

10   *Aircraft Accident Investigation Report. China Airlines, Airbus Industrie A300B4-622R, B1816, Nagoya Airport, April 26, 1994*, JTSB-AAAR-96-5, July 1996

11   *Aviation Accident Final Report, Westerly, RI, McDonnell Douglas MD-11, July 13, 1996*, NTSB-File NYC96LA148, March 1999

12   "Flight Guidance System," *Airworthiness Standards; Transport Category Airplanes*, 14 CFR Part 25, §25.1329, amendment 25-119, effective May 11, 2006

13   *NTSB Aviation Accident Final Report*, NTSB File ATL90FA099, March 1993

14   "High-intensity Radiated Fields (HIRF) Protection," *Airworthiness Standards; Transport Category Airplanes*, 14 CFR Part 25, §25.1317, amendment 25-122, effective September 5, 2007

15   *In-flight Breakup Over The Atlantic Ocean, Trans World Airlines Flight 800, Boeing 747-131, N93119, Near East Morishes, New York, July 17, 1996*, NTSB-AAR-00-03, August 2000

16   *Review of Federal Program for Wire System Safety*, National Science and Technology Council, Final Report. November 2000

17   *Fuel Tank System Fault Tolerance Evaluation Requirements*, SFAR 88, effective December 9, 2002

18   *Airworthiness Standards; Transport Category Airplanes*, 14 CFR Part 25, amendment 25-123, effective December 19, 2007

19   *Continued Airworthiness and Safety Improvements for Transport Category Airplanes*, 14 CFR Part 26, effective December 10, 2007

20   "Installed systems and equipment for use by the flightcrew," *Airworthiness Standards; Transport Category Airplanes*, 14 CFR Part 25, §25.1302, amendment 25-137, effective July 2, 2013

21   *Aircraft Accident Report: In-Flight Icing Encounter And Loss Of Control, Simmons Airlines, dba American Eagle Flight 4184, Avions De Transport (ATR) Model 72-212, N401AM, Roselawn, Indiana, October 31, 1994*, NTSB-AAR-96-01, July 1996

22   *Airworthiness Standards; Transport Category Airplanes*, 14 CFR Part 25, §15.1420 and 14 CFR Part 25 Appendix O, amendment 25-140, effective January 5, 2015

23   *Airworthiness Standards; Normal, Utility, and Acrobatic Category Airplanes*, 14 CFR Part 23, amendment 23-0, effective February 1, 1965

24   "Wing and Associated Structure," *Airworthiness Standards; Normal, Utility, and Acrobatic Category Airplanes*, 14 CFR Part 23, §23.572, amendment 23-7, effective September 14, 1969

25   FAA TCDS-A16EA, Revision 16, December 12, 2013; (Grumman American AA5)

26   "Seats, Berths, Safety Belts, and Harnesses", Airworthiness Standards: Normal, and Utility Category Airplanes, 14 CFR Part23, §23.785, amendment 23-7, effective September 14, 1969

27   "Arrangement and Visibility," *Airworthiness Standards; Normal, Utility, and Acrobatic Category Airplanes*, 14 CFR Part 23, §23.1321, amendment 23-14, effective December 20, 1973

28   "Seats, Berths, Safety Belts, and Harnesses", Airworthiness Standards: Normal, Utility, and Acrobatic Category Airplanes, 14 CFR Part23, §23.785, amendment 23-29, effective July 18, 1977

29   "Seats, Berths, Safety Belts, and Harnesses", Airworthiness Standards: Normal, Utility, Acrobatic, and Commuter Category Airplanes s, 14 CFR Part23, §23.785, amendment 23-32, effective December 12, 1985

30   "Special Retroactive Requirements", Airworthiness Standards: Normal, Utility, Acrobatic, and Commuter Category Airplanes, 14 CFR Part23, §23.2, amendment 23-32, effective December 12, 1985

31   "Structure. General", Airworthiness Standards: Normal, Utility, Acrobatic, and Commuter Category Airplanes, 14 CFR Part23, §23.561, amendment 23-36 effective September 14, 1988

32   "Emergency Landing Dynamic Conditions", Airworthiness Standards: Normal, Utility, Acrobatic, and Commuter Category Airplanes s, 14 CFR Part23, §23.562 amendment 23-36, effective September 14, 1988

33   "50th Percentile Male," *Anthropomorphic Test Devices*, 49CFR Part572, Subpart B, §§572.5 through 572,11

34   "Electronic Display Instrument Systems," *Airworthiness Standards; Normal, Utility, and Acrobatic Category Airplanes*," 14 CFR Part 23, §23.1311, amendment 23-41, effective November 26, 1990

35 "Electronic display instrument systems," *Airworthiness Standards; Normal, Utility, and Acrobatic Category Airplanes*, 14 CFR Part 23, §23.1311, amendment 23-41, effective November 26, 1990

36 "Emergency landing dynamic conditions," *Airworthiness Standards; Normal, Utility, and Acrobatic Category Airplanes*, 14 CFR Part 23, §23.562(d), amendment 23-44, effective August 18, 1993

37 *Airworthiness Standards; Normal, Utility, and Acrobatic Category Airplanes,*" 14 CFR Part23, §23.1308, amendment 23-49, effective March 11, 1996

38 "Structure. General", Airworthiness Standards: Normal, Utility, Acrobatic, and Commuter Category Airplanes, 14 CFR Part23, §23.561, amendment 23-48, effective March 17, 1996

39 "Arrangement and Visibility," *Airworthiness Standards; Normal, Utility, and Acrobatic Category Airplanes,*" 14 CFR Part23, §23.1321, amendment 23-49, effective March 11, 1996

40 *Airworthiness Standards; Normal, Utility, and Acrobatic Category Airplanes,*" 14 CFR Part23, amendment 23-50, effective March 11, 1996

41 "High-intensity Radiated Fields (HIRF) Protection," *Airworthiness Standards; Normal, Utility, and Acrobatic Category Airplanes*, 14 CFR Part 23, §23.1308, amendment 23-57, effective September 5, 2007

42 "Electrical and Electronic System Lightning Protection," *Airworthiness Standards; Normal, Utility, and Acrobatic Category Airplanes*, 14 CFR Part 23, §23.1306, amendment 23-61, effective August 8, 2011

43 "Certification of Part 23 Turbofan- and Turbojet-Powered Airplanes and Miscellaneous amendments," *Airworthiness Standards; Normal, Utility, and Acrobatic Category Airplanes*, 14 CFR Part 23, amendment 23-62, *Federal Register*: 76 [232], December 2, 2011; effective January 31, 2012, pp. 75736-75769

44 *Revision of Airworthiness Standards for Normal, Utility, Acrobatic, and Commuter Category Airplanes, Final Rule*, Docket No. FAA-2015-1621; 14 CFR Part 23, amendment 23-64, effective August 30, 2017

45 *Small Airplanes Capable of Carrying More Than 10 Occupants*, SFAR No. 23, effective January 7, 1969, Termination December 17, 1969

46 SFAR No. 41, effective August 12, 1982, Termination September 13, 1983

47 "Certification of Part 23 Turbofan- and Turbojet-Powered Airplanes and Miscellaneous amendments," *Airworthiness Standards; Normal, Utility, and Acrobatic Category Airplanes*, 14 CFR Part 23, amendment 23-62, effective January 31, 2012, pp. 75736-75769

48 *Airworthiness Standards: Transport Category Rotorcraft*, 14 CFR Part 29, amendment 29-0, effective February 1, 1965

49 "Applicability," *Airworthiness Standards: Transport Category Rotorcraft*, 14 CFR Part29, §29.1, amendment 29-21, effective March 2, 1983

50 "Notice of Proposed Rulemaking," *Federal Register*, 45, pp. 83424-ff, December 18, 1980

51 "Rotorcraft Regulatory Review Program. amendment No. 1," *Airworthiness Standards; Transport Category Rotorcraft*, 14 CFR Part 29, amendment 29-21, effective March 2, 1983

52 "Seats, Berths, Safety Belts, and Harnesses", *Airworthiness Standards: Normal Category Rotorcraft*, 14 CFR Part27, §27.785, amendment 27-21, effective December 6, 1984

53 "Seats, Berths, Safety Belts, and Harnesses", *Airworthiness Standards: Transport Category Rotorcraft*, 14 CFR Part29, §29.785, amendment 29-30, effective December 6, 1984

54 S P Desjardins *et al.*, *Aircraft Crash Survival Design Guide, Volume IV, Aircraft Seats, Restraints, Litters, and Cockpit/Cabin Delethalization*, USAAVSCOM TR-89-D—22D, December 1989

55 J W Coltman, Bolukbasi, A O, and D H Laananen, Analysis of Rotorcraft Crash Dynamics for Development of Improved Crashworthiness Design Criteria, FAA-CT-85-11, June 1985

56 "Occupant Restraint in Normal and Transport Category Rotorcraft," *Federal Register*: Volume 52, Number 106, June 3, 1987, pp. 20938-ff.

57 "General", *Airworthiness Standards: Normal Category Rotorcraft*, 14 CFR Part27, §27.561, amendment 27-25, effective December 13, 1989

58 "General", *Airworthiness Standards: Transport Category Rotorcraft*, 14 CFR Part29, §29.561, amendment 29-29, effective December 13, 1989

59 "Emergency Landing Dynamic Conditions.", *Airworthiness Standards: Normal Category Rotorcraft*, 14 CFR Part27, §27.562, amendment 27-25, effective December 13, 1989

60 "Emergency Landing Dynamic Conditions.", *Airworthiness Standards: Transport Category Rotorcraft*, 14 CFR Part29, §29.562, amendment 29-29, effective December 13, 1989

61 "General", *Airworthiness Standards: Normal Category Rotorcraft*, 14 CFR Part27, §27.561, amendment 27-30, effective November 2, 1994

62 "General", *Airworthiness Standards: Transport Category Rotorcraft*, 14 CFR Part29, §29.561, amendment 29-38, effective November 2, 1994

63 "General", *Airworthiness Standards: Normal Category Rotorcraft*, 14 CFR Part27, §27.561, amendment 27-32, effective June 11, 1996

64 "General", *Airworthiness Standards: Transport Category Rotorcraft*, 14 CFR Part29, §29.561, amendment 29-38, effective June 11, 1996

65 *Airworthiness Standards; Transport Category Rotorcraft*, 14 CFR Part 29, §29.1317, amendment 29-49, effective September 5, 2007

66 *Airworthiness Standards: Normal Category Rotorcraft*, 14 CFR Part 27, amendment 27-0, effective February 1, 1965

67 "Rotorcraft Regulatory Review Program. Amendment No. 1," *Federal Register*: **48** [21], 4374-4400

68 Airworthiness Standards; Normal Category Rotorcraft, 14 CFR Part 27, amendment 27-19, effective March 2, 1983

69 *Airworthiness Standards; Normal Category Rotorcraft*, 14 CFR Part 27, amendment 27-33, effective August 8, 1996

70 *Airworthiness Standards; Normal Category Rotorcraft*, 14 CFR Part 27, amendment 27-42, effective September 5, 2007

71 "Electrical and Electronic System Lightning Protection," *Airworthiness Standards; Normal Category Rotorcraft*, 14 CFR Part 27, §27.1316, amendment 27-47, effective August 8, 2011

72 "Electrical and Electronic System Lightning Protection," *Airworthiness Standards; Transport Category Rotorcraft*, 14 CFR Part 29, §29.1316, amendment 29-53, effective August 8, 2011

73 *Airworthiness Standards: Aircraft Engines*, 14 CFR Part 33, amendment 33-0, effective February 1, 1965

74 *Airworthiness Standards: Aircraft Engines*, 14 CFR Part 33, amendment 33-15, effective December 28, 1995

75 *Airworthiness Standards: Aircraft Engines*, 14 CFR Part 33, amendment 33-17, effective July 5, 1996

76 *Airworthiness Standards: Aircraft Engines*, 14 CFR Part 33, amendment 33-18, effective August 19, 1996

77 *Airworthiness Standards: Aircraft Engines*, 14 CFR Part 33, amendment 33-19, effective April 30, 1998

78 "Certification Standard Atmospheric Concentrations of Rain and Hail," Appendix B to 14 CFR Part 33, amendment. 33-19, effective April 30, 1998

79 *Aircraft Accident Report – Overseas National Airways, Inc., Douglas DC-10-30, N1032F, John F Kennedy International Airport, Jamaica, New York, November 12, 1975*, NTSB AAR-76-19, December 1965

80 National Transportation Safety Board Safety Recommendation A-76-64, April 1976

81 *Airworthiness Standards: Aircraft Engines*, 14 CFR Part 33, amendment 33-20, effective December 13, 2000

82 *Airworthiness Standards: Aircraft Engines*, 14 CFR Part 33, amendment 33-26, effective October 20, 2008

83 *Airworthiness Standards: Aircraft Engines*, 14 CFR Part 33, amendment 33-34, effective January 5, 2015

84 *Airworthiness Standards: Propellers*, 14 CFR Part 35, amendment 33-0, effective February 1, 1965

85 *Airworthiness Standards: Propellers*, 14 CFR Part 35, amendment 33-8, effective December 23, 2008

86 "Notice of Invitation to Submit Proposals"; Notice No. 74-5; Issued on 2/14/74. Accessed from http://rgl.faa.gov/

87 *Airworthiness Standards: Propellers*, 14 CFR Part 35, amendment 33-9, effective March 19, 2013

88 *Airworthiness Standards: Propellers*, 14 CFR Part 35, amendment 33-9A, effective July 26, 2013

PART III
EQUIPMENT APPROVAL AND
CERTIFICATION PROCEDURES

Chapter 7: Equipment Certification

Primum non nocere*

How does the designer decide what equipment is to be installed on the aircraft and how is it approved? The designer decides some of the equipment choices based on the design. The airworthiness requirements specify some equipment. The operating rules may specify additional equipment.

Required Equipment

Equipment Required by Certification Rules

The pertinent certification rules can be found Subpart F-Equipment. In particular,

§2x.1303 Flight and navigation instruments
§2x.1305 Powerplant instruments;
§2x.1307 Miscellaneous equipment;.

§2x.1303 lists the basic instruments, such as airspeed indicators, altimeters, attitude indicators, compasses, etc. §2x.1305 covers tachometers, oil pressure and temperature gauges and other engine instruments. §2x.1307 lists circuit breakers, radios, etc.

Table 8 shows a list of some required instruments required under the certification rules for Normal, Commuter, and Transport Categories,†[1] cross-referenced with some operating rules.

Table 8 Instruments Required by Certification and Operating Rules

Instrument	Part 23[a]		Part 23[a]		Part 25	
Aircraft Category	Normal		Commuter		Transport	
Operating Rule	FAR 91	FAR 91	FAR 91	**FAR 121**	FAR 91	**FAR 121**
VFR or IFR	VFR	IFR	IFR	Both	IFR	Both
Attitude		✓	✓		P	P
Airspeed	✓	✓	✓		P	P
Altimeter	NS	✓	✓		P	P
Turn & Bank		✓			P	P
Direction		✓	✓	✓	P	P
Vertical Speed			✓	✓	P	P
Standby Attitude			✓[b]	✓	✓	✓
Wet Compass	✓	✓	✓	✓	✓	✓
Clock		✓	✓	✓	✓	✓
Air Temperature				✓	✓	✓
Pitot Heat			✓	✓	✓	✓
Communication Radio(s)		✓		2	✓	2
Navigation System(s)		✓		2	✓	2
Legend ✓ Required.						
NS .on-sensitive altimeter, need not be adjustable for barometric pressure.						
P Required at each pilot station.						
2 Two systems required.						
Notes (a) Prior to amendment 23-64 (amendment 5 to CS-23)						
(b) Unless Type Certificate restricted to visual flight rules.						

---

* Latin, translated as "First, do no harm." from the Hippocratic Oath.
† Normal and Commuter Categories, prior to amendment 23-64.[1]

Equipment Required by Operating Rules

The pertinent certification rules can be found Part 91, Subpart C-Equipment. Instrument, and Certificate Requirements, or in particular,

§91.205(a) Visual flight rules (day);

§91.205(c) Visual flight rules (night);

§91.205(d) Instrument flight rules;

§121.305 Flight and navigational equipment;

§135.149 Equipment requirements:

A significant differences between equipment required by Certification Rules and Operating Rules happens with a requirement changes following the issuance of the Type Certificate. Normally, changes to the Certification Rules do not affect a certified aircraft. Changes to the Operating Rules will require compliance regardless of the certification data.

To illustrate, the early CAR 3 versions did not require an attitude indicator for any operations. Does that mean, that we don't need one to fly a Cessna 150 IFR? No, since the Operating Rules have been changed and require such an installation to fly in jurisdictions where Part 91 is the Operating Rule.

Recently, changes to the Operating Rules have required the installation of mode S transponders, ADS-B equipment for many operators.

Design Specified Equipment

Some equipment is included in the design because the designer (or the marketing department) wants it. Many small airplanes with controllable pitch propellers have manifold pressure gauges. This is not required by operating rules unless the engine is supercharged. It may be required by the manufacturer and will be so indicated on the equipment list.

The Lockheed C-130J included dual head-up displays, not for any operational requirement (In fact, at the time, there was no operational benefit). The marketing department thought that potential customers would want it.

The reader will note that Normal Category* airplanes only require three instruments for visual flight: airspeed indicator, altimeter, and a compass (often called a wet compass†). The altimeter need not be adjustable for changes in the barometric pressure. However, a sensitive altimeter adjustable for barometric pressure is required by current Part 91.

Customer Specified Equipment

Some airlines will equip their fleets with passenger entertainment equipment. Some small airplane owners may wish to install moving map displays. Normally such equipment will be approved on a *"no hazard"* basis.

However, installed equipment must be approved. It may be approved by the Type Certificate (TC) or by a Supplemental Type Certificate (STC). It may also be approved by a Technical Standard Order (TSO). There is a third choice for add-on equipment installations: a Field Approval (*vide infra*).

---

*   Prior to amendment 23-64.[1]

†   Sometimes called a "whiskey" compass because magnetic compasses often use alcohol to dampen oscillations.

Airworthiness Requirements for Installed Equipment

Many types of installed equipment are discussed specifically in the Airworthiness Requirements: Flight Instruments, Airspeed Indicators, Powerplant Instruments, Electrical Systems, and many others. The specific requirements are listed in Subpart F of FARs 23, 25, 27, and 29.

There are three sections that apply in general to equipment approvals: §§2x.1301, 25.1302, and 2x.1309.

§2x.1301
Originally, identical language was used for FARs 23, 25. 27, and 29:

Sec. §2x.1301 <u>Function and installation</u>.

Each item of installed equipment must:
(a) Be of a kind and design appropriate to its intended function;
(b) Be labeled as to its identification, function, or operating limitations, or
any applicable combination of these factors;
(c) Be installed according to limitations specified for that equipment; and
(d) Function properly when installed.

In 2007, Section §25.1301 added a new line, stating that that electrical wiring must meet §25.1701 through §25.1733. This extra verbiage seems unnecessary. It essentially has one rule saying that the design must comply with another

This paragraph seems straight forward. Yet in the late 1990s and early 2000s, many applicants introduced complex electronic displays with the intended function of providing "situation awareness." This was an attempt to avoid having to certify these displays on the ground that they weren't required. Current FAA policy deems that vague expressions such as *situation awareness* are not acceptable intended functions.[6]

§25.1302
Paragraph §25.1302, the "human factors" rule, uses 323 words to expand on the nature of equipment requiring pilot intervention. At the time the rule was added, we thought the rule was unnecessary and could be inferred from §25.1301 and §25.1309. We still find the rule unnecessary.

§2x.1309
Paragraph §2x.1309 could be known as the "First, do no harm" rule.* A paraphrased version of §27.1309 (a) through (c) expresses this succinctly (in eighty-seven words):[2]

Sec. §2x.1309 <u>Equipment, systems, and installations (paraphrased)</u>.
(a) The equipment, systems, and installations whose functioning is required by this subchapter,
must be designed and installed to ensure that they perform their intended functions under
any foreseeable operating condition.
(b) The equipment, systems, and installation of a multiengine aircraft must be designed to prevent
hazards to the aircraft in the event of a probable malfunction of failure.
(c) The equipment, systems, and installation of a single-engine aircraft must be designed to prevent minimize hazards to the aircraft in the event of a probable malfunction of failure….

Unfortunately, there is a tendency to add verbiage. In the various iterations, there were instances where a statement was added saying that the deign must comply with paragraph such and such. We don't think that it's necessary to have some section say that the design must comply with another.

Fortunately, later versions seem to be written more concisely while still making sure that the critical items are covered – at least for transport rotorcraft. For example, the 2011 version of §29.1309 is 494 words, but there is no "fluff."

Sec. §29.1309 <u>Equipment, systems, and installations</u>.
(a)   The equipment, systems, and installations whose functioning is required by this
subchapter must be designed and installed to ensure that they perform their intended

---

* Hence, the quote at the beginning of this chapter.

functions under any foreseeable operating condition.

(b) The rotorcraft systems and associated components, considered separately and in relation to other systems, must be designed so that--

    (1) For Category B rotorcraft, the equipment, systems, and installations must be designed to prevent hazards to the rotorcraft if they malfunction or fail; or

    (2) For Category A rotorcraft--

        (i) The occurrence of any failure condition which would prevent the continued safe flight and landing of the rotorcraft is extremely improbable; and

        (ii) The occurrence of any other failure conditions which would reduce the capability of the rotorcraft or the ability of the crew to cope with adverse operating conditions is improbable

(c) Warning information must be provided to alert the crew to unsafe system operating conditions and to enable them to take appropriate corrective action. Systems, controls, and associated monitoring and warning means must be designed to minimize crew errors which could create additional hazards.

(d) Compliance with the requirements of paragraph (b)(2) of this section must be shown by analysis and, where necessary, by appropriate ground, flight, or simulator tests. The analysis must consider--

    (1) Possible modes of failure, including malfunctions and damage from external sources;

    (2) The probability of multiple failures and undetected failures;

    (3) The resulting effects on the rotorcraft and occupants, considering the stage of flight and operating conditions; and

    (4) The crew warning cues, corrective action required, and the capability of detecting faults.

(e) For Category A rotorcraft, each installation whose functioning is required by this subchapter and which requires a power supply is an "essential load" on the power supply. The power sources and the system must be able to supply the following power loads in probable operating combinations and for probable durations:

    (1) Loads connected to the system with the system functioning normally

    (2) Essential loads, after failure of any one prime mover, power converter, or energy storage device.

    (3) Essential loads, after failure of--

        (i) Any one engine, on rotorcraft with two engines; and

        (ii) Any two engines, on rotorcraft with three or more engines.

(f) In determining compliance with paragraphs (e)(2) and (3) of this section, the power loads may be assumed to be reduced under a monitoring procedure consistent with safety in the kinds of operations authorized. Loads not required for controlled flight need not be considered for the two-engine-inoperative condition on rotorcraft with three or more engines.

(g) In showing compliance with paragraphs (a) and (b) of this section with regard to the electrical system and to equipment design and installation, critical environmental conditions must be considered. For electrical generation, distribution, and utilization equipment required by or used in complying with this subchapter, except equipment covered by Technical Standard Orders containing environmental test procedures, the ability to provide continuous, safe service under foreseeable environmental conditions may be shown by environmental tests, design analysis, or reference to previous comparable service experience on other aircraft.

The equivalent paragraph for transport airplanes (§25.1309) is lengthy, but in our opinion, not as well written. It leaves more to the reader to figure out what is needed.

The corresponding section for small rotorcraft (§25.1309) is quite short and does distinguish single from multiengine aircraft. The corresponding section for small airplanes is gone following the rewriting of Part 23.

Nevertheless, the reader can see the intent of the rule: "Make sure the equipment is safe and cannot result in loss of the airplane". This approach results in very subjective compliance finding. However, each directorate has provided Advisory Circulars (AC's) which provide acceptable means of compliance which allow consistent approval processes.[3][4][5][6]

Aircraft failures are defined in terms of their effect on the aircraft and on individuals. *Catastrophic* hazards involve multiple fatalities, loss of the aircraft, or incapacitation of the flight crew.* A *hazardous* event is one that involves a serious or fatal injury to an aircraft occupant, a large reduction in the functional capabilities of aircraft, a large reduction in safety margins or physical distress or excessive workload that impairs ability of crew to perform tasks, A *major* hazard involves physical distress for passengers, significant reduction in safety margins, or significant increase in crew workload. A *minor* hazard involves physical discomfort for passengers, slight reduction in safety margins, or slight increase in crew workload.

None of the hazards in the previous paragraph consider property damage to property on the ground.

In order to assess the risk of system operation, it is customary to categorize hazards into increasing levels of severity. MIL-STD-882D[7] suggests categories ranging from *Negligible* through *Catastrophic*. It also suggests frequencies ranging from Frequent through Improbable. When examined, certain pairs such as *Frequent-Catastrophic* are clearly unacceptable, but any improbable frequency or *Negligible* hazard should be acceptable. Table 9 shows the notional matrix, adapted from Bahr.[8]

Table 9 Notional Hazard Risk Assessment Matrix

Frequency	Hazard Category			
	1	2	3	4
	Catastrophic	Critical	Marginal	Negligible
(A) Frequent	1A	2A	3A	4A
(B) Probable	1B	2B	3B	4B
(C) Occasional	1C	2C	3C	4C
(D) Remote	1D	2D	3D	4D
(E) Improbable	1E	2E	3E	4E
Green: Acceptable; Yellow: Undesirable; Red: Unacceptable				

Roland and Moriarty[9] use these adjectives in to describe the qualitative probabilities of hazard occurrences. Table 10 shows the probability adjectives used by the FAA in their various system safety Advisory Circulars (ACs).

Table 10 General Mishap Probability Levels

Description		Individual Aircraft	Entire Fleet of a Model
Military[7][8][9]	Civilian[3][4][5][6]		
Frequent		Likely to occur frequently	Continuously experienced
Probable	Probable(a) Frequent(b)	Will occur several times in the life of a single aircraft	Will occur frequently
Occasional	Remote	Likely to occur once the life of a single aircraft	Will occur several times
Remote	Extremely Remote	Unlikely, but possible to occur in the life of a single aircraft	Unlikely, but can reasonably be expected to occur
Improbable	Extremely Improbable	It can be assumed occurrence will not happen	Unlikely to occur, but possible
Notes (a) *Probable* for fixed-wing aircraft.[2][3] (b) *Frequent* for rotary-wing aircraft.[4][5]			

---

* The definition shown is for Normal Category airplanes[3] or Transport Category helicopters.[6] For transport airplanes[4] or Normal Category helicopters,[5] the definition of a *catastrophic* hazard is one that would prevent continued safe flight and landing. It is unfortunate that the different directorates cannot agree on common definitions.

Quantitative allowable probabilities for aircraft hazards are shown in Table 11. It should be noted that the allowable probabilities should not be read with too much precision. The intent is to treat the goal as a probability of the order of the listed value.

Table 11 Allowable Probabilities for Aircraft Hazards

Failure Condition	MTOW	Minor	Major	Hazardous	Catastrophic
Probability Term		Probable[a]	Remote	Extremely Remote	Extremely Improbable
Small Airplane (SRE)[b]	≤6,000 lb	$<10^{-3}$	$<10^{-4}$	$<10^{-5}$	$<10^{-6}$
Small Airplane (non SRE)[b]	≤6,000 lb	$<10^{-3}$	$<10^{-5}$	$<10^{-6}$	$<10^{-7}$
Small Airplane[b]	≤12,500 lb	$<10^{-3}$	$<10^{-5}$	$<10^{-7}$	$<10^{-8}$
Commuter Airplane[b]	≤19,000 lb	$<10^{-3}$	$<10^{-5}$	$<10^{-7}$	$<10^{-9}$
Transport Airplane	Any				
Small Helicopter	≤7,000 lb				
Transport Helicopter	Any				

Notes: (a) The term "frequent" applies to rotorcraft
(b) Prior to Amendment 23-64

The reader will note that the allowed probabilities are consistent, except for small airplanes which have greater allowable probabilities for catastrophic and hazardous outcomes. The rationale for reducing these values for Part 23 airplanes is outlined in AC23.1309-1E, citing historical accident data indicating the probability of a fatal accident from operational and airframe causes was approximately one per 10,000 flight hours for single-engine airplanes under 6,000 lb. The AC concludes that system failures would be approximately one tenth of this rate or 10^{-5} per flight hour. In our opinion, this assessment is far too optimistic. In truth, the techniques described in a number of system safety texts should be followed.[8][9][10][11] Nevertheless, the AC has been published and "It is what it is" -- an acceptable means of compliance.

Customer Specified Equipment

Some airlines will equip their fleets with passenger entertainment equipment. Some small airplane owners may wish to install moving map displays. Normally such equipment will be approved on a "no hazard" basis.

Approval by Technical Standard Order

A Technical Standard Order (TSO) is a minimum standard for specified materials, parts, and appliances used on civil aircraft.* TSOs were originally developed to allow for substitution of standard parts without having to specifically approve each part. Typical TSOs include

- ✈ TSO-C2d   Airspeed Indicators,
- ✈ TSO-C10c   Altimeters,
- ✈ TSO-C16b   Pitot Tubes,
- ✈ TSO-C22g   Safety Belts,
- ✈ TSO-C25a   Seats and Berths,
- ✈ TSO-C62e   Aircraft Tires,
- ✈ TSO-C100c   Aviation Child Safety Devices (ACSD)
- ✈ TSO-C114   Torso Restraint Systems.

---

*   https://www.faa.gov/aircraft/air_cert/design_approvals/tso/ accessed 7/26/2017

➤ TSO-C119e   TCAS Systems,
➤ TSO-C145e   GPS Sensors,
➤ TSO-C151d   Terrain Awareness and Warning Systems (TAWS), and
TSOs can be issued for non-aircraft installed items, such as
➤ TSO-C13f   Life Preservers,
➤ TSO-C23f   Personnel Parachutes, and
➤ TSO-C164a   Night Vision Goggles.

Many TSOs are based on industry standards. For example, TSO-C10c is based on SAE AS-8009C.[12] A list of current TSOs may be found on RGL.*

A manufacturer may be issued a Technical Standard Order Authorization (TSOA) which is both a design approval and a manufacturing approval. Aircraft Certification Offices will be responsible for TSOA applications from manufacturers in the ACO's geographic region. The ACO will review the applicant's technical competence to certify conformance with the TSO. The appropriate MIDO will evaluate the applicant's quality control system.

In summary, TSO is a certification process in which the certification and production is placed in the hands of the manufacturer with oversight by the local ACO.

Field Approval

A Field Approval is a process intended for a single aircraft installation. The modification is inspected by a Flight Standards inspector. Presumably, the inspector is to use the same certification standards as the original aircraft TC. This approval is usually called a "337 approval" from the FAA form used to report major aircraft repairs or modification.

Figure 51 shows a ferry tank installation approved by an FAA maintenance inspector for a flight to Brazil in a Cessna 150. This modifications are supposed meet the same standards as the original certification. We used the FAR 23 requirements for fuel tanks and included a pressure test and loads analysis. The installation was approved by an FAA maintenance inspector..†

Figure 51 Cessna 150 Ferry Tank Approved Using FAA Form 337‡

---

* http://rgl.faa.gov/ accessed 7/26/2017
† The FAA inspector who was a maintenance inspector was surprised at the engineering report presented. He said, "I've never seen anything like this. Can I keep a copy of the report?"
‡ Photograph by the author (RLN) prior to flying the airplane to Brazil.

What Does "Installed" Mean?

On December 8, 2005, a Boeing 737 overran a snow-covered runway at Chicago's Midway Airport and collided with an automobile on the street.* A small boy in the automobile was killed.[13] As is often the case, this mishap had a number of causes and factors. One of the factors was the improperly coded software in the pilots' personal electronic device (PED). The accident investigation found that if the maximum allowable tailwind was exceeded, the software in the PED would calculate the stopping distance using an assumed maximum tailwind of 5 knots, which is non-conservative. Because of this, the PED gave a computed stopping margin of 40 feet (based on the programmed maximum tailwind, even though the pilots had entered 8 knots). Had the software used the actual wind, the algorithm would have clearly indicated an overrun.

The issue is that, at the time, the software was not certified by an acceptable means (DO-178). This has since been rectified for airline (Part 121) operations.

Does a hand-carried electronic device need to be certified for airworthiness? The answer is unclear. We have heard conflicting opinions stated by knowledgeable persons within the FAA, both while employed there and after retirement. This is particularly troubling since we have been asked to approve such devices as a consulting DER.

Operational Approval

There are several areas where equipment is not required for certification, but which may be required for specific types of operations. While the equipment isn't required, the operator (and the operational inspectors) will need to know if the equipment can perform its intended function and what the operating and maintenance procedures are to ensure that it can. Examples of such equipment include

✈: Category III ILS Approaches
✈: ETOPS
✈: RNP Operations
✈: RVSM Operations

Each of these has some combination of reliability verification and stringent performance criteria, and may require extensive testing.

ETOPS

Early airline operation rules required that twin-engine airplanes be operated within 1 hour (at single-engine cruise speed) of a suitable alternate airport at all times. This effectively precluded airlines from operating twin-engine airplanes over most oceanic routes. In 1985, US airlines were allowed to operate twin-engine turbojet airplanes on routes up to 120 minutes from a suitable alternate airports.[14]†

In 2007, the FAA amended the certification and operating rules for "the design, operation, and maintenance for airplanes that operated on flights that fly long distance from an adequate airport".[15] The air carrier operating rules now required specific ETOPS approval for:

✈ Twin-engine airplanes operating beyond 60 minutes from a suitable alternate airport; and

✈ Three or four engine passenger airplanes operating beyond 180 minutes from a suitable alternate airport.[16]

For twin-engine airplanes, there can be diversion times up to three hours.‡

---

* This accident is described in Appendix 4 on page 390.
† We were unable to locate the amendment to Part 121 in RGL.
‡ ETOPS-180–Imagine three hours on a single-engine over the North Pacific in winter!

Appendix K, Extended Operations (ETOPS) was added to Part 25 in 2007.[17] To qualify, there are extensive reliability requirements on the engines and systems. This may, in turn, result in special maintenance and equipment requirements. For example, an Auxiliary Power Unit may be inoperative for domestic flights, but would certainly be on the Minimum Equipment List for ETOPS operations.

While the original ETOPS only considered twin engine operations, the new rule considered passenger carrying airplanes with three or four engines, treating issues such as decompression or major loss of systems.*

Simulated diversions will be performed during the approval period to actually shut down an engine and simulate a 90 minute (or longer) diversion to an airport.†

Category III ILS Approaches

The technique for approving Category III (CAT III) ILS approaches dates back to the late 1960s with automatic landing approvals.[18] In summary, the touchdown dispersion must be determined based on a variety of wind conditions and turbulence. This distribution must show that the 2 sigma distribution:

→ Longitudinal dispersion of the main landing gear should not exceed 1500 ft total, but need not be symmetrical about the nominal point.

→ Lateral dispersion of the aircraft centerline should not exceed 27 ft either side of the centerline of the runway.

This may be demonstrated by a combination of analysis and flight demonstrations as appropriate to verify the result of the analysis. It must be remembered that AC-20-57A[18] was intended for automatic landing (Auto Land) approvals. As such much of the analysis was in the forms of repeated simulations using only hardware-in-the-loop with no pilots. Thus, thousands of simulations could be accomplished in relatively short time. Only a relatively few representative flights need be accomplished to show that the airplane-plus-system matched the simulations.

As the head-up display (HUD) began to appear in civil aviation, the question of how to approve such a display for CAT III arose. An FAA test pilot, Berk Greene,‡ decided to use a similar approach with thousands of manned simulation replications and hundreds of manned airplane landings. This was adopted by the FAA in a Flight Standards Advisory Circular AC-120-28.[19] AC-20-57 was folded into AC-120-28 and then cancelled.

Until recently, two visibility conditions were flown in flight test: "severe clear" or solid IMC. Sensor based flight, such as synthetic-vision or enhanced vision, creates new problems. Degraded visual environments (DVE) may be more critical than solid instruments. Hoh[20] evaluated "foggles" that are goggles with ground surfaces to block high-spatial-frequency images allowing low-spatial-frequency information to pass. This technique is promising for daytime degraded visual conditions, but has not yet been tried under night conditions. In a Degraded Visual Environment (DVE) the limited cues when combined with instrument cues via a HUD may be more challenging.[21]

An FAA/USAF technology demonstration program for synthetic vision flew actual low visibility approaches in a *Gulfstream-II* to determine the acceptability of using FLIR and MMWR sensors to create runway images.[22] The approaches were flown in actual low-visibility conditions

---

*   While in initial training with a major airline, I asked one of the instructors: "What is our procedure if you lose cabin pressure midway between Hawaii and Seattle?" His reply was "We try not to think about that."

†   Some test directors have insisted that this test be conducted over the ocean, presumably in the mistaken belief that ocean air is different from land air. We feel such a test would show extremely poor judgment on the part of the test director.

‡   Howard B. "Berk" Greene (1940-1994), a test pilot with the FAA.

down to 50 ft DH. As the image was recorded, the airplane carried air-sampling sensors to characterize the inflight liquid water content, water droplet size and distribution, air temperature, etc. The airport environment was surveyed and calibrated to establish radar cross sections for ground features. This allowed correlation of the actual atmospheric conditions with the recorded image quality and pilot responses. It also supported correlation of empirical data with theoretical performance prediction. To our knowledge, this was the only structured flight test program studying degraded visual conditions in actual weather.

RNP Operations

Required Navigation Performance (RNP) is a performance-based navigation allowing an aircraft to fly a specific path between two defined points in space. It differs from Area Navigation (RNAV) in that RNP requires an on-board navigation performance monitoring and pilot alerting indication. RNP-X refers to a performance level of the system where X is a distance to be maintained by the aircraft on a two-sigma basis. That is, 95% of the time, the aircraft will be within 1 NM for an RNP-1 system. The aircraft will be within 2 NM 99.999% of the time (four-sigma).[23]

It goes without saying that RNP approval will require sufficient testing over representative areas to demonstrate both the two- and four-sigma containment. This will be particularly significant if the deviations are non-Gaussian, such as with Pathway-in-the Sky (PITS) or Tunnel-in-the-Sky displays.

RVSM Operations

Reduced Vertical Separation Minima (RVSM) refers to the reduction in vertical separation from 2,000 ft to 1,000 ft at higher altitudes. Traditionally, vertical spacing was based on 1,000 ft. At higher altitudes, a standard barometric setting of 29.92 in Hg was used to minimize altimeter setting errors.* Because the change of pressure with altitude (pressure lapse rate) becomes less at high altitudes, the altimeters are less sensitive. In the late 1990s, RVSM airspace was implemented in specific airspace with 1,000 ft vertical separation. Generally, RVSM has been implemented between FL290 and FL410 in high traffic volume airspace, such as the North Atlantic.

To be RVSM compliant, the airplane must demonstrate rigorous static pressure altimetry standards as well as two independent altimeters, an altitude alerting system, an autopilot to control altitude, and an altitude reporting system for the ATC transponder.

References

1   "Applicability and Definitions," *Airworthiness Standards: Normal Category Airplanes*, 14CFRPart 23, §23.2005, amendment 23-64, effective August 30, 2017

2   "Equipment, Systems, and Installations," *Airworthiness Standards; Transport Category Airplanes*, 14 CFR Part 25, §25.1309, amendment 25-41, effective September 1, 1977

3   *System Safety Analysis and Assessment for Part 23 Airplanes*, FAA Advisory Circular 23.1309-E, 11/17/2011

4   *System Design and Analysis*, FAA Advisory Circular 25.1309-1A, June 21, 1988

5   "§27.1309 (amendment 27-21) Equipment, Systems, and Installations," *Certification of Normal Category Rotorcraft*, FAA Advisory Circular 27-1B, Change 8, June 29, 2018

6   "§29.1309 (amendment 29-40) Equipment, Systems, and Installations," *Certification of Transport Category Rotorcraft*, FAA Advisory Circular 29-2C, Change 4, May 1, 2014

7   *System Safety*, MIL-STD-882E, May 11, 2012

---

* In such airspace, Flight Levels (FL) show the indicated pressure altitude in hundreds of feet: FL295 is equivalent to 29,500 ft.

8   N J Bahr, *System Safety Engineering and Risk Assessment: A Practical Approach*, New York: Taylor and Francis, 1997

9   H E Roland and B Moriart, *System Safety Engineering and Management*, New York: Wiley, Second Edition, 1990

10   D B Brown, *Systems Analysis and Design for Safety*, Englewood Cliffs, NJ" Prentice-Hall, 1976

11   W Hammer, *Handbook of System and Product* Safety, Englewood Cliffs, NJ: Prentice-Hall, 1972

12   *Pressure Altimeter Systems*, Society of Automotive Engineers SAE AS-8009C, May 24, 2016

13   *Runway Overrun and Collision, Southwest Airlines Flight 1248 Boeing 737-7H4, N471WN, Chicago Midway International Airport, Chicago, Illinois, December 8, 2005*, NTSB AAR-07-06, October 2007

14   J A DeSantis, "Engines Turn of Passengers Swim: A Case Study of How ETOPS Improved Safety and Economics in Aviation," *Journal of Air Law and Commerce*: 77, 2013, 3-68

15   "Extended Operations (ETOPS) of Multi-Engine Airplanes," *Federal Register*, Docket No. FAA-2002-6716, amendments 1-55, 21-89, 25-120, 33-21, 121-329, and 135-108

16   *Extended Operations (ETOPS) and Polar Operations*, FAA-AC-120-42B, dated June 13, 2008, canceling AC-120-42A, dated December 30, 1988

17   "Extended Operations (ETOPS) of Multi-Engine Airplanes," *Airworthiness Standards: Transport Category Airplanes*, 14 CFR Part 25, Appendix K, amendment 25-120

18   *Automatic Landing Systems (ALS)*, FAA Advisory Circular AD-20-57A, dated January 12, 1971

19   *Criteria for Approval of Category III Weather Minima for Takeoff, Landing, and Rollout*, FAA Advisory Circular AC-120-28D, July 13, 1999

20   R H Hoh, *Investigation of Outside Visual Cues Required for Hover*, AIAA Paper 85-1808, 1985

21   P J Garman and J A Trang, "In Your Face! The Pilot's/Tester's Perspective on Helmet-Mounted Display (HMD) Symbology," *Proceedings Helmet- and Head-Mounted Displays and Symbology Requirements Symposium*, April 1994, p. 274-280

22   M A Burgess *et al.*, *Synthetic Vision Technology Demonstration, Executive Summary*, DOT/FAA-RD-93-40, December 1993

23   D Ellis, G Limesand, and B Syblon, "Equipping a Fleet for Required Navigation Performance," *Aero*, 4th Quarter 2011, pp. 25-28; published by Boeing Commercial Airplanes

Chapter 8:
Seat and Safety Belt Approval

While researching for accidents caused by fuel system icing, we came across a reference to an accident to a Pilatus PC-12 in March 2009.[1] While enroute, the fuel flow from the left tank became restricted as a result of ice accumulation at the fuel filter. About 1:21 into the flight, fuel supplied to the engine was being drawn solely from the right tank. A left-wing heavy condition resulted which continued to increase. The airplane was controllable in static flight with the left-wing heavy fuel imbalance that existed at the time of the accident, but the pilot lost control of the airplane in the traffic pattern.*

The airplane was equipped with eight passenger and two pilot seats; however there were a pilot and thirteen passengers on board. Seven of the passengers were children under the age of 9 years.

At least four of the seven children were not restrained or were improperly restrained. Except for the pilot and the adult passenger in the right cockpit seat, the NTSB was unable to determine the seating position for the airplane occupants. The airplane owner who organized the trip stated that adults could hold the children on their laps. However, only one of the seven children was under the age of 2 years and thus permitted (by the regulations) to be held on the lap of an adult. The other six children ranged in age from 3 to 9 years. Because the bodies of four of these children were found farthest from the impact site, the NTSB concluded that at least four of the seven children were not restrained or were improperly restrained.

The mental image of the children being thrown from the airplane prompted us to ask the NTSB for more details. This, in turn, led us to review crashworthiness and occupant restraint.

Crashworthiness Design

The basic principles of crashworthiness design are straightforward. An acronym, CREEP, has been proposed as a reminder of these principles.[2]

- Container: The structure surrounding the occupied space should not collapse. It should also prevent external objects from entering and injuring occupants.
- Restraint: The seat/restraint design should support the occupant during the deceleration and prevent her/him from contacting injurious objects.
- Environment: There should be no potentially injurious objects within each occupant's head and extremity strike zone. (See Figure 54.)
- Energy: The structure may absorb energy by proving "crush zones" to reduce the forces applied to the occupants.
- Post-Crash: Finally, the design should protect against post crash-hazards. The most serious post-crash hazard is fire. The design should also facilitate egress.

---

* This accident is described in Appendix 4 on page 339.

Human Tolerance to Crash Impact

Human tolerance of acceleration to a large extent depends on the orientation of the human to the acceleration vector. Figure 52 shows standard human coordinate system.[3]

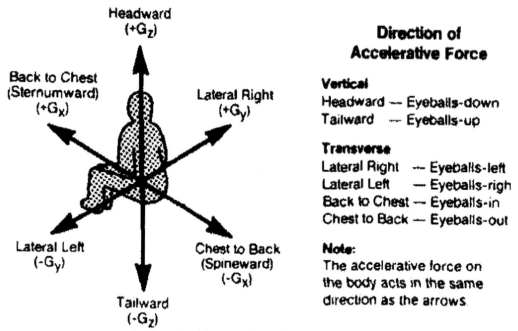

Figure 52 Human Coordinate System*

A limiting factor is the individual occupant's tolerance to abrupt decelerations.[4] Within the extremely short duration range of abrupt accelerations commonly experienced in an aircraft crash (0.2 sec and below), the effects on the human body are limited to mechanical overloading (skeletal and soft tissue stresses), there being insufficient time for functional disturbances due to fluid shifts.

It is difficult to find the limits. It is unlikely we will find human subjects willing to see how close to fatal injury we can come. Snyder[4] cites Stapp as defining tolerance as "the limit beyond which either the subject or the experimenter fears to go lest there be serious injury."

In 1959, Eiband performed a literature search of human tolerance to impacts.[5] The summary was based on human and animal experiments and these are presented on the basis of a trapezoidal pulse which was representative of much of the data at the time. Eiband displayed these data on log-log plots.[5] Figure 53 shows the Eiband Curve for spineward acceleration $(+G_Z)$. Several data points are shown for human and animal exposures.

There have been many additional reviews. In a 1966 FAA report, Mohler and Swearingen recommended a 25 g design goal for general aviation designs.[7]

In 1973, Snyder described the understanding of human impact tolerance at the time.[4] For example, he stated the -Gx (eyeballs-out) tolerance at about 50 g, provided there is adequate upper torso restraint. Yet he described 38 g at higher onset rates producing severe shock. Snyder said that rearward facing (+Gx, eyeballs-in) tolerance has not been clearly established. The accepted limits for +Gz (eyeballs-down) were stated as approximately 20 g, based on ejection seat design. Lateral tolerance appears to be considerably lower.

---

*   Figure from FAA-AC-21-22.[3]

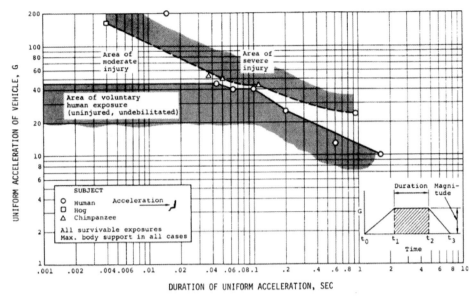

Figure 53 Eiband Curve for -G$_X$ (eyeballs out) Acceleration*

In a review of transport airplane crashworthiness, Snyder[8] cites a 1969 FAA memorandum,[9] which recommended the following

Direction	Deceleration
Upward	20 g
Forward	20 g
Sideward	10 g
Downward	20 g
Rearward	20 g.

The memorandum also recommended using a design passenger weight of 225 lb for seat design purposes.†[9] This would ensure that occupant protection would not be reduced for passengers heavier than the 170 lb weight used today.

In a review of general aviation crashworthiness, Snyder[10] recommended installation of shoulder harnesses at all seats and design crash loads of

Direction	Deceleration
Upward	16 g
Forward	25 g (applied 20 deg either side of the longitudinal axis)
Downward	15 g
Rearward	05 g

The forward direction applied 20 deg to either side was apparently to apply any sideward crash load simultaneously with a forward crash load, rather than a side crash load alone.

Desjardins *et al.*[11] made design recommendations for military helicopters. Crash load recommendations were

Direction	Deceleration
Rearward (+Gx)	12 g
Downward (+Gz)	23 g
Forward (-Gx)	35 g for crew seats
	30 g for passenger seats
Upward (-Gz)	8 g
Sideward (±Gy)	20 g.

---

*   The only available copy of Eiband's report was rather poor. Figure 53 was copied from Shanahan.[6]

†   Presumably a 75th percentile man.

Rearward-facing seats were recommended for passengers. Multi-unit seats were not recommended if energy-absorption were incorporated into the seat design. This would avoid situations where partially filled seats could render the energy-absorbing mechanisms ineffective.*

Based on these reviews, human tolerances for fully restrained occupants are summarized in Table 12.

Table 12 Fully Restrained Human Crash Tolerances

Symbol	Direction	Vernacular	Example	Tolerance
$-G_Z$	Tailward	Eyeballs-up	Downward ejection seat	~8-15 g
$+G_Z$	Headward	Eyeballs-down	Ejection seat	~15-25 g
$-G_X$	Chest-to-back	Eyeballs-out	Fly into vertical cliff	~25-45 g
$+G_X$	Back-to-chest	Eyeballs-in	Same with rear-facing seat	~30-45 g
$\pm G_Y$	Lateral	Eyeballs-left/right	Sideways skid into building	~10-15 g

One way to reduce the peak deceleration is to allow the aircraft structure deformation in a crash. Allowing the engine mounts to give way in a frontal crash can create a "crush zone." This can be used in small airplanes or in automobiles.

In addition to whole body tolerances in Table 12, we must consider localized secondary impacts where, for example, the occupant's head my strike an object, such as an instrument panel of flight control in the cockpit.

An occupant's head and extremities can flail about during a crash sequence. Hazardous objects should not intrude on space within the extremity-strike envelope. Figure 54 shows the envelope for an occupant's head (solid line), foot (dotted line), and hands (dashed line) for a chest-to-back (eyeballs-out) deceleration.[12] Figure 54(a) shows occupant restrained by a lap belt alone. Figure 54(b) shows the reduction with the use of a lap belt plus a shoulder harness. Clearly, removal of dangerous objects from these extremity-strike envelopes, particularly the head strike envelope is important.

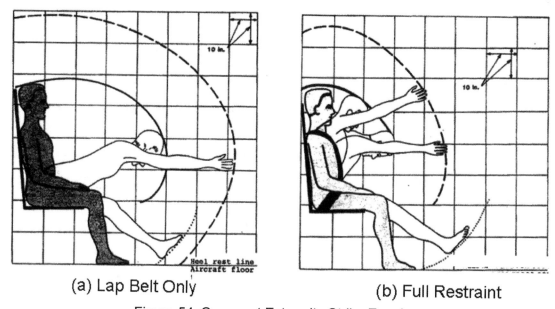

(a) Lap Belt Only   (b) Full Restraint

Figure 54. Occupant Extremity Strike Envelope

---

* Insufficient weight to activate the energy-absorbing mechanism at loads within human tolerance limits.

Application to Agricultural Application

Agricultural aerial application (crop dusting) is a particularly hazardous aviation segment. Bruggink et al.[13] reviewed that accident history of crop dusters with crashworthy designs compared to previous models. In 1949 and 1950, Texas A&M University developed a prototype incorporating a crashworthy design, the Ag-1.

Features of the design included locating the pilot station behind the payload and providing a large deformable "crush zone" in the event of a crash. The design crash loading was reported as to be 40 g.[14]

Following an accident which destroyed the prototype with the pilot surviving with negligible injuries, the industry began to produce new generation crop dusters with crashworthy designs. Bruggink et al. intentionally used this terminology to avoid commercial interpretation. Table 13 summarizes their data.* The new generation crop dusters had 13.8 percent of the accidents, but only 1.5 percent of the fatalities!

Table 13 Crash Experience of Old and New Generation Crop Dusters

Generation	Accidents	Occupants[a]		Injuries					
				Fatal		Serious		Minor/None	
		N	%	N	%	N	%	N	%
Old	460	461	86.2	66	98.5	50	92.6	345	83.3
New	074	074	13.8	01	01.5	4	07.4	069	16.7
Total	534	535	100	67	100	54	100	414	100
Note: (a) One mishap in an old generation aircraft had two crew members aboard.									

Crashworthiness Standards Since 1965

How has the research about human impact tolerance affected the airworthiness requirements? In 1965, the FAA published FARs 23, 25, 27, and 29. The initial publication reorganized the CAR requirements with no technical changes.†

1965--1987

During this period, most of the significant regulatory changes dealt with shoulder harnesses.

In 1969, Part 23 required the head-strike envelope to be free of injurious objects or that the seat be equipped with shoulder harnesses. In 1977, shoulder harnesses were required in "front" seats and, in 1985, in all seats.

In 1967, Part 25 now required shoulder harnesses or equivalent alternatives for sideway-facing seats. In 1969, these seats were redefined to be seats oriented more than 18 degrees to the vertical plane containing the aircraft centerline.

In 1984, Parts 27 and 29 required shoulder harnesses in crew seats and seats alongside crew seats.‡

1988--1989

During these two years, perhaps the most significant changes to crashworthy regulations were made. §§23.562 and 25.562 in 1988 and §§27.562, and 29.562 in 1989 now required seats approved for use during takeoff and landing to successfully complete dynamic testing or be demonstrated by a rational analysis based on dynamic tests on a similar seat. These tests were to use an approved 170 lb Anthropomorphic Test Device (ATD) seated in the normal, upright position.

---

*   One of the old generation accident had two crewmembers aboard, presumably for training purposes.

†   Details of the development of seat and restrain requirements are covered in Chapter 8.

‡   Avoiding the awkward terminology of *front seats*.

Sidebar: Crash Tests

Anthropomorphic Test Devices (ATD)

Anthropomorphic Test Devices (or Dummies) are used as surrogates for human bodies to measure the effectiveness of restraints during aircraft and automotive crashes.[15] They are designed to replicate the dynamic response to crash forces. In addition to mimicking the response of humans, they can be instrumented to record the forces that can produce injury.

The US National Highway Traffic Safety Administration (NHTSA) has published standards for ATDs for use by the automobile industry.[16] These standards have been adopted by the FAA and other aviation authorities for approval of airplane seats and restraints. Ruhle *et al.* describe the evaluation and approval process for including these dummies into these standards.[17]

In 1988, §§23.562 and 25,562 required dynamic airplane seat tests or rational analysis supported by dynamic tests. In 1989, the helicopter rules followed suit and §§27.562 and 29,562 also required dynamic testing. In each case, the tests were to use an ATD base on a 170 lb male (50th percentile male). This ATD was to be defined by 49CFR572, Subpart B unless otherwise approved.[18]

Other Subparts define child ATDs, such as Subpart C: 3-Year Old Child,[19] and Subpart J: 9 Month Old Child.[20] These are used for testing of Child Restraint Systems. Subpart O: 5th Percentile Female Test Dummy has not been used, to our knowledge in any aviation application.[21] Other subparts appear to be for works in progress.

The dynamic tests on an instrumented ATD can determine Head Injury Criterion (HIC) by measuring the acceleration if the ATD's head contacts an object. HIC is determined by the following equation

$$HIC = \left\{ (t_2 - t_1) \left[\frac{1}{t_2 - t_1} \int_{t_1}^{t_2} a(t)\, dt \right]^{2.5} \right\}_{max} \tag{3}$$

where, t1 and t2 are the initial and final integration time; a(t) is the acceleration versus time curve for the head strike. The units of t are seconds and the units of a multiples of gravity (g).[22 23 24 25]

HIC must not exceed 1000 units. The number 1000 is an arbitrary standard division between a "safe" and a "dangerous" ATD head response.[26]

In 1989, the design crash loads were increased for Normal and Transport Category rotor-craft

Direction	From	To
Upward	1.5 g	4 g
Downward	4 g	20 g
Forward	4 g	16 g
Sideward	2 g	8 g

It is interesting to note that only the 1989 changes to the helicopter design requirements cited published research findings. The US Army's *Crash Survival Design Guide*[11] and the FAA's *Analysis of Rotorcraft Crash Dynamics for Development of Improved Crashworthiness Design Criteria*[27] were cited in any Notice of Proposed Rulemakings. We also note that FAA's own study, *Cockpit Design for Impact Survival*,[7] which called for a seat design for a 25 g crash, was never mentioned in any of the NPRMs or final rule preambles.

<u>1994-on</u>
During this period, §23.561 now added a 6 g downward crash load; §25.561 increased the side-ward crash load to 3 g on the airframe and 4 g on the seats; and §§27.561 and 29.561 added a 1.5 g rearward crash load. §§27.561 and 29.561 added a requirement for increased protection for be-low floor fuel tanks.

Summary of Changes
Figure 55 traces the significant development of crashworthiness requirements from the initial publication of FARs in the mid-1960s and 1997. Table 14 summarizes the overall changes in crash load requirements during the same period.

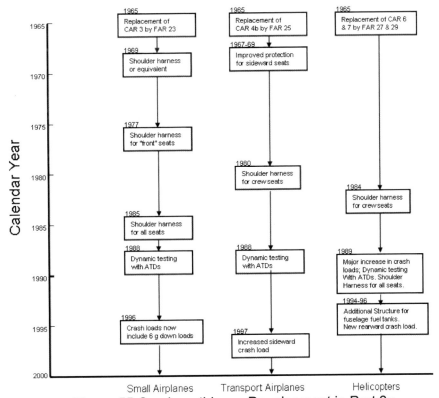

Figure 55 Crashworthiness Development in Part 2x

Table 14 Civil Airworthiness Requirements for Crash Loads

Force Direction (forward facing seats	Calendar Year	
	1965	1997
Small Airplanes		
Upward (eyeballs up)	3 g[a]	3 g[a]
Downward (eyeballs down)	-	6 g
Forward (eyeballs out)	9 g	9 g
Rearward (eyeballs in)	-	-
Sideward	1.5 g	1.5 g
Shoulder Harness Required	no	yes
Dynamic Testing Required	no	yes
Transport Airplanes		
Upward (eyeballs up)	2 g	3 g
Downward (eyeballs down)	4.5 g	6 g
Forward (eyeballs out)	9 g	9 g
Rearward (eyeballs in)	-	1.5 g
Sideward	1.5 g	4 g
Shoulder Harness Required	no	yes
Dynamic Testing Required	no	yes
Rotorcraft		
Upward (eyeballs up)	1.5 g	4 g
Downward (eyeballs down)	4 g	20 g
Forward (eyeballs out)	4 g	16 g
Rearward (eyeballs in)	-	-
Sideward	2 g	8 g
Shoulder Harness Required	no	yes
Dynamic Testing Required	no	yes
Note (a) 4.4 g for acrobatic		

Crashworthiness in the Operating Rules

The benefit of restraints are predicated on the proper use of seat belts and harnesses. This leads to an obvious question: "What are the requirements for using safety belts and harnesses?" These are found in the Operating Rules, FARs 91, 121, and 135 (See Chapter 2.

General Aviation Operating Rules

The initial publication of Part 91, in 1965, stated that safety belts were required to be installed,[28] but safety belt use was not required. During our review, we were surprised to find that there were requirement to actually fasten safety belts or harnesses.

In 1971, §91.14 was added stating "During the takeoff and landing of U.S. registered aircraft (except an airship) each person on board that aircraft must occupy a seat or berth with a safety belt properly secured about him. However, a person who has not reached his second birthday may be held by an adult who is occupying a seat or berth".[29] The preamble stated that "It is not intended that separate seats nor [sic] safety belts be required for operations under Part 91."*

In 1991, §91.14 morphed into §91.107.[30] It appears that the rule was modified at this time to include shoulder harness, if installed.

§91.107 was amended in 1996 to include provision for a child to occupy an approve Child Restraint System (CRS).[31]†

Air Carrier Operating Rules

---

* No amendments to Part 91 prior to amendment 91-211 in 1990 are available in RGL

† It is not clear from the RGL file exactly what the changes were or when they were made.

The RGL website does not show any Part 121 or Part 135 paragraph dealing with safety belts at its initial publication.[32] The earliest references in RGL we found were amendments 121-155 and 135-62 dated 1996. It is not clear when §121.311 was issued, but it was already in place in 1971.[33]

In 1996, the air carrier rules (§§121.311 and 135.128) required an approved seat or berth for each person on board the airplane who has reached his second birthday. It also requires a separate safety belt for each person who has reached his second birthday. It did allow that two persons may occupy a berth, lounge or divan seat and share one approved safety belt during cruise flight only.

A child may occupy an approved CRS if the child is accompanied by a parent, guardian, or attendant., A child may also be held by an adult, provided that child has not reached his second birthday.[34][35] There were several amendments to the CRS specifications during 2005--2015.

The current version of §121.311 requires transport airplanes manufactured after October 2009 to use passenger and flight attendant seats meeting the crashworthiness testing required by §25.562, amendment 25-64 in passenger-carrying operations.[36]

Lap Children

The practice of allowing small children to be held by an adult rather than being restrained has long been criticized. There have been a number of papers in various law reviews generally making strong arguments against the practice.[37][38][39][40] These arguments were generally based on a lack of equal protection for small children. They were the same things we felt when reviewing accounts of unrestrained babies in our review of airplane accidents.*

Two recent accidents made this more apparent to us. On June 3, 2008, a single-engine TBM airplane flying an *Angel Flight* mission† crashed on takeoff from Iowa City, IA airport. The airplane stalled during the downwind takeoff and the unrestrained child was expelled from her mother's arms and struck her head.‡ The cause of death was blunt force head trauma. The two occupants, the pilot and the child's mother received minor injuries.

The second accident occurred on December 22, 2012, when a commuter Swearingen SA227AC crashed during an attempted go-around in Sanikiluaq, Nunavut, Canada. An unrestrained infant was expelled from her mother's arms and struck the entry door mechanism and was found next to the captain's rudder pedals. The cause of death was blunt force head trauma. The two pilots and one passenger received serious injuries. Five passengers, including the child's mother, received minor injuries.§

In both accidents, the two infants were the only fatalities. Both of your authors have lost children; we can only imagine the additional trauma this would create to the surviving parent having the baby torn from your arms.

It seems obvious to us that few, if any, adults could possibly hold a two-year old during even a minor crash. Mohan and Schneider studied the ability of adults to restrain small infants.[41] They measured the ability of adults to withstand the forces to be expected in a crash situation. Three male and three female parents were in their subject pool. They concluded

> **"... children (under 2 yeas old) in airplanes are being exposed to undue risks of injury by the requirement that they be transported on adult laps."**

---

* Several accidents involving these issues are described in Appendix 4,

† *Angel Flight* is the name used by a number of groups whose members provide free air transportation for passengers in need of medical treatment far from home and perform other missions of community service.

‡‡ This accident is described in Appendix 4 on page 377.

§ This accident is described in Appendix 4 on page 378.

Fife *et al.* examined the relative mortality of infant passengers being held and unbelted non-infant passengers in US and non-US air carriers.[42] They reviewed accidents with both passenger fatalities and surviving passengers. Each passenger cabin was treated separately. Cabins with no survivors or no infants were excluded from the analysis. They concluded that unrestrained infants had a greater risk than restrained non-infants. Unrestrained infants were five times as likely to be killed than restrained infants in US air carriers. For non US air carriers, they were 9.6 times as likely.

Bil *et al.* also reviewed the safety of lap-held infants.[43] They discussed two cases, infants held solely by an adult's arms and in a supplementary loop belt. Using simulated conditions, they concluded that the infant held by the adult was likely to be "projected through the aircraft cabin". They also reported abdominal injury or head injury when restrained by a supplementary belt.

If we shouldn't hold small infants in our laps, why can't we just have them use the existing safety belts in the airplane (or automobile)? Burdi *et al.* reviewed this issue.[44] They point out differences between adults and children in terms of size, proportions, structural aspects, center-of-mass, and biomechanical properties.

In terms of restraint design, Burdi *et al.* state that particular attention must be paid to supporting the child's head because it is relatively massive compared with an adult's. They argue against a lap belt alone. The say that the most effective restraints for children should be designed to distribute the impact loads over a large portion of the body. While written for automotive applications, their observations clearly apply to aviation applications as well.

In seems to us that it is unconscionable that we are not allowed to carry unrestrained infants in automobiles, but we are allowed to in airplanes. The arguments for allowing such practice seem specious:

- Parents would have to pay for a seat.
- They might not own an approved seat. or
- To save money, they might drive
  (which, we all know, is more dangerous).

We must require all occupants on airplanes to be properly restrained.

Recommendations

1) Amend FARs 91, 121, and 135 to require proper restraint for all children in age-appropriate child seats and eliminate the practice of holding children in an adult's arms.
2) Amend Part 91 to eliminate the use of shared safety belts in all airplanes.
3) Amend Part 23 and 25 to increase the design crash loads for airplanes to

Direction	Deceleration
Upward	16 g,
Forward	25 g,
Downward	15 g,
Sideward	8 g, (no change for Part 25)
Rearward	5 g.

4) Amend FARs 23, 25, 27, and 29 to design seats based on the weight of the 75th percentile adult male.

References

1   *Aircraft Accident Report. Loss of Control While Maneuvering, Pilatus PC-12/45, N128CM, Butte, Montana, March 22, 2009*, NTSB AAR-11-05, Corrected copy, June 2012

2	D F Shannahan, *Basic Principles of Helicopter Crashworthiness*, USAARL Report 93-15, February 1993

3	*Injury Criteria for Human Exposure to Impact*, FAA Advisory Circular AC-21-22, June 20, 1985

4	R G Snyder, "Impact," *Bioastronautics Data Book*, Parker, J F and West, V R (eds.),NASA SP-3006, 1973, pp. 221-295

5	A M Eiband, *Human Tolerance to Rapidly Applied Accelerations: A Summary of the Literature*, NASA Memo 5-19-59E, June 1959

6	D F Shannahan, "Human Tolerance and Crash Survivability," *Pathological Aspects and Associated Biodynamics in Aircraft Accident Investigation*, RTO-EN-HFM-113, August 2005, Paper 6

7	S R Mohler and J J Swearingen, *Cockpit Design for Impact Survival*, FAA AM-66-03, February 1966

8	R G Snyder, *Advanced Techniques in Crash Impact Protection and Emergency Egress from Air Transport Aircraft*, AGARD-AG-221, June 1976

9	S R Mohler, "Crash Protection in Survivable Accidents," Memorandum, Staff Study, Office of Aviation Medicine, FAA, Washington, DC, 4 February 1973; cited by Snyder, *supra*

10	R G Snyder, *General Aviation Aircraft Crashworthiness: An Evaluation of FAA Safety Standards for Protection of Occupants in Crashes*, University of Michigan Highway Safety Research Institute UM-HSRI-81-10, May 1981

11	S P Desjardins *et al.*, *Aircraft Crash Survival Design Guide, Volume IV, Aircraft Seats, Restraints, Litters, and Cockpit/Cabin Delethalization*, USAAVSCOM TR-89-D-22D, December 1989

12	J W Coltman *et al.*, *Aircraft Crash Survival Design Guide, Volume II, Aircraft Design Crash Impact Conditions and Human Tolerance*, USAAVSCOM TR-89-D-22B, December 1989

13	G M Bruggink, A C Barnes, and L W Gregg, "Injury Reduction Trends in Agricultural Aviation," *Aerospace Medicine*, 1964, 472-475

14	H De Haven, *Development of Crash Survival Design Personal, Executive and Agricultural Aircraft*, Cornell-Guggenheim Safety Center, 1953; DDIC: TR-015100

15	H Mertz, "Anthropomorphic Test Devices," *Accidental Injury*, A M Nahum and J M Melvin (eds.) New York: Springer, 1993, Chapter 4, pp. 66-84

16	49CFR 572, as of October 1, 2011

17	D Rhule, H, Rhule, and B Donnelly, "The Process of Evaluation and Documentation of Crash Test Dummies for Part 572 of the Code of Federal Regulations," *19th International Technical Conference on the Enhanced Safety of Vehicles*, Washington DC, 2005.

18	"50th Percentile Male," *Anthropomorphic Test Devices*, 49CFR Part 572, Subpart B, §§572.5 through 572,11, *supra*

19	"3-Year-Old Child," *Anthropomorphic Test Devices*, 49CFR Part 572, Subpart C, §§572.15 through 572,25, *supra*

20	"6-Month-Old Infant," *Anthropomorphic Test Devices*, 49CFR Part 572, Subpart J, §§572.80 through 572,86, *supra*

21	"Hybrid III 5th Percentile Female Test Dummy",*Anthropomorphic Test Devices*, 49CFR Part 572, Subpart O, §§572.130 through 572,137, *supra*

22	"Emergency Landing Dynamic Conditions", *Airworthiness Standards: Normal, Utility, Acrobatic, and Commuter Category Airplanes*, 14 CFR Part 23, §23.562 amendment 23-36, effective 09/14/1988

23	"Emergency Landing Dynamic Conditions", *Airworthiness Standards: Transport Category Airplanes*, 14 CFR Part 25, §25.562, amendment 25-64, effective 06/16/1988

24	"Emergency Landing Dynamic Conditions", *Airworthiness Standards: Normal Category Rotorcraft*, 14 CFR Part27, §27.562, amendment 27-25, effective 12/13/1989

25	"Emergency Landing Dynamic Conditions", *Airworthiness Standards: Transport Category Rotorcraft*, 14 CFR Part29, §29.562, amendment 29-29, effective 12/13/1989

26	J A Newman, "Head Injury Criteria in Automotive Crash Testing," *SAE Transactions*: **89**, 1980, 4098-4115

27	J W Coltman, A O Bolukbasi, and D H Laananen, D H, *Analysis of Rotorcraft Crash Dynamics for Development of Improved Crashworthiness Design Criteria*, FAA-CT-85-11, June 1985

28	"Powered Civil Aircraft with Standard Category US Airworthiness Certificates; Instrument and Equipment Requirements," *General Operating and Flight Rules*, 14 CFR Part91, §91.33,

29	"Fastening of Safety Belts," *Federal Register*: 36, No. 127, July 1, 1971, 12511

30	"§91.14," email from T. Dempsey, Office of Chief Counsel, FAA, May 28, 2021

31 "Use of Safety Belts, Shoulder Harnesses, and Child Restraint Systems," *General Operating and Flight Rules*, 14 CFR Part91, §91.107, amendment 91-250,, effective 09/03/1996

32 Rgl.faa.gov, accessed 07/29/2021

33 "Clarification of Prior Interpretation of the Seat Belt and Seating Requirements for General Aviation Flights," *Federal Register*, **76**, No 121, June 23, 2011

34 "Seats, Safety Belts, and Shoulder Harnesses," *Operating Requirements: Domestic, Flag, and Supplemental Operations*, 14 CFR Part121, §121.311, amendment 121-255, effective 09/03/1996

35 "Use of Safety Belts and Child Restraint Systems," *Operating Requirements: Commuter and On-Demand Operations and Rules Governing Persons On Board Such Aircraft*, 14 CFR Part135, §135,128, amendment 135-62, effective 06/04/1996

36 "Seats, Safety Belts, and Shoulder Harnesses," *Operating Requirements: Domestic, Flag, and Supplemental Operations*, 14 CFR Part121, §121.311, amendment 121-315, effective 10/27/2005

37 B C Barksdale, "Child Safety Restraints: A Controversy Over Safe Infant Air Travel," *Journal of Air Law and Commerce*: 57, 1991, 201-255

38 A N Taylor, "What Is the Value of a Life: Child Restraint Systems on Commercial Airlines, *Journal of Air Law and Commerce*: 64, 1998, 307-336

39 S E Thompson, "Why, After All this Time, Is the FAA Just Now Taking Steps to Mandate Child Restraint Systems on Aircraft," *Gonzaga Law Review*: 37, 2001, 533-561

40 D E Rapoport, J Brown. and L A Epstein, "Babies Have a Right to a Safe Seat with Proper Restraints – The Infant Seat Exception Should Be Abandoned," *DePaul University Issues in Aviation Law and Policy*: 12, Autumn 2012, 67-96

41 D Mohan and L W Schneider, "An Evaluation of Adult Clasping Strength for Restraining Lap-Held Infants," *Human Factors*: 21, 1979, 635-645

42 D Fife, B Rosner, and W McKibben, "Relative Mortality of Unbelted Infant Passengers and Belted Non-Infant Passengers in Air Accidents with Survivors," *American Journal of Public Health*: 71, 1981, 1242-1246

43 C Bil, A Shrimpton, and G Clark, "Safety of Lap-Held Infants in Aircraft," *Procedia Engineering*: 99, 2015, 1311-1316

44 A R Burdi *et al.*, "Infants and Children in the Adult World of Automobile Safety Design: Pediatric and Anatomical Considerations for Design of Child Restraints, *Journal of Biomechanics*: 2, 1969, 267-280

Chapter 9: Type Certification Procedures

Type Certification in the US is covered in Subpart B of Part 21.[1] We will discuss the organizations within the FAA and then discuss the process.

Any interested person may apply for a Type Certificate. The applicant must make an application and within a prescribed time show compliance with the applicable requirements. To issue the TC, FAA, for its part, finds compliance and determines that there are no features or characteristic that renders the aircraft unsafe. The reader should review FAA Order 8110.4C.[2]

The TC application is of key interest, not because of the content, but because of the date. Beyond the name, etc of the applicant, the application need only contain a three-view drawing of the aircraft and "available preliminary data". An application for an engine TC must include "a description of the engine design features, the engine operating characteristics, and the proposed engine operating limitations". An exemplar application for a TC is shown in Figure 56.*

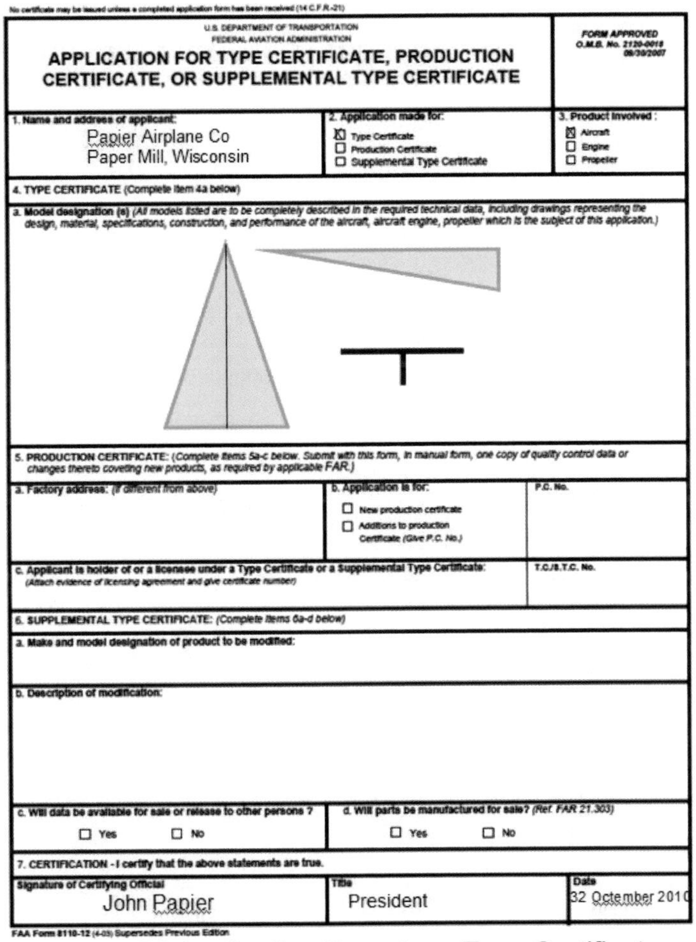
No certificate may be issued unless a completed application form has been received (14 C.F.R.-21)

U.S. DEPARTMENT OF TRANSPORTATION
FEDERAL AVIATION ADMINISTRATION

APPLICATION FOR TYPE CERTIFICATE, PRODUCTION CERTIFICATE, OR SUPPLEMENTAL TYPE CERTIFICATE

FORM APPROVED O.M.B. No. 2120-0018 09/30/2007

1. Name and address of applicant:
Papier Airplane Co
Paper Mill, Wisconsin

2. Application made for:
- [x] Type Certificate
- [ ] Production Certificate
- [ ] Supplemental Type Certificate

3. Product Involved:
- [x] Aircraft
- [ ] Engine
- [ ] Propeller

4. TYPE CERTIFICATE (Complete item 4a below)

a. Model designation (s) *(All models listed are to be completely described in the required technical data, including drawings representing the design, material, specifications, construction, and performance of the aircraft, aircraft engine, propeller which is the subject of this application.)*

5. PRODUCTION CERTIFICATE: *(Complete items 5a-c below. Submit with this form, in manual form, one copy of quality control data or changes thereto coveting new products, as required by applicable FAR.)*

a. Factory address: *(if different from above)*

b. Application is for:
- [ ] New production certificate
- [ ] Additions to production Certificate *(Give P.C. No.)*

P.C. No.

c. Applicant is holder of or a licensee under a Type Certificate or a Supplemental Type Certificate: *(Attach evidence of licensing agreement and give certificate number)*

T.C./S.T.C. No.

6. SUPPLEMENTAL TYPE CERTIFICATE: *(Complete items 6a-d below)*

a. Make and model designation of product to be modified:

b. Description of modification:

c. Will data be available for sale or release to other persons ?
- [ ] Yes
- [ ] No

d. Will parts be manufactured for sale? *(Ref. FAR 21.303)*
- [ ] Yes
- [ ] No

7. CERTIFICATION - I certify that the above statements are true.

Signature of Certifying Official	Title	Date
John Papier	President	32 Octember 2010

FAA Form 8110-12 (4-03) Supersedes Previous Edition

Figure 56 Application Form for a Type Certificate (FAA Form 8110-12)

* This exemplar application from the Papier Aircraft Company (*sic*) was used during our initial FAA course in aircraft certification taught to new employees.

The application date is the key element in determining what the applicable regulations are that will form the *Certification Basis* for the TC. The date will determine the applicable regulations – in general, these are the regulations in place on the date of the application. The application is effective for five years for Transport Category aircraft and three years for other aircraft types. If the TC is not completed in that period, the applicant runs the risk of seeing later revisions to the certification basis.*

The Organization

The FAA's certification standards are maintained by four aircraft directorates who report to the Compliance and Airworthiness Division (as opposed to the pre-2017 organization in which they reported to one of four regions). Each directorate is responsible for specific aircraft types.

> ✈ Engine/Propeller Directorate (Burlington, MA)
>> Engines
>> Propellers;
> ✈ Rotorcraft Directorate (Fort Worth, TX)
>> Normal Category Rotorcraft
>> Transport Category Rotorcraft
>> Powered Lift Aircraft;
> ✈ Small Aircraft Directorate (Kansas City, MO)
>> Normal Category Airplanes
>> Small Airplanes (Primary and Very Light Airplanes)
>> Gliders and Motorgliders
>> Airships; and
> ✈ Transport Airplane Directorate (Seattle, WA)
>> Transport Category Airplanes.

Aircraft Certification Offices

The applicant's point of contact during a certification project is the Aircraft Certification Office (ACO). At this time, there are thirteen ACO's reporting to the Compliance and Airworthiness Division.

> ✈ Anchorage ACO Branch;
> ✈ Atlanta ACO Branch;
> ✈ Boston ACO Branch;
> ✈ Chicago ACO Branch;
> ✈ Delegation Systems Certification Office (DSCO) Branch;
> ✈ Denver ACO Branch;
> ✈ Engine Certification Office (ECO) Branch;
> ✈ Fort Worth ACO Branch;
> ✈ Los Angeles ACO Branch;
> ✈ Military Certification Office (MCO) Branch;
> ✈ New York ACO Branch;
> ✈ Seattle ACO Branch; and
> ✈ Wichita ACO Branch.

---

*   As a consulting engineer in the early 1990s, we had a client with a display certification project. This client wanted to delay the application until it was complete. We pointed out the possibility of an adverse change in the requirements and recommend an early application. The client delayed for several months and finally agreed to an application. Unfortunately there was an amendment to §§23.1309 and 23.1311 which had become effective and added about a year to their program.

Military Certification Office

One of these offices seems out of place – the Military Certification Office (MCO). We saw at the very beginning of Chapter 1 that military aircraft are not under the FAA's jurisdiction. This is a result of recent policy for the military services to use FAA airworthiness approvals as a basis for military approval.

Normally, US military and public use aircraft do not require Type Approval by the FAA. In recent years, many certified civil aircraft have been procured by military organizations for non-combat roles, such as transport or training aircraft. Transport airplanes have also been modified for some military missions, such as maritime patrol etc. In recent years, military procurement programs have been encouraged to use "commercial off-the shelf" (COTS) aircraft where possible, rather than follow traditional military development programs.

The FAA has developed the Military Certification Office (MCO) to deal specifically with approvals of these COTS aircraft. We are not sure this was a wise policy.

We have been involved with these programs both as a DER and as a government engineer. In our experience, the organizations involved felt that the cost savings would be significant and that there would be little downside. In practice these programs have not been as successful as their promise. This was because of lack of understanding of the difference between airworthiness certification and operational certification.

In one case,* the military organization wanted to operate the transport airplanes using night vision goggles (NVGs). The civil requirements for cockpit warning, caution, and advisory (WCA) light colors. These colors are generally not compatible with NVGs. The designers decided to design the cockpit with NVG-compatible lights. This raised red flags with the FAA ACO. As a DER, we recommended using standard WCA colors and let the military operators change the lenses after delivery. The system designers decided to make the system compatible by developing a software warning systems. Rather than change the lenses for a small non-recurring charge, the spent millions of dollars to develop the new system. In the end, they ended up changing the lenses to civil standards and let the military operators make the changes after delivery.

The major issues with this approach to military procurement is the lack of understanding of the roles of the military agency procuring the aircraft and the aircraft vendor with the FAA's role.

Aircraft Evaluation Group (AEG)

In the 1990s, the FAA tried to bring some operational expertise into aircraft certification. The result was the creation of several Aircraft Evaluation Groups (AEG's) within the FAA. These groups work with, but are not part of the aircraft certification organization. Rather they are part of Flight Standards with a background in pilot licensing, or maintenance. There are two specialties in the AEG: Operations (pilots) and Maintenance (mechanics).

There are four AEG offices:
- ✈ Fort Worth AEG: Rotorcraft;
- ✈ Kansas City AEG: Small and Commuter airplanes;
- ✈ Los Angeles AEG: Transport airplanes; and
- ✈ Seattle AEG: Transport airplanes.

---

* This program predated the MCO concept.

AEG has the primary responsibility to evaluate

> ✈ Aircraft and systems for operational relevance;
> ✈ Continued Airworthiness
>   (Maintenance Review Board, MRB);
> ✈ Flightcrew type rating requirements
>   (Flight Standardization Board, FSB); and
> ✈ Minimum Equipment List (MEL) requirements
>   (Flight Operations Evaluation Board, FOEB).

The AEG also provides consultation and assistance in developing Airworthiness Directives (ADs). Normally, AEG will not become involved with small aircraft certification.

Manufacturing Inspection District Offices (MIDO's)

Manufacturing Inspection District Offices (MIDO's), another Flight Standards support aircraft certification. In several ways:

> ✈ Performing conformity inspections;
> ✈ Production approval and certification; and
> ✈ Support to ACOs during design approval.

Chief Scientific Technical Advisors

The FAA has several Chief Scientific Technical Advisors (CSTAs) to provide technical support in several specialized areas: Aircraft Controls, Flight Management, Human Factors, and Simulation, among other specialties. The CSTAs are available to assist ACO personnel as needed.

Issue Papers (IPs)

Issue Papers (IPs) are the formal means by which the applicant and the FAA resolve differences.[3] An IP is a formal means to develop

> ✈ Applicable Rules;
> ✈ Equivalent Level of Safety (ELOS) Findings;
> ✈ Means of Compliance (MOC);
> ✈ Potentially Unsafe Conditions; and
> ✈ Special Conditions.

Issue Papers are of several types:

> ✈ G: General Issues
>   G-1: Type Certification Basis;
> ✈ A: Airframe Issues;
> ✈ F: Flight Test Issues;
> ✈ M: Maintenance Issues;
> ✈ O: Operations Issues;
> ✈ P: Propulsion Issues; and
> ✈ S: Systems and Equipment Issues.

Figure 57 shows the Issue Paper process.

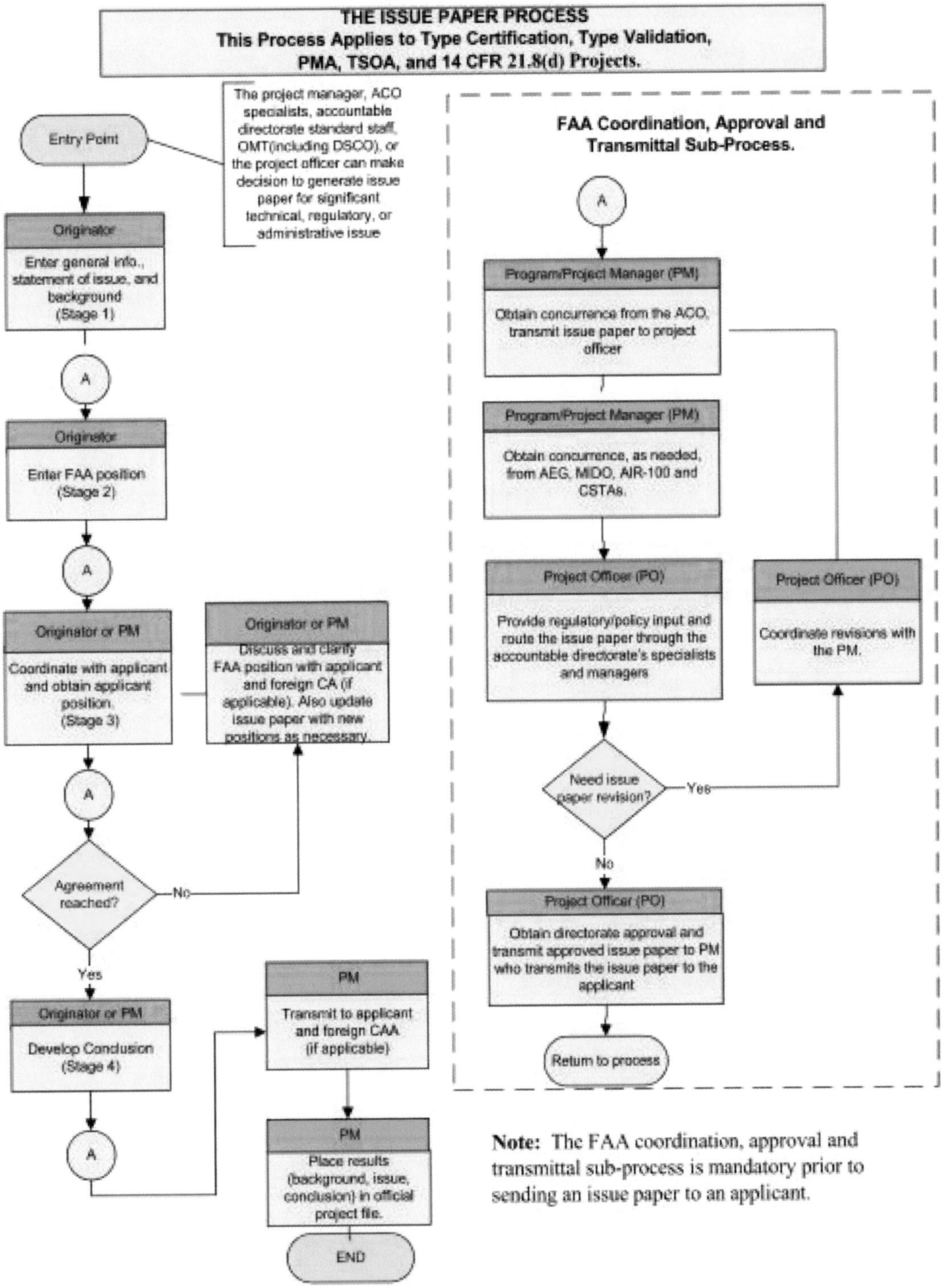

Figure 57 Issue Paper Process.
From FAA Order 8110.112A[3]

Normally, IPs begin with the originator (normally the Project Manager or an ACO engineer) preparing a statement of the issue and background information. (Stage 1*). The ACO staff then prepares an FAA position in consultation with the Directorate, FAA Headquarters, AIG, CSTAs, etc... This FAA coordinated position is referred to as Stage 2. If the involves an importing/Exporting Civil Aviation Authority (FCAA), the issue paper can include positions from that FCAA.

The Project Manager discusses the FAA position with the applicant and obtains the applicant's position (Stage 3). If there is agreement, the process moves forward. If not, revised Stage 2 and Stage 3 positions are proposed until there is agreement.

In some cases, the applicant will initiate the process. In this case, the Stage 3 will precede the Stage 2 position. This happens when the applicant is seeking an exemption or is proposing an Equivalent Level of Safety (ELOS) finding.

Once an agreement is reached (Stage 4), the issue paper will be prepared by the FAA Project Manager, transmitted to the applicant (and FCAA, if appropriate) and placed in the project files. The IP is not official until signed by the Directorate. The files are not releasable to the public until the IP is final.

The Type Certification Process

Following the acceptance of the application, the FAA will establish a project and appoint a Project Manager at the appropriate Aircraft Certification Office (ACO) and a Project Officer at the ACO.

Normally, the applicant and the ACO staff will have a preliminary meeting usually with a presentation by the applicant on the aircraft.

The applicant may petition for an exemption[4] or request an alternative means of compliance providing an Equivalent Level of Safety (ELOS). In addition, the applicant normally prepares a compliance checklist listing all the paragraphs believed to be relevant. This will have to be agreed to by the FAA.

The FAA may prescribe special conditions if it determines that existing regulations do not adequately contain safety standards dealing with novel or unusual characteristics of the aircraft.

The G-1 Issue Paper covers the certification basis of the aircraft and almost certainly be the first IP prepared. It will be based on the application date. Figure 23 shows representative issue papers. Most are resolved by the Issue Paper process, such as a Means of Compliance (MOC) IP. In such a case, the applicant might say, "The method in Advisory Circular such and such requires equipment we don't have". They would then propose to use another method or modify the method in that Advisory Circular. If the Stage 4 Issue Paper agrees, then that MOC will be acceptable for the current project.

As an example, the applicant might propose an airplane with a digital engine tachometer. They might say, "We can't easily comply with the requirement that an engine tachometer have a moving needle. In our design, the engine speed is controlled automatically and a digital display is just as safe". If the FAA agrees that it is, in fact, just as safe, they will issue an ESF--allowing a digital tachometer for that project. This ESF was used successfully in the Textron Model 3000 (T-6A) trainer.[5]

Two type of changes will require additional rulemaking. If the applicant petitions for an exemption, this petition must be published in the Federal Register and ample time allowed for public comments and disposition of those comments. Similarly, if the FAA wishes to propose a Special Condition (SC), the proposed SC will have to be justified as required because a "new and

---

* `Stage 1, 2, etc. are used within the FAA organization and specifically mentioned in the Order.[3]

novel" design feature that wasn't anticipated when the existing regulations were written. The SC will have to be published and time allowed for the comment period and disposition of the comments. This timeline is shown in Figure 58.

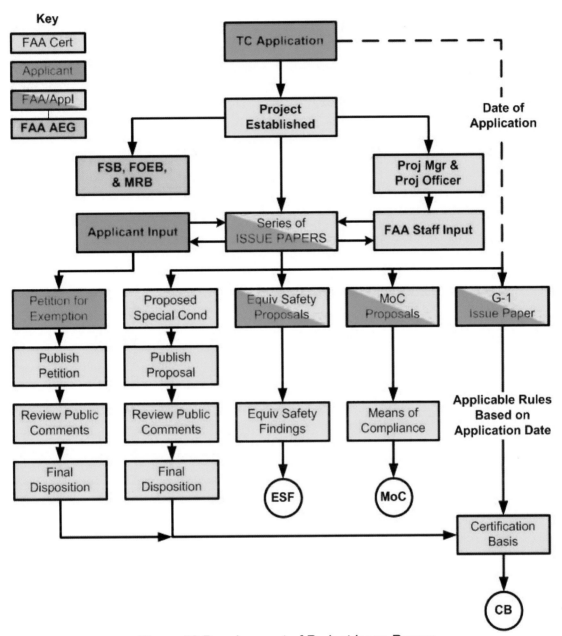

Figure 58 Development of Project Issue Papers

Once the Certification Basis, and the several Methods of Compliance or Equivalent Safety Findings are established, the applicant must show compliance with the regulations.

Compliance data includes engineering analyses, and engineering tests. A few examples of engineering certification tests include system function, static load tests, structural fatigue tests landing gear drop tests, ground vibration tests, and electro-magnetic interference test.[6] These engineering tests must be accomplished successfully before conducting any FAA certification flight test to validate the applicant's showing of compliance of any test requirement.

Compliance finding is discretionary on the part of the FAA. This means that the FAA can use discretion in deciding which parts of the applicant's submissions it reviews. This was decided in *United States v. Varig Airlines*.[7] In this case the CAA inspector did not inspect a heater installation.

The next step is for the applicant to submit engineering data and ground test data. The ACO engineers will review this data and approve it if it shows compliance with the regulations. The applicant may employ Designated Engineering Representatives (DER's) to review and, in some cases, approve the data

What follows is submission of lots of engineering data, both analyses and test results.

A major milestone is to test fly the aircraft. It is almost certain that applicant will have flown the aircraft even before submitting the TC application. These company test flights are flown under an Experimental Airworthiness Certificate issue for research and development flights. (See discussion beginning on page 15). However, the "official" flight tests are conducted under an Experimental airworthiness certificate issued to show compliance".

This step requires the signing of a Type Inspection Authorization (TIA) which allows for official ground and flight testing compliance tests. Before issuing the TIA, the applicant must present the airplane for a conformity inspection (show that it matches the drawings, etc.). The applicant must also state that all of the required flight tests have been performed by a company pilot.

Once the TIA is issued, the FAA test pilots may fly the aircraft. Does the FAA ever fly the aircraft before TIA? Of course they do, but these are properly called "familiarization flights." Also, many times the "FAA pilot" will be a designee – a company pilot with a Flight Test Pilot DER approval.

The outcome is a Type Inspection Report (TIR) which is the final flight test report. The process is shown in Figure 59.

Post-TC Activities

Following completion of the TIR, a review of all engineering data, and a draft Aircraft Flight Manual (AFM), there will be a meeting to review all of the data. It will be determined if the submissions are acceptable in light of the Certification Basis, the Methods of Compliance (MOC) and the Equivalent Safety Findings (ESF). If this is all satisfactory, the Type Certificate will be signed.

However, before entry into service, several documents are required from Flight Standards. The AEG staff will convene several boards as shown in Figure 60.

The Flight Standardization Board will issue a report describing the type rating requirements for the model. Now the aircraft type can enter service.

The Maintenance Review Board (MRB) will review the Instructions for Continued Airworthiness (ICA) and approve the maintenance program(s) for the aircraft.

The Flight Operations Evaluation Board (FOEB) will issue a report covering the Master Minimum Equipment List (MMEL) for the aircraft type. This forms the basis for operators to create their own Minimum Equipment List (MEL). It has been our observation that in many new highly integrated computer intensive airplanes, the FOEB did not consult with certification engineers and ensure the MEL procedures were agreed with the assumptions in the system safety analyses required by §2x.1309.

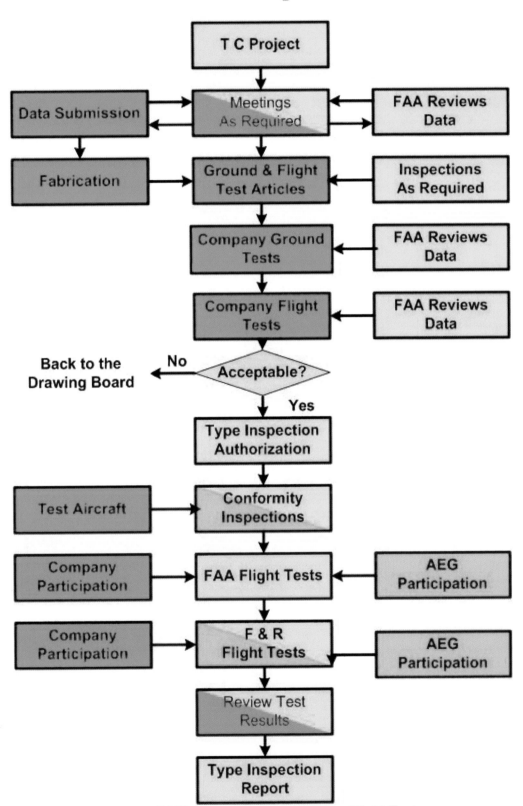

Figure 59 Type Certification Process: Flight Test

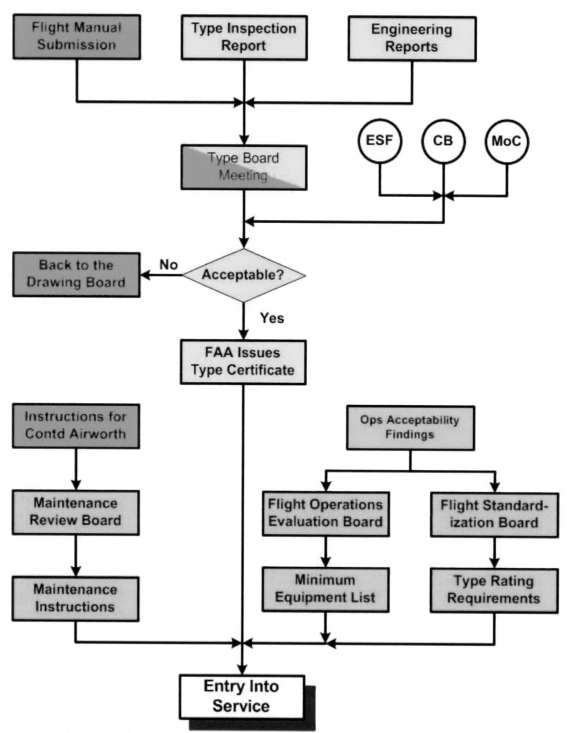

Figure 60 Type Certification Process: Completion and Post-TC Activities

Furthermore, dispatching an airplane with an inoperative, redundant piece of equipment may be of no consequence—until a second item or system fails. In 1969, a Boeing 727 flew three days with an inoperative generator. Three days later, the airplane had an engine fail on climb out from Los Angeles. This failure left the airplane with a single operating generator which soon failed from overload. The flight engineer was unable to switch the captain's instruments to a backup inverter. With no available attitude gyros, the crew lost control and crashed into Santa Monica Bay.[8] *

In June 2005, an Airbus A320 was dispatched on a flight from Milan to London-Heathrow with an inoperative Air Data Inertial Reference Unit (ADIRU). The inoperative ADIRU was located in the number 3 position per the MEL. On approach to Heathrow, the inertial part of the number 1 ADIRU failed. As a result, the autopilot and autothrottle both disconnected, the Captain's Primary Flight Display and Navigation Display were lost. In addition the flight control laws degraded to Direct Law. In addition, the number 1 yaw damper and the EGPWS were lost. The captain handed over control to the first officer. At about 250 ft AGL, the crew received several EGPWS alerts. ATC directed the flight to go around because the preceding flight had not cleared the runway.

Because of deteriorating weather, the captain decided to expedite the approach rather than troubleshoot the problem. The landing gear had to be lowered manually which disabled the nosewheel steering. Fortunately, the airplane landed successfully and was towed to parking.[9] †

References

1 "Type Certificates," *Certification Procedures for Products and Articles*, 14 CFR Part, Part 21 , Subpart B, as amended through amendment 21-92, effective April 14, 2010

2 *Type Certification*, FAA Order 8110.4C, change 5 dated December 20, 2011

3 *Standardized Procedures for Usage of Issue Papers and Development of Equivalent Levels of Safety Memorandums*, FAA Order 8110.112A, October 3, 2014

4 *General Rulemaking Procedures*, 14 CFR Part §11.15, amendment 11-46, effective September 20, 2000

5 *Type Certificate Data Sheet*, TCDS A00009WI, Revision 25, December 13, 2019; (Textron T-6A)

6 "Type Certification Processes," *Type Certification*, FAA Order 8110.4C, Chapter 2, March 28, 2007,

7 M E F Plave,, "United States v. Varig Airlines: The Supreme Court Narrows the Scope of Government Liability Under the Federal Tort Claims Act," *Journal of Air Law and Commerce*: 51 1985, 197-239

8 *Aircraft Accident Report. United Air Lines, Inc. Boeing 727-22C, N7434U Near Los Angeles, California January 18, 1969*, NTSB AAR-70-06, Dated March 1970

9 *Incident to Airbus A320-200, I-BIKE, 25 June 2005, On Approach to Runway 09L at London Heathrow Airport*, AAIB Bulletin 6/2006, 2006

---

*   This accident is described in Appendix 4 on page 352.
†   This incident is described in Appendix 4 on page 353.

PART IV:
CURRENT STATUS OF
AIRCRAFT CERTIFICATION

Chapter 10: New Technology in Transport Airplanes Certification

"What's it doing now?"*

How should we certify new technology? We can't rely on long experience with similar equipment. By definition, new technology is just that—new. We shouldn't rely solely on the claims of advocates. Let us examine some examples of certifying new technology. We will look at the experience from transport airplane certification and then from light airplane certification.

In the 1960-s, many large transports required a minimum crew of three (two pilots and a flight engineer). The flight engineer was a carryover from the era of large piston-powered four engine transports. These engines required the attention of a dedicated crew member—the flight engineer. With the advent of turbojet engines, the flight engineer's duty shifted from engine control to systems monitoring.

In the 1980s, there was a push from operators to reduce the crew complement to save costs. As a result, more automation was introduced. Computer-based monitoring systems (such as EICAS or ECAM) began to appear. These systems in one respect improved upon human monitoring, but often provided arcane warnings. At least the captain could ask the flight engineer: "What do you mean?"

As fly-by-wire (FBW) became more common, the flight controls would change mode without warning, and often without clear annunciation. Let's look at some examples of introduction of new technology to transport airplanes.

Ground Proximity Warning Systems

"You know, according to this dumb sheet it says thirty-four hundred to Round Hill --- is our minimum altitude."†

On December 1, 1974, TWA 514, a B-727, crashed into Mt Weather on an approach to Dulles Airport.[1]‡ This was a classic Controlled Flight into Terrain (CFIT) accident. All ninety-two souls on-board were killed. The outcome of this accident was a mandate for all US airlines to install Ground Proximity Warning Systems (GPWS) on their passenger jet transports.

At the time, we thought the rule was a rushed mistake and filed a five-page adverse comment to the NPRM. Our assessment was a mistake. We had been concerned that the black box would "Cry Wolf!" and that the pilot might be about to make a mistake. We were given the opportunity to spend three days in Seattle flying many CFIT scenarios in the Sundstrand *King Air* test bed. After flying many flight profiles near real mountains, our conclusion became that the box was actually telling the pilot that "You have already made a mistake and are about to die!"[2]

The original GPWS provided no preliminary advisories and started with a loud warning: "Whoop, Whoop, Pull Up!" An updated version provided advisories in some cases, alerting the crew that the system was approaching safe limits. They might hear a preliminary warning "Too

---

*   Spoken by a copilot in a hypothetical modern cockpit. The captain is alleged to have replied "Look! It's doing it again". The check pilot in the jump seat nodded wisely and said, "Ah yes, it does that sometimes".

†   Spoken by the Captain on TWA 514 about two minutes before colliding with Mt Weather at an elevation of 1,670 ft.[1]

‡   This accident is described in Appendix 4 on page 265.

Low – Gear", followed by the pull up command. This type of sequence provided confidence that the system was working by providing a rationale for the final alert.

While GPWS improved things greatly, it could not help in situations where the airplane was flying level towards a vertical cliff. This happened on January 1, 1985 to an Eastern B-727 approaching La Paz, Bolivia.* The airplane was apparently deviating around weather at night and flew into a mountain. While no accident report was published by the Bolivian investigators, the US Department of Transportation studied the available data and published a profile of the accident trajectory, shown in Figure 61.[3]

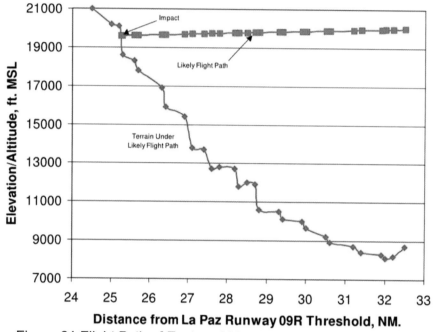

Figure 61 Flight Path of Eastern 980 at La Paz, January 1, 1985

The La Paz accident pointed out the problem was the steep approach into near vertical terrain. GPWS could not provide enough warning to avoid the mountain. The accident gave impetus to the development of world-wide digital terrain databases. In 2000, NASA flew the Shuttle Radar Topography Mission (SRTM) aboard the Space Shuttle *Endeavour*. This mission used dual radar antennas to acquire interferometric data processed to digital topographic data at 1 arc-sec resolution.[4]

Data from the SRTM allowed the enhancement of the GPWS to identify terrain around the airplane's flight, not just the airplane's vertical height above the ground. This led directly to Terrain Awareness and Warning System (TAWS). The renaming came about because Enhanced Ground Proximity Warning System (EGPWS) was a trademark.

As of March 29, 2002, Class A TAWS was now required on all newly manufactured US airline turbojet and turboprop airplanes.[5] These TAWS units require the following alerts:

➔ Reduced required terrain clearance,
➔ Imminent terrain impact,
➔ Excessive rate of descent,
➔ Excessive closure rate to terrain,
➔ Negative climb rate or altitude loss after takeoff,
➔ Flight into terrain when not in landing configuration,

---

* This accident is described in Appendix 4 on page 266.

✈ Excessive downward deviation from an ILS glideslope, or

✈ Descent to 500 ft above terrain or runway elevation.[6]

The rule was phased in with compliance required in all US Airline turbine powered airplanes by 2005.

The safety record shows a remarkable improvement in safety. If we examine the number of CFIT accidents to scheduled jet transport flights there is a remarkable change: We examined the Aviation Safety Network (ASN) database[7] for three 10-year periods:

Before the GPWS mandate;

Following the GPWS mandate; and

Following the TAWS mandate (2005 through 2014).*

Table 15 shows the results of this survey for US airlines for jet transports only.

Table 15 Effect of GPWS on Airline Turbojet CFIT Accidents

Before or After GPWS Mandate	Years	Accidents			Fatalities		
		No	No/yr	Pct	Fat	Fat/yr	Pct
Before GPWS Mandate	1966-1975	13	1.3	100%	897	89.7	100%
After GPWS Mandate	1979-1988	2	0.2	17%	33	3.3	3.8%
After TAWS Mandate	2005-2014	1	0.1	8.3%	2	0.2	0.2%

GPWS is the success story of what modern systems can do to improve aviation safety—a system that eliminated 99.8% of fatalities!

Automatic Ground Collision Avoidance System

Military attack aircraft have a similar, but more stringent need. They routinely fly single pilot aggressive low altitude maneuvering. A warning system would not be appropriate. Warning systems generally are designed to point out a hazard that the crew may not be aware of. In the attack mission the pilots are seeking to fly close to terrain to minimize enemy threats. A warning system would be constantly sounding. The solution is Automatic Ground Collision Avoidance System (Auto GCAS).

Auto GCAS uses a terrain data, but instead of warning of its proximity, continuously computes the aircraft trajectory and calculates the effect of performing a 5-g escape maneuver in a few seconds. As soon as the system concludes that this is insufficient, it immediately commands a 5 g escape maneuver to a safe altitude. Simultaneously, it sounds an unmistakable warning—considered essential to indicate to the pilot that the system is working and not a flight control malfunction. After the introduction of the system, losses were reduced dramatically.[8] [9]

Navigation Interfaces

The development of new, two-man cockpits provided changes in navigation interfaces. When one of us (RLN) started airline flying in 1966, VOR navigation radios were tuned by knobs. We had to identify the station aurally by Morse code. Newly hired pilots took a navigation standardization course in a Link trainer.† The airplanes' autopilots seemed advanced to this new copilot at the time, but very limited by today's standards guidance.

---

* We did not include the period when only part of the fleet was equipped to avoid confusing the data between equipped or non-equipped aircraft during the transitions.

† This was in the days before extensive use of simulators. We did all of our flight training in the airplane, not in the simulator. We did use a Link trainer to learn standard navigation procedures including ensuring a course change was reasonable before executing a turn. The Columbian accident report reviewed the American flightcrew's training which appeared similar.

So-called area navigation (RNAV) allowed direct flight between two points, and did not force the pilot to fly directly to or from the VOR station (or along arcs around a DME station). The capability improved with long-range navigation, such as LORAN. Fortunately for the industry, the global positioning system (GPS) came along. GPS had much better accuracy than any of the preceding navigation systems. Most GPS receivers have a vast number of fixes available to the pilot simply by a few letters to identify the fix.

With the ease of selecting a fix comes the opportunity for the wrong fix being selected. This resulted in the B-757 accident at Cali, Columbia on December 20, 1995.[10]* The crew was offered a short-cut to the airport – a slight right turn to the *Rozo* radio beacon. They entered the fix identifier, "R", into the navigation system. Unfortunately, *Rozo* radio beacon was not in the system's database. Instead, "R" was the *Romeo* beacon near Bogota. The pilot saw the "R" and entered "direct"; the airplane turned sharply left while continuing the descent.†

Clearly, such an interface can lead to errors. Who is responsible? The pilot? The agency that allowed two "R" named fixes near to each other? The designer of the interface? The customer who accepted it? The training department? Or the FAA engineer (or DER) who approved it? The short answer is, they all are to some degree.

While the human factors rule (§25.1302) was introduced since the Cali accident, there has been no regulatory action to address the issue of ensuring agreement between cockpit paper charts and an electronic navigation database. The "human factors" rule, §25.1302, uses generic wording, such as "predictable" and "appropriate to the task." As we said earlier (page 85),

"In our opinion, it is too general to be useful and is, as a result, unnecessary."

Crew Alerting Systems

Prior to the introduction of the new, two-man cockpits, crew alerting was generally based on a few warning sounds and warning lights, which were complemented with paper emergency check-lists. Our old B-727 Operating Manual,[11] showed the following aural warnings:

✈ Bell   Engine, Wheel-Well, or APU Fire;
✈ Steady Horn   Unsafe Landing Gear;
✈ Intermittent Horn   Takeoff Configuration or Depressurization; and
✈ Clacker   Excessive Airspeed or Mach.

In the Helios B-737 depressurization, apparently confused by the intermittent warning horn, the crew spent some effort in checking the various takeoff settings instead of donning their oxygen masks. After the crew became unconscious, the airplane flew on under the control of the autopilot until it exhausted its fuel.[12]‡

Warning and caution lights were generally located around the cockpit. This design was state-of-the-art back in the day. It was impractical, at that time, to integrate the systems. Besides, we had a dedicated systems monitor—the Flight Engineer.

However, the development of new, two-man cockpits required improvements in crew alerting. The industry has developed integrated warning, caution, and advisory (WCA) systems. Two examples are ECAM and EICAS.

---

*   This accident is described in Appendix 4 on page 268.
†   We saw a similar problem when our navigation computer thought Nashville (BNA) was in Europe.
‡   This accident is described in Appendix 4 on page 331.

Electronic Centralized Aircraft Monitor (ECAM)

ECAM is a system on Airbus aircraft for monitoring and displaying engine and aircraft system information to the pilots. In the event of a malfunction, it will display the fault and may also display the appropriate steps of the remedial action.[13] ECAM presents data on the Engine/Warning Display (E/WD) and the System Display (SD) including:

✈ Primary engine indications, fuel quantity, flap and slat position;
✈ Warning and caution alerts, or memos;
✈ Synoptic diagrams of aircraft systems, and status messages; or
✈ Permanent flight data.

Engine Indicating and Crew Alerting Systems (EICAS)

EICAS is an aircraft system for displaying engine parameters and alerting crew to system configuration or faults.[14] EICAS systems are found on Boeing, Embraer and many other aircraft types.

EICAS systems display engine parameters and, depending upon manufacturer and model, may display other information such as fuel quantity, cabin pressure or landing gear and flap/slat position. It will also alert the crew to aircraft configuration issues such as open passenger or cargo doors and will, in conjunction with a Master Warning or Master Caution light and aural alert, indicate system faults and failures by displaying the Quick Reference Handbook (QRH) checklist title of the appropriate remedial action.

Comparison of Central Warning Systems

Since many "high tech" installations have effects that are not clearly obvious, any approval must ensure that pilots are adequately trained and that the warning systems provide warnings that assist properly trained pilots to identify problems.

Singer and Dekker evaluated four different commercial transport warning systems.[15] Specifically, they examined how line pilots reacted to each of four different warning philosophies representing four different warning philosophies certified by four different aircraft designs. These philosophies are described in Table 16.

Table 16 Warning Philosophies

Display Philosophy	Display Features	Modeled on
"Show All"	All failed systems displays RED/AMBER priority only Red only if single failure critical Direct subsystem failures not shown Failures shown chronologically	Boeing 777 certified 4/1995
"Sort & Show"	Predetermined clutter when no crew action RED/AMBER priority Direct subsystem failures not shown Clear primary failure shown on top Automatic reconfigurations not shown	Saab 2000 certified 4/1994
"Sort & Guide"	Full predetermined priority RED/AMBER priority Guidance for immediate action follows failure message Subsystems presented as status only Automatic reconfigurations not shown	Airbus 320 certified 12/1988
"Do and Show"	Full predetermined priority Combined failure effect stated Automatic system reconfigurations—information to crew Guidance where crew action required	Boeing MD-11 certified 11/1990

The test matrix used line pilots, each flying 16 approaches and presented with one out of four system failures. Their task was to keep flying the approach within normal tolerances and identify the primary failure as quickly as they could. When they believed that they were confident they had identified the primary fault, they indicated this by pressing a button. At the conclusion of each run, the evaluation pilots' accuracy was checked by means of a brief questionnaire. To minimize learning, no feedback to the pilots was given until all 16 approaches were completed.

The test results are shown in Figure 62. All warning philosophies had been certified and were in service. Presumably, all had been vetted by engineers, human factors specialists, and test pilots. We question whether or not such evaluations by test pilots would be sufficient. We will discuss this later, in the chapter summary.

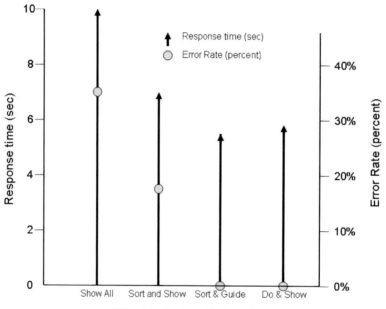

Figure 62 Results from Singer & Dekker

Fly-by-Wire (FBW)

Fly-by-wire is likely the most significant change in the airplane-pilot interface. FBW is the description of airplane flight controls where there is no direct mechanical connection between the pilot's stick and rudder and the flight control surfaces, such as the ailerons or elevators. Most, if not all, new transport airplanes have FBW flight control systems. For the manufacturers, elimination of the cable and pulleys means a significant weight savings which translates to increased payload or fuel savings. It also greatly reduces the manufacturing and maintenance man-power requirements. Any airplane mechanic will tell you that control rigging can be time consuming.

Modern FBW flight controls use on-board digital computers to modify the pilot's control inputs before sending the signal to the actual control surfaces. The first operational use of FBW flight controls was on the military F-16 fighter. By programming the flight control computers to compensate for the less basic airframe stability when the operational center of gravity is moved aft, the airplane drag can be reduced and the maneuverability can be increased. In transport airplanes, such as the Airbus A320 or Boeing 777, this can mean burning less fuel or being able to carry more payload.

In addition, the flight control system can be designed to make the airplane's handling qualities appear the same to the pilot across the range of speed, altitude, and aircraft loading. Different airplanes can be designed to fly with virtually identical handling qualities, thus reducing pilot training costs.

Most FBW flight control system designs also include flight envelope protection features. If present, envelope protection can help prevent the pilot from reaching unsafe flight conditions, such as stalling, overspeeding, overstressing, or overbanking the airplane. As one NASA test pilot said, "This results in carefree handling."* In other words, with full authority envelope protection, you just fly the airplane and don't worry about losing control.

Certification Basis for FBW Systems

Certification requirements for airplane flight controls were developed for traditional mechanical systems. They have not been updated to cover FBW designs. The certification rules still speak of stick-and-rudder motion and forces in terms of direct mechanical systems. None of the current transport FBW airplane designs meet all of the Part 25 requirements. All models to date have employed special conditions for the flight control system.

With such systems, one cannot examine accidents, but must search for precursors. In 2010, Newman and Lambregts[16] reviewed the service history of civil FBW airplanes, using several databases for FBW accidents and serious incidents. They found 9 accidents and 18 serious incidents involving FBW as a cause or as a factor. The results are summarized in Table 17.

Table 17 FBW Events in Civil Aircraft

Model	Accidents	Serious Incidents	Total	Rate[a]
A-320	7	13	20	3.0
A330/340	2	2	4	2.6
B-777		2	2	1.2
E-170/190	9	18	27	1.6
Note: (a) Rates are events per flight hour.				

Uncommanded Control Actions

These are generally caused by sensor errors, such as incorrect angle-of-attack (AOA). incorrect airspeed data, or inertial platform errors. They can also be caused by faulty processing of the sensor data. Representative events include: an A340 pitch-up caused by inappropriate stall prevention;[17] an A330 pitch-down caused by faulty AOA signal;[18] and a B-777 pitch-up caused by failed accelerometer.[19] These occurrences are all described in Appendix 4.

Abrupt Maneuvers

These have been the result of both pilots applying control inputs to the flight controls simultaneously. An early control logic added the second input to the first input, potentially doubling the result. Representative events include: an A340, caused by both pilots responding to TCAS[20]† and an A320, caused by both pilots' simultaneous control inputs.[21]‡

Other control arrangements have been used, such as giving more weight to the most recent input, a cut-out button to allow the pilot to takeover from the copilot, etc. These incidents have become less common with the introduction of better training and new procedures. Some manufacturers (such as Boeing) avoid this problem by mechanically connecting the control sticks or yokes mechanically.

---

*   Rogers Smith (NASA Dryden Flight Research Center), May 2004
†   This incident is described in Appendix 4 on page 311.
‡   This accident is described in Appendix 4 on page 313.

We are reluctant to mandate reversible control inceptors. There was an incident in which an A320 was released from maintenance with the Captain's sidestick wired backwards. *[22] At liftoff, the Captain corrected for a slight wing drop. Since his sidestick was reversed, the wing dropped further. The First Officer instinctively grabbed the stick and applied corrections – effectively canceling the Captain's input. They were able to reach a safe altitude where they decided to return to the airport – with the First Officer flying. Having irreversible pilot controls allowed a minor outcome replace a likely catastrophic accident. Generally such an aileron reversal results in a catastrophic crash on takeoff, such as Northwest Flight 706.†

Pilot Induced Oscillations

A Pilot-Induced Oscillation (PIO) is an oscillatory response when a pilot attempts to precisely control the airplane. These differ from aircraft oscillations caused by deliberate control motions that are open loop in nature. Most of the severe PIOs are characterized by rate limited responses of control surfaces. In most cases pilots report the onset was sudden, unexpected, and cliff-like.[23] Representative events include: An A320 accident during takeoff caused by mistrimmed stabilizer;[24] and two A320 roll PIOs caused by ice accretion on unprotected surfaces.[25]‡

Finally, PIOs will continue to be an issue, particularly when there are sudden changes in flight control gain. Current flight test evaluations are addressing this by requiring evaluation in those areas where PIO is likely. Various prevention approaches have been proposed to develop system design attributes to reduce susceptibility to PIO or to detect PIO and change gains accordingly.

Sensor Error Management

Many of the issues with FBW systems, indeed for many automated functions in non-FBW airplanes, involve sensor errors. It is instructive to discuss the common types of sensor errors: air data sensors.. One of the problems with air data sensors in general is their location. By necessity, these sensors are located external to the airframe and are exposed to weather and other hazards. There have been a number of mishaps in which multiple airspeed probes, angle-of-attack vanes either froze or were damaged by hail, bird strikes, and so on.[26 27 28]

Many of these FBW occurrences involved angle-of-attack (AOA) errors. There are several approaches to sensor redundancy: use one, two, or three AOA sensors. Clearly with one sensor the system may not even detect if there is an error. With two, the system can detect an error but may not be able to tell which one is wrong. Let us examine three separate approaches to using AOA sensors to detect and correct an unsafe situation.

Three AOA Sensors

We will look at the Airbus A330 as an exemplar.§ This model has layers of redundancy. Sensor data is collected from three separate sensors (for AOA, as well as other air date, such as pitot pressure and static pressure). Inertial data is obtained from three sets of gyroscopes and accelerometers. The triplicated data are sent to three Air Data Inertial Reference Units (ADIRUs), labels ADIRU-1, -2, and -3. Included with the data is a validity flag.

---

*   This incident is described in Appendix 4 on page 311.
†   This accident is described in Appendix 4 on page 307
‡   These incidents are described in Appendix 4 beginning on page 309.
§   One should not infer too much from our use of the A330. Most of the details of flight control systems are proprietary and not generally available. We are able to obtain the figures and details from a public source. We acknowledge the use of the ATSB figures and descriptions.[18]

There are five flight control computers, each with a separate command function and monitor function. The FCS has three identical primary computers (labeled PRIM) and two identical secondary computers (labeled SEC). The hardware and software for the PRIM and SEC computers are different. In addition, the software for the command and monitor functions in each computer were developed by different teams. Figure 63 shows the logic flow for processing AOA data. (Other data are treated similarly.)

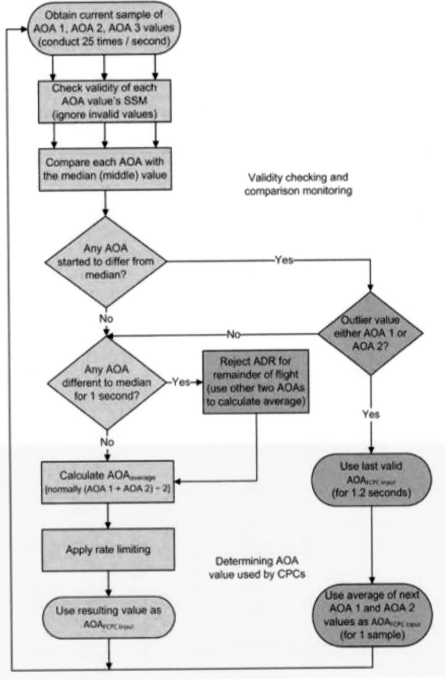

Figure 63 A330 Algorithm for Processing AOA Data*

_____

* From ATSB Report on A330 upset.[18]

The A330 algorithm for processing AOA data has been able to detect and manage almost all situations involving incorrect or inconsistent AOA data being received as "valid" data. Figure 64 shows some typical examples.

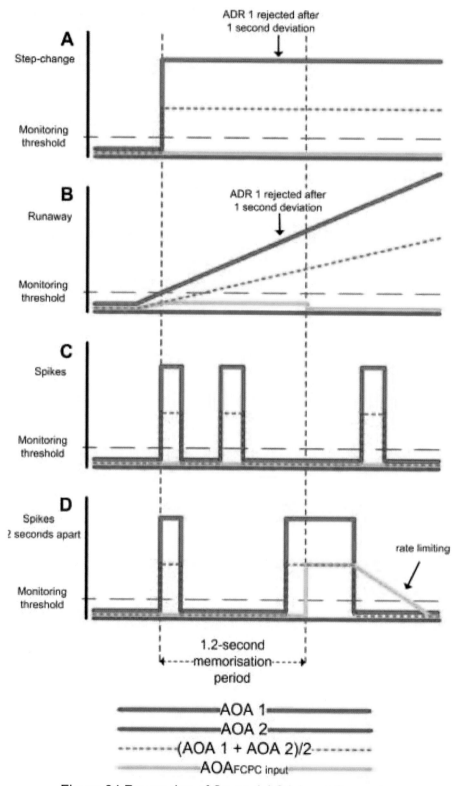

Figure 64 Processing of Several AOA Input Scenarios

Ramp-Change

A ramp change, a steady increase or decrease in AOA-1 (or -2) would trigger the 1.2 sec memorization period once it reaches the monitoring threshold. During this period, the FCS computers would use the average of AOA-1 and -2 from that point. Again after 1 sec, AOA-1 would be rejected and the value used by the FCS computers would be the average of AOA-2 and -3. (See Case B in Figure 64.)

Data Spike(s)

A single, short-duration spike in AOA-1 (or -2), would trigger a 1.2 sec memorization period, with the last valid value used. The presence of other spikes during this 1.2 sec period would not trigger another memorization period. A steady increase or decrease in either AOA-1 or -2 would trigger the 1.2 sec memorization period once it reaches the monitoring threshold. During this period, the FCS computers would use the average of AOA-1 and -2 from that point. Again after 1 sec, AOA-1 would be rejected and replaced by AOA-3. (See Case C in Figure 64.)

Data Spikes 1.2 Seconds Apart

Following the occurrence to A330, VH-QPA,[18] the aircraft manufacturer identified a scenario in which data spikes spaced 1.2 sec apart could trigger an incorrect AOA response. This requires that the first spike to AOA-1 (or -2) last less than 1 second. In this case the spike would end before the 1 second rejection interval. The system would treat AOA-1 data as valid. If it then spikes 1.2 sec after the first spike, it would be accepted as valid and be averaged with (in this case AOA-2). (See Case D in Figure 64.) After the second spike ends, it would slowly decay.

This was determined to be the cause of the upset to A330, VH-QPA. After the 1.2 sec memorization period, a single, short-duration spike in AOA-1 (or -2) would trigger another 1.2 sec memorization period, with the last valid value used. The presence of other spikes during this 1.2 sec period would not trigger another memorization period. A steady increase or decrease in either AOA-1 or -2 would trigger the 1.2 sec memorization period once it reaches the monitoring threshold. During this period, the FCS computers would use the average of AOA-1 and -2 from that point. Again after 1 sec, AOA-1 would be rejected and replaced by AOA-3. (See Case C in Figure 64.)

Two AOA Sensors

The Airbus A320 uses three AOA sensors and a two-out-of-three voting to detect sensor errors. However, the airplane's Minimum Equipment List (MEL) allows dispatch with two ADIRUs. We have already discussed an incident (in Chapter 8, page 139) in which an airplane was dispatched with one inoperative. The crew was forced to land with no instruments for the captain and the first officer making an approach in direct law.[29]

One AOA Sensor

The B-737(MAX) uses one AOA sensor. While the airplane has two sensors installed, only one is used at a time. As a result, the crew must

      (1)   Detect a disagreement between the two;

      (2)   Determine which sensor is incorrect or if both are; and

      (3)   If one sensor is incorrect, select the other, or
            if both are incorrect, shut the system off.

This would be extremely difficult at best. The decision to require no additional training makes it nigh impossible. The accidents to Lion Air 610[30] and Ethiopean 302[31] were caused by an erroneous angle-of-attack (AOA) sensor.* In this case it drove the controls to reduce the angle

---

* These accidents are described in Appendix 4 on pages 315-ff..

of attack. Since these two accidents happened just after takeoff, the result was a collision with the ground. While many of the details are still not public, there have been reports that the safety analysis was based on a preliminary design. Having no internal data checking would require the crew detect the error and take steps to correct it. At best, this would be difficult. With the manufacturer's apparent policy to require no differences training for the pilots, it would be well nigh impossible. This provides further impetus for the need for additional crew training for novel systems.

Latent Sensor Failures

Another upset was Malaysian 124,[32] caused by a faulty accelerometer which had failed four years previously! It had not been replaced in the intervening period. When a second accelerometer failed, the airplane pitched up abruptly. It seems obvious, but improvements in system fault detection and isolation are required. At the same time, more attention should be paid to the effect of dispatch with faulty or inoperative system components (allowed by MEL) on the probability of successfully coping with subsequent additional failures (e.g. correct handling of a second failure).

The issue of Dual Control Inputs should be studied to determine if simple summing of the pilot inputs is the best solution for designs using sidesticks with passive feel forces. This will require some research study and careful assessment of the consequences of dual inputs. More research is also needed to establish sidestick safety requirements related to stick maneuver command sensitivity, the scheduling of the maneuver command authority for large stick deflections appropriate to the flight condition and harmonization between the displacement maneuver commands and the required deflection forces. Similar issues with respect to FBW rudder control system designs also need to be addressed.

Automatic Emergency Descent

Following two depressurization accidents, there have been some efforts to develop a system to mitigate such mishaps, mostly for small airplanes. Recently, Airbus has developed such a system.[33][34] In the Airbus version, the system automatically triggers when the cabin altitude exceeds 14,500 ft and the airplane is at an altitude above 25,000 ft. When triggered, the Airbus system automatically:

- ✈ Activates an "Emergency Descent" aural warning;
- ✈ Engages the autopilot if not engaged already;
- ✈ Engages the auto-throttle if not engaged already;
- ✈ Reduces thrust to idle;
- ✈ Sets an safe altitude, usually 14,000 to 15,000 feet as altitude select;
- ✈ Start a rapid descent;
- ✈ Engages heading mode and flies the plane's present heading;
- ✈ Resets the transponder code to 7700;
- ✈ Turns the seat belt sign to ON; and
- ✈ Keeps TCAS engaged.

The system will be standard for all A350-1000 aircraft.

Small airplanes have taken the lead in developing these systems and have already fielded systems. These systems are discussed in the following chapter.(see page 157).

A Final Point

It may seem counterintuitive, but safety equipment can reduce safety, particularly if the flight crews do not receive clear training material. This point must be emphasized over and over again.

References

1 *Aircraft Accident Report: Trans World Airlines, Inc., Boeing 727-231, N54328, Berryville, Virginia, December 1, 1974*, NTSB AAR-75-16, November 1975

2 R L Newman, "Ground Prox Revisited," *Professional Pilot*, August 1977, pp. 92-94

3 R O Phillips, *Investigation of Controlled Flight into Terrain: Descriptions of Flight Paths for Selected Controlled Flight into Terrain (CFIT) Aircraft Accidents, 1985-1997*, Volpe National Transportation Systems Center Report DOT-TSC-FA9D1-99-01, March 1999

4 T G Farr *et al.*, "The Shuttle Radar Topography Mission", *Reviews of Geophysics*: 45, June 2007, RG2004,doi:10.1029/2005RG000183;http://agupubs.onlinelibrary.wiley.com/doi/epdf/10.1029/2005RG000183, accessed November 5, 2020

5 "Terrain Awareness and Warning System," *Operating Requirements: Domestic, Flag, and Supplemental Operations*, §121.354, amendment 121-273, effective March 29, 2001

6 *Airworthiness Criteria for the Installation Approval of a Terrain Awareness and Warning Systems Approved (TAWS) Approved for Part 25 Airplanes*, FAA Advisory Circular AC-25-23, Dated May 22, 2000

7 *Aviation Safety Network*, http:// aviation-safety.net /; accessed October 27, 2020

8 D Swihart *et al.* "Results of a Joint US/Swedish Auto Ground Collision System Program", Paper ICAS-98-1,8,2 presented at 21st ICAS Congress, Melbourne, Australia, September 1998

9 N Ho *et al.*, "A Longitudinal Field Study of Auto-GCAS Acceptance and Trust: First Year Results and Implications," *Journal of Cognitive Engineering and Decision Making*: 11, 2017, pp. 239-251

10 *Aircraft Accident Report. Controlled Flight Into Terrain, American Airlines Flight 965, Boeing 757-223, N103AA, Near Cali, Columbia, December 20, 1995*, Aeronautica Civil of the Republic of Columbia Final Report, September 1996

11 *727 Cockpit Operating Manual, Northwest Airlines*, Vol 727, Revised April 12, 1966, Superseding October 12, 1965

12 *Accident of the Aircraft 5B-DBY of Helios Airways, Flight HCY522 on August 14, 2005, in the Area of Grammtiko, Attikis, 35 km Northwest of Athens International Airport*, Ministry of Transport & Communications AAR-11/2006, November 2006

13 "Electronic Centralized Aircraft Monitor (ECAM)", https://www.skybrary.aero/index.php/Electronic_Centralized_Aircraft_Monitor_%28ECAM%29

14 "Engine Indication and Crew Alerting System (EICAS)", http://skybrary.aero/index.php/Engine_Indicating_and_Crew_Alerting_System_(EICAS), accessed November 9, 2020

15 G Singer and S Dekker, "Pilot Performance During Multiple Failures: An Empirical Study of Different Warning Systems," *Transportation Human Factors*: 2 [1], 63-76, 2000

16 R L Newman and A A Lambregts, "A Review of Fly-by-Wire Accidents," *ISASI Forum*, April-June 2010, Presented at 40th ISASI Annual Seminar, Orlando

17 *Airbus A330 C-GGWD and Airbus A340 TC-JDN*, 2 October 2000, AAIB Bulletin No. 6/2001

18 *In-Flight Upset, 154 km West of Learmonth, WA, 7 October 2008*, VH-QPA, Airbus A330-303, ATSB Aviation Occurrence Investigation AO-2008-070, 2011, October 2019

19 *In-Flight Upset Event, 240 km North-West of Perth, WA, Boeing Company 777-200, 9M-MRG, 1 August 2005*, ATSB Aviation Occurrence Investigation Report AO-200503722, 2007

20 *Aviation Accident Final Report*, NTSB File FTW96LA269, dated March 1998

21 *Accident of aircraft Airbus A-320-214, registration EC-HKJ, at Bilbao Airport on 7 February 2001*, CIAIAC Technical Report A-006/2001, 2006

22 *Airbus A-340-200, Registration D-AIBE*, NTSB File FTW96LA269

23 D T McRuer *et al.*, *Aviation Safety and Pilot Control*, Washington: National Academy Press, 1997

24 *Aviation Accident Final Report*. NTSB File CHI01FA104

25   *Roll Oscillations on Landing, Airbus 321-211, C-GJVX and C-GGIUF, Toronto/Lester B Pearson International Airport, Ontario, 07 December 2002*, TSB Report A02O0406, Report Date July 4, 2005

26   *Aircraft Accident Report, Northwest Airlines, Incorporated, Boeing 727-25, N264US, Near Theills, New York, December 1, 1974*, NTSB AAR-75-13, August 1975

27   *Accident on 27 November 2008 off the coast of Canet-Plage (66) to the Airbus A320-232 registered D-AXLA operated by XL Airways Germany*, BEA d-la081127, report undated

28   *Final Report on the Accident on 1st June 2009 to the Airbus A330-203 registered F-GZCP, Operated by Air France, Flight AF 447 Rio de Janeiro - Paris*. BEA Report f-cp090601en, Published July 2012

29   *Incident to Airbus A320-200, I-BIKE, 25 June 2005, On Approach to Runway 09L at London Heathrow Airport*, AAIB Bulletin 6/2006, 2006

30   *Aircraft Accident Investigation Report. PT. Lion Mentari Airlines Boeing 737-8(MAX); PK-LQP, Tanjung Karawang, West Java, Republic of Indonesia, October 29, 2018*, KNKT Final Report 19.10.35.04

31   *Interim Investigation Report on Accident to the B-737-8 (MAX) Registered ET-AVJ Operated by Ethiropian Airlines on 10 March 2019*, Aircraft Accident Investigation Bureau Report AI-01/19, March 2020

32   *In-Flight Upset Event, 240 km North-West of Perth, WA, Boeing Company 777-200, 9M-MRG, 1 August 2005*, ATSB Aviation Occurrence Investigation Report AO-200503722, 2007

33   http://simpleflying.com/airbus-a220-emergency-descent-mode/, accessed November 8, 2021

34   http://www.airbus.com/en/newsroom/news/2018-03-airbus-developed-a350-xwb-safety-feature-enables-automated-emergency-descents, accessed November 8, 2021

Chapter 11: New Technology in Small Airplanes Certification

The general aviation safety record is so bad, we encourage certifying innovative instruments and equipment because it can only help improve the safety record.*

The Small Airplane Directorate has a very difficult job. Most of the aircraft certified by the directorate encompass what is known as General Aviation, which includes are civil aircraft not flown by a scheduled or non-scheduled airline. It should not come as a surprise that this encompasses a wide variety of airplane sizes as well as a wide variety of pilot skill and experience. As a result, Part 23 had become very complicated.

As can be expected with this range of airplanes, the pilots have a wide variety of experience and training. The safety record can best be described as "spotty." The Directorate encouraged manufacturers and owners to install advanced aircraft systems to improve safety.

To formalize this encouragement, a policy for Non-Required Safety Enhancing Equipment (NORSEE) was formalized.[1] To address the wide variety of airplanes certified under Part 23, Small Airplane Directorate completely rewrote Part 23.[2] We shall look at both of these changes, the promotion of advanced aircraft systems and the complete rewrite of Part 23 in this chapter.

This philosophy was well-meaning but, in our opinion, flawed because pilots, and to some extent regulators don't fully understand the implications of the new technology.

We will discuss the implications of introducing advanced technology to light airplanes: the Good, the Bad and the Ugly:

The Good:	Emergency Descent Mode (EDM), and
	Ballistic Recovery Systems (BRS), and
	Automatic Emergency Landing;
The Bad:	Angle-of-Attack Displays, and
	Technically Advanced Airplanes;
The Ugly:	NORSEE; and
	The New Part 23.

We based our choice of adjectives based on the relative success of the technology. Generally, "Bad" systems were implemented without regard for ensuring that pilots were adequately trained and the systems had well thought-out procedures. The "Ugly" refers to policies and procedures we feel are detrimental to safely introducing new technology.

Emergency Descent Mode

In October 1999, a Lear 35 lost pressurization and flew for several hours at or above FL390. At this altitude, all occupants died from hypoxia in a short time. In any event, the airplane flew on and crashed when it ran out of fuel. In August 2005, a Boeing 737 also flew for several hours at FL340 before crashing. In this accident 121 passengers and crew died.†

In a more recent accident, a privately owned TBM 700 on an IFR flight from Rochester, NY to Naples, FL, requested an emergency descent from FL280 to a lower altitude. When cleared

---

* Frequent comment by one of our former colleagues in the Small Airplane Directorate.
† Hypoxia accidents are described in Appendix 4 beginning on 331.

to descend, the pilot became unresponsive and did not descend. The airplane continued past the Florida coast. It was intercepted by Air Force fighters and escorted until the accident airplane crossed into Cuban airspace. The airplane ultimately crashed into the sea north of Jamaica. The flight time was consistent with fuel exhaustion after departing with a full fuel load.*

To mitigate this problem, several manufacturers have produced autopilots that monitor cabin altitude and detect potential hypoxia. These systems then program the autopilot to descend the airplane to a safe breathing altitude. All such systems require integrated avionics systems.

Honeywell has designed systems for a number of business jets as well as Part 23 propeller airplanes, such as the Pilatus PC-12.

Garmin has produced an autopilot with a "hypoxia detection" mode which monitors the cabin altitude and provides an *Are You Alert?* message if no activity has been detected in a time interval.[3] Any sign of activity, such as pressing an avionics system key resets the time interval. The time interval increases with increasing cabin altitude, presumably to match the physiological time-of-useful consciousness (TUC). If no response is received, the autopilot is programmed to descend to a safe altitude

When triggered, these systems automatically:

> ✈ Activate an "Emergency Descent" aural warning
> (usually with a delay to allow the pilot to cancel the mode);
> ✈ Engage the autopilot if not engaged already;
> ✈ Set a safe altitude, usually 14,000 to 15,000 feet as altitude selected;
> ✈ Start a rapid, descent;
> ✈ Engage the heading mode and fly the airplane's present heading
> (Some systems turn the airplane off present heading); and
> ✈ Reset the transponder code to 7700.

We are not aware of any "saves" with these systems.

Airbus has also developed a similar system for their aircraft described in the previous chapter. (See page 154.)

Ballistic Recovery Systems

Using a parachute to lower the entire airplane safely to the ground has been an idea for years. Snyder discussed the concept in 1976.[4] It was generally thought to be impractical for large aircraft, but became relatively common with the introduction of ultralight vehicles using the Ballistic Recovery Systems (BRS) parachutes deployed by a ballistic rocket. There is an aftermarket BRS parachute system for some general aviation airplanes, such as the Cessna 150.

Cirrus Aircraft incorporated such a system from the outset. Their system, Cirrus Airframe Parachute System (CAPS) also uses a ballistic rocket to deploy the recovery parachute.[5] CAPS is the only approved spin recovery method for the Cirrus SR-20 and SR-22.

The system is deployed using a handle located in the cabin ceiling. This causes a rocket to fire which has a cable to extract the parachute. A slider ring on the suspension lines allows the canopy to partially inflate. As the canopy loads increase, the slide ring moves down and allows the canopy to open fully. The airplane will descend under the canopy at less than 1700 ft/sec. The "landing" is the equivalent of dropping from a height of 13 ft.

As January 2014, there have been 85 lives by the CAPS system.

Sadly, 120 fatalities have resulted in loss-of-control accidents where the pilot did not deploy the parachute before the crash. Similar fatalities have resulted when military pilots fail to use their ejection seats, leading to the expression "Better dead than look bad."

---

*   This accident is described in Appendix 4 on page H.

Emergency Automatic Landing

In 2020, Garmin announced certification of an Autoland function on their G3000 integrated flight deck on the Piper M600. The system requires an on-board terrain and airport database as well as GPS navigation capability. It also requires interfaces with aircraft systems and available weather information. Because of the need for extensive interactions between the autopilot and the aircraft systems, these systems are not currently available for aftermarket installations.

In the event of an emergency or pilot incapacitation, the pilot or a passenger may activate the system using a guarded switch on the instrument panel or on an overhead panel. The system could be automatically activated. When activated, the system determines the nearest appropriate airport based on proximity and weather (primarily surface winds and thunderstorms) and navigates there while maintaining adequate terrain clearance.

The autopilot flies a standard instrument approach or standard pattern and lands. It automatically broadcasts its intentions to air traffic control. It also displays this information to passengers. It lands the airplane, stops on the runway, and secures the engine. Shortly after writing the first draft of this system, an article appeared in *AOPA Pilot* describing the operation.[6]

The Garmin system is available on the Cirrus *Vision Jet* as well the Piper M600.

Transport Airplane automatic landing systems are designed for routine operations and require extremely small probabilities of a catastrophic outcome. This generally requires many simulated replications of landings to ensure that the autoland itself does not lead to such a catastrophic outcome. They also do not include the routing algorithm, flap and gear control, or engine shut down after landing.

In this light airplane application, it was argued that the catastrophic event (pilot incapacitation) has already happened, allowing a relaxed statistical criteria. This seems quite reasonable to us.

Angle-of-Attack Displays

In the 1960's the FAA began to promote research to show that angle-of-attack (AOA) displays would improve pilot training and safety. Studies conducted by two university flight training departments* failed to show any significant improvement.[7][8] Part of the problem with these two studies may have been the need to meet the no AOA final check requirement. A NASA study[9] in the same time period concluded that using the AOA as control parameter was unsatisfactory because of lightly damped phugoid oscillations.† Recent studies have been inconclusive, yet continue to recommend AOA installations.[11][12][13]

Many head-up displays show flight path angle relative to aircraft pitch attitude, this difference is the angle-of-attack.[14][15] Some airplanes show this relationship head down as well.[16]

$$\alpha \quad = \quad \theta - \gamma \tag{3}$$

where θ = airplane pitch attitude

γ = flight path angle

We are not against AOA displays. We are, however, against installing them without a clear understanding on what they show and how pilots will use them. This must be followed by adequate pilot training.

---

*   Ohio State University and Embry-Riddle Aeronautical University
†   Phugoid oscillations are typically lightly damped and occur at nearly constant AOA.[10]

Technically Advanced Airplanes

What is a Technically Advanced Airplane (TAA)? There is no definition or description of a TAA in where we would expect to find it, in FAR 1. The closest we can find is in the experience requirements for a single-engine airplane rating on a commercial pilot certificate. Until October 2009, applicants for a commercial pilot certificate (airplane) needed ten hours of training in an airplane with retractable landing gear, flaps, and a controllable pitch propeller (commonly called a "complex airplane".[18] Following amendment 61-142, the ten hours of training could be done in a "complex airplane", a turbine-powered airplane or a TAA.[19]

The intent of the Part 61 definition is an attempt to require new commercial pilots to have experience in airplanes beyond simple fixed-gear, fixed-pitch propeller airplanes.

In 2003, an FAA-industry safety study team used TAA to mean one with an IFR-certified GPS navigation system with a moving map or a display that could show weather, airborne traffic, or terrain. It also had to have an integrated autopilot which was integrated with these systems. The report stated that a significant concern was that the airplane be intended for long-distance trips flown by a lone private pilot.

In their review of the safety record of airplanes with modern systems, the NTSB described the instruments as "integrated, computerized displays commonly referred to as glass cockpits".*

> **Technically Advanced Airplane (TAA):** An airplane equipped with an electronic Primary Flight Display (PFD), an electronic Navigation Display (ND), and an autopilot capable of maneuvering the airplane about the three axes.†

This definition is not solely concerned with the display format (computer screen or round dials) but the integration with the aircraft systems.

Safety Record of Glass Cockpit Light Airplanes

The NTSB compared the accident rates for thirteen models of light single-engine aviation airplanes, manufactured between 2002 and 2006.[21] Of these aircraft models, the distribution of instrument configuration was:

- Glass Cockpit   5,516 built
- Round Dials   2,848 built
- Total   8,364 built

The NTSB compared the accident data from this cohort from 2002 through 2008. To quote the report: "The study aircraft fleet included aircraft manufactured between 2002 and 2006; therefore, the distribution of glass cockpit aircraft in the study aircraft fleet remained constant from 2006 through 2008".‡ Table 18 shows the total and fatal accident rates for the two instrument configurations. Figure 65 from the NTSB Glass Cockpit report shows this data graphically

---

* Conventional instruments are usually referred to as "round dials" or "steam gauges." We will use "round dials" to denote conventional flight displays.
† This is the wording used in §135.105(c)(1).[20]
‡ The is a direct quote from the NTSB report.[21]

Table 18 Accident Rates by Cockpit Configuration*

Year	Accident Rate Per 100,000 Flight Hours					
	Total Accidents			Fatal Accidents		
	Round Dials	Glass Cockpit	All General Aviation[a]	Round Dials	Glass Cockpit	All General Aviation[a]
2006	3.70	4.10	6.33	0.51[b]	1.37	1.28
2007	3.71	3.46	6.92	0.36[b]	0.72[b]	1.20
Combined	3.71	3.77	6.63	0.43[b]	1.03	1.25
Standard Error	15.3%	12.7%	-	44.2%	19.2%	-
Notes (a)	From *NTSB Aviation Accident Statistics, Fatalities, Rates, 1989-2008, U. S. General Aviation,*					
(b)	Rate based on fewer than 10 events.					

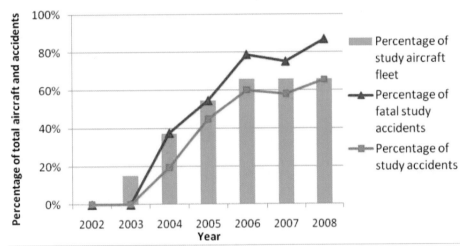

Figure 65 Distribution of Glass Cockpit Cohort Aircraft and Accidents per Year†

The NTSB also reported that glass cockpit airplanes were involved in higher percentages of loss-of-control in flight and collision-with-terrain events, and round dial airplanes were involved in more loss-of-control on ground and more takeoff and landing accidents. This is consistent with the use of the airplanes with glass cockpit airplanes being used in more personal and business transportation and round dial airplanes being used for training and local flights. A summary of the accident types is shown in Figure 66 from the NTSB study.‡ The totals do not round to 100 percent because of rounding.

The NTSB concluded that higher percentage for collisions-with-terrain compared with all other events for the glass cockpit cohort was the only statistically significant difference between the two cohorts in the accident events: χ^2 (1, N=255) = 3.980. p = 0.046.

There had been anecdotal reports of pilots using the red terrain color in their moving map displays to allow them to fly closer to mountainous terrain at night than they would have otherwise. These anecdotal reports provide a possible explanation for the NTSB conclusion.

The NTSB concluded that these advanced displays did not show a significant improvement in safety for glass cockpit general aviation airplanes. They also raised concerns about a lack of knowledge and training requirements for glass cockpit displays. This is significant in view of the overwhelming success of TAWS in transport airplanes.

---

* From the NTSB report.[21]

† From the NTSB report.[21]

‡ One accident (NTSB File SEA03LA180) was excluded from the NTSB study analysis because it occurred during the factory flight test and therefore not representative of normal, day-to-day operations.

In our opinion, this supports the absolute need to consider pilot behavior during certification and to ensure adequate training and operational procedures for new technology equipment.

This is troubling. The lack of required operational training for pilots in the GA fleet has led to an increased occurrence of CFIT accidents compared with transport operations. We saw earlier in Table 15 that the introduction of TAWS almost completely eliminated CFIT mishaps in transport airplanes. Why, then, do we see an increase in GA airplanes? We have heard anecdotal stories of pilots who used the terrain display in the mountains at night to navigate through passes by descending to make the terrain display just turn red. In any event, introducing new technology without proper training and procedures is fraught with issues.

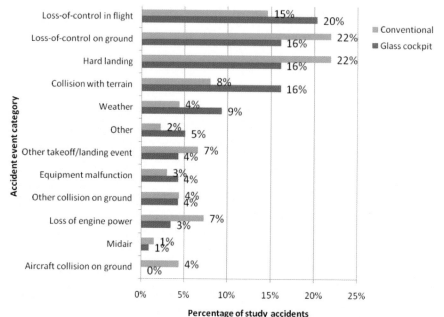

Figure 66 Comparison of Accident Event Type by Instrument Configuration (NTSB Aviation Accident Data 2002-2008, N=255)

Non-Required Safety Enhancing Equipment (NORSEE)

Recently, the FAA promulgated a policy relaxing the standards for "equipment that is not required by any Federal Regulation with the intent to measurably increase aircraft safety".[22] The basic premise of NORSEE is that non-required equipment which has a safety rationale will have a "minor" safety effect. As defined in the policy memorandum, a minor safety effect is:

> **A faulty condition, caused by equipment, that has no safety effect on the certificated aircraft (for example, structural integrity, electrical, mechanical, mechanical, pneumatic, hydraulic, navigation, communication, stability and control, or performance), and is within the pilot's capability and control.**

It should be noted that this definition differs from the definition of a "minor effect" in AC-23.1309-1E.[23]

It is difficult to reconcile this definition with, for example, an increased number of CFIT results* identified in the NTSB Safety Study.[21] In our opinion, the NORSEE approach is fraught with "gotcha's.". We conclude that the NORSEE initiative is a well-intentioned, but misguided

---

* TAWS equipment is specifically listed as being within the scope of NORSEE.[22]

program that will not improve general aviation safety. This reminds us of a large number of Transport Category equipment whose intended function was "increase situation awareness."

The "New" Part 23

At the start of this chapter, we discussed how complicated and diverse the general aviation fleet is. Not only do they deal with many types of aircraft from gliders through Commuter Category, the flight crew members range from student pilots through airline pilots. As a result, Part 23 had become very complicated. Table 19 shows the wide range of fixed-wing aircraft that are certified through the Small Airplane Directorate.*

Table 19 Fixed-Wing Aircraft Types Certified by the Small Airplane Directorate (Pre-Amendment 23-64)

Type of Aircraft	Maximum Weight [lb]	Maximum Seats	Rule	Equipment Cert Level
Light Sport	1,320	2[a]	§21.190	
Glider	1,650	2[a]	§21.17	
Motorglider	1,750	2[a]	§21.17	
Primary Category	2,700	4[a]	§21.24	
Single Reciprocating Engine (SRE)	6,000	9[b]	FAR-23	Level 1
Single Turbine Engine (STE), Multi-Engine	6,000	9[b]	FAR-23	Level 2
Single & Multi Engine	12,500	9[b]	FAR-23	Level 3
Commuter	19,000	19[b]	FAR-23	Level 4
Notes: (a) Including pilot seats. (b) Excluding pilot seats.				

There are seven "subsets" of fixed-wing certification standards" – the four equipment levels, and three external standards, EASA SC-22, Light Sport consensus standards, and primary aircraft standards. We feel there are two points to be made:

→ Difficulty in predicting how divisions by weight will last; and
→ Artificiality of separating aircraft by engine type.

Divisions Based on Aircraft Weight

It is difficult to set weight limits to separate different classes of aircraft. In the 1930s, 12,500 lb was a good separation point between small airplanes from large. The Beech 18 was representative of the larger small airplanes while the Lockheed 18 was representative of the smaller transport. So 12,500 lb was a reasonable choice:

Beech 18   8,750 lb   6 passengers[24]
Lockheed 18   17,500 lb   15 passengers[25]

This worked well until engineering design became more sophisticated materials improved so that the same payload could be carried in a much lighter airframe, such as the DeHavilland DHC-6-100[26] (11,579 lb and 19 passengers). It is easy to see how difficult it is, in the long term, to set weight limit between certification categories. The 6,000 lb division between Levels 1 and 2 and Level 3 may blur in the future.

---

* In addition, the Small Airplane Directorate is responsible for Manned Free Balloon and Airship certification.

Divisions Based on Engine Type

The second point is that the difference between engine types seems to us to be artificial. In general, turbine engines are simpler for pilots to operate than high-performance reciprocating, turbocharged engines. It seems illogical that replacing a piston engine with a turboprop engine should require an increase in certification level, all else being equal. We cannot help but feel that the underlying reasons were concern that an increase in certification requirements would increase the certification cost and that this increase would be proportionally higher in reciprocating–powered airplanes than turbine-powered airplanes.

As can be expected with this range of airplanes, the pilots have a wide variety of experience and training. The safety record can best be described as "spotty." The Directorate encouraged manufacturers and owners to install advanced aircraft systems to improve safety.

To address the wide variety of airplanes certified under Part 23, Small Airplane Directorate completely rewrote Part 23.[27]

In our opinion, dividing certification classes by weight or engine type, make about as much sense as dividing classes by list price.

Complete Rewrite of Part 23

The most noticeable change in amendment 23-64 is the abandoning of the common paragraph number scheme that served the certification team so well since 1965. The *New Part 23* also eliminates Acrobatic, Utility, and Commuter Categories. All newly certified Part 23 airplanes will be Normal Category with limits of 19,000 lb MTOW and 19 passengers.[28]

§23.2005 defines four Certification Levels, based on the number of passengers. These Certification Levels are

> ✈ Level 1: Passenger Seating ≤ 1
> ✈ Level 2: Passenger Seating   2–6
> ✈ Level 3: Passenger Seating   7–9
> ✈ Level 4: Passenger Seating   10–19

These levels are similar to the equipment levels in Table 11 (on page 110) but appear to apply to the entire aircraft, not just to installed equipment as used previously.

The Certification Levels are based on the number of passengers. The rationale was stated in the preamble was that "passenger count has historically correlated to risk tolerance". This seems overly simplistic and is probably not measurable. Our first observation is that safety, while it may be related to the passenger count, is also related to the risk to the general public. It is easy to envision a 19,000 lb single-engine airplane designed to carry cargo only. Such an aircraft could pose significant risk on the non-flying public.

A second point is that the number of passenger seats does not measure the risk to the occupants. Besides ignoring the risk to crew members, the number of passenger seats does not limit the number of passengers. The FAA has long interpreted Part 91 as permitting occupants to share seats and safety restraints. In 2008, A Pilatus PC-12 crashed near Butte, MT with 13 passengers aboard an airplane with 8 passenger seats.[29]

"Occupant" versus "Passenger"

As discussed in the preamble, the term *occupant* was consciously rejected and replaced by *passenger*, probably in an attempt to exclude flight crew members. We requested a legal opinion from the FAA legal counsel on the definition of a "passenger"? The reply was: "The Office of the Chief Counsel has determined that your question, at least at this time does not require a legal interpretation".[30]

We concluded from this that the appropriate definition would be similar to that definition found in CAR 43, the General Operating Rules: A passenger is an occupant of the aircraft in flight other than a crew member.[31]

This agrees with our office dictionary.[32] In other words, anybody on board other than crew members. We were unable to identify when or why this definition was dropped from the regulations. If it was still in effect in three years later in 1962, it should have been brought over to the initial version of Part 91.

In our letter to the Office of Chief Counsel, we asked if other supernumerary crew, such as loadmasters, meteorologists on a weather survey flight, or cargo handlers, would be considered "passengers" or crew members. Their reply was that they could not answer the question.

We find this vagueness at odds with precise language usually associated with aircraft certification. We will continue this discussion in a later chapter (on page 174).

Lack of Objective Standards

We have heard the *New Part 23* summarized as "Do good and avoid evil".* The requirements are now written as high-level safety objectives.

It is now up to the applicant to provide a means of compliance,[33] such as a consensus standard, on how the aircraft will comply. A side effect is that compliance finding will now be much more subjective.

The end result, we believe, will be extremely subjective compliance findings lacking transparency. We do not believe that relying on so-called consensus standards is an acceptable means of providing regulation.

We shall discuss this in more depth in Chapter 14.

The *New Part 23* will be especially hard for new entrants to the industry. We have personally heard complaints from new entrant companies about an inability to "tell them exactly what the requirements were. After several of these, we decided to terminate our DER consulting.

Recommendations

Since the Commuter Category developed into form that was extremely close to Part 25, the Commuter Category should be contained within Part 25.

The distinction between reciprocating and turbine engines should be eliminated. If a particular design has special issues, Special Conditions or Equivalent Levels of Safety are always options.

The Primary Category should be formalized with a slightly higher weight limit, perhaps of 4,000 lb. and codified as part of Part 23 or as another rule.

References

1   *Approval of Non-Required Safety Enhancing Equipment (NORSEE)*, FAA Policy PS-AIR-21.8-1602, March 31, 2016

2   *Revision of Airworthiness Standards for Normal, Utility, Acrobatic, and Commuter Category Airplanes, Final Rule*, Docket No. FAA-2015-1621; 14 CFR Part 23, amendment 23-64, effective August 30, 2017

3   *GFC 700 Automatic Flight Control System*, Cirrus Design, Pilot's Operating Handbook and FAA Approved Airplane Flight Manual Supplement, SR22/SR22T, P/N 13772-135, revised September 8, 2014

---

*   The speaker is employed in the FAA's Aircraft Certification Service and will not be identified.

4   R G Snyder, *Advanced Techniques in Crash Impact Protection and Emergency Egress from Air Transport Aircraft*, NATO-AGARD-AG-221, 1976

5   CAPS. Guide to the Cirrus Airframe Parachute System (CAPS™) , Cirrus Aircraft Corporation. 2013, http://cirrusaircraft.com/wp-content/uploads/2014/12/CAPS_Guide.pdf

6   T A Horne and M Fizer, "Pilotless Precision," *AOPA Pilot*, December 2021, pp. T-10- T-16

7   Gandelman, J H, *Evaluation of Angle of Attack Instrumentation in the Training of Student Pilots to Private Pilot Certification*, FAA DS-68-19, August 1968

8   F G Forrest, *Angle of Attack Presentation in Pilot Training*, FAA DS-69-6, March 1969

9   S W Gee, H G Gaidsick, and E K Enevoldson, *Flight Evaluation of Angle of Attack as a Control Parameter in General Aviation Aircraft*, NASA TN-D6291, March 1971

10 B W McCormick, *Aerodynamics, Aeronautics, and Flight Mechanics*, New York: Wiley, 1979, pp. 572-576

11 M Bromfield and B Dillman, "The Effects of Using an Angle of Attack System on Pilot Performance and Workload During Selected Phases of Flight," *Sixth International Conference on Applied Human Factors and Ergonomics*, AHFE 2015, 2015, pp 3222-3229

12 B Dillman *et al.*, *Effect of Angle of Attack Displays on Approach Stability*, AIAA-2017-4064, June 2017

13 L R Le Vie, *Review of Research on Angle-of-Attack Indicator effectiveness*, NASA TM-2014-218514, August 2014

14 R L Newman, *Head-Up Displays: Designing the Way Ahead*, Aldershot, UK: Ashgate, 1995

15 G Klopfstein, "Rational Study of Aircraft Piloting," *Intrados*, 1966; reprinted by Thomson-CSF

16 *A330 & A340 Flight Crew Training Manual*, Cathay Pacific, July 2004, downloaded on October 1, 2020 from http://connect.cathaypacific.com/flightcrew/0,8385,14771,00.html

17 *Definitions and Abbreviations*, 14 CFR Part 1, §1.1, amendment 1-73, June 27, 2018

18 "Aeronautical Experience," *Certification: Pilots, Flight Instructors, and Ground Instructors*, 14 CFR Part61, §§61.129(a) and 61.129(j), amendment 61-124A, effective October 20, 2009

19 *ibid.*, amendment 61-142, effective July 27, 2018

20 "Exception to second in command requirement: Approval for use of autopilot system," *Operating Requirements: Commuter and On Demand Operations and Rules Governing Persons on Board Such Aircraft*, 14CFRPart135, §135. 105, amendment 135-58, effective December 29, 1995

21 *Introduction of Glass Cockpit Avionics into Light Aircraft*, NTSB Safety Study NTSB-SS-01-10, March 2010

22 *Approval of Non-Required Safety Enhancing Equipment (NORSAA)*, FAA Policy PS-AIR-21.8-1602, March 31, 2016

23 *System Safety Analysis and Assessment for Part 23 Airplanes*, FAA Advisory Circular 23.1309-E, 11/17/2011

24 *Type Certificate Data Sheet, Model 18D*, FAA-TCDS-A-684, Revision 6, November 27, 2017

25 *Type Certificate Data Sheet, Lockheed Model 18*, FAA-TCDS-723, Revision 23, July 23, 2012

26 *Type Certificate Data Sheet, Viking Air Limited DHC-6-100*, FAA TCDS-A9EA, Revision 21, February 28, 2018

27 *Revision of Airworthiness Standards for Normal, Utility, Acrobatic, and Commuter Category Airplanes, Final Rule*, Docket No. FAA-2015-1621; 14 CFR Part 23, amendment 23-64, effective August 30, 2017

28 "Applicability and Definitions," *Airworthiness Standards: Normal Category Airplanes*, 14CFRPart 23, §23.2005, amendment 23-64, effective August 30, 2017

29 *Aircraft Accident Report. Loss of Control While Maneuvering, Pilatus PC-12/45, N128CM, Butte, Montana, March 22, 2009*, NTSB AAR-11-05, Corrected copy, June 2012

30 "Re: Request for Clarification on Whether Certain Persons Onboard Aircraft Would be Considered Passengers," Letter from FAA Office of the Chief Counsel, January 24, 2018

31 "Definitions," *General Operation Rules*, CAM 43, §43.70, effective September 1959

32 Webster's New World Dictionary," New York: Simon and Schuster, 1989, p. 987

33 "Accepted Means of Compliance," *Airworthiness Standards: Normal Category Airplanes*, 14 CFR Part23, §23.2010, amendment 23-64, effective August 30, 2017

Chapter 12: Current Issues in Transport Airplane Certification

Pitot-Static System Issues*

Airplane pitot-static (P-S) systems provide airspeed and altitude data for pilot displays. In modern aircraft, they also provide air data to allow autopilots, flight directors, and autothrottles to operate correctly. Loss of air data has been the cause of many loss-of-control (LOC) and other mishaps. A number of these mishaps were reviewed and show the pilots have difficulty understanding the indications of P-S problems. The major problem is diagnosing the multiple, apparently independent systems of the common cause of a blocked pitot tube or static port.[1]

A Tale of Two Airplanes

It was the best of flights; it was the worst of flights.†

This is the story of two B-757's that departed tropical airports at night after the airplanes had been parked for several days. During the takeoff rolls, with the captain flying, the left airspeed never responded. Both captains decided to continue the takeoff. These two P-S mishaps had virtually identical series of precursors, differing only at the end

Birgenair TC-GEN,‡ February 6, 1996;[2]§
Astraeus G-STRZ, January 28, 2009.[3]**

If we examine the progression of each mishap, we see that the sequence of events is virtually identical. In both cases, the pitot tubes were left uncovered in a tropical area, the airplane departed at night with no ground lights available, and the captain failed to reject the takeoff in spite of a lack of airspeed indication. Figure 67 shows a side-by-side comparison of the precursors for the two flights.

What was the same for the two flights? A breakdown in procedure when the captain decided to continue the night takeoff contrary to best practices, followed by confusion caused by misleading, contradictory indications. The captain's airspeed appeared to work again, then increased to show excessive airspeed with attendant warnings. As the captain tried to bring the indicated airspeed under control, the first officer's airspeed showed slow airspeed. The stall warning triggered.

What was different between the two flights. Basically the only difference was in Crew Resource Management (CRM). The Astraeus first officer followed best CRM practices; the Birgenair first officer didn't. Possibly it was a cultural issue; perhaps he was awed by the difference between the other two pilots (captain and relief pilot) experience and his own.

---

* This section was adapted from a research report submitted to NASA.[1]

† With apologies to Charles Dickens

‡ Birgenair Airlines (Turkey, ICAO code BHT) was operating flight 301 for Alas Nacionales (Dominican Republic, ICAO code ALW) on February 6, 1996. The accident report and most references cite ALW 301 as the flight number. Both airlines ceased operations in 1996.

§ This accident is described in Appendix 4 on page 361.

** This serious incident is described in Appendix 4 on page 362.

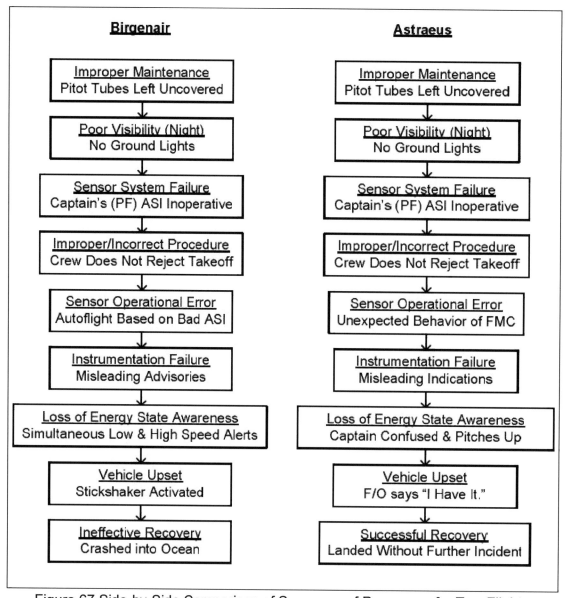

Figure 67 Side-by-Side Comparison of Sequence of Precursors for Two Flights

Other Pitot-Static Accidents

Another accident was called "The Poster Child" of pitot-static accidents.* Northwest 6231 has special significance to the writers.† This was a pitot-icing accident to a Boeing 727 on December 1, 1974. The accident occurred when the airplane climbed through icing conditions, and the pitot tubes became blocked. As the airplane continued to climb, the static pressure became less and the indicated airspeed became larger. Ultimately, the airplane's stall warning triggered which confused the flight crew leading to a loss of control.[5] ‡

Part of this confusion was the simultaneous high and low-speed warnings. During these events, crews forget that the high-speed warnings are triggered by indicated airspeed, and the

---

* Bertorelli referred to Northwest 6231 as the *Poster Child* of pitot-static accidents.[4]

† The captain was hired by Northwest Airlines in the class before one of your authors (RLN). Efforts of trying to understand how this confusion could arise led to becoming an accident investigator.

‡ This accident is described in Appendix 4 on page 360.

low-speed warnings by angle-of-attack. This is a recurring theme in many of the P-S scenarios reviewed.

Table 20 lists a representative sample of P-S related mishaps.

Table 20 Pitot-Static Related Mishaps

Date	Model	Flight ID	Fat	Occurrence	Result	Ref
Dec 1974	B727	NWA 6231	003	Pitot Icing (heat off)	Stall, Uncontrolled Descent	5[a]
May 1986	Tu154	AFL 327		Pitot Icing (heat off)	Stall, Improper Recovery	6
Feb 1996	B757	ALW 301	189	Blocked Pitot Insect	Stall, Uncontrolled Descent	2[a]
Oct 1996	B757	PLI 603	070	Blocked Static Ports	Disorientation, CFIT	7[a]
Oct 1997	DC-9	AUT 2553	075	Pitot Icing (heat off)	Improper Response	8
Apr 1999	B737	THY 5904	006	Pitot Icing (heat off)	Loss of Control	9
Oct 1999	MD-11	FDX 87		Blocked Pitot Drain	Excessive Speed, Overrun	10
Oct 2002	B757	ICE 662		Blocked Pitot	Stall, Safe Recovery	11
May 2005	B717	MEP 490		Pitot Icing	Improper Control Input	12
Jan 2009	B757	AEU TRZ[b]		Blocked Pitot Insect	Stall, Safe Recovery	3[a]
Jun 2009	A330	AFR 447	228	Pitot Icing[c]	Uncontrolled Descent	13[a]
Feb 2013	A340	ETD EHF[b]		Pitot Icing[c]	Safe Diversion	14[a]
Oct 2013	B757	UAL 140		Pitot Icing	Stall, Improper Recovery	15
Notes	(a) Mishap listed in Appendix 4. (b) Last three characters of registration used vice flight number. (c) Icing level beyond requirements for pitot probes.					

<u>Causes:</u> Six mishaps were caused by pitot icing. Pitot icing typically affects all on-board air data systems, including the pilot/copilot airspeed indicators and all systems that use airspeed data. These systems include autopilots, flight directors, and some flight control functions. Five mishaps were caused by inoperative pitot heat (either switched off or failed). One was caused by atmospheric conditions beyond pitot heat design requirements.

Five mishaps were triggered by a single blocked pitot tube A blocked pitot tube affects a single air data system, usually the pilot's or copilot's systems. In this case, there will be disagreement between the cockpit indications. Two were caused by leaving pitot covers off overnight. Two were caused by an internal blockage, One was caused by ingestion of ice crystals in a single pitot probe, although it could have been caused by pitot icing on that probe.

One mishap was caused by static ports taped over by maintenance crew. Blocking all static ports affects all on-board air data systems, the pilot, copilot, and standby systems and all flight control systems that use air data. These include autopilots, flight directors, and some flight control functions.

<u>Confusion:</u> Seven events resulted in flight confusion with misleading or conflicting cues and warnings. It was not clear to the flight crews what was happening. Attempts to isolate the failed systems appeared to be ineffective because the selection logic for airspeed/altitude input to autopilots or flight directors was not clear. The pilots did not understand what the effect of changing altitude had on their indications. Confusing warnings, such as MACH/SPD, TRIM, and RUDDER RATIO were shown with no or insufficient training. Simultaneous overspeed and stall warnings were presented.

Four of these were accidents. The crews were still trying to sort the situation out when they crashed, killing all on board. The other three showed the same confusion on the flight deck with the same indications. Fortunately for all on board, the crew finally reverted to basic pitch and power control and safely recovered the airplanes.

Sidebar: Pitot-Static Background

Airspeed Measurements

Airspeed is measured in aircraft by measuring the delta pressure between static air pressure and the total pressure. The total pressure (P_T) is measured using a forward-facing pitot tube located in undisturbed airflow to measure the stagnation pressure. Static pressure (P_S) is usually measured by averaging the pressure sensed at flush static ports on either side of the fuselage.

The subsonic airspeed equation shows the velocity through the air (True Airspeed, TAS) as a function of total pressure (P_T), ambient (static) pressure (P_A), and ambient air density (ρ_A):

$$V_T = \sqrt{\frac{2\gamma}{\gamma-1}\frac{P_A}{\rho_A}\left[\left(\frac{P_T-P_A}{P_A}+1\right)^{(\gamma-1)/\gamma}-1\right]} \tag{4}$$

Airspeed indicators are designed as differential pressure gauges, calibrated to show airspeed. Because the indicator only senses $\Delta P = (P_T - P_A)$, all other ambient air conditions are assumed to be sea level, standard-day conditions. This allows for the construction of a differential pressure gauge measuring the ΔP between total pressure and static pressure. This airspeed is called Calibrated Airspeed (CAS) and shown by Equation (5)

$$V_C = \sqrt{\frac{2\gamma}{\gamma-1}\frac{P_0}{\rho_0}\left[\left(\frac{P_T-P_A}{P_0}+1\right)^{(\gamma-1)/\gamma}-1\right]} \tag{5}$$

Or, in standard British units (with speed in knots),

$$V_C = 1478\sqrt{\left[\left(\frac{P_T-P_A}{2116}+1\right)^{0.2857}-1\right]} \tag{5a}$$

For completeness, Indicated Airspeed (IAS) is found by substituting measured pitot pressure (P_{PT}) for P_T and measured static port pressure (P_{SP}) for P_A

$$V_I = \sqrt{\frac{2\gamma}{\gamma-1}\frac{P_0}{\rho_0}\left[\left(\frac{P_{PT}-P_{SP}}{P_0}+1\right)^{(\gamma-1)/\gamma}-1\right]} \tag{6}$$

These equations are based on Livingston and Gracey,[16] *SFTE Reference Handbook*,[17] and Kimberlin.[18]

Altimeter Measurements

Discussion of altimetry must begin with the standard atmosphere. The standard atmosphere is a mathematical derivation treating air as a dry, perfect gas in hydrostatic equilibrium and based on agreed-to boundary conditions (shown in Appendix 1, page 240).

The pressure reduces with altitude as

$$dP/dH \quad = -g\rho \tag{7}$$
$$= -g\,P/(R(T_0-\lambda H)) \tag{7a}$$
and
$$P \quad = P_0(1-\lambda H)^{g/RL} \tag{8}$$
$$= -P_0(1 - 6.87535\times10^{-6}\,H)^{5.2561} \tag{8a}$$

Inappropriate Stall Recovery: Three events resulted when the PF, saw zero airspeed and immediately applied an aggressive stall recovery. In these cases, the pilot over-stressed the airplane causing structural damage. In one case, the pilot extended the slats, causing one to depart the airplane which precluded recovery. In the other two cases, the aircraft was damaged, but landed safely.

Erroneous Airspeed Display: In one accident, in incorrect airspeed was displayed. The pilot used that airspeed that was too high. The airplane was unable to stop and overran the runway on landing at the destination.

Discussion

With P-S problems the pilot must first identify the problem, then either change the P-S source or change the control strategy. If the pilot wishes to use the autopilot, he/she must also change the autopilots P-S source. Changing the autopilot source is often different from changing the airspeed/altimeter display source. What could possibly go wrong?

Effect of Blocked Pitot: The pitot tube may become blocked by ice accretion* or by internal contamination, such as insects entering during ground stops.† When this happens, the pressure sensed by the airspeed indicator (ASI) pitot fitting will register a reduced pressure. The airspeed will usually read low, often zero. Other air instruments (altimeters or vertical speed indicators) are not affected.

However, if the airplane is climbing, the pressure in the pitot line remains near the static pressure at the altitude where the blockage occurred, but the static pressure will continue to drop. This can cause an apparent increase in airspeed. For Northwest 6231, the airplane was climbing at an indicated airspeed of 305 knots at 16,000 ft when the pitot tube iced over. At these conditions, the static pressure is 16.89 in Hg and the total pressure is 20.69 in Hg. Therefore the total pressure will be trapped in the pitot lines as the static pressure falls as the airplane climbs. As the airplane passes 20000 ft, the displayed airspeed will read 357 knots; as it passes 25000 ft, 421 knots.

Similarly, if a single pitot tube is blocked at sea level, the static pressure at sea level will be trapped. The airspeed will not increase during the takeoff run. However, as the airplane climbs, the sea level pressure trapped will indicate increasing airspeed. Pilots may think the problem has been fixed,. However, this may not be true. The Birgenair captain commented that his ASI "began to operate."[2]

Effect of Blocked Static Port: The static port may become blocked by ice accretion, although this is rare for fuselage-mounted ports. The more common cause is covering the ports during maintenance as happened to AeroPerú Flight 603 in 1996.‡ Given blockage at the start of flight, there will be no effect during the takeoff roll. The airspeed will show the correct value. Once the aircraft starts to climb, however, the displayed airspeed will decrease while the displayed altitude will remain constant. An airplane flying at 250 knots will have the displayed airspeed decrease to zero somewhere between 2000 and 3000 ft.

Pitot-Static Arrangements: In addition to displaying flight data to each pilot, the P-S system also provides data to the autopilot, flight director, and other flight guidance and flight management systems. To accomplish this, the following arrangement is typical.

Three separate P-S systems are typically installed. These are often referred to as Captain's, F/O's, and Standby. Additional systems are sometimes installed to provide data for the flight control system. Figure 68 shows a typical arrangement.

---

*   Such as in the accident to Northwest 6231[5] or Air France 447.[13]

†   See, for example, the mishap reports to Birgenair 301[2] or Astraeus G-STRZ.[3]

‡   This accident is described in Appendix 4 on page 361.

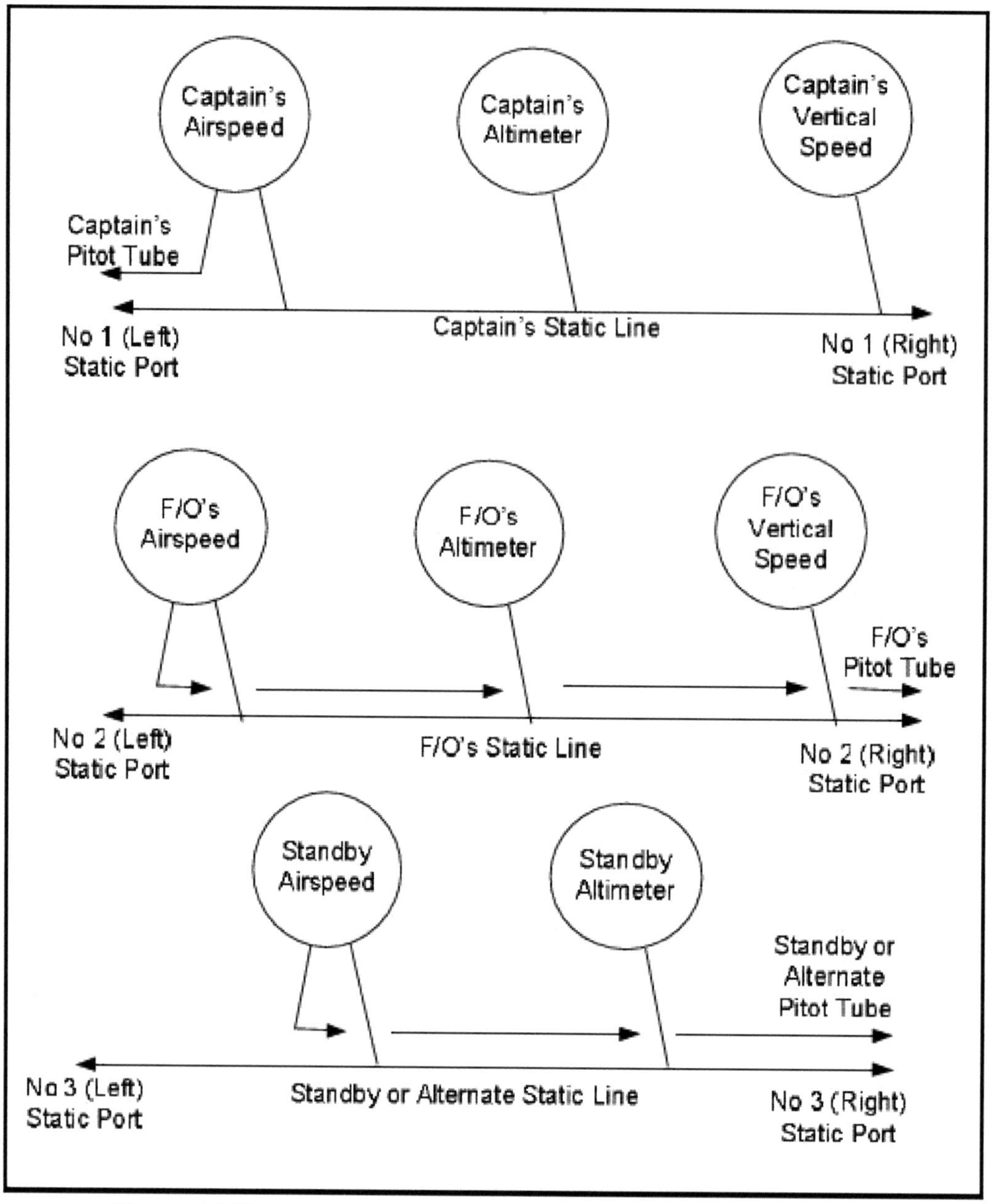

Figure 68 Typical Arrangement of Triplicated Air Data Systems

<u>Display Source Switching</u>: Most P-S arrangements allow the captain's airspeed and altimeter to switch between the captain's static line and pitot tube (shown in the figure) and the standby static line and pitot tube. The first officer's instruments can likewise be switched to the standby sources. This is allows the crew to compensate for a single pitot-tube or static port problem. It should have allowed the crews in the Birgenair 301, Astraeus G-STRZ, or the Icelandair 662 mishaps to select an appropriate pitot source. What actually happened in each of these mishaps is described below on page 174.

Effect on Warning Indications

One thing that is evident in these mishap reports is the confusion as the airplane climbs higher. In the stress of the situation, the pilots seem to forget that overspeed warnings are triggered by airspeed (or Mach) indications, while low speed warnings are generally triggered by angle-of-attack.

Many airplanes do not show an indication of airspeed or altitude disagreement. This seems strange to the writers in view of their importance. Many of the airplanes in this study detect an airspeed discrepancy. Unfortunately, they indicate this to the crew by displaying of MACH/SPD TRIM and RUDDER RATIO simultaneously. This arcane warning does not appear to adequately warn the flight crew. It should be corrected in all airplanes where it is found.

We feel the problem is a diagnostic one. These situations are examples of multiple failure indications with a common cause. Most pilots could describe the symptoms for a given common cause, such as what would happen of the static ports were blocked prior to takeoff. What is more difficult is being presented with the multiple symptoms and then trying to deduce the single cause.

When faced with difficulty, humans develop a hypothesis to explain the symptoms. There is a tendency for the initial diagnosis to take hold, for the person to seize on corroborating symptoms and to reject those that disagree with the initial diagnosis. In medicine, this tendency is called "diagnosis momentum."[19]

In the case of Northwest 6231, the crew became convinced they were encountering "unusual weather, which included strong updrafts." The NTSB report says their analysis must have been "strongly influenced by these factors and by the fact that *both airspeed instruments were indicating essentially the same value*.[Emphasis added] When the stickshaker activated, the crew interpreted this as Mach buffet.[5]

Air Data Source Switching

In addition to providing data sources for the pilots' ASI, altimeters, and VSI, the P-S system also provides data for the autopilot, flight directors, and autothrottle. Different airplanes have different arrangements. A common arrangement is three P-S inputs to two air data computers (ADCs). Such an arrangement is found in the B-757. The following description was taken from the accident report for Icelandair 662. This is not intended to make a point of the B-757's design.*

> "Each of the B-757's three flight control computers (FCCs) receives air data from both the left and right air data computers (ADCs). The source of air data utilized by the FCCs is dependent upon the autopilot or flight director engagement configuration. The FCCs will utilize air data from the right ADC if the right FCC is engaged first or if only the first officer's flight director is switched on. Otherwise, the FCCs will utilize data from the left ADC.

> "During the altitude hold, flight level change, and vertical speed autopilot pitch modes, the FCCs generate pitch commands using ADC information. For VNAV modes, the FCCs receive VNAV steering commands directly from the Flight Management Computer (FMC). The FMC receives air data from both ADCs and selects the left ADC as the primary source.

> "During a VNAV climb, the FMC generates pitch steering commands to control the airspeed to the FMC target speed. With the autothrottle engaged, the FMC requests the autothrottle to control thrust to the target thrust setting (climb thrust). VNAV will command an increase of the pitch attitude to manage excess airspeed until the FMC target airspeed is reached or a 15 degree pitch attitude limit is reached. During VNAV PATH mode the FMC generates pitch steering commands to control the airplane to the FMC target altitude, while the autothrottle manages the thrust to control airspeed to the FMC target speed."[11]

---

* This is not a criticism of the B-757. There are many similar idiosyncrasies in many other airplane designs. Since we have three B-757 mishaps, use of the excerpt from one of the accident reports is appropriate.

What actually happened in the three mishaps was described below:

Birgenair   The crew did not make any source selections. The captain stated at one point that his ASI "began to operate."

Astraeus   The crew believed that these systems would use the *alternate* data if it was selected as the captain's source.

Icelandair   The crew switched to the right autopilot believing it would use the data from the right ADC instead of possible erroneous data from the left ADC.

Comments on Air France 447

Air France 447 also involved pitot icing. However it differed from the other accidents it two significant ways. First, the crew quickly recognized the loss of airspeed. The PNF said "We've lost the speeds." The most significant difference was that the loss of airspeed triggered a degraded mode in the flight control system. The crew was now hand-flying an airplane with degraded flight controls, no airspeed, and at high altitude in turbulence.*

While pitot icing did play a part, a more significant factor was that the junior member of the crew was the PF and very likely hand-flying an A330 at altitude for the first time in extremely difficult conditions.

An Etihad Airways A340 encountered similar conditions to AFR 447 over the Indian Ocean during cruise. The flight encountered moderate to heavy turbulence and experienced airspeed oscillations. The autopilot, autothrust, and flight directors disconnected, and the flight control changed from Normal to Alternate twice, then remained in Alternate law. The autopilots could not be re-engaged. The crew flew the airplane manually and safely diverted to Singapore.[14]

Supernumerary Crew Members

As discussed in the preamble to the Part 23 rewrite, the term *occupant* was consciously rejected and replaced by *passenger*, in an attempt to exclude flight crew members. We requested a legal opinion from the FAA legal counsel on the definition of a "passenger" versus a "supernumerary," but were told that there was no FAA definition.†

This all started as an attempt at humor aimed at inconsistencies within the FAA legal department. Then one of us (MK) pointed out the significant effort within the Transport Directorate dealing with applicants petitioning for exemptions for supernumeraries for cargo transports. At the time, we (RLN) had been surprised that supernumerary crew members were not simply treated as crew members, having been assigned duties during the flight.

Our office dictionary's definition of supernumerary" is "beyond the regular or prescribed number."[20] FAA Advisory Circular AC-120-85A essentially defined supernumeraries as those persons identified in §121.583:[21][22]

> **Supernumeraries**: Persons that are not members of the crew who are carried on board all-cargo aircraft that do not comply with all passenger-carrying requirements of Part 121.

Further research found this excerpt from CAR 8:

> **"The minimum crew specified in section 8.33 includes those persons necessary to navigate the aircraft, such as pilot, copilot, and flight engineer, and such other persons as may be required to perform the special purpose operations. For example, a multi-engine[sic] airplane engaged in and agricultural operation of dis-**

---

* The crew was also at their circadian low point.

† This was discussed in the previous chapter, on page 164.

pensing poison bran might be navigated by a pilot and copilot, and also have as part of the crew persons engaged in the dispensing of the bran. All of these persons would be considered crew members since each has a specific job to perform in connection with the special purpose operation."[23]*

What we have is a hodgepodge of definitions in the regulations, definitions in Advisory Circulars (ACs); these differ from document to document. Why the wording from CAR 8 was dropped, we don't know.

We reviewed the RGL website for exemptions for "supernumeraries" and found 98 cases over a 20-year span. These exemptions were generally for personnel on cargo airplanes to be seated in "crew" seats, not "passenger" seats. Approximately 98 per cent were granted. The requests generally were for relief from the following Part 25 paragraphs:

§25.785(j) Relief from passenger seat requirements for certain personnel;

§25.812(e) Relief from requiring escape path marking to provide emergency evacuation guidance and lighting for passengers;

§25.855(e) Relief from the requirement for automatic drop-down oxygen mask at passenger seats; or

§25.857 Does not allow passengers on airplanes with a Class E cargo compartment.

Each such action would have required several FAA staff preparing these petitions for public comments, issuing the announcement through the Federal Register and responding to comments received over a 60-day period.

It seems clear to us, that the FAA needs to fix regulations that require so many exemptions with so few denials.

Proposed Supernumerary Definition

We started with the list from §§121.583(a) and 135.85 that meet the definition of a *crew member*: and added additional ones from §§91.313(d) and 135.601(b).[21 24 25 26]

Therefore, we recommend the following definitions be added to the Part 1:

> **Supernumerary Crew Member**: A crew Member with special skills and training that is aboard an aircraft to perform special functions, in addition to the required minimum crew. The specialized functions are associate with any operation that is within the scope of the airworthiness certificate of the aircraft. Supernumerary functions include, but are not limited to the following:
>
> ✈ A courier necessary for the security of valuable or confidential cargo; A courier necessary for the security of valuable or confidential cargo;
> ✈ A Medical Crew Member;
> ✈ A Loadmaster responsible for the loading, securing, unloading of heavy or outsized cargo, and the operation of cargo handling devices.
> ✈ A person necessary for the safe handling of animals;
> ✈ A person necessary for the safe handling of fragile or perishable cargo;
> ✈ A person necessary for the safe handling of hazardous materials;

---

* We note that this was published after the founding of the FAA before the conversion from CAR to FAR.

↷ A person necessary for the safety of flight;

↷ A person required by an operating rule;

↷ A person to accomplish the work activity directly associated with an operation for which the airplane is certified;

↷ A person who performs an essential function in connection with an operation for which the airplane is certified;

↷ A person performing duty as an honor guard accompanying a shipment made by or under the authority of the US Government; and

↷ An inspector from the FAA, DOD, or NTSB performing an inspection or evaluation.

Simulator Approval

Negative Training: When the Simulator Lies*[27]

Accident to American Flight 587

In November 2001, American Flight 587, an Airbus A300B4, suffered an in-flight structural failure when the vertical tail departed the airplane after a series of roll-yaw excursions following a wake encounter after takeoff.† The investigation revealed that the first officer, who was the PF, had been trained to aggressively use the rudder pedals during such a recovery in spite of concerns expressed by the airplane manufacturer.[28]

The probable cause of the accident was the in-flight separation of the vertical stabilizer was a result of loads beyond ultimate design that were created by the copilot's unnecessary and excessive rudder pedal inputs.

During the course of the investigation, it was found that a member of the airline's training cadre had taken a course in unusual attitude training. This course, which used a military trainer (Beech T-34), emphasized aggressive use of the rudder during upset recovery with the slogan "step on the sky" to denote using pushing on the rudder away from the bank.

To force company pilots to use this method, the airline training staff modified the simulators' control response during upset training to force the pilots into aggressive rudder inputs. This modification placed the simulator under a preprogrammed maneuver:

> **"Roll the airplane about 10° either left or right and the roll the airplane in the opposite direction past 90° The simulator inhibited the use of the ailerons, spoilers, and rudder until the airplane reach a bank angle of 50° or a maximum bank angle of 50° or a maximum of 10 seconds had elapsed. After the airplane's bank angle reached 50°, yaw and roll control were phased back in during the next 1.3 seconds."**   **from the NTSB report[28]**

Airbus, as well as other manufacturers of transport airplanes, had criticized this modification and recommended that it not be implemented. They also expressed concerned about the types of maneuvers being flown in simulators and the conclusions that were being drawn from them.‡

---

*   The chapter title on the American 587 accident from <u>Why Planes Crash. Case Files 2001</u> by Sylvia Wrigley.[27]

†   This accident is described in Appendix 4 on page 374.

‡   Section 1.17.1.2.5 in the American Flight 587 accident report.[28]

New Regulation: Part 60

In 2006, a new rule was proposed: *Flight Simulation Training Device Initial and Continuing Qualification and Use*.[29]* The new part is extremely thorough in terms of the process of obtaining simulator data but is vague on the actual data requirements. It seems to us, with a background in certification from a flight test background, the main effect of the Part 60 will be documentation of any changes without meaningful technical requirements on the data.

We are concerned with the feeling that we can use ground-based simulation for any and all aspects of pilot training. While simulators can be extremely valuable for procedural training, we must remember that they are not sufficient for some areas – such as replicating the effects of highly dynamic motion cues.

Supporting Data

The data to support a Flight Simulation Training Device (FTSD) must include aircraft manufacturer flight test data. It may or may not contain additional flight test data. The qualification requirements for additional data are quite vague. However the procedures to obtain this data and the qualifications of the test personnel are vague. There is no discussion of resolving conflicting data from different sources. The simulator data package may also contain extrapolated or predicted data. There is no mention of validating such extrapolation.

Validation of Simulator Data: The performance and handling qualities of the FTSD must be validated by pilots qualified in the aircraft involved. Presumably, this could mean qualified line pilots. There is no mention of flight test qualifications of such pilots.

Lack of Motion Cues: The major limitation of any ground-based simulator is the generation of motion cues to the flight crew. Typically, such simulators can briefly apply short duration forces by small servo-driven motion with a washout. They can simulate sustained forces only by tilting the cockpit. While there are obvious limitations to these techniques, ground-based simulators can be acceptable for most tasks within the normal operating envelope.

The upshot of the creation of Part 60 is standardization in approving simulators or modifications to existing simulators. The modification to the American Airlines simulators discovered following the Flight 587 accident would likely be flagged under Part 60.

Rudder Use Survey

The FAA circulated a questionnaire to transport pilot intended to document pilots' experiences and understanding of lateral/directional issues.[30] Several observations were made:

1) Rudder inputs are reported to be more common than expected.
2) Rudder is used in ways not always trained.
3) Erroneous and accidental inputs have occurred.
4) Some pilots report making rudder reversals.
5) Some pilots are not clear on appropriate use of rudder – many felt they needed more training.
6) Wake vortices were reported to be the most common initiator of upsets; most commonly during approach.
7) Pilots were not concerned about differences between control systems across aircraft.

Clearly better training is needed. The report also recommended incorporating more error-tolerant designs in control systems.

Another Alternating Rudder Upset

On January 10, 2008, an Air Canada A319 was enroute between Victoria, BC, and Toronto, ON when, while climbing through FL366, it encountered the wake of a B-747 at FL 370.† During the recovery, pilot rudder and sidestick control inputs resulted in vertical accelerations between -0.77

---

* The NPRM makes no reference to the American 587 accident, but the timing suggests otherwise.
† This accident is described in Appendix 4 on page 375.

and +1.57 g and lateral accelerations between 0.49 (R) and 0.46 (L). The crew used extensive rudder inputs which were out of phase with the yaw oscillations.

While crew had been exposed to training regarding cyclic control inputs, it appeared that it was inadequate.

Limitation of Existing Training Simulators

The data requirements for FTSDs make them suitable for training conducted within the normal flight envelope and absent large dynamic motion. Unfortunately, their use has become almost the only way pilots are trained. In fact, many copilots have virtually all of their "flight" experience in simulators before they are hired by an airline. Their first flight in a transport airplane might well be with paying passengers on board.

In a 2013 report,[31] the BEA studied a number of accidents, incidents, and serious incidents they described as loss of "aeroplane state awareness during go-around" (ASAGA). A common theme of the these mishaps was high gravito-inertial forces on the pilots during all-engine go-arounds, particularly after a long-range flight. This created an illusion, called the somatogravic illusion, in which the direction of the gravito-inertial forces is perceived as vertical. This leads to the illusion that the airplane is tilting up.

Twin engine airplanes have greater thrust with all-engines operating than three or four engine airplanes. This is particularly apparent at the end of a long-range flight where the airplane is much lighter than usual. The BEA identified this as an issue because most flight training go-arounds are made with one engine inoperative. This was confirmed in a later paper.[32]

The major limitation with relying solely on ground based simulators is the inability to adequately replicate highly dynamic flight situations, such as in Loss of Control Accidents, or high energy go-arounds (ASAGA). This leads to the next section on Loss of Control accidents.

Loss of Control Accidents

Loss of Control (LOC) have become the predominant transport airplane accident cause, particularly with the successful reduction in Controlled Flight Into Terrain (CFIT). During the period 2100-2020, there were eight fatal LOC accidents worldwide with 695 fatalities.[33] By comparison, the second-most accident category during the same period was CFIT with fatal accidents with 229 fatalities worldwide.

Case Study: The Pittsburgh Scenario

A case in point was the B-737 accident at Aliquippa, PA in September 1994. While on approach to Pittsburgh, the airplane departed controlled flight and crashed with 132 fatalities.*

The NTSB determined that the probable cause was a loss of control of the airplane resulting from the movement of the rudder surface to its blow down limit. The rudder surface most likely deflected in a direction opposite to that commanded by the pilots as a result of a jam of the main rudder power control unit servo valve secondary slide to the servo valve housing offset from its neutral position and over travel of the primary slide.[34]

Much effort was spent developing training scenarios to train pilots in the event the problem recurred. One of your authors (RLN) has flown *The Pittsburgh Scenario* many times in simulators (like many other pilots). Subjectively, this became a very mechanical exercise: the airplane begins to roll, leading to full opposite aileron and rudder with no effect. The drill was then to reduce the AOA to make the ailerons more effective and thus be able to recover.

---

* This accident is described in Appendix 4 on page 301.

Later, we had the occasion to fly the same scenario in the Calspan variable stability *Learjet*.[35] This is an In-Flight Simulator (IFS) that replicates the response of another aircraft (the simulated aircraft). Instead of using hydraulic jacks to provide motion cues, it flies the *Learjet* IFS through the same trajectory as the aircraft being simulated which provide the same gravito-inertial cues as that aircraft. This allows the in-flight simulator to be used in highly dynamic scenarios, such as training for unusual attitudes or upset events requiring adoption of an alternative control strategy.

Figure 69 Calspan Variable Stability *Learjet* Inflight Simulator*

Prior to flying the *Learjet* IFS, we flew a preliminary sortie in an acrobatic *Bonanza* to refresh our subjective sensation with g-forces. Unlike most transports, this *Bonanza* had a g-meter installed to help us calibrate our senses. We also practiced recovery from a nose-low spiral by reducing the g's by lowering the nose, which to one not practiced in these situations might appear counterintuitive. The instructor used a mantra of "half g, roll wings level, two g's." At this point, we realized that we had not been able to unload to ½ g in previous simulator exercises since we had neither the feel of the reduced g nor a g-meter in the simulator. We also appreciated the fact that, as an aeronautical engineer, we knew all this intellectually, we hadn't really applied it to flying the airplane. Parenthetically, fighter pilots are quite aware of the benefit of reduced g on rolling performance because of practice.

During the training sortie in the *Learjet*, we were given many different upsets, usually based on actual operational accidents or incidents. For each scenario, we were instructed to verbally call our assessment. One event appeared to be *The Pittsburgh Scenario*, but which was actually an aileron reversal. When that event happened, the learning effect of all the previous simulator experiences led to our reducing the AOA. The result was an instant roll to an inverted attitude. After several uncomplimentary comments from the safety pilot, we continued with the exercise.

Sidebar: Definition of "Loss of Control"

The definition of loss of control is often confused with "upset" or "unusual attitude." These terms need to be defined unequivocally.

The first definition should be a departure from the parameters normally experienced in line operations or training: Pitch attitude more than 25° nose-up; Pitch attitude more than 10° nose-down; Bank angle greater than 45°; or Airspeed inappropriate for the conditions. Various authorities have called such a condition an "Airplane Upset";[36][37][38] or a "Jet Upset".[39] We dislike the term "Jet Upset" since some of the memorable accidents involved Turboprop airplanes.* We would prefer to use "Unusual Attitude."

Unusual Attitude is an aircraft in flight unintentionally exceeding the parameters normally experienced in routine operations, typically:

+·· Pitch attitude more than 25° aircraft nose up (ANU);
+·· Pitch attitude less than 10° aircraft nose down (AND);
+·· Bank angle greater than 45 deg; or
+·· Airspeeds inappropriate for the conditions.

We also recognize that different aircraft and altitudes can have a major effect. For example, Lambregts et al.[40] pointed out that a 10° nose down attitude at 500 ft overwater at night or a 45° bank at FL 390 would be extremely problematic.

Unusual attitude training ensures that the pilot is able to quickly use his instruments to recognize the situation and make proper recovery control inputs. This is different than loss of control.

Lost of control definitions are often circular, such as "accidents where the crew was unable to maintain control". The US Navy describe out of control as "when unexpected results occur from normal control inputs."[41] We would paraphrase this as

Loss-of-Control: The point when normal control inputs produce unexpected-aircraft responses.

Finally, combining the two, we have"

Aircraft Upset Event: The situation when normal control inputs produce unexpected aircraft responses and the aircraft enters or is already in an Unusual Attitude.

---

*   Such as the Roselawn loss of control accidents.

When we finally experienced the rudder hardover, the lateral g-forces were very large, Our verbal call was "engine failure."* The lateral forces were very large. Pilots in multi-engine airplanes seldom feel large lateral forces except following an engine failure. The large lateral g-force led us to the "engine failure" call and to use asymmetric thrust to balance the lateral force. (Parenthetically, when we returned to the office, we were criticized by our co-workers for using the wrong recovery.)

Gawron in her review of the training states that the only pilots who recovered from the Pittsburgh Scenario used asymmetric thrust.[42] It is worth noting that NTSB records shown only one line pilot experienced such an upset and recovered -- using asymmetric thrust.†

These observations were clearly different from previous simulator effects in these scenarios. Ground-based training simulators can only produce brief forces of limited magnitude. We realized what had happened and split the throttles to aid our application of rudder. This technique uses one engine to balance the yaw caused by the rudder hardover.

How Should We Train For Upset Events?

An ideal simulator should (1) have an accurate representation of flying like the real airplane; (2) have the same cockpit presentation and controls that are in the real airplane, and (3) provide the same gravito-inertial cues as the real airplane. Typical ground-based training simulators generally have (1) and (2); although we have experienced training simulators examples that did not have the same cockpit presentation and secondary controls as the real airplane.

No ground-based training simulator has the motion capability to replicate the motion cues which are important to training pilots for Upset Event Recovery (UER). The options, listed in increasing cost, are:

➤ Do nothing; continue to use existing training simulators.

➤ Continue to use existing training simulators but augment with classroom training to make pilots aware of these differences.

> This is unlikely to work well. Using classroom training for hypoxia training doesn't work well. Pilots need to actually experience the physiological experience. This also applies to spatial disorientation training.

➤ Use In-Flight Simulators for initial UER training, and existing training simulators for recurrent UER training.

> This is likely to be the best approach.

➤ Use line aircraft UER training.
Use In-Flight Simulators for all UER training,

> This is likely to be more expensive.

➤ Use a research simulator, such as the NASA Vertical Motion Simulator (VMS), for all UER training.

> Cost and scheduling issues rule out using existing research simulators with large motion capabilities. These are useful for research but not practical for training organizations.

In our opinion, the best approach would be to use an IFS for initial UER training and use conventional simulators for recurrent training. This would require an assessment to determine the retention of the IFS training. A sample of IFS students should be evaluated at intervals to determine the retention of the training. The initial assessment could be made in short order evaluating those pilots who have already completed their IFS UER training.

---

* We were supposed to make a verbal call as to our first assessment of the problem.

† Eastwind Airlines upset incident June 9, 1996 near Richmond, VA. This accident is described in Appendix 4 on page 302.

Uncontrolled Fire

Historical Air Carrier Fire Accidents

Several inflight fire accidents to air carrier flights have been listed in Appendix 4.* One thing that appears in these accidents is the fairly short time the flight crew has to land safety. Table 21 shows the times between the discovery of the fire and the crash. Based on this sample, the available time is less than twenty minutes. Even if we disregard the Varig and ValuJet accidents, the average time is still under 25 minutes.

Table 21 Time Available to Extinguish an Uncontrolled Fire

Date	Mishap	Source	Time Fire Detected	Time of Crash	Elapsed Time
07/11/1973	VRG 820	Lavatory	13:58	14:03	~5 min
0/02/1983	ACA 797	Lavatory	23:00	23:20	~20 min
11/28/1987	SAA 295	Unknown	23:45	00:07	~22 min
05/11/1996	VJC 592	HAZMAT	18:10	18:14	~4 min
09/02/1998	SWR 111	Electrical	01:13	01:31	~18 min
09/03/2010	UPS 006	HAZMAT[a]	15:13	15:41	~28 min
07/28/2011	AAR 991	HAZMAT[a]	03:53	04:11	~18 min

Note: (a) Lithium-Ion batteries suspected, but not proven.

What did we learn from these accidents?

Changes Following Varig 820 Accident

In the Varig 820 accident in 1973, we learned that lavatory fire detectors were needed. Unfortunately, we still hadn't accomplished this until after the Air Canada fire in 1983.

Changes Following ValuJet 592 Accident

After the ValuJet 592 accident, we converted all Class D cargo compartments to Class C. Class C use fire extinguishers, while Class D does not admit fresh air and thus smothers any fires. Unfortunately, the combustible materials in this accident were chemical oxygen generators. Recent FAA research reports conclude that fire extinguishers might not be effective against some chemical fires.

Changes Following UPS 006 Accident

Lithium batteries were suspected, but not proven, to have caused this fire. To date, all of the changes have involved safety notices with little regulatory action. Many recent fires are caused by lithium batteries. This should be of concern to the industry.

Lithium Batteries

On November 9, 2010, a B-787 prototype experienced an in-flight battery fire during a flight test and diverted to Laredo, Texas. Several electrical problems were experienced and the ram air turbine (RAT) deployed to ensure electrical power to critical systems. The crew evacuated the airplane after landing using deployable slides.†

Since then, the FAA reports 343 lithium battery fires on airline flights reported.[43] Most were either in passenger checked baggage or carry-on articles.

There have been two main or APU lithium battery fires during or following B-787 airline flights. The B-787 Main and APU batteries are the same model.

On January 7, 2013, an incident involving the main battery occurred aboard a 787 airplane operated by Japan Airlines following a flight from Tokyo.‡

On January 16, 2013, an All Nippon Airways flying from Yamaguchi to Tokyo, Japan was forced to divert to Takamatsu Airport (TAK), Takamatsu, Japan, shortly after takeoff.* The

---

* Summaries are shown in Appendix 4 beginning on page 287.
† This incident is described in Appendix 4 on page 295.
‡ This incident is described in Appendix 4 on page 296.

Japan Transport Safety Board (JTSB) investigated this incident with support from the National Transportation Safety Board (NTSB).

Boeing was responsible for the overall integration and certification of the equipment in the 787's electrical power conversion subsystem, which is part of the airplane's electrical power system (EPS). Boeing contracted with Thales Avionics Electrical Systems to design the 787 electrical power conversion subsystem, which includes the main and APU batteries. Thales then subcontracted with various manufacturers for the main and APU battery system components, including GS Yuasa Corporation, which developed, designed, and manufactured the main and APU batteries.

From our perspective, it seems that an American company contracting with a French company who then subcontracts with a Japanese supplier is fraught with problems. The language issues, as well as the logistical issues seem overwhelming.

Boeing was required to demonstrate that the 787's design complied with a Special Condition,[44] which detailed nine specific requirements regarding the use of these batteries on the airplane. As part of the compliance demonstration with these requirements and 14 CFR Part 25 airworthiness standards, Boeing performed a safety assessment to determine the potential hazards that various failure conditions of EPS components could introduce to the airplane and its occupants. Boeing determined that the rate of occurrence of cell venting for the 787 battery would be about 1 in 10 million flight hours. However, at the time of the BOS and TAK incidents (both of which involved cell venting), the in-service 787 fleet had accumulated less than 52,000 flight hours. After the BOS and TAK events and the FAA's subsequent grounding of the US 787 fleet, Boeing modified the 787 main and APU battery design and its installation configuration to include, among other things, a stainless-steel enclosure for the battery case and a duct that vents from the interior of the enclosure to the exterior of the airplane to prevent smoke from entering the occupied space of the airplane. In April 2013, the FAA issued an airworthiness directive mandating the installation of the modified battery aboard US 787 airplanes before they could return to service.

The 787 main and APU battery design was also modified to mitigate the most severe effects of an internal short circuit (that is, cascading, cell-to-cell thermal runaway of other cells within the battery; excessive heat; flammable electrolyte release; and fire). The recommendations resulting from the safety issues identified during this investigation could help prevent such effects from occurring in future battery designs. The NTSB identified the following safety issues as a result of this incident investigation: Cell internal short-circuiting and the potential for thermal runaway of one or more battery cells, fire, explosion, and flammable electrolyte release. This incident involved an uncontrollable increase in temperature and pressure (thermal runaway) of a single APU battery cell as a result of an internal short circuit and the cascading thermal runaway of the other seven cells within the battery. This type of failure was not expected based on the testing and analysis of the main and APU battery that Boeing performed as part of the 787 certification program.

Boeing indicated in certification documents that it used a draft version of FAA Advisory Circular (AC) 25.1309-1B, (the Arsenal draft), as guidance during the 787 certification program.† However, the analysis that Boeing presented in its EPS safety assessment did not appear to be consistent with the guidance in the draft AC. In addition, Boeing and FAA reviews of the EPS safety assessment did not reveal that the assessment had not (1) considered the most severe effects of a cell internal short circuit and (2) included requirements to mitigate related risks. The FAA did not recognize that cascading thermal runaway of the battery could occur as a result of a cell internal short circuit. As a result, FAA certification engineers did not require a thermal runaway test as part of the compliance demonstration (with applicable airworthiness regulations and

---

*   This serious incident is described in Appendix 4 on page 296..
†   No official version of this draft has been published. A version can be found at http://en.wikipedia.org/wiki/AC_25.1309-1#AC_25_1309%E2%80%931B

lithium-ion battery special conditions) for certification of the main and APU battery. Guidance to FAA certification staff at the time that Boeing submitted its application for the 787 type certificate, including FAA Order 8110.4, *Type Certification*, did not clearly indicate how individual special conditions should be traced to compliance deliverables (such as test procedures, test reports, and safety assessments) in a certification plan.

It seems to us that neither Boeing nor the FAA was up to the challenges that certifying these batteries posed. It also seems that having these batteries supplied by a contractor (Thales) in another continent directing a subcontractor (GS Yuasa) in a second continent can be fraught with problems, not to mention each speaking a different language.

Effect on proposed electric airplanes

These challenges will be pushed further with the advent of electric aircraft, which almost certainly will rely on lithium batteries. Even with that, they will certainly be limited in flying time for each sortie.

The first certified electric airplane is the Pipistrel SW-121.[45] We note that its airplane's endurance is 80 minutes, making the maximum legal VFR mission 50 minutes.

Emergency Automatic Landing for Transports

On September 3, 2010, a UPS B-747 experienced an on-board fire during the climb. After takeoff from Dubai, the crew advised air traffic control (ATC) of the main deck fire and that they needed to land as soon as possible. ATC advised the crew that Doha was the closest airport at 100 nm. The captain elected to return to Dubai (at 148 nm). The cockpit became filled with smoke. The captain's oxygen supply was not working and he left his position to get a portable oxygen bottle. He apparently become overcome since there was no further CVR recording from him.

The FO tried to fly by "feel" since he couldn't see any of the instruments or controls. Ultimately, the airplane crashed.

It seems to us that a single *Land at Nearest Suitable Airport* button on the autopilot control would be helpful in this situation. Most transports have such a capability. This would be certified as a "last resort." In the UPS case, pressing the button would have directed them to an airport 48 nm closer which might have produced a better outcome.

Even if the crew could see, the system would get them started in the right direction. A crew in mid-ocean could use such an autoland mode to get started in the right direction. If it could work for a light airplane, it should work for a transport. As in the light airplane version, the system could be turned off at any time.

One final thought could be a *hijack mode*. Instead of directing the flight to the nearest suitable airport, this *hijack mode* would divert to the nearest military airport and squawk the appropriate code. This mode could only be reset on the ground. If the existence of this mode were made public, it might have a deterrent effect on potential hijackers.

System Safety Issues

Pilots today complain that their airplanes are too complicated. Most new transport airplanes require many days, if not weeks, of classroom training followed by hours of simulation and a satisfactory check by a pilot examiner. At this point, they are allowed to fly the airplane under the supervision of a check pilot. Complicated systems make it difficult to know what the outcome of a particular action will be. For pilots, the only possible way to cope is with lots of practice and careful checklists.

The same problem arises with design engineers. It is difficult in complex systems to see intuitively what the outcome of any given action or any individual component failure will be.

With the advent of ballistic missiles in the 1950s, the US Air Force started to develop the concepts of, what is known today as, system safety.[46]

System safety analyses are performed routinely, yet we have seen otherwise capable engineers who didn't understand the principles.* For example, accident report for the faulty AOA signal that sent a QANTAS A330† into a -0.8 g dive incorrectly stated that one such failure in 28 million A330 flight hours world-wide was consistent with the relevant certification requirements.[47] Clearly, the writer of the accident report didn't understand that such a pitch-over must be designed for the worst credible outcome. For a pitch down, that would, of course, be striking the ground. This would be clearly a catastrophic outcome.

In fact, a single failure in a cumulative flight exposure of 28 million flight hours is 3.6×10^{-8}, which is a little over 35 times too large. We should expect no more than one such event in a billion hours, fleet wide ($p < 10^{-9}$).

Lest the reader think treating such a pitch down as having a catastrophic outcome is excessively conservative, consider the B-737(MAX) accidents.‡ Fortunately the QANTAS A330 was flying at FL370 (37,000 ft). In the case of the MAX, the system may have been have been designed considering cruise flight and neglected to consider flight at low altitude.

In October 2000 over the North Atlantic, an Airbus A340 had the stall protection triggered in cruise flight by a gust.[48] § In stall protection mode, sidestick commands alpha directly with neutral sidestick commanding *alpha-prot* (~4.5 deg in this case) When the protection logic triggered the airplane was being controlled by the autopilot, the command immediately cancelled the A/P and the commanded AOA went from cruise (AOA about 2.5 deg) to 4.5 deg. This resulted in a 1.8 g zoom, causing a near miss. When we examined the report in the mid-2000s, we felt that the design implicitly assumed this would only happen in the airport terminal area; why else would you allow a stall prevention measure to *increase* AOA.

The problem seems to be that the system safety engineers did not consider any off-nominal cases in either the B-737(MAX) or the Airbus stall prevention software.

We also saw a lack of proper system safety considerations in the B-787 lithium battery design. We feel that program managers (and certification managers) are inclined to downplay system safety estimates.

References

1   R L Newman, *Pitot-Static Accidents: A Tale of Two Airplanes*, Crew Systems Report TN-16-10, July 2016

2   *Final Aviation Accident Report: Birgenair Flight ALW-301, Puerto Plata, Dominican Republic, February 6, 1996*, Junta Investigadora De Accidentes Aeros, Director General Aeronautics, DGAC-final-report, ALW-301, September 1997

3   *Serious Incident: Boeing 757-258, G-STRZ, Following Departure from Accra, Ghana, January 28, 2009*, AAIB Bulletin 12/2009

4   Bertorelli, P, "Pitot-Static Systems," *Aviation Safety Magazine*, December 2013

5   *Aircraft Accident Report: Northwest Airlines, Inc., Boeing 727-251, N274US, Near Thiells, New York, December 1, 1974*, NTSB AAR-75-13, August 1975

6   *Accident to AFL 327, TU-154, Registration CCCP-85327, on May 21, 1986*, http://aviation-safety.net/database, accessed January 24, 2021

7   *Accident of the Boeing 757-200 Aircraft Operated by Empresa de Transporte Aéreo del Perú S A Aeroperú Property of the Cintra Company with its Head Office in Mexico City, Base of Operations*

---

*   Including a program manager who, after the prototype airplane had flown three sorties totaling seven hours, said, "Well, now we've demonstrated 10^{-9} reliability."

†   This accident is described in Appendix 4 on page 314.

‡   This accident is described in Appendix 4 on page 315.

§   This incident is described in Appendix 4 on page 312.

Lima, Perú, Date: 2 October 1996; Place:: Lima, Perú, Location: Off the Coast of Lima Department, Position: 48 DME Miles to the North-West of the Lima VOR, Time of Impact: 06:11:30 UTC, Equivalent to 01:11:30 Local Time, Accident Investigation Board, Directorate General of Air Transport

8   *Accident to AUT 2553, DC-9-32, Registration, LV-WEG, on October 10, 1997,* http://aviation –safety-net/database, accessed January 24, 2021

9   *Accident to THY 5904, B-737-4Q8, Registration TC-JEP, on April 7, 1999,* http://aviation –safety-net/database, accessed January 24, 2021

10   *Accident to FDX 87, MD-11F, Registration N-581FE, on October 17, 1999,* http://aviation –safety-net/database, accessed January 24, 2021

11   *Factual Report: Registration: TF-FII,* NTSB Factual Report File DCA03IA005, 2005

12   *Aviation Incident Final Report, Boeing 717-200, Registration N-910ME, Union Star, MO,* 05/1/1005, NTSB File NYC05MA083, January 14, 2009

13   *On The Accident On 1st June 2009 To Airbus A-330-203 Registered F-GZCP Operated By Air France Flight AF 447 Rio De Janeiro - Paris,* BEA-f-cp090601-en, July 2012

14   *Serious Incident Investigation Unreliable Airspeed Indication,* UAE GCAA-Final Report NQ. AIFN/0005/2013, Dated September 13, 2015

15   *Serious Incident. Boeing 757-224, N41140, 80 nm Southwest of Dublin, Ireland, 20 October 2013,* AAIU Report 2016-007, May 10, 2016

16   S P Livingston and W Gracey, *Tables of Airspeed, Altitude, and Mach Number Based on Latest International Values for Atmospheric Properties and Physical Constants,* NASA TN-D-822, August 1961

17   *SFTE Reference Handbook,* Lancaster, CA: Society of Flight Test Engineers, 2007

18   R D Kimberlin, *Flight Testing of Fixed-Wing Aircraft,* Reston, VA: American Institute of Aeronautics and Astronautics, 2003

19   P Croskerry, "The importance of cognitive errors in diagnosis and strategies to minimize them," *Academic Medicine*: 78, 2003, 775-780.

20   "Re: Request for Clarification on Whether Certain Persons Onboard Aircraft Would be Considered Passengers," Letter from FAA Office of the Chief Counsel, January 24, 2018

21` "Definitions Related to Air Cargo Operations," Air Cargo Operations, FAA Advisory Circular AC-120-85A, Appendix C, dated 6/25/15,

22   "Carriage of persons without compliance with the passenger carrying requirements of this part", *Operating Requirements: Domestic, Flag, and Supplemental Operations,* 14 CFR Part 121, §121.583, as amended through amendment 121-373, September 30, 2015, effective October 30, 2015

23   "Passengers Prohibited During Special Purpose Operations (CAA Interpretations Which Apply to §8.33)," *Aircraft Airworthiness – Restricted Category,* CAM 8, Second Edition, March 1959, §8.33-1, p 15

24   "Carriage of Persons Without Compliance With the Passenger Carrying Requirements of this Part", *Operating Requirements: Commuter and On Demand Operations and Rules Governing Persons on Board Such Aircraft,* 14 CFR Part 135, §135.85, as amended through amendment 135-88, July 10, 2003

25   "Restricted Category Civil Aircraft: Operating Limitations," *General Operating and Flight Rules,* 14 CFR Part 91, §91.313(d), as amended through amendment 91-346, effective August 30, 2017

26   "Definitions", *Operating Requirements: Commuter and On Demand Operations and Rules Governing Persons on Board Such Aircraft,* 14 CFR Part 135, §135.601(b), as amended through amendment 135-129, February 21, 2014, effective April 22, 2014

27   S Wrigley, *Why Planes Crash – Case Files:2001,* EQP Books, 2013

28   *Aircraft Accident Report. In-Flight Separation of Vertical Stabilizer, American Airlines Flight 587, Airbus Industrie A300-606R, N14053 Belle Harbor, New York, November 12, 2001,* NTSB AAR-04-04, October 26, 2004

29   *Flight Simulation Training Device Initial and Continuing Qualification and Use,* 14 CFR Part 60, amendment 60-1, effective October 30, 2007

30   L S Peterson *et al. An International Survey of Transport Airplane Pilots' Experiences and Perspectives of Lateral/Directional Control Events and Rudder Issues in Transport Airplanes (Rudder Survey),* FAA-AM-10-14, October 2010

31   *Study on Aeroplane State Awareness During Go-Around* , BEA, August 2013

32   R L Newman and A H Rupert, "The Magnitude of the Spatial Disorientation Problem in Transport Airplanes," *Aviation Medicine and Human Performance*: **91**, 2020, 65-70

33 *Statistical Summary of Commercial Jet Airplane Accidents. Worldwide Operations, 1959-2020*, Boeing, https://www.boeing.com/resources/boeingdotcom/company/about_bca/pdf/statsum.pdf, Accessed October 21, 2021

34 *Aircraft Accident Report. Uncontrolled Descent and Collision with Terrain, USAir Flight 427, Boeing 737-300, N513AU Near Aliquippa, Pennsylvania, September 8, 1994*, NTSB AAR-99-01, March 24, 1999, corrected November 4, 1999 and February 16, 2000

35 J Priest, *Research into Reduction of Loss-of-Control Accident Rate in a Meaningful and Measurable Way Through the Use of Innovative Pilot Training Techniques*, AIAA Paper 2014-1005, Presented at AIAA Modeling and Simulation Technologies Conference, National Harbor, MD, January 2014

36 Kochan, J A, Priest, J E, and Moskal, M, "Human Factors Aspects of Upset Recovery Training", *17th Annual European Aviation Safety Seminar*, Flight Safety Foundation and European Regional Airline Association, Warsaw, March 2005

37 *Upset Prevention and Recovery Training*, FAA Advisory Circular AC-120-111, Change 1, January 4, 2017

38 *Guidance Material and Best Practices for the Implementation of Upset Prevention and Recovery Methods*, IATA Publication ISBN-978-92-9252-713-6, 2015

39 M F Belcher and T M Luallen, *Advanced Maneuver & Upset Recovery Training (AN-URT)*, Calspan Report TM-FRG-GEN-0123-R00, April 2011

40 A A Lambregts *et al.*, *Airplane Upsets: Old Problem, New Issues*, AIAA 2008-6867, presented at AIAA Modeling and Simulation Technologies Conference, Honolulu, August 2008

41 *Flight Training Instruction, Out-of-Control Flight, T-45 Strike*, CINATRA P-1216, 2012, Revision 5-12

42 V. J. Gawron, *Airplane Upset Training Evaluation Report*, NASA SR-2002-211405, May 2002

43 *Events with Smoke, Fire, Extreme Heat or Explosion Involving Lithium Batteries*, http://www.faa.gov/hazmat/resources/lithium_batteries/media/Battery_incident_chart.pdf, dated October 29, 2021, accessed November 10, 2021

44 *Boeing Model 787-8 Airplane; Lithium-Ion Battery Installation*, FAA Special Conditions 25-359-SC, November 13, 2007

45 *Type Certificate Date Sheet for Virus SW 121*, EASA TCDS No EASA.A.533, 2021

46 N J Bahr, *System Safety Engineering and Risk Assessment*, New York: Taylor and Francis, 1997

47 *In-Flight Upset, 154 km West of Learmonth, WA, 7 October 2008*, VH-QPA, Airbus A330-303, ATSB Aviation Occurrence Investigation AO-2008-070, 2011, October 2019

48 *Airbus A330 C-GGWD and Airbus A340 TC-JDN, 2 October 2000*, AAIB Bulletin No. 6/2001

Chapter 13:
Proposed UAV Certification

Background and Approach*

The proven record of manned aircraft shows that the industry and regulators know how to design aircraft in terms of structures, systems, propulsion, etc. They also know how to tailor the requirements for different type of aircraft: small light airplanes, helicopters, etc. The requirements grew from the safety problems in the past. The set of regulations grew from nothing to extensive requirements in the 1920s—*based on the accident record.*

We will use the terms "Unmanned Aerial Vehicle" (UAV) to denote the airborne vehicle and "Unmanned Aerial System" (UAS) when discussing the entire system including the ground control station

To develop appropriate requirements for unmanned aircraft mishaps, we should first look at the problems encountered to date in terms of accidents, injuries, and fatalities and then address the problems that exist. The present report has documented such a study of the problems to date for UAVs.

Ground Rules for UAS Airworthiness Requirements.

The following summarize the ground rules† proposed to adapt airworthiness requirements for manned aircraft to unmanned.[3]

1. Adopt the manned aircraft Normal Category requirements (FARs 23 and 27 as appropriate) for UAS
2. Do not consider ownship damage to the UAS in safety assessment
3. Ensure containment of the UAV (i. e., ensure UAV remains within specified airspace or in uncontrolled airspace)
4. Relax the requirements for UAV models below a specified weight
5. Establish hazard and estimate risk to non-participating people on the ground
6. Establish hazard and estimate risk to non-participating manned aircraft (MA)
7. Compare risk to established requirements for FARs 23 and 27
8. If necessary, establish operational rules to achieve safety for people on the ground.

Unmanned Aircraft Accident Review

What should the upper weight limit be for the relaxed standards (item 4 above)? There have been proposals between 55 lb and 300 lb. This weight limit should be harmonized to agree with the accident definition for considering damage in UAV accidents.

This accident definition[4] lists two criteria for determining if an event is an accident or an incident. For both manned and unmanned aircraft, the injury criterion is the same: if there is a serious injury or a fatal injury to any person, the event is an accident. Substantial aircraft damage is considered for MA of any weight and for UAV over 300 lb, MTOW. Under the accident defini-

---

* This chapter has been adapted from a NASA-sponsored report[1] which in turn was based on a paper presented at NATO Meeting STO-MP-265 in Vilnius, Lithuania.[2]

† Ground Rules: Basic or governing principles of conduct in any situation or field of endeavor.[3]

tion, *substantial damage* is based on the damage effect on aircraft airworthiness, *not* on the monetary cost of the damage. Property damage other than aircraft damage is disregarded in determining whether a mishap is an accident.*

Unmanned Aircraft Weight Classes

It is convenient to group UAV into several weight classes. Table 22 shows several classes from the literature. We have made one change, using 300 lb for the lower limit for large UAV. Most references have set this at 330 lb (150 kg). This was done to harmonize the weight classes with the NTSB accident definition.

Table 22 UAV Categories from the Literature

Weight Class	W_{MAX}	V_{MAX}		Kinetic Energy	References
	[lb]	[ft/sec]	[kts]	[ft lb]	
A: Micro-UAV	4.4	010.3	060	2.26×10^4	NATO Micro[5]
B: Mini-UAV	20	146.7	087	2.15×10^5	DOD Group 1[6]
C: Small UAV	55	146.7	085	5.92×10^5	DOD Group 2[6] NASA Category 1[7 8] NATO Small[5]
D: UAV	300[a]	337.8	200	1.71×10^7	CAA Light UA[9]
E: Large UAV	1320	422.3	250	1.18×10^8	DOD Group 3[6]
F: Very Large UAV	No limit	No limit	No limit	No limit	DOD Group 4 or 5[6]
Note: (a) This value was harmonized from 330 lb to agree with NTSB accident criterion.[4]					

Table 23 shows UAV accidents distributed by weight classes for civilian mishaps. The astute reader may note two property damage accidents for Weight Class A in the table.

Table 23 Civilian Operations UAV Mishap Distribution by Weight and Severity

Incidents and Accidents	UAV Weight Class						Total
	A	B	C	D	E	F	
Incidents	72	36	16	16	4	36	180
Property Damage Accidents	2				4	10	16
Injury Accidents	4						4
Fatal Accidents	2	2	1	1	1		7
Total Civilian Operations	80	38	17	17	9	46	207

Effect of Unmanned Aircraft Configuration

Table 24 show configurations for UAV models and the distribution in our mishap sample.

Table 24 Configuration Distribution of UAV Mishaps

Configuration	Small UAV W≤300 lb	Large UAV W>300 lb	Total
Fixed-Wing (FW)	56	53	109
Multi-Rotor (MR)	77		77
Rotary-Wing (RW)	12	2	14
Tilt-Rotor (TR)	5		5
Thrust Vector	1		1
Not Reported	1		1
Total	152	55	207

The following tables show the primary causes for the various mishaps as well as the types of outcomes. Table 25 summarizes primary mishap cause for unmanned aircraft populations and Table 26 summarizes the mishap results for each UAV population.

---

*   The definition of military accident damage is based on the total dollar amount of the damage to the aircraft and other property.

Table 25 Primary Mishap Cause for Each UAV Population

Primary Mishap Cause	Small UAV W≤300 lb	Large UAV W>300 lb	Total
Lost Link	15	22	37
Flight Crew	18	7	25
Flight Control System	14	2	16
MAC or NMAC	16		16
Propulsion System	10	3	13
Validation/Verification	11	1	12
Sensors	4	1	5
Remote Control Station	2	2	4
Airframe Structure	1	3	4
Other	19	6	25
Undetermined	42	8	50
Total	152	55	207

Table 26 Mishap Results for Each UAV Population

Mishap Result	Small UAV W≤300 lb	Large UAV W>300 lb	Total
Collision with Terrain	21	8	29
Loss of Control	54	5	59
Lost Link	20	18	38
Collision with Obstacle	17	2	19
ATC Event	14	2	16
Near Midair Collision[a]	13	3	16
Flyaway	12	5	17
Collision with Person(s)	13		13
Midair Collision[a]	07		07
Navigation Error	2	2	04
CFIT on Lost Link Profile[a]	3		03

Note (a) Confirmed MAC or NMAC
Included in Lost Link and Collision with Obstacle

Unmanned Aircraft Hazard Categories

Manned aircraft failures were discussed in Chapter 7 in terms of the effect on the aircraft and occupants. We will adapt these definitions for UAS to omit any reference to aircraft occupants or damage to the UAV itself. We will retain effects on crew workload. For UAS, proposed hazard categories are shown in Table 27.

Table 27 Proposed Unmanned Aircraft Hazard Categories

Hazard Categories	Injuries	Safety Margins	Crew Workload
Minor	None	Slight Decrease	Slight Increase
Major	None	Significant Decrease	Significant Increase
Severe Major	Serous Injuries	Large Decrease	Compromises Safety
Catastrophic	Fatalities		

Allowable Hazard Probabilities

Qualitative probabilities for hazards were also discussed in Table 10 in Chapter 7. We will continue to use these same terms for unmanned aircraft: *Probable, Remote, Extremely Report,* and *Extremely Improbable*. We will have to determine how to apply these to unmanned aircraft. The most likely choice is the Part 23 level 1 requirements for fixed-wing UAV models and the Part 27 requirements for MR and RW UAV models. For midair collisions, typical practice is to use the Part 25 or Part 29 requirements.[10] These choices are public policy issues, not technical issues.

Estimated Severity of Unmanned Aircraft Hazards

Using the outcome data in our sample of UAV mishap data shown in, we can create a list of hazards associated with UAV mishaps. These are shown in Table 28.

Table 28 Hazards Associated with Unmanned Aircraft Mishaps

Hazard	Result	Factors
Uncontrolled Descent to Ground	Injury to People on Ground	Mass and Speed of UAV Trajectory of UAV Fragmentation following In-flight Breakup Trajectory of Fragments
	Fire after Impact Injures People on Ground	Mass and Speed of UAV UAV Fuel Type and Quantity
Collision with Terrain of Water Flight Trajectory onto the Surface.	Injury to People on Ground	Mass and Speed of UAV
	Fire after Impact Injures People on Ground	Mass and Speed of UAV UAV Fuel Type and Quantity
Collision with Obstacle Flight Trajectory into an Obstacle or Structure	Fragments Injure People on Ground	Mass and Speed of UAV Altitude Fragmentation Following Collision Trajectory of UAV or any Fragments
	Fire after Impact Injures People on Ground	Mass and Speed of UAV UAV Fuel Type and Quantity
	Damage to Structure	Not Considered-See Note (a)

Collision with Person or Persons
 	
Flight Trajectory into a Person or Group of People
	Injury to People on Ground	Mass and Speed of UAV
		Unguarded Rotor/Propeller
		    Helicopter Rotor Injury
		    Tractor Propeller Injury

Collision with Surface Vehicle Includes Boats or Ships	Injury to People on Ground	Mass and Speed of UAV
	Fire after Impact Injures People on Ground	Mass and Speed of UAV Other Vehicle Fuel Type Other Vehicle Fuel Quantity UAV Fuel Type and Quantity
	Damage to Vehicle	Not Considered-See Note (a)
Accident within Runway Safety Area	Fire after Impact Injures People on Ground	UAV Fuel Type and Quantity People Present (Fire Crew)
	Damage to Vehicle	Not Considered-See Note (a)
MAC or NAMC	Loss of Other Aircraft	Other Aircraft Type Mass of UAV-See Note (b) Relative Speed of UAV and Other Aircraft
Lost-Link or Lost Communications	Midair Collision	UAV Flight Control Programming
Forced Landing or Flight Termination	Injury to People on Ground	Mass and Speed of UAV
Flight Termination Using a Deployed Parachute	Injury to People on Ground	Mass of UAV
Airport Ground Collision Collision with another aircraft or service vehicle on the ground and an airport	Injury to People in Other Aircraft	Other Aircraft Type Mass of UAV Relative Speed
	Fire After Collision Injures People on Ground	Other Aircraft Type UAV Fuel Type and Quantity
Notes (a) Property damage is not normally considered in aircraft certification risk assessment. (b) UAV velocity considered negligible compared to manned aircraft velocity.		

Injury to Person(s) on Ground

To evaluate the relationship between out-of-control aircraft striking and injuring persons on the ground, we can use the methodology used for assessing debris from spacecraft in-flight breakups used by the FAA in assessing launch safety.[11] This requires the risk of injury caused by falling debris.

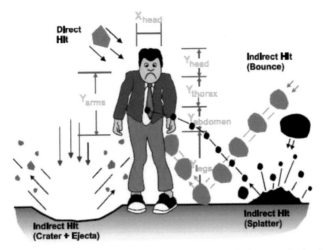

Figure 70 Casualty Producing Events from Debris Impacts in the Open*

For errant UAV models, particularly for the lighter weights, we would normally consider direct hits or secondary (indirect) hits from striking buildings. This approach was used in the investigation of space shuttle *Columbia* accident in 2003.[12 13]

Figure 71 shows the relationship between fatality and non-fatality as a function of speed and mass (kinetic energy) of an object striking a person. Collisions are assessed by the mass of the UAV along the x-axis and the speed on the y-axis. The blue line is a median that represents injuries of an AIS Level 3. The area above the median line indicates where the injuries will occur at AIS level 3 or higher. Conversely, the area below the median line indicates where the injuries, if any are expected to be less than AIS level 3.[14 15]

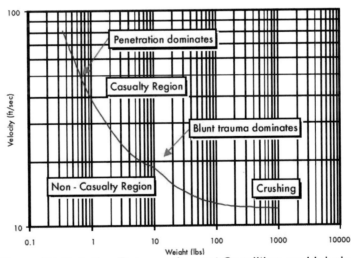

Figure 71 Relation Between Impact Condition and Injuries

The following examples are intended to outline the approach.

This does not mean that all UAV models in the same weight class have the same hazard level. Each individual aircraft must be assessed separately. In the following discussion, when we talk about a UAV weight class, we will be considering the maximum kinetic energy for the class

---

* from FAA *Flight Safety Analysis Handbook*[11]

based on mass and speed. Furthermore, for collisions with persons, we must consider each physical arrangement separately.

Case 1: Uncontrolled Descent to Ground

The type of trajectory for uncontrolled descent to the ground is different for fixed-wing, for multi-rotor, and for helicopter UAV models.

Multi-Rotor UAV Models

In multi-rotor UAV models, the most common uncontrolled descent results from failure or one or more propulsion units. With complete power loss, usually by battery depletion, the UAV simply falls. Recent simulations and flight test research describe the trajectory of various MR UAV models following engine failure as steep trajectories to the ground.[16] Unpublished NASA wind tunnel results indicate that the terminal conditions for a failure at 400 ft would be a velocity of 75 ft/sec and a flight path angle of -70 to -80 deg.*

For all multi-rotor UAV models greater than 4.4 lb, an uncontrolled descent has the potential for fatal injuries. Therefore outcomes are classed as *Severe Major* for Weight Class A and *Catastrophic* for heavier MR UAV models.

Fixed-Wing UAV Models

Typically, a FW UAV will depart in either a stall/spin or in a high-speed spiral dive. FW UAV outcomes are classed as *Severe Major* for Weight Class A, and *Catastrophic* for heavier UAV models.

Rotary-Wing UAV Models

For RW UAV models, uncontrolled descent can be caused by mast-bumping, ring vortex state (power settling or loss-of-tail-rotor effectiveness), or the loss of one or more main rotor blades. Outcomes are *Catastrophic* for all weight classes.

Case 2: Undesired Trajectory into Person or Persons

The causes of the undesired trajectory are common to all types of unmanned aircraft -- Loss of guidance or control systems aboard the UA or pilot error. While this may be referred to as a subset of loss of control. It is actually guidance error caused by the flight crew or by the navigation system. It might also be triggered by loss of control link. These go beyond the vehicle itself and must be considered for the UAS

Multi-Rotor UAV Models

As with the loss of control case, Weight Class A would have a hazard level of *Severe Major*. Heavier MR UAV models have considerably more kinetic energy and the outcome would be *Catastrophic*.

Fixed-Wing UAV Models

Again, we will assume that the UAV collides with a person at the maximum kinetic energy of the UAV class. By the same token, fixed wing UAV models will result in *Severe Major* outcomes. for Weight Class A, and *Catastrophic* for heavier UAV models.

Rotary-Wing UAV Models

For RW UAV models, the chief hazard is in the main rotor. The size and speed are almost certainly lethal. In addition, multiple fatalities are possible. The hazard level for all RW UAV models is *Catastrophic*.

---

* Preliminary results: John Foster (NASA Langley) personal communication November 18, 2018

Flight Termination

This type of outcome includes forced landings and intentional crashes. It also includes parachute deployments. We will consider these separately. Property damage is not considered as a hazard.

Forced Landings/Intentional Crashes

Generally, the hazards to people are similar to "collision with terrain" outcomes and similar hazard assessments can be made.

Parachute Deployment

The deployment of a parachute would reduce the threat to people on the ground reducing the hazard level to *Minor* for Weight Class A UAV models; to *Major* for Weight Classes B and C; and *Severe Major* for Weight Class D UAV models. In fact, deployable parachutes are a significant mitigating strategy for the Uncontrolled Descent to Ground cases.

Midair Collision (MAC)

This type of outcome includes MAC, NMAC, and lost-link cases. While the NMAC and lost-link cases to date have been without injury, the potential outcome is definitely *Catastrophic*.

With the exception of a midair collisions between a 22 lb model aircraft and a *Grob* motorglider,[17] no other midair collision have resulted in loss of the manned aircraft. In 2011, a 450 lb *Shadow* collided with a C-130 *Hercules* which landed safely.[18][19] One could say that we have been quite lucky with respect to midair collisions.

The outcome of a collision with a Class A UAV is likely to be survivable to a transport. Transports are currently required to be capable of surviving a collision with an 8 lb bird. Collisions between heavier UAV models and transports or any UAV with general aviation airplanes or with helicopters are likely to be *Catastrophic*. In spite of this prediction, a recent MAC between a DJI *Phantom* and an Army Sikorsky UH-60M, *Black Hawk* resulted in a safe return to base for the helicopter.[20]

Table 29 shows the summary of the hazard level assessments.

Table 29 Summary of Hazard Level Assessments

Outcome	FW	MR	RW	FW	MR	RW	FW	MR	RW	FW	MR	RW
Uncontrolled Descent	SM	SM	Cat	Cat	Cat	Cat	Cat	Cat	Cat	Cat	Cat	Cat
Collision w/Persons	SM	SM	Cat	Cat	Cat	Cat	Cat	Cat	Cat	Cat	Cat	Cat
Mid-Air Collision	Maj	Maj	Maj	Cat	Cat	Cat	Cat	Cat	Cat	Cat	Cat	Cat
Flight Termination (no parachute)	SM	SM	Cat	Cat	Cat	Cat	Cat	Cat	Cat	Cat	Cat	Cat
Flight Termination (with parachute)	Min	Min	Min	Maj	Maj	Maj	Maj	Maj	Maj	SM	SM	SM

Assessment of Risk

We can now arrive at overall system reliabilities for selected UAV mishaps. We will use the following are generic system failure rates:

- Small Electric Motors:[21] $p \approx \bullet .0 \times 10^{-5}$
- Small Piston Engines:[22] $p \approx 1.5 \times 10^{-4}$
- FCS (*Predator* data):[23] $p \approx 2.5 \times 10^{-4}$

In practice, one would obtain analyze the system with known failure rates using failure modes and effects analysis (FMEA).[24][25]

We are forced to use generic failures and estimate the outcomes from our mishap review. Nonetheless, this will illustrate the methodology.

We have not considered loss of GPS signals, which can be quite significant in urban settings. We were unable to locate any representative data. The loss-of-control conclusions must be tempered with this limitation, which would make urban estimates non-conservative.

Flight Control Failures

The outcome of FCS failures (N = 49) shows:

- Undesired trajectory   20%
- Uncontrolled descent   80%

Propulsion Failures

There was a marked difference with the type of UA. The outcome of propulsion failures (N=110) shows

- Undesired trajectory 100%   (FW)
                           0%   (MR)
                          50%   (RW)
- Uncontrolled Descent   0%   (FW)
                         100%   (MR)
                          50%   (RW)

Probability of Ground Fatalities

We will examine mishaps involving striking persons on the ground as a result of a system fault or failure. Such failures are likely the result of a fault in the flight control system (FCS). These mishaps are classed in Table 28 as *Uncontrolled Descent to Ground*, as *Collision with Person or Persons*, or as *Forced Land*ing. In this review, all fatal UAV to person collisions in our database have involved a UAV striking a person on the head.

Urban (7700 persons/sq mi) and Suburban (2700 persons/sq mi) density figures were obtained from Demographia.[26] Rural population densities were estimated to be ten percent of suburban densities and remote areas were estimated to be 5 persons per square mile. The population density of congested areas was based on an assumed one person per fifteen square feet.

Unmanned Aerial Vehicles

For vertical descents of small objects,* the approximate cross-section of a person is about $A_P = 1.5$ ft^2. For a shallow-angle descent, the approximate cross-section is about $A_P = 11$ ft^2. In either case, the approximate cross-section of a person's head is $A_H = 0.56$ ft^2. The chance of a person being struck by a UAV is shown in Table 30.

What do the data in Table 30 mean? For a vertical trajectory, we would expect that a UAV would strike a person (causing an injury) 10% of the time in a congested area, but only 0.4% in a general urban environment, etc. The figures for striking a person in the head (presumably fatal) would be 3.75% in a congested area and 0.3% in a general urban environment, etc.

Table 30 Fraction of Unmanned Aircraft Striking Persons Excluding Unmanned Helicopters

Population Environment	Density [ft^{-2}]	Steep Trajectory		Shallow Trajectory	
		Striking Head	Striking Body	Striking Head	Striking Body
Congested	0.066667	0.037333	0.100000	0.037333	0.733333
Urban	0.000276	0.000155	0.000414	0.000155	0.003038
Suburban	9.68E-05	0.000054	0.000145	0.000054	0.001065
Rural	9.68E-06	~0	~0	~0	~0
Remote	1.79E-07	~0	~0	~0	~0

---

* This would exclude rotary-wing unmanned aircraft which will be considered separately.

Rotary-Wing UAV Models

For RW-UAV Models, the problem is slightly different. Instead of the UAV striking the person, it only has to strike within the radius of its main rotor. Thus the percent of impacts is greater. Table 31 shows the percent of impacts striking the head or body of a person.

Table 31 Fraction of Unmanned RW Impacts with Rotor Striking Person

Population Environment	Density [ft^{-2}]	Steep Trajectory		Shallow Trajectory	
		Striking Head	Striking Body	Striking Head	Striking Body
Congested	0.066667	~1	~1	~1	~1
Urban	0.000276	0.005800	0.015536	0.005800	0.113932
Suburban	9.68E-05	0.002034	0.005448	0.002034	0.039950
Rural	9.68E-06	0.000203	0.000545	0.000203	0.003995
Remote	1.79E-07	~0	~0	~0	~0

Risk of Striking Person on Ground

Flight Control Failure

Table 32 shows the estimated probability of a UAV striking a person in the head following a flight control system failure resulting in a fatal collision. Note that as the population density becomes less, the probability becomes less and ultimately is not a factor.

Table 32 Predicted Rate of Fatal Collision with Person After FCS Failure

Weight Class	Weight Class A			Weight Classes B - D		
UAV Arrangment	FW	MR	RW	FW	MR	RW
Certification Basis	Part 23	Part 27	Part 27	Part 23	Part 27	Part 27
Hazard Level	SM	SM	Cat	Cat	Cat	Cat
Allowable Probability	1.00E-05	1.00E-07	1.00E-09	1.00E-06	1.00E-09	1.00E-09
Populstion Description						
Congested	9.33E-06	9.33E-06	2.38E-04	9.33E-06	9.33E-06	2.38E-04
Urban	6.90E-08	3.87E-08	1.45E-06	6.90E-08	3.87E-08	1.45E-06
Suburban	2.42E-08	1.36E-08	5.08E-07	2.42E-08	1.36E-08	5.08E-07
Rural	2.42E-09	1.36E-09	5.08E-08	2.42E-09	1.36E-09	5.08E-08
Remote	4.48E-11	2.51E-11	9.42E-10	4.48E-11	2.51E-11	9.42E-10
FCS Failure Rate:	2.50E-04					

Color Code	2.42E-09	Meets Part 23 or 27 criteria as appropriate
	1.36E-09	Data Is of the order of the Part 27 criteria.
	9.33E-06	Does not meet Part 23 criteria
	9.33E-06	Does not meet Part 27 criteria

In Table 32 and subsequent tables, we have indicated results that are marginal for compliance with the level of safety for FARs 23 or 27 (i. e., within a factor of two of the values shown in Table 11, on page 110).

Propulsion System Failure

The next two tables show the estimated probability of a fatal UAV–person collision following a propulsion system failure. Based on these generic failure rates neither MR nor RW unmanned aircraft meet the requirements for overflying congested areas of people, such as at a sports event or while watching a parade. The lack of probabilities for FW models is based on mishap data showing no loss of control following a propulsion failure. All FW aircraft were able to make a controlled forced landing.

Table 33 Predicted Fatal Collision Rate Following Small Electric Motor Failure

Weight Class	Weight Class A			Weight Classes B - D		
UAV Arrangement	FW	MR	RW	FW	MR	RW
Certification Basis	Part 23	Part 27	Part 27	Part 23	Part 27	Part 27
Hazard Level	SM	SM	Cat	Cat	Cat	Cat
Allowable Probability	1.00E-05	1.00E-07	1.00E-09	1.00E-06	1.00E-09	1.00E-09
Population Description						
Congested	no data	1.19E-05	1.49E-06	no data	1.19E-05	1.49E-06
Urban	no data	4.95E-08	6.19E-09	no data	4.95E-08	6.19E-09
Suburban	no data	1.74E-08	2.17E-09	no data	1.74E-08	2.17E-09
Rural (a)	no data	1.74E-09	2.17E-10	no data	1.74E-09	2.17E-10
Remote (a)	no data	3.21E-11	4.02E-12	no data	3.21E-11	4.02E-12
Propulsion Failure Rate	8.00E-05 (electric motors)					
Color Code	1.74E-09 Meets Part 27 criteria as appropriate.					
	1.74E-08 Data Is of the order of the Part 27 criteria.					
	1.19E-05 Does not meet Part 27 criteria.					

Table 34 Predicted Fatal Collision Rate Following Small Piston Engine Failure

Weight Class	Weight Class A			Weight Classes B - D		
UAV Arrangement	FW	MR	RW	FW	MR	RW
Certification Basis	Part 23	Par 27	Part 27	Part 23	Part 27	Part 27
Hazard Level	SM	SM	Cat	Cat	Cat	Cat
Allowable Probability	1.00E-05	1.00E-07	1.00E-09	1.00E-06	1.00E-09	1.00E-09
Population Description						
Congested	no data	2.24E-05	7.13E-05	no data	2.24E-05	7.13E-05
Urban	no data	9.28E-08	4.35E-07	no data	9.28E-08	4.35E-07
Suburban	no data	3.25E-08	1.53E-07	no data	3.25E-08	1.53E-07
Rural	no data	3.25E-09	1.53E-08	no data	3.25E-09	1.53E-08
Remote	no data	6.03E-11	2.82E-10	no data	6.03E-11	2.82E-10
Propulsion Failure Rate	1.50E-04 (small piston engine)					
Color Code	9.28E-08 Meets Part 27 criteria.					
	2.24E-05 Does not meet Part 27 criteria					

Allowable Risk for Midair Collision

Collisions with Transport Airplanes

The assumption is that operations of civil UAV models will be confined to low altitudes, of the order of 400 ft AGL or less. It is also assumed that no operations will be permitted near airports. If unmanned aircraft are to operate outside of these limits, they should be designed and certified as airplanes or rotorcraft with appropriate safeguards in place.*

The likely cause of a collision risk with a transport airplane is a "fly-away" scenario. In this scenario, the unmanned aircraft is no longer being controlled and is flying on an unpredictable, random trajectory. This approach is not new. Weibel and Hansman[27] used a similar analysis.

The transport flies a distance, d, through a particular airspace segment. The transport has a cross-sectional area, A_{XP}, exposed to a midair collision. Weibel and Hansman used the cross-section of a Boeing 757, approximately 560 ft^2 The UAV cross-section is much less and will be treated as a point. Figure 72 shows the geometry.

If we consider an airspace segment 100 NM by 100 NM and 1000 ft thick, we have a volume of 3.69×10^{14} ft^3. The area swept by the transport is 560 ft^2 by 100 NM or 3.40×10^8 ft^3. However, we have to consider that there will be n transports as shown in Figure 73. Thus the combined area swept by all transports will be n times 3.40×10^8 ft^3. We will consider the headway between the two transports as 12-1/2 NM, then n = 8. Thus the probability of a collision is $= 8 \times 3.4 \times 10^8 / (3.69 \times 10^{14})$.

---

* As an acceptable substitute for certification, these aircraft could be operated in segregated airspace, such as Restricted Area.

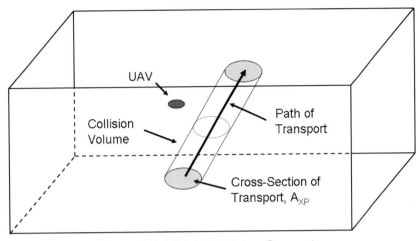

Figure 72 Midair Collision Geometry

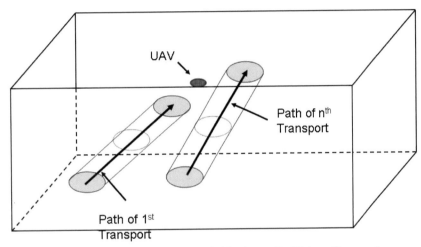

Figure 73 Multiple Transport Airplane Collision Geometry

Thus the combined area swept by all transports will be n times 3.40×10^8 ft^3. We will consider the headway between the two transports as 12-1/2 NM, then n = 8. Thus the probability of a collision is = $8 \times 3.4 \times 10^8 / (3.69 \times 10^{14})$.

The probability of a collision is the ratio of the two volumes or 7.37×10^{-6} based on the time for a transport airplane to fly 100 NM - about 1/5th of an hour. On a per hour basis, the collision probability is $P_{COLL} = 1.47 \times 10^{-8}$ collisions per hour.

This analysis is based on enroute transport operations. The approach for transport operations at airports is similar. The speed will be less (250 knots), but the traffic density is likely to be higher (6 NM headway). The resulting probability for terminal operations is $P_{COLL} = 6.14 \times 10^{-8}$ collisions per hour.

The risk of a "flyaway" resulting in a collision with a transport airplane is $P_{FCSFAIL} \times P_{COLL}$ or $2.50 \times 10^{-4} \times 1.47 \times 10^{-6}$ or 3.69×10^{-10}. The level of safety for a collision with a transport airplane is 1.0×10^{-9}, based on the data in Table 11.* Thus, civil unmanned aircraft appear to meet the required level of safety.

This analysis assumes that the "flyaway" rate equals $P_{FCSFAIL}$ and that every flyaway reaches airspace where transports are flying. These are very conservative assumptions

Collisions with Small Aircraft

We have not analyzed the risk to general aviation aircraft nor to helicopters. This should be performed in the near future. The risk to helicopters is likely to be large. Unmanned Aircraft share the same low altitude airspace (below 400 ft) and small helicopters are not designed to withstand bird strikes.

Proposed UAS Certification Rules

We have attempted to show an approach to approving UAV models for routine operation by adapting the principles of the FARs 23 and 27 safety analyses to the propulsion, control and guidance systems of the UAS and only considering possible human casualties.[28][29] Implicit in this approach is having no human occupants aboard the aircraft.

We envision a three tier system for UAS approval.

First: All UAV over 300 lb, would be certified as Normal Category Aircraft under FARs 23 or 27 as appropriate. Aircraft operated for Research and Development or for Test and Evaluation could operate under an Experimental Approval much as is done today with manned aircraft.

Second: UAS with an AV under 300 lb could be certified as unmanned aircraft under the guidelines in this section.

Third: uncertified UAS qualifying as Model Aircraft could operate in rural or remote areas. They could also operate in dedicated areas in Suburban Airspace on a case by case basis.

Propulsion System

All certified manned aircraft have to demonstrate that an engine failure should not cause loss of control. Even if they are unable to maintain altitude, all aircraft must maintain control. It is our recommendation that the UAV, "must be capable of a controlled emergency landing.† This requirement should be part of any UAS certification.

Guidance and Control Systems

The major remaining issue for approving unmanned aircraft is the guidance and control system. This includes the flight control system, the navigation system, the related sensors, the up- and down-links between the UAS and the control station, and the control station hardware and software.

It is appropriate to approve UAS systems using Technical Standard Order developed for UAS. This would be similar to that for manned aircraft accessories, such as autopilots, navigation systems, etc.

Software

Flight control software can mitigate many flight control problems. Many proposed mitigations have recommended that on-board software be better prepared to deal with issues, such as loss of

---

* In midair collision cases, we should consider the effect on the likely most critical aircraft, as is done during certification of collision avoidance systems or cockpit display of traffic information.[10] Normally, this would be a Transport Category aircraft.

† As described in the EASA proposed rulemaking for VTOL with distributed left/thrust units.[30]

GPS, or have recommended "geofencing" to force unmanned aircraft to avoid certain areas, such as the White House, airports, or high altitude. Current UAS flight control software, even if coded with these functions, does not have the integrity to avoid *Catastrophic*, or even *Severe Major* events. At a minimum, flight control and guidance software should have an appropriate software assurance level.[31] This would be level A or B for fixed-wing and level A for multi-rotor or rotary-wing UAS.

Combination of Mitigations

If the UAS can detect an out-of-control condition, many of the *Catastrophic* outcomes can be mitigated to a *Major* or *Minor* outcomes if the flight is terminated with a parachute system, analogous to the Cirrus ballistic parachute. This is could reduce a *Catastrophic* outcome to a *Severe Major* or even to a *Major* outcome.

Comments on Unmanned Aircraft Configuration

In the preceding discussions, the unmanned rotorcraft (MR or RW) often have higher predicted casualties than fixed-wing aircraft. This is partly because of the aircraft dynamics and partly an effect of the loss of a single-engine for multi-rotor aircraft. Fixed-wing aircraft are generally more benign.

In addition, the astute reader will have noted that the allowable levels of safety are quite different for fixed-wing and for rotary-wing aircraft. Indeed, this is true. Part 23 predecessor requirements have always been relaxed from those for other aircraft types. As a result, small rotorcraft would have to meet a higher safety standard than small fixed-wing aircraft. Holding rotorcraft to a higher standard is a policy decision, not an engineering decision.

References

1   R L Newman, *Review of Military and Civilian UAS Mishaps*, Crew Systems TR-18-08, December 2018

2   R L Newman, "Review of Unmanned Aircraft Mishaps," *Proceedings: Approval of RPAS Operations: Airworthiness, Risk-Based Methods, Operational Limitations*, NATO STO Specialists' Meeting, Vilnius, Lithuania, May 2017

3   Random House Webster's *Unabridged Dictionary*, New York: Random House, Second Edition, 1999, p. 844

4   "Definitions," *Notification and Reporting of Aircraft Accidents or Incidents and Overdue Aircraft and Preservation of Aircraft Wreckage, Mail, Cargo, and Records*, Title 49, Code of Federal Regulations, Part 830.2

5   L Brooke-Holland, *Overview of Military Drones Used by the UK Armed Forces*, UK House of Commons Briefing Paper 06493, June 2015

6   *Unmanned Systems Integrated Roadmap FY2013-2038*, Office of the Secretary of Defense Memorandum, 2013; Reference 14-S-0553

7   J M Maddalon *et al.*, *Perspectives on Unmanned Aircraft Classification for Civil Airworthiness Standards*, NASA TM-2013-217969, February 2013

8   *Aircraft Operations Management*, NASA Procedural Requirement 7900.3C, July 2011; cited by Maddalon *et al. supra*, Reference 7

9   *Unmanned Aircraft System Operations in UK Airspace-- Guidance.* CAA (UK) CAP-722, 6th Edition, March 2015

10 *Airworthiness Approval of Automatic Dependent Surveillance - Broadcast (ADS-B) Out Systems*, FAA AC-20-165A, November 7, 2012

11 *Flight Safety Analysis Handbook*, FAA Office of Commercial Space Transportation, Version 1.0, September 2011

12   "Determination of Debris Risk to the Public Due to the Columbia Breakup During Reentry," *Columbia Accident Investigation Board, Report Volume II*, Appendix D-16, October 2003, based on ACTA, Inc. Technical Report 03-517-01[13]

13   M Y Y Lin, E W F, Larson, and J D Collins, *Determination of Debris Risk to the Public Due to the Columbia Breakup During Reentry*, ACTA, Inc. TR-03-517-01, September 2003

14   W S Copes *et al.*, "Progress in Characterizing Anatomic Injury," *Journal of Trauma*: 31, 1990, 1200-1207

15   S P Baker *et al.*, "The Injury Severity Score: A Method for Describing Patients with Multiple Injuries and Evaluating Emergency Care," *Journal of Trauma and Acute Care Surgery*: 14, 1974, 187-196

16   N Goli *et al.*, "Modeling and Simulation of sUAS Impact Characteristics Following In-Flight Failures," presented at *74th Annual AHS Forum*, Phoenix, May 2018

17   *Collision between Grob G-109B and Model Airplane, near Schopfheim, 3 August 1997*, BFU Report 3X306-1/2/97, March 1999

18   K Feltman, *UAS Accidents* (Unclassified), December 13, 2017

19   M Peck, "New Details Released about 2011 Drone Collision," *Aerospace America*, May 2016, p. 7

20   *National Transportation Safety Board Aviation Incident Final Report*, NTSB File DCA17IA202

21   R E Shafer *et al.*, *Nonelectronic Reliability Notebook*, RADC-TR-85-194, October 1985

22   *Engine Failures and Malfunctions in Light Aeroplanes 2009-2014*, ATSB-AR-2013-107, March 2016

23   *Unmanned Aerial Vehicle Reliability Study*, Office of the Secretary of Defense, February 2003.

24   *Guidelines and Methods for Conducting the Safety Assessment Process on Civil Airborne Systems and Equipment*, SAE ARP-4761, December 1996

25   W Hammer, *Handbook of System and Product Safety*, Englewood Cliff, NJ: Prentice-Hall, 1972

26   *Demographia*, http://www.demographia.com/db-intlsub.htm, downloaded March 29. 2016 at 1728 PT

27   R E Weibel and R J Hansman, *Safety Considerations for Operation of Unmanned Aerial Vehicles in the National Airspace System*, MIT Report ICAT-2005-1, March 2005

28   *System Safety Analysis and Assessment for Part 23 Airplanes*, FAA AC-23.1309-1E, November 17, 2011, Cancels AC-23.1309-1D, dated January 16, 2009

29   "Equipment, Systems, and Installations," *Certification of Normal Category Rotorcraft*, FAA AC-27-1B, Change 8, June 29, 2018

30   *Special Condition. Vertical Take-Off and Landing (VTOL) Aircraft*, EASA Document No. SC-VTOL-01, Issue 1, October 15, 2018

31   *Software Considerations in Airborne Systems and Equipment Certification*, RTCA DO-178, December 1992

Chapter 14:
Other Aircraft Certification Issues

New Technology: General Issues

It may seem counterintuitive, but safety equipment can reduce safety, particularly if the flight crews do not receive clear training material. This point must be emphasized. We need to consider the ability of the average user pilot to use these systems. If the average untrained user pilot has difficulty, then we must determine what training is required prior to introduction into service.

We have been involved with many certification/test programs with many displays and systems. Quite often civilian and military certification engineers miss the point about operational issues.[1][2][3] Further, since many of the "high tech" installations have effects that are not clearly obvious, any approval must use system safety tools to identify non-obvious solutions.

Let us examine these two issues: first, to incorporate system safety tools to identify non-obvious solutions and, second, to have a test philosophy to ensure that user pilots can safely use the systems.

System Safety Philosophy

System safety is a well-known discipline designed to identify non-obvious solutions. The biggest shortcoming seems to us that engineering managers downplay non-obvious results. There also seems to be a tendency not to consider off-nominal situations. For example, a system designed to prevent upsets at cruise altitude should also be examined at, for example, during the landing approach or initial climb and vice versa.

An example of the first case is, of course, the B-737(MAX). It would seem that the issue was a high-altitude loss of pitch stability. In the two instances, the system triggered during climb, shortly after takeoff. The second example is the Airbus alpha-protection case. When reviewing the North Atlantic "zoom" climb of the A340, it seemed to us that the designers were focusing on the approach to landing case where the AOA is slowly increased. In this case, the stall protection was sudden triggering of the stall prevention routine at relatively low AOA. The logic caused the system to command an increase the AOA with a resulting zoom and an altitude bust.

In both cases, somebody should have been asking what happens if it triggers at the wrong time?

Of course, any system safety assessment must be verified by an independent party. (But isn't that what the certification process is all about?)

Systems Testing Philosophy

The essential features of our test and evaluation (T&E) philosophy are easy to state:
- ✈ Clear Definition of Intended Function;
- ✈ Valid Test Scenarios;
- ✈ Valid and Objective Test Completion Criteria; and
- ✈ The Ultimate Tester is the User.

Clear Definition of Intended Function

The intended function should be clearly stated and be clearly defined. Normally, this will be determined by the customer operational requirements. In addition, there are implicit intended functions, such as unusual attitude recognition and recovery or the need for the pilot to monitor the various systems.

Valid Test Scenarios

The flight tasks flown during the test should reflect the intended function description and the customer's requirements. These should be agreed to by the design team, by the customer, and by the certification authority. These flight tasks should be clearly defined and based on operational usage. Both nominal and off-nominal points should be assessed.

Valid and Objective Test Completion Criteria

There should be valid and objective test completion criteria. The use of subjective criteria should be avoided. We feel that objective test completion criteria make the ultimate design specification. The objective criteria must be based on performance measurements. The final tests (and to some extent the important tests) are operational scenarios and situation awareness tests. All tests to that point are for risk reduction or proof of concept.

This is important enough to repeat... The test criteria must be based on objective performance measures.

The Ultimate Tester Is the User

A frequent question is "Test pilots or operational pilots?" We strongly believe that the objective performance measures mentioned above must be achieved by operational pilots–the end users. For this reason, we feel that test pilots should not act as evaluators during operational and situation awareness testing.

Final operational testing must use line pilots. Test pilots should certainly participate in this testing, but as test directors, not as evaluators. If the test pilot community has concerns with aspects of a cockpit system, they need to design test scenarios to address these concerns.

For line pilots, we must use a variety of experience levels and abilities. Newman[4] recommended 6 to 10 operational pilots. These pilots should be current in the mission or a related one and have a range of experience. Low-experience pilots for general aviation were recommended to have 100–200 hours of experience. Low experience pilots for transport copilots were recommended to have 200–500 hours with abilities commensurate with just passing their flight check for their pilot certificate.

The low flight hours for low-experience pilots in transports reflects the entry-level for many foreign carriers using multi-crew pilot certificates (MCP).

A Final Point

It may seem counter intuitive, but safety equipment can reduce safety, particularly if the manufacturers do not develop clear operating procedures or flight crews do not receive clear training material. We have seen this in the results of some advanced display systems in general aviation. This point must be emphasized over and over again.

Common Paragraph Numbering

One of the major advantages of the change from CARs 3, 4b, 6, and 7 to Parts 23 to 29 was the adoption of a common organization and use of identical paragraph numbers. We discussed this earlier in Chapter 1 (page 12) and in Chapter 6 (page 79). This worked well for when initially implemented, particularly for those engineers who worked on certification projects in various categories of aircraft. Then, it gradually deteriorated.

In 2007, the High Intensity Radiated Field rules were promulgated. They were all given paragraph numbers §2x.1317–except for Part 23 which was given §23.1308. Then four years later, the Lightning Rule followed with §2x.1316 used, again except for Part 23 which used §23.1306. We have no idea why Part 23 had to be different or why the FAA lawyers in Washington, DC allowed this to happen.

Then, of course, Amendment 23-64 did away with the previous numbering scheme.

We had an extremely workable organization of airworthiness regulations; why in the world did we abandon it?

Common Definitions For the Same Things

Similarly, there seems to be a lack of standardized in nomenclature. The primary case in point can be seen in the system safety advisory circulars for the various aircraft types.

There are two terms for hazards that will happen several times in the life of a particular aircraft: "Probable" for airplanes,[5][6] and "Frequent" for helicopters.[7][8]

Pick one, please.

The definitions for a catastrophic hazard differ also, but in a bizarre way:

(1) Hazards involving multiple fatalities, loss of the aircraft
   or incapacitation of the flightcrew[5][8]
   for Small airplanes and Transport rotorcraft and
(2) Hazards preventing continued safe flight and landing[6][7]
   Transport airplanes and Small rotorcraft.

Again, pick one, please.

At several points in this book, we have lamented on conflicting definitions. We need consistent definitions. It strikes us that in an industry thought to be rigorous, we are so sloppy in out definitions.

Performance-Based Certification Standards

We have heard the *New Part 23* summarized as "Do good and avoid evil".* The requirements are now written as high-high level safety objectives.

It is now up to the applicant to provide a means of compliance,[9] such as a consensus standard, on how the aircraft will comply. A side effect is that compliance finding will now be much more subjective.

The end result, we believe, will be extremely subjective compliance findings lacking transparency. We do not believe that relying on so-called consensus standards is an acceptable means of providing regulation.

Industry Working Groups

We have served on several industry committees and working groups generating these standards as committee members and as committee chair. We are still active on several now. The individuals who serve on these committees perform a service to the industry.

While these committees do not act in secrecy, neither are they completely open. Some charge registration fees to attend meetings. Some committees are unwilling to allow nonmembers to have any input.† There is no representation from the public at large.

We do not believe that the use of industry consensus standards meet either the intent or the letter of the Administrative Procedures Act (APA).[10] While such standards are available to the public after their adoption, there is generally no opportunity for public comment. In some cases the copies have significant charges or restrictions on their use. In our opinion, to make the *New*

---

* Response when I asked a colleague what he thought of the *New Part 23*. The speaker is employed in the FAA's Aircraft Certification Service and will not be identified.

† In the 1970s, one of us (RLN) was the chairman on an engineering society's committee. We were pursuing a document which would standardize aircraft manuals for small airplanes. Another industry committee was pursuing a similar effort. Our committee decided to terminate our effort and give the other committee access to our files and notes. The answer to receiving our offer was a telephone call: "You [deleted] aren't going to tell us what to do!" This was followed by a disconnection.

Part 23 or any other Performance-Based Rule compliant with the APA, every certification basis would have to be published and public comments solicited much as other regulatory changes are handled.

A typical industry group sells copies of their standards at exorbitant prices. These are frequently sold as encrypted PDF files which may not be copied and may only be printed once. In one case, we had one that could not be printed and could not be transferred to another computer and was lost when the initial computer had to be recycled.

In addition, the approval or disapproval of any such standard may not be known with certainty until the application for a Type Certificate is made and the G-1 Issue Paper finalized.

This is a significant issue for any engineering design. Engineers are accustomed to designing to specifications: customer specifications, company policies, certification standards and certification requirements. In many cases, the customer specifications are the strictest. With the advent of this new approach, the actual certification requirements will not be known until the issuing of the G-1 Issue Paper. This means that the design cannot safely proceed past the Preliminary Design Review before making an application. This places the design team under a strict three-year* time limit.[11]

Existing Rules <u>Are</u> Performance-Based Standards

We already have performance-based standards. The flexibility of consensus standards is not needed. The traditional requirements, FARs 25 through 29 and the Old Part 23 already had enough flexibility. Any applicant could deviate from existing standards by making a case that his alternative was "equally safe" as the published rule. At the same time, the FAA could make a case that the rule did not provide adequate safety and required a special condition. This has worked well for the industry, for both applicants and authorities.

We note that, at this time, *none* of the existing Transport Category airplanes with Fly-by-Wire (FBW) controls meet the stability and control requirements in Part 25. All were certified using a combination of equivalent safety findings or special conditions.

The arguments for "performance-based standards" ignore the fact that existing rules allow an applicant to bypass a rule if it can be shown the proposed design is at least as safe as the rule. This has allowed many innovative designs. At the same time, the certification authorities can rule that the existing rule does not provide adequate safety and propose a special condition. In either case the proposed certification basis, either for equivalent safety or for special condition will be published for public comment before certification.

The *New Part 23* will be especially hard for new entrants to the industry. We have personally heard complaints from new entrant companies about an inability to "tell them exactly what the requirements are." After several of these, we decided to terminate our DER consulting.

Recommendations

We feel that the Certification Authorities must reconsider their drive to "Performance-Based Rules."

References

1   L A Haworth and R L Newman, *Test Techniques for Evaluating Flight Displays*, NASA TM-103947, Army USAATCOM-TR-92-A-006, February 1993

---

*   Five years for transport airplanes.

2   R L Newman and K W Greeley, *Cockpit Displays: Test and Evaluation*, Aldershot, UK: Ashgate, 2001

3   R L Newman, *Display Evaluation Protocol*, Sabbatical Summary, Embry-Riddle Aeronautical University, May 2003

4   R L Newman, *Advanced Display Certification Issues*, Sabbatical Project Report, Operator Performance Laboratory, University of Iowa, May 2003

5   *System Safety Analysis and Assessment for Part 23 Airplanes*, FAA Advisory Circular 23.1309-E, 11/17/2011

6   *System Design and Analysis*, FAA Advisory Circular 25.1309-1A, June 21, 1988

7   "§27.1309 (amendment 27-21) Equipment, Systems, and Installations," *Certification of Normal Category Rotorcraft*, FAA Advisory Circular 27-1B, Change 8, June 29, 2018

8   "§29.1309 (amendment 29-40) Equipment, Systems, and Installations," *Certification of Transport Category Rotorcraft*, FAA Advisory Circular 29-2C, Change 4, May 1, 2014

9   "Accepted Means of Compliance," *Airworthiness Standards: Normal Category Airplanes*, 14 CFR Part23, §23.2010, amendment 23-64, effective August 30, 2017

10   *Administrative Procedures Act*, Public Law 79-404, enacted June 11, 1946

11   "Designation of Applicable Regulations," *Certification Procedures for Products and Articles*, 14 CFR Part 21, §21.17(c), as amended through amendment 21-100, dated December 10, 2018

Chapter 15:
Certification Management

If you think safety is expensive, try an accident.*

Within the FAA, aircraft certification has become very process oriented. There are detailed procedures and organization charts showing who is to approve the data, in what form the data needs to be, and how to document the approval process. There is little guidance on the actual approval criteria. There needs to be substantial criteria on what the data is and how the calculations or analyses are to be carried out. We discussed this earlier, beginning on page 177.

The unfortunate trend is characterized by emphasis on *business* or *management* rather than *engineering*. The effect is an emphasis on schedule and cost, in particular as they apply to certification and manufacturing. We take exception to the idea that engineers aren't concerned about cost or schedule. Some of the more challenging courses in our engineering education dealt with project management: cost and schedule.

The overall effect is to vigorously attack both development schedule and development cost. The undesirable effect is reduced examination of life-cycle costs and downplaying of system safety analysis.

Certification Management: FAA

When your authors joined the FAA in the early 2000s, we were both surprised to see the frequent references to project management. Many certification projects had schedules with due dates over which there was limited control.

One of us (MK) had frequent assignments to shepherd exemption petitions for allowing cargo handlers on cargo flight.† These petitions had to be put in proper format, coordinated with other FAA personnel, published, and comments from the public disposed. Often, required information was needed from the applicant–which was often delayed. If the applicant were to be late with the data, an adverse personnel performance evaluation might result.

As a DER, we (RLN) found our performance was evaluated based on the numbers of approvals. We once pointed out that we had turned down several data submissions because of noncompliance before receiving an acceptable submission. Our FAA contact said, "We're not interested in data rejections–we want to see data approvals!"

It should be clear to most who study certification that it is hard to schedule findings; it is even harder to schedule approvals.

Certification Scheduling

There was no doubt that FAA management had project completion goals. When the project would be completed on-time, bonuses would appear in our paychecks. Most of us were able to resist any pressure to accelerate approvals.

Eclipse Type Certificate Approval
On September 30, 2006, the FAA signed the Type Certificate for the Eclipse EA-500, Very Light Jet (VLJ). The date was interesting, not only was it the last day of the government fiscal year, but

---

* Common saying among safety engineers.

† See the discussion on page 174.

it was a Saturday. At the time, many of us in our office at the FAA commented on the timing and the travel to the Eclipse office in Albuquerque on a Saturday. The consensus was that that the EA-500 was a highly visible FAA project and that one or more officials were in line to receive a large bonus.

Two years later, in Congressional testimony, the Inspector General (IG) examined the processes by which this airplane was certified. The IG stated that FAA's desire to promote the use of VLJs led to many shortcomings.[1] The testimony focused on three main points:

> ✈ "FAA allowed Eclipse to use alternate means of compliance to meet design certification requirements despite unresolved design problems. Users continue to report similar problems after certification;
>
> ✈ "FAA awarded Eclipse a Production Certificate despite known deficiencies in the company's supplier and quality control systems; and
>
> ✈ "FAA's desire to promote the use of VLJs may have contributed to its decision to accelerate the Eclipse Certification process."

The IG cited a mishap involving an 500 when a throttle failure resulted in uncontrollable maximum thrust from one of the engines. After consulting the emergency procedures, the pilot shut the engine down. This caused the second engine to roll back to idle power. Fortunately, the crew was able to perform a flame-out landing on the runway with minor damage and no injuries.*

Certification Management: Industry

In reading the reports of the B-737(MAX)'s introduction into service, we were struck by the lengths to which company management went to cut costs. We couldn't help notice that the company would charge the airlines extra for the AOA indicators and avoided doing anything that might add to training on the type. The additional cost would have been approximately $1,000,000 per airplane.[2] The savings would be approximately 1 percent of the list price. Based on our market estimate of about 5000 airplanes, the total would be five billion dollars.[3]

Certainly safety is expensive. However, to put this into perspective, consider the cost related to the two B-737(MAX) accidents. Reports put Boeing's cost for the two accidents at $19 billion.[4] We'd like to ask Boeing,

"How's that working for you?"

We had seen cost cutting with our clients in our pre-FAA days. However, they were generally receptive to reconsider extreme cost reduction. We were able to convince them that this was a cost of doing business in aviation. The clients who were adamant, we gave contact information for our competitors and parted on good terms. Some came back after interviewing other prospective consultants.

We have also found that new entrant organizations had difficulty in understanding the difference between manufacturing a new product (such as a washing machine) and manufacturing an airplane. Even engineers, experienced in light airplanes, often do not appreciate the extensive testing requirements for a Part 25 airplane. On one Part 25 airplane program, an experienced Part 23 engineer asked "Why did we need perform takeoff tests with and without antiskid, with and without auto-feather, etc.? The short answer was "If we don't test a particular configuration, the operator will not be able to use it in service." – That's still the answer today.

---

* This incident is described in Appendix 4 on page 276.

Conclusion

Having focused on these certification shortcomings, we should offer our opinion on the degree of responsibility of the industry and FAA management. When one of us had to face a similar issue being asked to approve something as a DER that wasn't right, we said that if we approved the form and the FAA actually reviewed it, they would think

> (1) We were careless in our review;
> (2) We didn't understand the requirements, or
> (3) We would sign anything.

We didn't want anybody to think any of these things about us. After all, a signature should mean something.

Within most large organizations, most of the problem seems to lie with the higher management. There is always pressure to compromise an engineering position to match organization policy While management may not be the cause, they are certainly enablers.

It is a terrible thing when professional judgment must give way to organization policy

References

1   C L Scovel, Jr, "FAA's Certification of the Eclipse EA-500 Very Light Jet," *Statement of The Honorable Calvin L Scovel III Inspector General of the US Department of Transportation*, US House of Representatives Committee on Transportation and Infrastructure, September 17, 2008

2   *The Design, Development, and Certification of the Boeing 737 Max*, Final Committee Report of the House of Representatives Committee on Transportation and Infrastructure, September 2020

3   *http://en.wikipedia.org/wiki/Boeing_737_MAX*, accessed February 28, 2022

4   http://www.theguardian.com/business/2020/jan/29/boeing-puts-cost-of-737-max-crashes-at-19bn-as-it-slumps-to-annual-loss, accessed January 11, 2022

Epilogue

"Half a thou* too small. The difference between Right and Wrong. Half a thou bigger, and it'd be Right. As it is, it's Wrong, and you can't cheat about it. Too bad when God gives the mind of an Inspector, isn't it?"†

The use of designees is essential for a smoothly work aerospace industry. The FAA (or other Certification Authority) simply do not have the manpower to adequately handle the engineering workload to adequately review data to arrive at a certification decision. But, how do we ensure that these designees have the engineering background, the regulatory knowledge, and, most importantly, the proper ethical character.

Recent aircraft programs have shown an increasing use of designated organizations. We submit: Organizations do not have ethics; only people do. The current ODA system isolates the certification engineer at the FAA or other agency from the engineer who does the review and approval. This engineer reports to a manager at his company, effectively isolating him from the engineering staff at the agency.

When we first became a DER, it was made quite clear that while we were approving a data for certification, we were working for the FAA, not for our employer or client. Our ACO supervisor said that if any pressure were to be brought to bear, we should immediately report that fact immediately to him.

We believe that while approving certification test plans, data, and designs all certification engineers and pilots should not be subject to any form of management pressure. This applies to both FAA and applicant personnel. It is a terrible thing when organization policies outweigh professional judgment.

We are troubled by recent reports of company certification project managers "hiding certification from the FAA. We are even more troubled by reports of a company technical pilot indicted for "lying to the FAA".

The Society of Experimental Test Pilots (SETP)‡ expects its members to strive to perfect professional integrity, excellence, and accountability:

- Hold paramount the safety, health, and welfare of the end-user, passenger, and public.
- Deliver accurate results and objective conclusions.
- Reject bribery, fraud, and corruption in all forms, including conflicts of interest.
- Properly credit the contributions of others.
- Treat all persons fairly, based solely upon their words and deeds.

As a member of SETP, we have endeavored to apply these principles in our professional life. Many professional societies have similar standards. Do we need one for certification professionals?

We think that we do. We must hold those in aircraft certification to a high standard.

---

* "Thou" equals a thousandth of an inch (0.001 in)
† Shute, N, <u>Round The Bend</u>, New York: William Morrow & Co, 1951
‡ Other engineering professional organizations have similar codes. We are most familiar with SETP's.

APPENDICES

Appendix 1:
Abbreviations and Definitions

Abbreviations

14 CFR	Title 14 of the Code of Federal Regulations (Aeronautics and Space)
49 CFR	Title 49 of the Code of Federal Regulations (Transportation)
A	Category (Part 23)
A/P	Autopilot
AAIB	Air Accident Investigation Board (United Kingdom)
AAIU	Air Accident Investigation Unit (Ireland)
AC	Advisory Circular
ACG	Airline Carrier-Goods (CAR 4)
ACO	Aircraft Certification Office
ACP	Airline Carrier-Passengers (CAR 4)
ACSD	Aviation Child Safety Devices
AD	Airworthiness Directive
ADAC	Avion à Décollage et Atterrissage Court* (Short Takeoff and Landing Aircraft)
ADIRU	Air Data Inertial Reference Unit
ADM	Air Data Module
AEG	Aircraft Evaluation Group
AFM	Airplane Flight Manual
AFML	Aircraft Flight Maintenance Log
AGL	Above Ground Level
AIR	Aircraft Certification Code (FAA organization code)
AIS	Abbreviated Injury Scale
AMC	Acceptable Means of Compliance
AND	Aircraft Nose Down
ANPRM	Advance Notice of Proposed Rule Making
ANU	Aircraft Nose Up
AOA	Angle-of-Attack
APA	Administrative Procedures Act
APU	Auxiliary Power Unit
AR	Authorized Representative
ARAC	Aviation Rulemaking Advisory Committee
ARTCC	Air Route Traffic Control Center
ASAGA	Aeroplane State Awareness During Go-Around
ASD	Accelerate-Stop Distance
ASTM	American Society for Testing and Materials
ATC	(1) Air Traffic Control (2) Approved Type Certificate
ATD	Anthropomorphic Test Device (Dummy)
ATIS	Automatic Terminal Information Service
ATSB	Australian Transport Safety Bureau
AW	Airways
AWC	Airworthiness Certificate
BAA	Bilateral Airworthiness Agreement.
BASA	Bilateral Aviation Safety Agreement.
BCAR	British Civil Airworthiness Requirements

---

* Some have humorously suggested ADAC might stand for Douglas DC-3, known in some countries as as a *Dakota* or *Dak*.

BEA	Bureau d'Enquêtes et d'Analyses pour le sécurité de l'aviation civile (France)
BFU	Bundesstelle für Flugunfalluntersuchung (Germany)
BRS	Ballistic Recovery System
C	(1) Commuter Category (Part 23)
	(2) Celsius or Temperature in Celsius
CA	Certificating Authority
CAA	(1) Civil Aeronautics Authority (USA, 1938-1940)
	(2) Civil Aeronautics Administration (USA, 1940-1966)
	(3) Civil Aviation Authority (generic for other countries)
CAB	Civil Aeronautics Board
CAM	Civil Aeronautics Manual
CAPS	Cirrus Airframe Parachute System
CAR	Civil Air Regulations
CAS	Calibrated Airspeed
CAT III	Category III ILS Approach
CB	Circuit Breaker
CEI	Critical Engine Inoperative
CFR	Code of Federal Regulations
CG	Center of Gravity
CREEP	Acronym for crashworthiness principles.* [1]
CRM	Crew Resource Management
CRP	Cushion Reference Point
CRS	Child Restraint Seat (System)
CS	Certification Specifications (EASA)
CVR	Cockpit Voice Recorder
DA	Density Altitude
DAH	Design Approval Holder
DAR	Designated Airworthiness Representative
DAS	Designated Alteration Station (replaced by ODA)
DER	Designated Engineering Representative
DFDR	Digital Flight Data Recorder
DH	Decision Height
DME	Distance Measuring Equipment
DMIR	Designated Manufacturing Inspection Representative
DOA	Delegation Option Authorization (replaced by ODA)
DOD	Department of Defense (US)
DOT	Department of Transportation (US)
DSCO	Delegation Systems Certification Office
DVE	Degraded Visual Environment
EAPAS	Enhanced Airworthiness Program for Airplane Systems
EASA	European Aviation Safety Agency
ECAM	Electronic Centralized Aircraft Monitoring
ECL	Engine Control Lever
ECO	Engine Certification Office
EDM	Emergency Descent Mode
EFB	Electronic Flight Bag
EGPWS	Enhanced Ground Proximity Warning System
EICAS	Engine Indication and Crew Alerting Systems
ELAC	Elevator Aileron Computer
ELOS	Equivalent Level of Safety
ELSA	Experimental Light Sport Aircraft
ELT	Emergency Locator Transmitter
EMI	Electromagnetic Interference
EOEA	Executive Office of Environmental Affairs
ESF	Equivalent Safety Finding

---

* Stands for Container, Restraint, Environment, Energy, and Post-Crash, see Page 117.

ETOPS	(1) Extended-range Operations
	(2) Extended-range Twin-engine Operations (original term)
ETOPS-N	Extended-Range Operations, with a maximum diversion time of N minutes
ETSO	Technical Standard Order (issued by EASA)
EWIS	Electrical Wiring Interconnection System
F	Fahrenheit or Temperature in Fahrenheit
F&R	Function and Reliability
FAA	(1) Federal Aviation Agency (1958-1966)
	(2) Federal Aviation Administration (since 1966)
FAR	Federal Aviation Regulation
FBW	Fly-by-Wire
FCAA	Foreign Civil Aviation Authority
FCS	Flight Control System
FDR	Flight Data Recorder
FFS:	Full Flight Simulator
FIR	Flight Information Region
FL	Flight Level
FLIR	Forward Looking Infrared
FMEA	Failure Modes and Effects Analysis
FMS	Flight Management System
FOEB	Flight Operations Evaluation Boards
FPA	Flight Path Angle
FSB	Flight Standardization Boards
FSII	Fuel System Icing Inhibitor
FSTD:	Flight Simulation Training Device
FW	Fixed Wing Aircraft
g	Acceleration of Gravity (32.2 ft/sec^2)
G	Glider Category
GA	General Aviation
GAMA	General Aviation Manufacturers' Association
GPS	Global Positioning System
GPWS	Ground Proximity Warning System
H/V	Height/Velocity Diagram
HERF	High Energy Radiated Fields (obsolete term)
HAZMAT	Hazardous Material
HIC	Head Injury Criteria
HIGE	Hover in Ground Effect
HIRF	High Intensity Radiated Fields
HOGE	Hover Out of Ground Effect
HP	High Pressure
HPa	HectoPascal (unit of pressure)
HPC	High Pressure Compressor
HUD	Head-Up Display
IAS	Indicated Airspeed
IATA	International Air Transport Association
ICA	Instructions for Continued Airworthiness
ICAO	International Civil Aviation Organization
IFEN	In-Flight Entertainment Network
IFR	Instrument Flight Rules
IFS	In-Flight Simulation (Simulator)
ILS	Instrument Landing System
IMC	Instrument Meteorological Conditions
IP	(1) Instructor Pilot
	(2) Intermediate Pressure
	(3) Issue Paper
ISA	International Standard Atmosphere
ISA+T	International Standard Atmosphere plus T degrees

JAA	Joint Aviation Authorities
JAR	Joint Airworthiness Requirements
JTSO	Joint Technical Standard Order - Issued by JAA
K	Kelvin or Temperature in Kelvin
KCAS	Knots, Calibrated Airspeed
KIAS	Knots, Indicated Airspeed
KTAS	Knots, True Airspeed
L	(1) Light Airplane Category (Bulletin 7 and CAR 4)
	(2) Limited Category (CAR 9)
lb	Pounds
LBA	Luftfahrt-Bundesamt (national CAA of Germany)
LORAN	Long Range Navigation System
LP	Low Pressure
LSA	Light Sport Aircraft
M/E	Multi-Engine
MAC	Midair Collision
MC	Maximum Continuous Power (Thrust)
MCO	Military Certification Office
METO	Maximum Except Takeoff Power (Thrust)
MIDO	Manufacturing Inspection District Office
MIL	Military Specification (Standard)
MMWR	Millimeter Wave Radar
MOC	Means of Compliance
MR	Multi-Rotor Aircraft
MRB	Maintenance Review Board
MRE	Multiple Reciprocating Engines (Part 23)
MSL	[Height Above] Mean Sea Level
MTE	Multiple Turbine Engines (Part 23)
MTOW	Maximum [Certified] Takeoff Weight
N	Normal Category (Part 23)
N/A	Not Applicable
NAC	Non-Airline Carrier (CAR 4)
NACA	National Advisory Committee for Aeronautics (predecessor to NASA)
NASA	National Aeronautics and Space Administration
NATO	North Atlantic Treaty Organization
Navaid	Navigation aid (such as a radio beacon or a VOR)
NDB	Non-Directional [Radio] Beacon
NGO	Non-Government Organization
NMAC	Near Midair Collision
NORSEE	Non Required Safety Enhancing Equipment
NPRM	Notice of Proposed Rule Making
NR	Normal Category (Rotorcraft)
NSPM	National Simulator Program Manager
NTSB	National Transportation-Safety Board
NVG	Night Vision Goggles
ODA	Organization Designation Authorization
OEI	One Engine Inoperative
OL	Operating Location
OSTIV	Organization Scientifique et Technique Internationale du Vol a Voile
P-S	Pitot-Static
PA	Pressure Altitude
Part 2x	Generic reference for Parts 23, 25, 27, and 29
PC	Production Certificate
PDF	Portable Document Format
PED	Personal Electronic Device
PF	Pilot Flying
PITS	Pathway-in-the-Sky [Display]

PM	Pilot Monitoring
PMA	Parts Manufacturer Approval
PNF	Pilot Not Flying
POB	Persons on Board
psf	Pounds per square foot
QANTAS	Queensland and Northern Territories Air Service
QPS:	Qualification Performance Standards
QRH	Quick Reference Handbook
R	(1) Restricted Category
	(2) Rankine or Temperature in Rankine
R&D	Research and Development
RA	Resolution Advisory (TCAS)
RAE	Royal Aircraft Establishment
RAT	Ram Air Turbine
RCO	Rotorcraft Certification Office
RGL	Regulatory and Guidance Library
RH	Relative Humidity
RNAV	Area Navigation
RNP	Required Navigation Performance
ROC	Rate of Climb
ROD	Rate of Descent
rpm	Revolutions Per Minute
RVSM	Reduced Vertical Separation Minima
RW	Rotary Wing Aircraft
RWSA	Runway Safety Area
S/E	Single-Engine
S/N	Serial Number
SFA	Special Flight Authorization
SFAR	Special Federal Aviation Regulation
SFP	Special Flight Permit
SLD	Supercooled Large Droplets
SLS	Sea Level Standard [Conditions]
SLSA	Special Light Sport Aircraft
SOC	Statement of Compliance
SOD	State of Design
SOM	State of Manufacture
SOR	State of Registration
SR	Special Regulation
SRE	Single Reciprocating Engine (Part 23)
SRP	Seat Reference Point
SRTM	Shuttle Radar Topography Mission
STC	Supplemental Type Certificate
STE	Single Turbine Engine(Part 23)
STOL	Short Takeoff and Landing
T	Transport Category (Part 25)
TA	Traffic Advisory (TCAS)
TAA	Technically (or Technologically) Advanced Aircraft
TAS	True Airspeed
TAWS	Terrain Awareness and Warning System
TC	(1) Type Certificate
	(2) Transport Canada (national CAA of Canada)
TCAS	Traffic Advisory and Collision Avoidance System
TCDS	Type Certificate Datasheet
TDP	Takeoff Decision Point
THS	Trimmable Horizontal Stabilizer
TIA	Type Inspection Authorization
TIR	Type Inspection Report

TO	Takeoff Power (Thrust)
TOD	Takeoff Distance
TOGA	Takeoff/Go Around
TOR	Takeoff Run
TR	Transport Category (Rotorcraft)
TRACON	Terminal Radar Approach Control
TSA	Transportation Security Administration
TSB	Transportation Safety Bureau (Canada)
TSO	Technical Standard Order
TSOA	Technical Standard Order Authorization
TUC	Time of Useful Consciousness
TWA	(1) Transcontinental and Western Airlines (before 1950)
	(2) Trans World Airlines (after 1950)
U	Utility Category (Part 23)
UA	(1) Unmanned Aircraft
	(2) Unusual Attitude
UAR	Unusual Attitude Recovery
UAS	Unmanned Aerial System
UAV	Unmanned Aerial Vehicle
UER	Upset Event Recovery
US	United States
USAF	United States Air Force
UTC	Coordinated Universal Time
VA	Validating Authority
VFR	Visual Flight Rules
VLA	Very Light Aircraft
VLJ	Very Light Jet (Aircraft)
VMC	Visual Meteorological Conditions
VMS	Vertical Motion Simulator (NASA)
VOR	VHF Omni Range (Navigation Radio)
WAT	Range of Weight, Altitude, and Temperature Corrections to Aircraft Performance
WCA	Warning, Caution, and Advisory [Display, System]
WFD	Widespread Fatigue Damage
WSC	Weight Shift Control
Z	Coordinated Universal Time

ICAO and IATA Airline Codes

Airlines and other operators are identified by ICAO and IATA airline codes. Throughout the book, we have referred to airlines using the three letter ICAO code. These are provided for some non-airline operators, such as AIB for Airbus, BOE for Boeing Company, or LAC for Lockheed. These are issued by ICAO upon recommendation of the appropriate national aviation organization (the FAA for US operators). Associated with the designator is a unique phonetic call sign.* Advisory Circular AC-120-26M provides more information.[2]

Two letter IATA codes are generally reserved for airlines and are the same abbreviations used for reservations and airport monitors.†

Table 35 ICAO and IATA Airline Codes

Operator	ICAO	IATA	Call sign	Operator	ICAO	IATA	Call sign
Aeroflot	AFL	SU	Aeroflot	Excellaire Services	XLS		Excellaire
Aeroperú	PLII	PL	Aeroperú	Federal Express	FDX	FX	Fedex
Air Canada	ACA	AC	Air Canada	Flash Airlines	FSH		Flash
Air France	AFR	AF	Airfrans	GOL Transportes Aeréos	GLO	G2	GOL
Air Inter	ITF	IT		Gulfstream	GLF		Gulfstream
Air South	KKB	WV	Khaki Blue	Helios Airways	HCY	ZU	Helios
Airbus Industrie	AIB		Airbus	Iberia Airlines	IBE	IB	Iberia
AirTransat	TSC	TS	AirTransat	Interflug	IFL	IF	
Alas Nacionales	ALW		Alnacional	Lauda Air	LDA	NG	Lauda Air
Alaska Airlines	ASA	AS	Alaska	Lion Air	LNI	JT	Lionsair
Alitalia	AZA	AZ	Alitalia	Lockheed Aircraft	LAC		Lockheed
All Nippon Airlines	ANA	NH	All Nippon	Lufthansa	DLH	LH	Lufthansa
Aloha Airlines	AAH	AQ	Aloha	Malaysian Airlines	MAS	MH	Malaysian
America West	AWE	HP	Cactus	Midwest Airlines	MEP	YX	Midex
American Airlines	AAL	AA	American	Northwest Airlines	NWA	NW	Northwest
American Eagle	EGF	MQ	Eagle Flight	Overseas National AW	ONA	OV	Liberty
Amtrak		2V		Pan American World AW	PAA	PA	Clipper
Astraeus Airways	AEU	5W	Flystar	Perimeter Aviation	PAG	JV	Perimeter
Atlantic Southeast	ASQ	EV	Acey	Petroleum Helicopters	PHM		Petroleum
Birgenair Airlines	BHY		Birgenair	QANTAS	QFA	QF	Qantas
Boeing Company	BOE		Boeing	Singapore	SIA	SQ	Singapore
Braniff	BNF	BN	Braniff	South African Airways	SAA	SA	Springbok
Bristow Helicopters	BHL	UH	Bristow	Southern Cross Aviation	SFS		Ferry
British Airways	BAW	BA	Speedbird	Southwest	SWA	WN	Southwest
British Antarctic Survey	BAN		Penguin	Swissair	SWR	SR	Swissair
China Airlines	CAL	CI	Dynasty	TACA	TAI	TA	Taca
Colgan Air	CJC	9L	Colgan	TAM Linhas Aéreas	TAM	JJ	TAM
Comair	COM	OH	Comair	Trans World Air	TWA	TW	TWA
Continental Connection	CJC	SL	Colgan	TransAsia Airways	TNA	GE	
Delta Air Lines	DAL	DL	Delta	Turkish Airlines	THY	TK	Turkish
Eastern Air Lines	EAL	EA	Eastern	United Air Lines	UAL	UA	United
Eastwind	SGR	W9	Stinger	United Parcel Service	UPS	5X	UPS
El Al	ELY	LY	ELAL	USAir	AEW	US	Cactus
Emery Worldwide	EWW	EB	Emery	ValuJet	VJA	J7	Critter
Ethiopian Airlines	ETH	ET	Ethiopian	Varig	VRG	RG	Varig
Etihad Airways	ETD	EY	Etihad	XL Airways Germany	GXL	X4	Stardust

---

* Our personal favorite call sign is for the British Antarctic Survey, ICAO Code BAS, with the call sign "Penguin."

† It is interesting to note that Amtrak has an IATA code, 2V. This is presumably to allow travel agents to sell tickets for mixed mode travel.

Symbols

a	Acoustic Velocity (Speed of Sound)
G_X	Inertial Resultant of Body Acceleration (positive: eyeballs in)
G_Y	Inertial Resultant of Body Acceleration (positive: eyeballs left)
G_Z	Inertial Resultant of Body Acceleration (positive: eyeballs down)
H	Altitude
H_Z	Geopotential Altitude
H_{TP}	Altitude of Tropopause
M_D	Design Dive Mach
M_{DF}	Demonstrated Dive Mach
M_{FC}	Maximum Mach for Stability Characteristics
M_{MO}	Maximum Operating Mach
n_-	Design Negative Load Factor
n_+	Design Positive Load Factor
P	Atmospheric Pressure
p	Probability of an event
P_{PT}	Total Pressure (measured at aircraft pitot tube
P_{SP}	Static Pressure (measured at aircraft static port
T	Atmospheric Temperature
V_1	Takeoff Engine Failure Decision Speed
V_2	Takeoff Safety Speed
V_A	Design Maneuvering Speed
V_B	Design Speed for Maximum Gust Intensity
V_C	Design Maximum Structural Cruise Speed
V_D	Design Dive Speed
V_{DF}	Demonstrated Dive Speed
V_{EF}	Takeoff Engine Failure Speed
V_{FC}	Maximum Speed for Stability Characteristics
V_H	Maximum Speed in Level Flight with Maximum Continuous Power
V_{MA}	Maneuvering Speed
V_{MC}	Minimum Control Speed with Critical Engine Inoperative (General term)
V_{MCA}	Minimum Control Speed with Critical Engine Inoperative (Airborne)
V_{MCG}	Minimum Control Speed with Critical Engine Inoperative (Ground)
V_{MO}	Maximum Operating Speed
V_{MU}	Minimum Unstick Speed
V_{NE}	Never Exceed Speed
V_{NO}	Maximum Structural Cruising Speed
V_R	Rotation Speed
V_{RA}	Rough Air Speed
V_{REF}	Approach Reference Speed
V_S	Stall speed
V_{S0}	Stall Speed in Landing Configuration
V_{S1}	Stall speed in a specific configuration
V_{SR}	Reference Stall Speed
V_{SR0}	Reference Stall Speed in landing configuration
V_{TOSS}	Takeoff Safety Speed (Category A Rotorcraft)
V_X	Airspeed for Best Angle of Climb
V_Y	Airspeed for Best Rate of Climb
W	Weight

α Angle of Attack

β Angle of Sideslip

χ^2 Chi square statistical measure

γ (1) Flight Path Angle

        (2) Ratio of Specific Heats (C_P/C_V)

λ Temperature lapse rate

θ (1) Aircraft Pitch Attitude

        (2) Atmospheric (absolute) temperature ratio (T_A/T_0)

        (3) Azimuth (in polar coordinates)

ρ (1) Atmospheric density

        (2) Distance (from a navaid)

        (3) Radius in polar coordinates)

Symbol Subscripts

$(\_)_0$ Sea Level Standard Conditions

$(\_)_A$ Ambient Atmospheric Conditions

$(\_)_T$ Total Pressure, Temperature, etc.

$(\_)_S$ Static Pressure, Temperature, etc.

Definitions

337 Approval: See **Field Approval**.

Abbreviated Injury Scale (AIS): An anatomically based. consensus derived, global severity grading system that classifies each injury to every body region according to its relative importance on a 6 point ordinal scale. Using AIS level 3 or greater is appropriate for describing a medical condition sufficiently to allow modeling of casualties for the purpose of determining whether a [space] launch satisfies the public risk criteria.[3]

Abrupt Acceleration: An acceleration of short duration primarily associated with crash impacts, ejection seat shocks, capsule impacts, etc. One second is generally accepted as the dividing point between abrupt and prolonged accelerations.[4]

Acceptable Means of Compliance (AMC): An EASA document containing the text and supporting policy and guidance for a specific CS.

Acrobatic Category: Airplanes intended for use without restrictions. The only operating restrictions will be those indicated by flight test. This Category was established in 1942. Since 1953, new designs are limited to a maximum takeoff weight of 12,500 lb or less. Since 1987, new designs have been limited to nine passengers or fewer. Part 23 Acrobatic Category certifications were discontinued following amendment 23-64 in 2017.[5]

Administrator: The Federal Aviation Administrator or any person to whom he has delegated his authority in the matter concerned. [6]

Advance Notice of Proposed Rule Making (ANPRM): A public notice that a government agency may choose to publish to announce its intention to establish a regulation or amendment, inviting comments from the public. An ANPRM is not required by the *Administrative Procedures Act* (APA).[7]

Advisory Circular (AC): An FAA document containing policy and guidance for complying with specific regulations.

Advisory: The level or category of alert for conditions that require flight crew awareness and may require subsequent flight crew response.[8]

Acrobatic (Aerobatic) Flight: An intentional maneuver involving an abrupt change in an aircraft's attitude, an abnormal attitude, or abnormal acceleration, not necessary for normal flight.*[9]

Aerodynamic Ceiling: The point at which the flight envelope boundary defined by a high incidence stall intersects with that defined by the critical Mach number. In other words, it occurs when, for a specific gross weight and G-force loading, the aircraft has climbed to an altitude where the speed differential between the onset of low speed stall buffet and the onset of high speed Mach buffet approaches zero.[10] Sometimes referred to as the *Coffin Corner* or the *Corner*.

Aircraft Accident: An occurrence associated with the operation of an aircraft which takes place between the time any person boards the aircraft with the intention of flight and all such persons have disembarked, and in which any person suffers death or serious injury, or in which the aircraft receives substantial damage.[11] See **Unmanned Aircraft Accident**.

---

* This rule is often confused with §91.307, "Parachutes and Parachuting," which requires the use of parachutes for maneuvers exceeding 60 degrees of bank or ±30 degrees of pitch, with some exceptions.

Aircraft Category: A grouping of aircraft based upon intended use or operating limitations. Examples include: transport, normal, utility, acrobatic, limited, restricted, and provisional.[6]

Aircraft Engine: An engine that is used or intended to be used for propelling aircraft. It includes turbosuperchargers, appurtenances, and accessories necessary for its functioning, but does not include propellers.

Aircraft Upset Event: The situation where normal control inputs produce unexpected aircraft responses and the aircraft enters or is already in an Unusual Attitude. Also called an Upset Event. Some use the term Jet Upset.*

Aircraft: A vehicle that is used or intended to be used for flight in the air.[6]

Airplane: An engine driven, heavier than air, fixed wing aircraft, that is supported in flight by the dynamic reaction of the air against its wings.[6]

Airworthiness Certificate (AWC): An approval for an individual aircraft indicating that the aircraft is in compliance with its Type Design and is in a safe condition to fly. Aircraft without Type Certificates are issued Permits to Fly or, in the United States, Special Airworthiness Certificates. Sometimes referred to as a Certificate of Airworthiness.

Airworthiness Directive (AD): Maintenance requirements necessary to correct an unsafe condition in an aircraft. They are part of Part 39 and are published in the Federal Register as amendments to §39.13. AD's are legally enforceable rules that apply to aircraft, aircraft engines, propellers, and appliances.[12]

Airworthiness Standards: Regulations promulgated by FAA. In the context of this discussion, the specific regulations dealing with aircraft certification, Parts 21 through 43.

Airworthy Aircraft: An aircraft that conforms to its Type Design and is in a condition for safe flight.[13]

Alert: A flight deck indication meant to attract the attention of and identify to the flight crew a non-normal operational or airplane system condition. Alerts are classified at levels or categories corresponding to Warnings, Cautions, and Advisories. Alert indications also include non-normal range markings (for example, exceedances on instruments and gauges).[8]

Alteration, Major: An alteration not listed in the aircraft, aircraft engine, or propeller specifications that might appreciably affect weight, balance, structural strength, performance, powerplant operation, flight characteristics, or other qualities affecting airworthiness; that is not done according to accepted practices; or that cannot be done by elementary operations.

Alteration, Minor: An alteration other than a major alteration.

Alternate Law (Mode): Flight Control Laws that are degraded from Normal Law (Mode) usually resulting from degraded sensor date, often with some envelope protections removed. See **Normal Law (Mode)** or **Direct Law (Mode)**.

Aviation Child Safety Device (ACSD): A Child Restraint Seat (*q. v.*) approved for use in an aircraft.[14]

Amended Type Certificate: A design approval for an aircraft modification when the applicant demonstrates that the aircraft or other product complies with the appropriate regulations or standards. Only the holder of a TC apply for an amended TC. Other applicants should apply for a Supplemental Type Certificate

---

* We dislike this term since many such accidents involve turboprop airplanes.

Anthropomorphic Test Device (ATD): A test device that simulates the dimensions, weight proportion, and articulation of the human body to a vehicle collision. Sometimes referred to as a Anthropomorphic [Test] Dummy, or as a *crash test dummy*.

Applicant: An individual or organization seeking FAA [or other agency] approval of a specific aircraft, component, or installation.[15]

Apron: A defined area on an airport intended to accommodate aircraft for purposes of loading or unloading passengers or cargo, refueling, parking, or maintenance.[16][17] See **Ramp**.

Attaboy: An interjection to express encouragement, approval, or admiration.* Made famous by a humorous certification the recipient, where one thousand "attaboys" qualifies the recipient "to be a leader of men, work overtime with a smile, explain assorted problems to management, and be looked upon as a local hero".†

Autorotation: The condition of flight during which the main rotor is driven only by aerodynamic forces with no power from the engine.[18]

Aviation Child Safety Device (ACSD): A Child Restraint Seat (q. v.) approved for use in an aircraft.[19]

Ballistic Recovery System (BRS): A recovery system using a rocket to deploy a parachute to recover an out-of-control aircraft.

Balanced Field Length: The length of the runway when the distance to takeoff with an engine failure at the critical point is the same as the distance to accelerate to that critical point, have an engine failure, decide to reject the takeoff and come to a complete stop.

Bilateral Airworthiness Agreement (BAA): Agreement between the aviation safety agencies of two countries to facilitate airworthiness approvals, export, and import of aircraft and related products. BAA's have been replaced by Bilateral Aviation Safety Agreements (BASA's).

Bilateral Aviation Safety Agreement (BASA): Agreement between the aviation safety agencies of two countries for technical cooperation. BASA's normally include implementation procedures for airworthiness. Prior to 1996, Bilateral Airworthiness Agreements (BAA's) were used to facilitate airworthiness approvals, export, and import of aircraft and related products.

Catastrophic Failure Condition: Failure conditions that are expected to result in multiple fatalities, in incapacitation or fatal injury to a flight crew member normally with the loss of the aircraft.

Category A: A Transport Category, multi-engine rotorcraft designed with engine and system isolation features specified in Part 29 and utilizing scheduled takeoff and landing operations under a critical engine failure concept which assures adequate designated surface area and adequate performance capability for continued safe flight in the event of engine failure.[6]

Category B: A Transport Category, single-engine or multi-engine rotorcraft which do not fully meet all Category A standards. Category B rotorcraft have no guaranteed stay up ability in the event of engine failure and unscheduled landing is assumed.[6]

Category: A grouping of aircraft based upon intended use or operating limitations, such as acrobatic, commuter, limited, normal, and transport categories.

---

*   https://www.merriam-webster.com/dictionary/Attaboy, accessed October 18, 2021
†   One "awshit" wipes out all "attaboys" and the recipient has to start all over again.

Caution: The level or category of alert for conditions that require immediate flight crew awareness and a less urgent subsequent flight crew response than a Warning alert.[8]

Certificate of Airworthiness: A document issued by the civil aviation authority of a given country (such as FAA or EASA) when a given aircraft is in conformity with its Type Certificate. It is sometimes called an Airworthiness Certificate.

Certificating Authority (CA): The Certificating Authority is the agency responsible for the original TC. The CA issues the final design approval (TC or STC) for products designed in its own country. The FAA is the CA for aircraft designed in the United States.

Certification Basis: The listing of airworthiness requirements which apply to a particular aircraft certification project.

Certification Standards (CS): Regulations promulgated by EASA. In the context of this discussion, the specific regulations dealing with aircraft certification, CS 21 through CS 29, CS VLA, etc. Some use number designations and some letter designations. These regulations are published with supporting policy and guidance (AMC) similar to the CAM.

Child Restraint Seat (CRS): A device capable of accommodating a child occupant in a sitting or supine position. It is so designed as to diminish the risk of injury to the wearer, in the event of a collision or of abrupt deceleration of the vehicle, by limiting the mobility of the child's body. Sometimes referred to as a Child Restraint System.

Child: For the purposes this book, an individual less than 12 years of age. (See **Infant**.)

Civil Aeronautics Administration (CAA): The aviation authority in the United States from 1938 until replaced by the FAA in 1958. From 1938 until 1940 it was known as the Civil Aeronautics Authority.

Civil Aeronautics Manual (CAM): A document containing the text and supporting policy and guidance for a specific CAR.

Civil Air Regulations (CAR): Regulations promulgated by the Civil Aeronautics Administration (and later by the Federal Aviation Administration). In the context of this discussion, the specific regulations dealing with aircraft certification, CAR 1 through CAR 10.

Civil Aircraft: Aircraft other than Public Aircraft.[6]

Clearway: An area beyond the runway which is unsuitable for takeoff or stopping, but is clear of obstacles and is under the control of the airport.

Code of Federal Regulations (CFR): A codification of the general and permanent rules published in the Federal Register. There are numerous sections, called titles within the CFR. Two titles of interest to aviation are Title 14: Aviation and Space and Title 49: Transportation.

Commuter Category: Multi-engine airplanes intended for regional commercial operations carrying passengers or cargo for hire. Commuter Category airplanes are limited to maneuvers incident to normal flying; carry 19 or fewer passenger seats; and have a maximum takeoff weight of 19,000 lb or less. This Category was established in 1987. Prior to 2012, Commuter Category was restricted to propeller-driven airplanes. Commuter Category certifications were discontinued following amendment 23-64 in 2017.[5]

Complex System: A system requiring structured methods of analysis for a thorough and valid safety assessment. A structured method is very methodical and highly organized. Failure modes and effects, fault tree, and reliability block diagrams are exampled of structured methods.[20]

Condition for Safe Operation: The condition of the aircraft relative to wear and deterioration, such as skin corrosion, wind delamination/crazing, fluid leaks, and tire wear.

Conformity: The aircraft configuration and the engine, propeller, and articles installed are consistent with the drawings, specifications, and other data that are part of the TC. This also includes and STCs, repairs, and alternations incorporated into the aircraft.

Consensus Standard (1): For the purpose of certificating light-sport aircraft, an industry-developed consensus standard that applies to aircraft design, production, and airworthiness. It includes, but is not limited to, standards for aircraft design and performance, required equipment, manufacturer quality assurance systems, production acceptance test procedures, operating instructions, maintenance and inspection procedures, identification and recording of major repairs and major alterations, and continued airworthiness.[6]

Consensus Standard (2): Any standard or modification thereof which
   (a) has been adopted and promulgated by a nationally recognized standards-producing organization under procedures whereby it can be determined by the [Agency] that persons interested and affected by the scope or provisions of the standard have reached sub have reached substantial agreement on its adoption,
   (b) was formulated in a manner which afforded an opportunity for diverse views to be considered, and
   (c) has been designated as such a standard by the Agency after consultation with other appropriate Federal agencies.[21]

Continued Safe Flight and Landing: The capability for continued controlled flight and landing at a suitable airport, possibly using emergency procedures, without requiring exceptional pilot skill or strength. Some airplane damage may be associated with a failure condition or upon landing.[22]

Control Column: An aircraft Control Wheel that is mounted on a column that pivots fore-and aft. Also called a Control Yoke.

Control Stick: An aircraft control normally mounted on the cockpit floor that controls aircraft pitch and bank. Pitch is controlled by a fore-and-aft motion and the bank by a left and right motion.

Control Wheel: An aircraft control that controls aircraft pitch and bank. Pitch is controlled by a fore-and-aft motion of the wheel and bank by a left-and-right rotary motion of the wheel.

Control Yoke: See **Control Column**.

Conventional Cockpit: An aircraft cockpit using analog gauges. Sometimes referred to as Steam Gauges or Round Dials.

Crashworthiness: The ability of an aircraft to protect occupants in the event of a crash.[23]

Crew Member: A person assigned to perform duty in an aircraft during flight time.[6]

Crew Member:
   (1) An individual assigned by an air carrier for the performance of duty on an airplane in flight.[24]
   (2) A person assigned to perform duty in an aircraft during flight time.[6]

Critical Altitude: The maximum altitude at which, it is possible to maintain a specified engine output (power, thrust, etc.).

Cushion Reference Point (CRP): The intersection of the plane of the uncompressed top of the seat bottom cushion with the plane of the uncompressed front of the back cushion. This was originally known as the Seat Reference Point (SRP).[25]

Design Eye Position (DEP): The point, specified by the airframe manufacturer, from which the pilot can view cockpit instrumentation, have adequate external view, and can reach cockpit controls.[26]

Design Eye Reference Point (DERP): The spatial position of the observer's eye relative to the optical axis of a head-up display designated by the HUD Manufacturer.[27][28] Note that the DEP is determined by the airframe manufacturer, while the DERP is determined by the HUD manufacturer.

Designated Airworthiness Representative (DAR): An individual designated by the FAA to perform examination, inspection, and testing services necessary to the issuance of certificates.

Designated Engineering Representative (DER): An individual designated by the FAA to approve technical data for a certification project.

Designated Manufacturing Inspection Representative (DMIR): An individual employed by a holder of a Production Certificate and designated by the FAA to perform examination, inspection, and testing services necessary to the issuance of certificates.

Direct Law (Mode): Flight Control Laws that result in direct matching of control surface position with stick, yoke, or pedal position. See **Normal Law (Mode)** or **Alternate Law (Mode)**.

Electronic Centralized Aircraft Monitor (ECAM): A system on Airbus aircraft for monitoring and displaying engine and aircraft system information to the pilots. In the event of a malfunction, it will display the fault and may also display the appropriate steps of the remedial action.[29]

Emergency Descent Mode (EDM): A system which automatically performs an emergency descent to an altitude which does not necessitate supplementary oxygen when the crew does not respond to warnings concerning cabin altitude.[30] Also referred to as *Automatic Descent Mode* or *Automatic Emergency Descent System*.

Engine Indicating and Crew Alerting Systems (EICAS): An aircraft system for displaying engine parameters and alerting crew to system configuration or faults.[31]

Enhanced Airworthiness Program for Airplane Systems (EAPAS): An FAA program to ensure the airworthiness of aging aircraft fuel systems and wiring. It is a result of the in-flight explosion of TWA 800 on July 17, 1996.

Equivalent Level of Safety (ELOS): A finding by the Certification Authority that literal compliance with a specific airworthiness requirement is impractical and that there are compensating factors that provide an equivalent level of safety. These findings may be documented in an Equivalent Safety Finding (ESF).

Equivalent Safety Finding (ESF): See **Equivalent Level of Safety (ELOS)**.

European Aviation Safety Agency (EASA): The European Union organization responsible for aircraft certification and operations.

Exemption: Immunity. Exemption implies release from some duty, tax, etc. [32]

Extended Operations (ETOPS): An airplane flight operation during which a portion of the flight is conducted beyond a time threshold specified in the operating rules. Cargo aircraft with three or four engines are excluded. Originally, ETOPS was an acronym for Extended Range Twin Operations.*

Failure Condition: The effects on the airplane and its occupants, both direct and consequential, caused or contributed to by one or more failures, considering relevant adverse operational or environmental conditions.

Failure: A loss of function, or a malfunction, of a system or a part thereof.

Federal Aviation Administration (FAA): The aviation authority in the United States since 1958. From 1958 until 1966 it was known as the Federal Aviation Agency. Flight Standards is the branch dealing with operations, pilot certification and operating regulations. Aircraft Certification is the branch dealing with aircraft certification and certification regulations.

Federal Aviation Regulations (FAR): Regulations promulgated by the Federal Aviation Administration. In the context of this book, these are the specific regulations dealing with aircraft certification, FAR 21 through FAR 45. Strictly speaking, the abbreviation FAR should not be used; rather these regulations should be cited as Title 14, Code of Federal Regulations (14 CFR Parts 1-199).

Ferry Permit: See **Special Flight Permit**.

Field Approval: An airworthiness approval for a modification to a single aircraft (identified by its serial number) by an FAA maintenance inspector in lieu of a STC. Field Approvals are sometimes referred to as "337 Approvals" after FAA Form 337 which is used to document major repairs or alterations.

Flightcrew Member: A pilot, flight engineer, or flight navigator assigned to duty in an aircraft during flight time.[6]

Flight Director: A guidance air overlaid on the attitude indicator showing the attitude [or control inputs] required to follow a certain trajectory.[34]

Flight Management System (FMS): An on-board computer system that uses a data base to allow routes to be pre-programmed and fed into the system by means of a data loader. The system is updated with aircraft position by reference to available navigation aids. The most appropriate aids are automatically selected during flight.[35]

Flight Simulation Training Device (FSTD): A Full Flight Simulator or a Flight Training Device.[36]

Flight Training Device (FTD): A replica of aircraft instruments, equipment, panels, and controls in an open flight deck area or an enclosed aircraft flight deck replica. It includes the equipment and computer programs necessary to represent aircraft (or set of aircraft) operations in ground and flight conditions having the full range of capabilities of the systems installed in the device as described in Part 60 of this chapter and the Qualification Performance Standard (QPS) for a specific FTD qualification level.[6]

Flyaway: A loss of control to an unmanned aircraft (UA) when the control signals from the ground stations is interrupted making it impossible to control the UA. The UA may crash or simply fly out of sight.

---

* Some have said, humorously, that ETOPS could stand for Engines Turn or Passengers Swim[33] or, alternatively for cargo flights, as Engines Turn or Packages Sink.

Fuel System Icing Inhibitor (FSII): An fuel additive designed to reduce the condensation of ice in the fuel tanks and line. These additives are added in a small amount, typically of the order of 0.10 to 0.15 percent by volume.[37]

Full Flight Simulator (FFS): A replica of a specific type, make, model, or series aircraft. It includes the equipment and computer programs necessary to represent aircraft operations in ground and flight conditions, a visual system providing an out-of-the-flight deck view, a system that provides cues at least equivalent to those of a three-degree-of-freedom motion system, and has the full range of capabilities of the systems installed in the device as described in Part 60 of this chapter and the Qualification Performance Standards for a specific FFS qualification level.[6]

Function and Reliability (F&R): A "Service Test" to ascertain whether there is reasonable assurance that a given airplane, its components and equipment are reliable and function properly.

General Aviation (GA): Aircraft operations other than airline, air taxi, or military operations.

Geofencing: A feature in a software program that uses the global positioning system (GPS) or other location system to define geographical boundaries for an unmanned aircraft and ensure containment.

Glass Cockpit: An aircraft cockpit using computer screens rather than analog gauges. cf. Technically Advanced Airplane.

Glider: A heavier-than-air aircraft, that is supported in flight by the dynamic reaction of the air against its lifting surfaces and whose free flight does not depend principally on an engine.[6]

Global Positioning System (GPS): A satellite-based navigation system operated by the US Space Force. It uses four (or more) satellites to provide timing information to a receiver. With knowledge of the satellites' position and time signals, the navigation system can solve four (or more) simultaneous equations for three dimensional coordinates and for time.

Gotcha: An unexpected usually disconcerting revelation or catch.*

Gravito-Inertial Forces: The sum of gravitational forces and force resulting from acceleration.

Ground Effect: A usually beneficial influence on aircraft performance that occurs while flying lose to the ground. It results from a reduction in upwash, downwash, and tip vortices, which provide corresponding decrease in induced drag.

Gyroplane: A rotorcraft whose rotors are not engine driven, except for initial starting, but are made to rotate by action of the air when the rotorcraft is moving; and whose means of propulsion, consisting usually of conventional propellers, is independent of the rotor system.[6] These are sometimes called *autogyros*.

Hazard: A potentially unsafe condition resulting from failures, malfunctions, external events, errors, or combinations thereof.

Hazardous Failure Condition: Failure conditions that would reduce the capability of the airplane or the ability of the crew to copy with adverse operating conditions occupants to the extent that there would be a large reduction in safety margins or functional capabilities, higher workload or physical distress such that the crew could not be relied on to perform its tasks accurately or completely, or serious or fatal injury to an occupant other than the flight crew.

---

* https://www.merriam-webster.com/dictionary/gotcha, accessed October 18, 2021

Head Injury Criterion (HIC): A measure of the severity of a head injury caused by the head striking an object in a crash. HIC is determined from dynamic crash tests using an instrumented ATD and measuring the acceleration of the dummy's head as it contacts an object. HIC is determined by the following equation

$$HIC = \left\{ (t_2 - t_1) \left[\frac{1}{t_2 - t_1} \int_{t_1}^{t_2} a(t)\, dt \right]^{2.5} \right\}_{max} \tag{3}$$

where, t_1 is the initial time and t_2 the final integration time, $a(t)$ is the acceleration vs. time curve for the head strike, and t is in seconds, and a is in units of gravity (g).[38]

Head-Up Display (HUD): A display which presents flight control symbols into the pilot's forward field of view. The symbols should be presented at a virtual image focused at optical infinity.[39]

Height-Velocity Diagram: A combination of low height and slow speed (including hover) from which a safe autorotation and landing cannot be made following an engine failure.[40] Sometimes referred to as the *dead man's curve*.

Helicopter: A rotorcraft that, for its horizontal motion, depends principally on its engine driven rotors.

High Energy Radiated Fields (HERF): A now unused alternative for HIRF.

High Intensity Radiated Fields (HIRF): Radiated energy of strength sufficient to cause damage to unprotected electronic circuits.

Hover-in-Ground-Effect (HIGE): A hover close to the surface (usually less than one rotor diameter above the surface) under the influence of ground effect.[18]

Hover-out-of-Ground-Effect (HOGE): A hover greater than one rotor diameter above the surface. Because induced drag is greater when hovering out of ground effect, it takes more power to achieve a hover out of ground effect.[18]

Infant: For the purposes this book, an individual less than 2 years of age. (See **Child.**)

In-Flight Simulation (IFS): The use of a variable-stability airplane as a surrogate for another using the control inputs by the pilots to create the response of the simulated airplane and move the controls of the variable-stability airplane to match that response. The result, from the viewpoint of the pilot, is that her seat moves exactly as it would were she were in the airplane being modeled.

In-Flight Simulator (IFS): The variable-stability airplane used as a surrogate for another.

Instrument Flight Rules (IFR): A set of operating rules to govern flight where the meteorological conditions may not allow for visual operation of aircraft by direct pilot vision.

Issue Paper: A formal document for describing and tracking the resolution of significant issues between the applicant, Certifying Authority (CA) and Validating Authority (VA).

Jet Upset: See **Aircraft Upset Event.***

Joint Airworthiness Authorities (JAA): JAA was an association of the civil aviation authorities of a number of European civil aviation regulating authorities. JAA began in 1970 producing common airworthiness requirements (JAR). In 1987, JAA became Joint Aviation Authorities and expanded its work to include operations and licensing. EASA replaced JAA.

---

* We dislike this term since many such accidents involve turboprop airplanes.

Joint Airworthiness Requirements (JAR): Regulations promulgated by the JAA. In the context of this discussion, the specific regulations dealing with aircraft certification: JAR 21 through JAR 29, JAR-VLA, etc.

Lap Baby (Child): A small child being carried on an adult's lap and not in a separate seat.

Large Aircraft: An aircraft of more than 12,500 lb, maximum certificated takeoff weight.[6]

Light Sport Aircraft (LSA): An fixed-wing aircraft that
  (1) has an MTOW ≤ 1320 lb for land operations or an MTOW ≤ 1430 lb for water operations;
  (2) has a V_H ≤ 120 kts CAS for an airplane or a V_{NE} ≤ 120 kts CAS for a glider;
  (3) has a V_{S1} ≤ 45 kts CAS with the use of lift-enhancing devices at maximum certificated takeoff weight and most critical center-of-gravity;
  (4) has a single reciprocating engine, if powered;
  (5) has a fixed pitch or ground adjustable propeller for an airplane or a fixed or feathering propeller if a glider;
  (6) is unpressurized;
  (7) has a fixed landing gear for a land airplane
       (a glider or a seaplane may have a retractable landing gear).
  There are other restrictions for gyroplanes. There is a significant restriction that the aircraft must have continued to meet these limitations since its original certification. For a complete definition, the reader is directed to Part 1.[6]

Limited Category: Aircraft which are surplus military aircraft. They may not be used to carry passengers or cargo for hire.

LORAN: An acronym for Long Range Navigation. LORAN is a hyperbolic radio navigation. The system used for civil aircraft was LORAN-C which became common in the 1970s. Global Position System (GPS) superseded LORAN in the 1990s for civil aviation. We use the upper case; some use lower case presentation.

Loadmaster: An individual responsible for performing airplane weight and balance calculations, inspecting cargo and pallets, and ensuring that items are secured properly with the provided restraints or supplemental restraints.

Major Failure Condition: Failure conditions which would reduce the capability of the airplane or the ability of the crew to cope with adverse operating conditions to the extent that there would be a significant reduction in safety margins or functional capabilities, a significant increase in crew workload or in conditions impairing crew efficiency, or some a discomfort to the flight crew, or physical distress to passengers or cabin crew, possibly including injuries.

Medical Crew Member: Person or persons with medical training, carried aboard an aircraft during air ambulance or search and rescue operations in order to provide medical care.

Means of Compliance (MOC): Analyses, tests, or inspections used by the applicant to demonstrate compliance with the certification airworthiness standards. MOC include descriptions of methodologies employed, assumptions used in applying the methodologies, and discussions of the procedures used to verify the methodologies. MOC's are documented in Issue Papers.

Minor Failure Condition (1): A failure condition that would not significantly reduce airplane safety and [would] involve crew actions that are within their capabilities. Minor failure conditions may include a slight reduction in safety margins or functional capabilities, a slight increase in crew workload (such as routine flight plan changes), or some physical discomfort to passengers or cabin crew.[26]

Minor Failure Condition (2): A faulty condition, caused by equipment, that has no safety effect on the certificated aircraft (for example, structural integrity, electrical, mechanical, pneumatic, hydraulic, navigation, communication, stability and control, or performance), and is within the pilot's capability and control.[41]

Mishap: An unplanned event or series of events resulting in death, injury, occupational illness, damage to or loss of equipment or property, or damage to the environment.

Must: A verb used in advisory documents to refer to an action that is required by regulation. See **shall** and **should**.

National Simulator Program Manager (NSPM): The FAA manager responsible for the overall administration and direction of the National Simulator Program (NSP), or a person approved by that FAA manager.[36]

Navaid: A ground radio aid allowing for radio navigations, such as a VOR or a radio beacon.

Navigation Display (ND): A display showing a graphical presentation of the geographic location and path of the aircraft.

Normal Category Airplane: An airplane intended for non-acrobatic, nonscheduled passenger and nonscheduled cargo operations and with a certification basis prior to August 30, 2017; or Any airplane certified under Part 23 with certification basis subsequent to August 30, 2017.

Normal Category Rotorcraft: A rotorcraft with MTOW limited to 6,000 lbs and limited to nine passenger sears. For rotorcraft with a certification basis subsequent to October 18, 1999, MTOW is limited to 7,000 lb.

Normal Law (Mode): Flight Control Laws that are the intended function of design control response and envelope protection. See **Alternate Law (Mode)** or **Direct Law (Mode)**.

Notice of Proposed Rule Making (NPRM): A public notice when a government agency wishes to add, remove or amend a rule or regulation as part of the rulemaking process. An NPRM is required by the Administrative Procedures Act (APA).[7]

Organization Designation Authorization (ODA): An organization authorized to perform specified functions on behalf of the FAA related to engineering, manufacturing, operations, airworthiness, or maintenance.

Ownship: During discussion of risk to an aircraft, ownship refers to risk to that particular aircraft and not to other aircraft.

Passenger: An occupant of the aircraft in flight other than a crew member.[42][43]*

Permit to Fly: An approval for an individual aircraft which either is not in compliance with its Type Design or which does not have an approved Type Design, but which is in a safe condition for flight and is allowed to fly for specific purposes, such as flight test or exhibition. This term is used in some countries. See **Special Airworthiness Certificate**.

Primary Flight Display (PFD): The display presenting information sufficient to maneuver the aircraft about all three axes, control its flight path, and accomplish a mission segment.[44]

Production Certificate (PC): An approval by the FAA or other approval authority to manufacture or alter an aircraft, aircraft engine, or a propeller.

---

* It is not clear why this definition was dropped from the initial version 14CFRPart 1[6] as the intent was to only change the arrangement, not add or subtract content.

Public Aircraft: (From legislation and Advisory Circulars)

Any of the following:[45] [46]

(a) Except with respect to an aircraft described in subparagraph (e), an aircraft used only for the United States Government, except as provided in §40125(b).

(b) An aircraft owned by the Government and operated by any person for purposes related to crew training, equipment development, or demonstration, except as provided in §40125(b).

(c) An aircraft owned and operated by the government of a State, the District of Columbia, or a territory or possession of the United States or a political subdivision of one of these governments, except as provided in §40125(b).

(d) An aircraft exclusively leased for at least 90 continuous days by the government of a State, the District of Columbia, or a territory or possession of the United States or a political subdivision of one of these governments, except as provided in § 40125(b).

(e) An aircraft owned or operated by the armed forces or chartered to provide transportation or other commercial air service to the armed forces under the conditions specified by §40125(c). In the preceding sentence, the term "other commercial air service" means an aircraft operation that

(i) Is within the United States territorial airspace;

(ii) The Administrator of the Federal Aviation Administration determines is available for compensation or hire to the public, and

(iii) Must comply with all applicable civil aircraft rules under Title 14, Code of Federal Regulations.

Public Aircraft: (From FAR definition))

Any of the following aircraft when not being used for a commercial purpose or to carry an individual other than a crewmember or qualified non-crewmenber:[6]

(1) An aircraft used only for the United States Government; an aircraft owned by the Government and operated by any person for purposes related to crew training, equipment development, or demonstration; an aircraft owned and operated by the government of a State, the District of Columbia, or a territory or possession of the United States or a political subdivision of one of these governments; or an aircraft exclusively leased for at least 90 continuous days by the government of a State, the District of Columbia, or a territory or possession of the United States or a political subdivision of one of these governments.

(i) For the sole purpose of determining public aircraft status, commercial purposes means the transportation of persons or property for compensation or hire, but does not include the operation of an aircraft by the armed forces for reimbursement when that reimbursement is required by any Federal statute, regulation, or directive, in effect on November 1, 1999, or by one government on behalf of another government under a cost reimbursement agreement if the government on whose behalf the operation is conducted certifies to the Administrator of the Federal Aviation Administration that the operation is necessary to respond to a significant and imminent threat to life or property (including natural resources) and that no service by a private operator is reasonably available to meet the threat.

(ii) For the sole purpose of determining public aircraft status, governmental function means an activity undertaken by a government, such as national defense, intelligence missions, fire-fighting, search and rescue, law enforcement (including transport of prisoners, detainees, and illegal aliens), aeronautical research, 241-242or biological or geological resource management.

(iii) For the sole purpose of determining public aircraft status, qualified non-crewmember means an individual, other than a member of the crew, aboard an aircraft operated by the armed forces or an intelligence agency of the United States Government, or whose presence is required to perform, or is associated with the performance of, a governmental function.

(2) An aircraft owned or operated by the armed forces or chartered to provide transportation to the armed forces if--

(i) The aircraft is operated in accordance with title 10 of the United States Code;

(ii) The aircraft is operated in the performance of a governmental function under title 14, 31, 32, or 50 of the United States Code and the aircraft is not used for commercial purposes; or

(iii) The aircraft is chartered to provide transportation to the armed forces and the Secretary of Defense (or the Secretary of the department in which the Coast Guard is operating) designates the operation of the aircraft as being required in the national interest.

(3) An aircraft owned or operated by the National Guard of a State, the District of Columbia, or any territory or possession of the United States, and that meets the criteria of paragraph (2) of this definition, qualifies as a public aircraft only to the extent that it is operated under the direct control of the Department of Defense.

Qualification Performance Standards (QPS): The collection of procedures and criteria used when conducting objective and subjective tests, to establish FSTD qualification levels. These are published in the appendices to this Part 60.[36]

Radial: A line of position from a radio fix measured from the angle from magnetic north. In some cases, such as near the earth's magnetic poles, the angle from true north is used.

Ramp: US term used for **Apron**.[16]

Reduced Vertical Separation Minima (RVSM): The standard vertical separation for aircraft under Instrument Flight Rules (IFR) is 1,000 ft from sea level to FL290 and 2,000 ft above. The increased separation was a result of the lower atmospheric gradient above FL290. RVSM requires more accurate pressure sensors, air data computers, and autopilots, to achieve comparable safety to that achievable at lower altitudes.

Remise des Gaz: French for Go Around (literally *Put the gas back on*).

Repair, Major: A repair that, if improperly done, might appreciably affect weight, balance, structural strength, performance, powerplant operation, flight characteristics, or other qualities affecting airworthiness; that is not done according to accepted practices; or that cannot be done by elementary operations.

Repair, Minor: A repair other than a major repair.

Restricted Category: Aircraft which are limited to special purpose operations, such as agricultural application, aerial surveying, etc.

Rho-Rho (ρ-ρ) Navigation: Navigation based on the intersections of two or more circles defined by their distance (radius or ρ) from fixes, such as DME or TACAN stations.

Rho-Theta (ρ-θ) Navigation: Navigation based on a distance and bearing from affix, such as a radial and DME from a VOR/DME or a TACAN station.

Risk: An expression of the effect and possibility of a mishap in terms of potential mishap severity and probability of occurrence.

Rotorcraft: A heavier-than-air aircraft that depends principally for its support in flight on the lift generated by one or more rotors.[6]

Round Dials: See **Conventional Cockpit**.

Safety: Freedom from those conditions than can cause death, injury, occupational illness, damage to or loss of equipment or property, or damage to the environment.

Seat Reference Point (SRP): See **Cushion Reference Point**.

Service Bulletin: Maintenance requirements recommended by the manufacturer.

Shall: A verb used in advisory documents to refer to an action that is required to be accomplished in satisfying the use of the advisory document. See **must** and **should**.

Should: A verb used in advisory documents to refer to an action that is recommended to be accomplished in satisfying the use of the advisory document. See **must** and **shall**.

Sidestick: An aircraft Control Stick mounted on one side of the pilot on the side opposite the engine controls.

Small Aircraft: An aircraft of 12,500 lb or less, maximum certificated takeoff weight.[6]

Small/Large Aircraft: Aircraft defined as "small aircraft" for the purposes of Parts 23, 36, 121, 135, and 139 and as "large aircraft" for the purposes of Parts 61 and 91. Examples of small/large aircraft include BE-200, BE-1900, Lear 25, EMB-110P, and PA-42.

Somato-Gravic Illusion: A false sensation of body tilt that results from perceiving as vertical the direction of non-vertical gravito-inertial acceleration or force.

Special Airworthiness Certificate: An FAA approval for an individual aircraft which either is not in compliance with its Type Design or which does not have an approved Type Design, but which is in a safe condition for flight and is allowed to fly for specific purposes, such as flight test or exhibition. See **Permit to Fly**.

Special Conditions: Specific safety standards issued by the CA when airworthiness requirements do not contain adequate or appropriate standards for an aircraft, aircraft engine, or other product because of a novel or unusual design feature not anticipated when the standards were written.

Special Federal Aviation Regulation (SFAR): Regulations, issued by the FAA, limited to a limited number of aircraft models or to specific operations, often with an established expiration date.

Special Flight Authorization (SFA): A document authorizing the operation within the United States of a foreign civil aircraft not having a current and effective registration and Airworthiness Certificate.

Special Flight Permit (SFP): A special airworthiness certificate issued for a US registered aircraft that may not currently meet applicable airworthiness requirements but is capable of safe flight. It authorizes flight within US airspace. This is often called a Ferry Permit.

Special Regulation (SR): Regulations, issued by the CAA, limited to a limited number of aircraft models or to specific operations, often with an established expiration date.

Sponsor: A certificate holder who seeks or maintains FSTD qualification and is responsible for the prescribed actions as prescribed in this part and the QPS for the appropriate FSTD and qualification level.

Standard Atmosphere: Atmospheric properties (pressure, temperature, etc.) based air being a perfect gas (obeying equation (9)

$$PV = nRT \tag{9}$$

and on boundary conditions based on observations. These boundary conditions are shown in. The calculated properties are published at intervals. The model calculates air temperature, pressure, etc. based on assumed boundary conditions and a standard temperature profile, shown in Table 36. The FAA uses the *US Standard Atmosphere 1976.*[47]

Table 36 Assumed Conditions for the Standard Atmosphere

Symbol	Parameter	Metric Units	British Units
T_0	Temperature at Sea Level	15 C	59 F
P_0	Pressure at Sea Level	1913.2 HPa	2116.23 psf
g_0	Acceleration of Gravity	9.807 m/sec^2	32.15 ft/sec^2
λ	Lapse Rate within Troposphere	-0.0065 C/m	-0.003562 F/ft
H_{TP}	Height of Tropopause	11,000 m	36,089 ft
T_{TP}	Temperature above Tropopause	-56.5 C	-69.7 F

There are several other "standard" atmospheres, published since the US standard. These include those published by ISO[48] and ICAO.[49] These are no significant differences between these standards in the range of altitudes for civil aviation. See Johnson, Roberts, and Vaughan for a more complete discussion.[50]

State of Design (SOD): The country or jurisdiction having regulatory authority of the organization responsible for the design and continued airworthiness of a civil aeronautical product or article.[51]

State of Manufacture (SOM): The country or jurisdiction having regulatory authority of the organization responsible for the production and airworthiness of a civil aeronautical product or article.[51]

State of Registry (SOR): The country or jurisdiction on whose register an aircraft is recorded.[51]

Steam Gauges: See **Conventional Cockpit**.

Stick: A generic term for a Control Column, Control Stick, Control Wheel, Control Yoke, or a Sidestick. *Stick* can refer to a pilot, as in *"He's a good stick." Stick and Rudder* skills means the ability to fly using manual controls.

Stickshaker: A stall warning device which rapidly vibrates the control yoke or stick or yoke to warn when the AOA nears stall.

Strike Envelope: The extent of space surrounding a restrained occupant defined by the flailing of extended body parts during a crash impact of the aircraft. Parts of the body may strike objects located within this envelope,

Stopway: An area beyond the runway which is unsuitable for continuing a takeoff ground run, but can be used for stopping following a decision to reject a takeoff.

Substantial Damage: Damage or failure which adversely affects the structural strength, performance, or flight characteristics of the aircraft, and which would normally require major repair or replacement of the affected component.[4] This excludes engine failure or damage limited to an engine if only one engine fails or is damaged, bent fairings or cowling, dented skin, small punctured holes in the skin or fabric, ground damage to rotor or propeller blades, and damage to landing gear, wheels, tires, flaps, engine accessories, brakes, or wingtips.

Supernumerary Crew Member: A Crew Member with special skills and training that is aboard an aircraft to perform special functions, in addition to the required minimum crew. Supernumerary functions include, but are not limited to the following:

- ✈ A courier necessary for the security of valuable or confidential cargo;
- ✈ A Medical Crew Member;
- ✈ A Loadmaster responsible for the loading, securing, unloading of heavy or outsized cargo, and the operation of cargo handling devices.
- ✈ A person necessary for the safe handling of animals;
- ✈ A person necessary for the safe handling of fragile or perishable cargo;
- ✈ A person necessary for the safe handling of hazardous materials;
- ✈ A person necessary for the safety of flight;
- ✈ A person required by an operating rule;
- ✈ A person to accomplish a work activity directly associated with an operation for which the airplane is certified;
- ✈ A person who performs an essential function in connection with an operation for which the airplane is certified;
- ✈ A person performing duty as an honor guard accompanying a shipment made by or under the authority of the US Government; and
- ✈ An inspector from the FAA, DOD, or NTSB performing an inspection or evaluation.

Supplemental Type Certificate (STC): A design approval for an aircraft modification when the applicant demonstrates that the aircraft or other product complies with the appropriate regulations or standards. STCs may be issued to any applicant. The holder of the original TC has the option to apply for either an Amended Type Certificate or an STC.

Survivable Accident: An accident in which the forces transmitted to the occupant through the seat and restraint system do not exceed the limits of human tolerance to abrupt accelerations.[52]

Takeoff Power (Thrust): The approved engine power (thrust) with respect to reciprocating, turbopropeller, and turboshaft engine type certification, means the approved brake horsepower that is developed statically under standard sea level conditions, within the engine operating limitations established under Part 33, and limited in use to periods of not over 5 minutes for takeoff operation.

Tarmac: British trademark for material used for surfacing roads or other outdoor areas, consisting of crushed rock mixed with tar. As a trademark, it should be capitalized. The term is derived from tar combined with macadam.

Technical Standard Order (TSO): A minimum performance standard issued by the FAA for specified materials, parts, processes, and appliances used on civil aircraft.

Technically (or Technologically) Advanced Airplane (TAA): An airplane equipped with an electronic Primary Flight Display (PFD), an electronic Navigation Display (ND), and an autopilot capable of maneuvering the airplane about the three axes.[53]

Time of Useful Consciousness (TUC): The period of elapsed time from the interruption of normal air supply or exposure to an oxygen-poor environment until the time when the ability to function usefully is likely to be lost at which point an affected individual would no longer be capable of taking normal corrective or protective action. Time of useful consciousness (TUC) is not the time to total unconsciousness.

Theta-Theta (θ-θ) Navigation: Navigation based on cross bearings from two fixes, such as the intersection of two VOR radials.

Translational Lift: The additional lift obtained when entering forward flight due to the increased efficiency of the rotor system.[18]

Transport Category: Aircraft intended for commercial airline operations carrying passengers or cargo for hire.

Tropopause: The boundary between the lowest portion of the atmosphere, the troposphere, and the next region, the stratosphere. The tropopause marks the change in the temperature lapse rate from decreasing with altitude below and isothermal above the tropopause.

Type Certificate (TC): A design approval issued by the FAA (or FCAA) when the applicant demonstrates that the aircraft or other product complies with the appropriate regulations or standards. The TC includes the Type Design, operating limitations, the TCDS, and other conditions or limitations prescribed by the FAA (or FCAA). The TC is the foundations for other approvals including STCs, Amended Type Certificates or individual Airworthiness Certificate.

Type Certificate Datasheet (TCDS): A published summary of design approval which summarizes the design restrictions, limitations, and certification basis. The certification basis will list the application and approval dates, and the amendment level of the airworthiness rules showing compliance. The TCDS will also list all exemptions, special conditions, and equivalent levels of safety findings.

Type Design: The engineering description of an aircraft, aircraft engine or propeller. The Type Design consists of the following:
  (1) Drawings and specifications
  (2) Dimensions, materials, and processes,
  (3) Airworthiness limitations,
  (4) Other data as needed.

Type Inspection Authorization (TIA): The approval to conduct required conformity and airworthiness inspections and conduct ground and flight tests required to complete an aircraft certification project.

Type Inspection Report (TIR): The report of the tests and inspections conducted under a TIA.

Type Validation: The process for an airplane design from one country to be granted a Type Certificate in a second country.

Unmanned Aerial System (UAS): An unmanned aircraft including the remote control system.

Unmanned Aerial Vehicle (UAV): An unmanned aircraft.

Unmanned Aircraft (UA): An aircraft with no crew on board. These aircraft may be referred to as an unmanned aerial vehicle (UAV)

Unmanned Aircraft Accident: An occurrence associated with the operation of any public or civil unmanned aircraft system that takes place between the time that the system is activated with the purpose of flight and the time that the system is deactivated at the conclusion of its mission, in which
  (1) Any person suffers death or serious injury; or
  (2) The aircraft has a maximum gross takeoff weight

weight of 300 pounds or greater and sustains
substantial damage.[54]   See **Aircraft Accident**.

Unusual Attitude: An aircraft in flight unintentionally exceeding the parameters normally experienced in routine operations, typically:
  ✈ ‥Pitch attitude more than 25° aircraft nose up (ANU);
  ✈ ‥Pitch attitude less than 10° aircraft nose down (AND);
  ✈ ‥Bank angle greater than 45 deg; or
  ✈ ‥Airspeeds inappropriate for the conditions.
Sometimes called an *Upset*.[55] [56] [57]

Upset Event Recovery (UER): The recovery from an Upset Event such that the pilot must adopt an alternate control strategy to recover and maintain controlled flight.[58]

Upset Event: See **Aircraft Upset Event**.[58] Some use the term *Jet Upset*.*

Upset: See **Aircraft Upset Event** or **Unusual Attitude**.

Utility Category: Applies to airplanes intended for normal operations and limited acrobatic maneuvers. These airplanes are not suited for snap or inverted maneuvers. This Category was established in 1946. Since 1953, new designs have been limited to 12,500 lb. or less. Since 1987, new designs have been limited to nine passengers or fewer. Utility Category certifications were discontinued following amendment 23-64 in 2017.

Validating Authority (VA): The agency responsible for compliance with its own airworthiness requirements based on approvals from the exporting country's airworthiness authority. The VA may issue a Type Certificate for the aircraft or product. This is sometimes called an import TC. Not all countries use the validation approach but have alternative methods.

Visual Flight Rules (VFR): A set of operating rules to govern flight where the meteorological conditions must allow for visual operation of aircraft by direct pilot vision.

Waiver: The act of relinquishing a right, interest, or intentionally or a written statement of such relinquishment.[32]

Warning, Caution, and Advisory (WCA): See **Alert**.

Warning: The level or category of alert for conditions that require immediate flight crew awareness and immediate flight crew response.[8]

References

1   D F Shannahan, "Human Tolerance and Crash Survivability," *Pathological Aspects and Associated Biodynamics in Aircraft Accident Investigation*, RTO-EN-HFM-113, August 2005, Paper 6
2   *Assignment of Aircraft Call Signs and Associated Telephonies*, FAA AC-120-26M, dated 09/10/2018
3   *Flight Safety Analysis Handbook*, FAA Office of Commercial Space Transportation, September 2011, Version 1.0
4   S P Desjardins, D H Laananen, and G T Singley, *Aircraft Crash Survival Design Guide, Volume I, Design Criteria and Checklists*, USARTL TR-79-22A, December 1980
5   *Revision of Airworthiness Standards for Normal, Utility, Acrobatic, and Commuter Category Airplanes*, amendment 23-64, effective August 30, 2017
6   *Definitions and Abbreviations*, 14 CFR Part 1, §1.1, amendment 1-73, June 27, 2018
7   *Administrative Procedures Act*, Public Law 79-404, enacted June 11, 1946
8   *Flightcrew Alerting*, FAA Advisory Circular AC-25.1322-1, December 13, 2010

---

*   We dislike this term since many such accidents involve turboprop airplanes.

9   'Aerobatic Flight," *General Operating and Flight Rules*, 14 CFR Part 91, §91.303, amendment 91-227, effective December 17, 1991

10   http://www.skybrary.aero/index.php/Coffin_Corner, accessed September 26, 2021

11   "Definitions," *Notification and Reporting of Aircraft Accidents or Incidents and Overdue Aircraft, and Preservation of Aircraft Wreckage, Mail, Cargo, and Records*, 49CFR Part830, §830.2, revised August 24, 2010, effective October 25, 2020

12   "Definition of Airworthiness Directives," *Airworthiness Directives*, 14 CFR Part 39, §39.3, amendment 39-9474, effective August 21, 2002

13   "Definitions," *General Requirements*, 14 CFR Part 3, §3.5(a) amendment 3-1, effective October 17, 2005

14   *Aviation Child Safety Device (ACSD)*, FAA-TSO-C100c, effective April 6, 2012

15   *The FAA and Industry Guide to Product Certification*, Published by Aerospace Industries Association, Aircraft Electronics Association, General Aviation Manufacturers Association, and Federal Aviation Administration, Third Edition, May 2017

16   *Surface Movement Guidance and Control System*, FAA Advisory Circular AC-120-57A, December 19, 1996

17   *Air Traffic Management*, ICAO-DOC-4444, Sixteenth Edition, 2016

18   *Helicopter Flying Handbook*, FAA-H-8083-21B, 2019

19   *Aviation Child Safety Device (ACSD)*, FAA-TSO-C100c, effective April 6, 2012

20   *System Design and Analysis*, FAA Advisory Circular AC-25.1309-1A, June 21, 1988; Cancels AC-25.1309-1, dated September 7, 1982

21   Adapted from "Definitions," *Occupational Safety and Health Standards*, 29 CFR, Subpart A, §1910.2(g)

22   *System Design and Analysis*, FAA Advisory Circular AC-25.1309-1A, June 21, 1988; Cancels AC-25.1309-1, dated September 7, 1982

23   D F Shannahan, *Basic Principles of Helicopter Crashworthiness*, USAARL Report 93-15, February 1993

24   "Applicability and Definitions, CAR 40, §40.1, As amended to November 16, 1964

25   *Transport Airplane Cabin Interiors Crashworthiness Handbook*, FAA Advisory Circular, AC-25-17A, Change 1, Dated 05/24/2018

26   *Pilot Compartment View for Transport Category Airplanes*, FAA AC-25,773-1, January 1966

27   R L Newman and K W Greeley, *Cockpit Displays: Test and Evaluation*, Aldershot, UK: Ashgate, 2001

28   *Minimum Performance Standard For Airborne Head-Up Display (HUD)*, SAE AS-8055,

29   *Electronic Centralized Aircraft Monitor (ECAM)*, https://www.skybrary.aero/index.php/Electronic_Centralized_Aircraft_Monitor_%28ECAM%29

30   *Automatic Emergency Descent System*, http://skybrary.aero/articles/automatic-emergency-descent-system, accessed November 19, 2021

31   http://skybrary.aero/index.php/Engine_Indicating_and_Crew_Alerting_System_(EICAS), accessed November 9, 2020

32   *Random House Webster's Unabridged Dictionary*, New York: Random House, 2nd edition, 1999

33   J A DeSantis, "Engines Turn of Passengers Swim: A Case Study of How ETOPS Improved Safety and Economics in Aviation," *Journal of Air Law and Commerce*: 77, 2013, 3-68

34   http://www.skybrary.aero/index.php/Flight_Director#:~:text=Definition.%20A%20flight%20director%20is%20a%20guidance%20aid,the%20attitude%20required%20to%20follow%20a%20certain%20trajectory.Accessed December 23, 2020

35   Adapted from http://www.skybrary.aero/index.php/Flight_Management_System, accessed October 30, 2020

36   "Definitions," *Flight Simulation Training Device Initial and Continuing Qualification and Use*, 14 CFR Part60, §60.3, and Appendix F, amendment 60-1, effective October 30, 2007

37   *Use of Aircraft Fuel Anti-Icing Additives*, FAA Advisory Circular FAA-AC-20-29B, January 18, 1972, cancels FAA-AC-20-29A

38   *Injury Criteria for Human Exposure to Impact*, FAA Advisory Circular FAA-AC-21-22, June 20, 1985

39   R L Newman, *Head-Up Displays: Designing the Way Ahead*, Aldershot, UK: Ashgate, 1995

40   *Helicopter Flying Handbook*, FAA-H-8083-21B, 2019

41 *System Safety Analysis and Assessment for Part 23 Airplanes*, FAA Advisory Circular 23.1309-E, 11/17/2011

42 "Definitions," *General Operation Rules*, CAR 43, §43.9(c), effective July 1, 1945

43 "Definitions," *General Operation Rules*, CAM 43, §43.70, Revision September 1, 1962

44 R L Newman, "Definition of Primary Flight Reference," *Journal of Aircraft*: 35, 1998, 497-500

45 "General Definitions," *Transportation,Part A, Air Commerce and Safety,* 49 USC, $40102, as published in FAA Advisory Circular AC-00-1.1B[46]

46 *Public Aircraft Operations – Manned and Unmanned*, FAA Advisory Circular AC-1.1B, Dated 9/21/18

47 *US Standard Atmosphere 1976*, NASA-TM-X-74335, October 1974

48 *Standard Atmosphere*, International Organization for Standardization, IS)-2533: 1975, 1975

49 *Manual of the ICAO Standard Atmosphere (extended to 80 kilometres (262599 feet))*, ICAO Doc 7488-CD, 1993

50 D E Johnson, B C Roberts, and W W Vaughan, "Reference and Standard Atmosphere Models," *10th Conference on Aviation, Range, and Aerospace Meteorology*, May 13, 2002, Portland, OR, May 13, 2002

51 *Airworthiness Certification of Aircraft*, FAA Order 8130.2J, July 21, 2017

52 S P Desjardins *et al.*, *Aircraft Crash Survival Design Guide, Volume IV, Aircraft Seats, Restraints, Litters, and Cockpit/Cabin Delethalization*, USAAVSCOM TR-89-D—22D, December 1989

53 "Exception to second in command requirement: Approval for use of autopilot system," *Operating Requirements: Commuter and On Demand Operations and Rules Governing Persons on Board Such Aircraft*, 14CFRPart135, §135. 105, amendment 135-58, effective December 29, 1995

54 *Definitions*, 49FCR830.2, effective October 25, 2010

55 J A Kohan, J E, Priest, and M Moskal, "Human Factors Aspects of Upset Recovery Training," Paper presented at 17th Annual European Aviation Safety Seminar, Warsaw, March 2005

56 B Wainwright *et al.*, *Aerodynamic Principles of Large Airplane Upsets*, Airbus and Boeing White Paper, 2002

57 R L Sumwalt, "Airplane Upset Recovery Training: A Line Pilot's Perspective," *Flight Safety Digest*, July-August 2003, pp. 1-18

58 J Priest, *Research into Reduction of Loss-of-Control Accident Rate in a Meaningful and Measurable Way Through the Use of Innovative Pilot Training Techniques*, AIAA Paper 2014-1005, Presented at AIAA Modeling and Simulation Technologies Conference, National Harbor, MD, January 2014

Appendix 2:
Aviation Regulation Bibliography

This section should help readers identify additional background references.

Required Reading for All Certification Personnel and Managers

J H Fielder and D Birsch (eds.), The DC-10 Case: A Study in Applied Ethics, Technology, and Society, Albany: State University of New York Press, 1962

C Haddon-Cave, *An Independent Review into the Broader Issues Surrounding the Loss of RAF Nimrod MR2 Aircraft XV230 in Afghanistan in 2006*, London: HMSO, 2009

D Vaughan, *The Challenger Launch Decision*, Chicago: University of Chicago, 1996

Aviation Regulation History

L F Schmeckebier, *The Aeronautics Branch Department of Commerce. Its History, Activities, and Organization*, Washington: The Brookings Institution, 1930, Service Monographs on the United States Government, No. 61

H G Hotchkiss *A Treatise on Aviation Law*, New York: Baker, Voorhis & Co., 1938

R W Fixel, *The Law of Aviation*, Charlottesville, VA. The Michie Company, 1945

F H Frederick, *Commercial Air Transportation*, Homewood, IL: Richard D Irwin, Inc., 1955

D R Whitnah,, *Safer Skyways, Federal Control of Aviation, 1926-1966*, Ames: Iowa State University Press, 1966

N A Komons, *The Cutting Air Crash: A Case Study in Early Federal Aviation Policy*, Washington: Superintendent of Documents, 1973; reprinted by University of Michigan Library

G Pucci, *Aviation Law. Fundamental Cases*, Dubuque, IA, Kendall/Hunt Publishing, Second Edition 1974

V F Rollo, *Aviation Law. An Introduction*, Lanham, MD: Maryland Historical Press, Third Edition 1975

S I Rochester, *Takeoff and Mid-Century, Federal Civil Aviation Policy in the Eisenhower Years 1953-1961*, Washington: Superintendent of Documents, 1976

Anonymous, *Safe Skies for Tomorrow: Aviation Safety in a Competitive Environment*, Washington: Office of Technology Assessment, 1980, reprinted by University of Michigan Library

J R M Wilson *Turbulence Aloft, The Civil Aeronautics Administration Amid Wars and Rumors of Wars 1938-1953*, Washington: Superintendent of Documents, 1979

R J Kent, *Safe, Separated, and Soaring, A History of Federal Civil Aviation Policy 1961-1972*, Washington: Superintendent of Documents, 1980

R E Bilstein, *Flight Patterns: Trends of Aeronautical Development in the United States, 1918-1929*, Athens: University of Georgia Press, 1983

D R Whitnah, *Government Agencies*, Westport, CT: Greenwood Press, 1983

M E F Plave,   *"United States v. Varig Airlines*: The Supreme Court Narrows the Scope of Government Liability Under the Federal Tort Claims Act," *Journal of Air Law and Commerce*: 51 1985, 197-239

H P Wolfe and D A NewMyer, *Aviation Industry Regulation*, Carbondale, IL: Southern Illinois University Press, 1985

R E Bilstein,   *Flight Orders of Magnitude. A History of the NACA and NASA. A History of the NACA and NASA, 1915-1990*, Washington: National Aeronautics and Space Administration, 1989

N A Komons,   *Bonfires to Beacons: Federal Civil Aviation Policy Under the Air Commerce Act, 1926-1938*, Washington: Smithsonian Institution Press, Reprinted 1989

D R Whitnah,   *U.S. Department of Transportation: A Reference History*, Westport, CT: Greenwood Press, 1998

E Preston *et al.*, *FAA Historical Chronology: Civil and the Federal Aviation Government*, Washington: Federal Aviation Administration, 1998

Anonymous,   *1997-2015 Update to FAA Historical Chronology: Civil Aviation and the Federal Aviation Administration*, Washington: Federal Aviation Administration, 1998

M K Sparrow,   *The Regulatory Craft. Controlling Risks, Solving Problems, and Managing Compliance*, Washington: Brookings Institution Press, 2000

J S Hamilton,   *Practical Aviation Law*, Ames, IA: Blackwell, 2005

F De Florio,   *Airworthiness. An Introduction to Aircraft Certification. A Guide to Understanding JAA, EASO, and FAA Standards*, Amsterdam: Elsevier, 2006

R Holanda,   *A History of Aviation Safety*, Bloomington, IN: AuthorHouse, 2009

C M Kerwin and S R Furlong, *Rulemaking. How Government Agencies Write Law and Make Policy*, Washington: CQ Press, Fourth Edition 2011

S E Dudley and J Brito, *Regulation. A Primer*, Arlington, VA: Mercatus Center and Washington: Regulatory Studies Center, 2012

C C Rodrigues and S K Cusick, *Commercial Aviation Safety*, New York: McGraw-Hill, Fifth Edition, 2012

Appendix 3: Sources of Documents

FAA Documents

These include FAA Orders, Notices, TSOs, Type Certificate Data Sheets, Regulations, NPRMs, Equivalent Levels of Safety, Exemptions, Special Airworthiness Information Bulletins (SAIBs), Special Conditions, Supplemental Type Certificates, etc.

Sources: FAA Regulatory and Guidance Library:* http://rgl.faa.gov
            FAA Dynamic Regulatory System: *http://drs.faa.gov.*
            The DRS site is available at this time . We have not had great success in locating out of date regulations on this site. This will be a serious issue for those who are adding equipment to older airplane models
            Summit Aviation, Digital Aviation Reference Library,
            (13997 S Minuteman Drive, Draper, UT 84020)
            There is a charge for a subscription.

EASA Documents

http://www.easa.europa.eu/document-library

ICAO Documents

http://www.icao.int/safety/lpr/Pages/Documents-and-Manuals.aspx
There are charges for the documents'

Aircraft Accident Data

These include accident and incident data. These can be found on the website of various countries' web sites. We have used all of the following sites:

Australia   ATSB: Australian Transport Safety Bureau
                     http://www.atsb.gov.au/

Aviation Safety Network (Sponsored by Flight Safety Foundation)
                     http://aviation-safety.net/

Brazil   CENIPA: Centro de Investigação e Prevenção de Acidentes Aeroãnauticos
                     http://www2.fab.mil.br/cenipa/

Canada   TSB: Transportation Safety Board
                     http://www.tsb.gc.ca

France   BEA: Bureau d'Enquêtes et d'Analyses pour la sécurité de l'aviation civile
                     http://www.bea.aero/

Germany   BFU:Bundesstelle für Flugunfalluntersuchung
                     http://www.bfu-web.de/DE/Home/homepage_node.html

Indonesia   NTSC: National Transportation Safety Committee (also KKNT)
                     http://knkt.dephub.go.id/knkt/ntsc_home/ntsc.htm

Ireland   AAIA: Air Accident Investigation Unit
                     http://www.aaiu.ie/

---

* The RGL Databases have a planned decommission by the end of 2022. All information is currently available on the Dynamic Regulatory System (DRS) which will replace RGL. The URL for DRS is drs.faa.gov We have not had much success in finding old regulations and NPRMs on the DRS website.

Japan	JTSB: Japan Transport Safety Bureau http://www.mlit.go.jp/jtsb/
Netherlands	OVV: Onderzoeksraad voor Veilgheid (Dutch Safety Board) http://www.onderzoeksraad.nl/
New Zealand	TAIC: Transport Accident Investigation Commission http://www.taic.org.nz
Norway	Norwegian Safety Investigation Authority http://havarikommisjonen.no/
Russia	MAK: Interstate Aviation Committee http://mak-iac.org/
South Korea	ARIAB: Aviation and Railway Accident Investigation Board http://araib.molit.go.kr/intro.do?gubn=english
Spain	CIAIAC: Comisión de Investigación de Accidentes e Incidentes de Aviación Civil http://www.mitma.es/organos-colegiados/ciaiac
Sweden	SFH: Statens Haverikommission http://www.havkom.se/
Switzerland	SUST: Sicherheitsuntersuchungsstelle http://www.sust.admin.ch/de/sust-startseite/
Taiwan	TTSB: Taiwan Transportation Safety Board http://www.ttsb.gov.tw/english/
United Kingdom	AAIB: Air Accident Investigation Board https://www.gov.uk/government/organisations/air-accidents-investigation-branch
United States	NTSB: National Transportation Safety Board https://www.ntsb.gov/Pages/default.aspx The NTSB has reconstructed their database. All mishaps since 1983 are now available using the Case Analysis and Reporting Online (CAROL). We have found CAROL less satisfactory than the previous system. However, data in Mictrosoft Access® format are available for downloading. These data are updated weekly.

Appendix 4.
Accidents That Affected Certification

> Aviation in itself is not inherently dangerous. But to an
> even greater degree than the sea, it is terribly unforgiv-
> ing of carelessness, incapacity, or neglect.*

Finding the underlying causes of aircraft accidents and incidents is the quality control of aircraft certification. This Appendix lists many such accidents. Each entry lists a brief description of the occurrence, the probable cause as determined by the investigating agency, and additional commentary as it affected aircraft certification. Table 37 provides an index to aircraft models in this Appendix

Most of these mishaps came from number of sources: NTSB files and the Aviation Safety Network[1] provided most of the listings. The reader is also directed to the FAA's "Lessons Learned" website[2] which shows 93 aviation accidents from 1950 through 2010. It points out many of the regulatory changes resulting from each. The "Lessons Learned" website seems sparse in its coverage of light airplanes and helicopters. Only one light airplane and six helicopter accidents are listed.

The descriptions and probable causes were copied directly from the source documents and have not been edited. These are shown in Arial font, while our comments are shown in Times New Roman font as in the rest of the book.

To aid the reader who is interested in specific models, Table 37 lists the mishaps by make and model with a reference to the page number.

---

* Anonymous, http://www.quotes.net/quote/36031, accessed October 24, 2020

Table 37 Accidents to Specific Aircraft Models

Aircraft	Date	Flight	Page	Aircraft	Date	Flight	Page
A300B4	26-Apr-94	CAL 140	357	B-767	23-Jul-83	ACA 143	327
	12-May-97	AAL 903	373		26-May-91	NG 004	278
	12-Nov-01	AAL 587	374	B-777	01-Aug-05	MAS 124	341
A310	11-Feb-91	IFL	373		17-Jan-08	BAW 38	338
A319	10-Jan-08	ACA 190	375		26-Nov-08	DAL 18	339
A320	17-Mar-91	NWA 985	309		08-Mar-14	MAS 370	356
	20-Jan-92	ITF 148	267		11-Dec-18	ACA 15	381
	22-Mar-98	PAL 137	279	B-787	09-Nov-10	N-787EX	295
	07-Feb-01	IBE 1456	313		07-Jan-13	JAL 008	298
	20-Mar-01	DLH	307		16-Jan-13	ANA 692	296
	28-Aug-02	AWE 794	280		12-Jul-13	ET 700	299
	18-Oct-04	GA 536	280	BE-35	03-Feb-59	N-3851N	384
	25-Jun-05	AZA Unkn	353	BE-99	06-Jul-69	KKB 168	311
	17-Jul-07	JJ 054	281	CE-152	18-Jul-98	C-GZLZ	303
	27-Nov-08	D-AXLA	322		11-Apr-05	N-24779	304
	15-Jan-09	USA 1549	258	CE-U206	24-May-85	N-1803Q	275
A321	07-Dec-02	ACA 1130	309	CE-310	16-Oc-72	N-1812H	354
	07-Dec-02	ACA 457	309	Concorde	25-Jul-00	AFR 4590	328
A330	30-Jun-94	F-WWKH	320	DC-10	12-Jun-72	AAL 96	255
	24-Aug-01	TSC 236	326		03-Mar-74	THY 981	256
	07-Oct-08	QFA 72	314		12-Nov-75	ONA 32	259
	27-Dec-08	QFA 71	315		25-May-79	AAL 191	344
	01-Jun-09	AFR 447	363		19-Jul-89	UAL 232	283
A340	21-Jun-96	DLH	311	DC-2	03-May-39	TWA 2	262
	02-Oct-00	THY	312	DC-7	30-Jun-56	UAL 718	350
	03-Feb-13	EDT 460	365	DC-8	05-Jul-70	ACA 621	367
A380	04-Nov-10	QFA 32	284		16-Feb-00	EWW 17	306
Airship600	15-Apr-90	N-602SK	330	DC-9	02-Jun-83	ACA 797	287
AS-332L	19-Jan-95	BHL 53C	343		02-Jul-94	USA 101	394
ATR 72	31-Oct-94	EGF 4184	334		11-May-96	VJC 592	290
B-377	16-Oct-56	PAA 6	274	Comet	10-Jan-54	G-ALYP	346
B-707	11-Jul-73	VRG 820	287		01-Apr-54	SAA 201	346
B-727	18-Jan-69	UAL 266	352	DHC-4A	27-Aug-92	N-400NC	319
	04-Sep-71	ASA 66	263	DHC-8	12-Feb-09	CJC 3407	337
	01-Dec-74	TWA 514	265	EA-500	05-Jun-08	N-612KB	276
	01-Dec-74	NWA6231	360	EMB-120	05-Apr-91	ASQ 2311	368
	01-Jan-85	EAL 980	266		11-Sep-91	BTA-2574	345
B-737	28-Apr-88	AAH 243	347		09-Jan-97	CCM 3272	336
	24-May-88	TAI 110	372	EMB-135	29-Sep-06	N-600×L	351
	25-Feb-90	THY 1951	370	Fo-10A	31-Mar-31	TWA 599	253
	03-Mar-91	UAL 5858	301	G650	02-Apr-11	N-652GD	324
	08-Sep-94	USA 427	301	LL-1049A	30-Jun-56	TWA 2	350
	09-Jun-96	SGR 517	302	L-1011	02-Aug-85	DAL 191	391
	03-Jan-04	ZHA 604	386	L-188	29-Sep-59	BNF 542	253
	14-Aug-05	HCY 522	331		17-Mar-60	NWA 710	254
	08-Dec-05	SWA 1248	390		04-Oct-60	EAL 375	258
	29-Sep-06	GLO 1907	351		17-Sep-61	NWA 706	307
	17-Feb-14	UAL 1676	379		30-Oct-74	CF-PAB	264
B-737MAX	29-Oct-18	LNI 610	315	L-382	03-Feb-93	N-130X	320
	10-Mar-19	ETH 302	316	LR-35A	25-Oct-99	N-47BA	331
B-747	24-Jun-82	BAW 009	389	M-20F	06-Feb-75	N-6371Q	341
	28-Nov-87	SAA 295	289	MD-11	13-Jul-96	AAL 140	358
	24-Feb-89	UAL-811	257		02-Sep-98	SWR 111	291,
	15-Dec-89	KLM 867	389	MD-82	16-Aug-87	NWA 255	392
	04-Oct-92	ELY 1862	349	MD-83	31-Jan-00	ASA 261	305
	17-Jul-96	TWA 800	396	PA-32	16-Jul-99	N-9253N	385
	06-Nov-01	9V-SPP	272	PC-12/45	22-Mar-09	N-128CM	339
	12-Jan-02	Unreported	272	S-61L	22-May-68	LAA 841	368
	03-Sep-10	UPS 006	293	S-76	04-Jan-09	N-748P	260,
B-757	20-Dec-95	AAL 965	268		26-Jan-20	N-72EX	387
	06-Feb-96	ALW 301	361	S-92	14-Mar-17	EI-ICR	270
	02-Oct-96	PLI 603	361	SA-227	22-Dec-12	PAG 993	378
	28-Jan-09	AEU	362	TBM-700	03-Jun-08	N-849MA	377
	15-Jun-19	UAL 627	382		05-Sep-14	N-900KN	333

Airframe Failure

March 31, 1931: Fokker F.10A, NC-999E

Location	Near Bazaar, KS
Flight	TWA 599
Operator	Transcontinental & Western Air
Operation	Scheduled Airline
Route	Kansas City, MO to Wichita, KS
On-Board Injuries	Fatal 8
Aircraft Damage	Destroyed

Description   Source: Aviation Safety Network[1]

A TWA Fokker F.10A crashed near Bazaar, Kansas, after one of the wings failed in-flight. The aircraft operated on a scheduled TWA service from Kansas City Municipal Airport, Missouri, to Los Angeles, California. En route stops were planned at Amarillo, Texas; Albuquerque, New Mexico; and Winslow, Arizona.

The flight departed Kansas City at 09:15. The aircraft encountered worsening weather conditions en route with fog, clouds and icing. When maneuvering near Bazaar, Kansas, one of the wings failed and the aircraft crashed out of control. All eight on board were killed. Among the passengers was Knute Rockne (43), coach of the "Fighting Irish" American football team of the University of Notre Dame.

Probable Cause   Source: Aviation Safety Network[1]

The probable cause of the accident was never made public by the Civil Aeronautics Branch. A requirement was issued for aileron counterbalancing on all F.10A's, suggesting aileron flutter may have been a factor.

Fifty-five years after the crash, a former TWA mechanic, came forward and claimed he saw that wing panels had come loose on the aircraft. He made this observation during an inspection of the aircraft a few days before he crash.

The wings of Fokker F.10 aircraft were manufactured out of wood laminate; in this instance, moisture had possibly leaked into the interior of one wing over a period and had weakened the glue bonding the structure. The failure of a spar could have caused the wing to develop uncontrolled flutter.

Comments

The main effect of this accident on the industry was the design of the Douglas DC-1 to meet a TWA specification for a three-engine replacement for the Fokker F.10A. Douglas proposed a twin engine airplane which met and exceeded the TWA requirements. While only one DC-1 was built, it was followed by 218 DC-2 and 13,300 DC-3 aircraft. [3]

September 29, 1959: Lockheed L-188, N-9705C

Location	Near Buffalo, TX
Flight	BNF 542
Operation	Braniff International Airways
Operation	Scheduled Airline
Route	Houston, TX to Dallas, TX
On-Board Injuries	Fatal 34
Aircraft Damage	Destroyed

Description   Source: Aviation Safety Network[1]

Braniff International Airways Flight 542, a Lockheed L-188A *Electra*, departed the ramp at Houston International Airport at 22:37, 22 minutes behind schedule. The delayed departure was due to a mechanical discrepancy involving No. 3 generator. This generator was inoperative on arrival of N9705C at Houston. Prior to departure from Houston the Nos. 3 and 4 voltage regulators were interchanged. The estimated time en route to Dallas was 41 minutes.

The flight was given an IFR clearance which was to the Leona omni, via Victor Airway 13 west to the Gulf Coast intersection, direct to Leona, to maintain 2,300 feet altitude to Gulf Coast, then to climb to and maintain 9,000. At approximately 22:40 the flight was cleared for takeoff and at 22:44 the crew reported airborne.

After takeoff Houston departure control advised that it had the flight in radar contact and requested it to report when established outbound on the 345-degree radial of the Houston omni. Flight 542 complied and subsequently was cleared to 9,000 feet and advised to contact San Antonio Center upon passing the Gulf Coast intersection.

At approximately 22:52 Flight 542 reported to San Antonio Center as being over Gulf Coast intersection at 9,000 feet. The flight was then issued its destination clearance to the Dallas Airport and it was cleared to climb to its cruising altitude of 15,000 feet. After the *Electra* had passed Leona at 23:05, the crew contacted company radio with a message for maintenance, advising that the generators were then OK but that there had been insufficient time for maintenance to insulate the terminal strip on No. 3 propeller at Houston and it would like to have it done in Dallas.

At 23:09 the left wing and the No. 1 gear box propeller separated. The horizontal stabilizer then broke up under the impact of parts coming from the wing; wing planking from the right wing tip came free; the No. 4 powerplant tore loose; and the right wing outboard of engine No. 4 separated. All of these events happened in a short period of time. Somewhat later, at much lower altitudes, the fuselage broke in two separate portions at a point about halfway back near fuselage station No. 570.

All 34 on board were killed in the accident.

Probable Cause   Source: Accident Report[4]

Structural failure of the left wing resulting from forces generated by undampened propeller whirl mode.

March 17, 1960: Lockheed L-188, N-121US

Location   Near Cannelton, IN
Flight   NWA 710
Operator   Northwest Orient Airlines
Operation   Scheduled Airline
Route   Chicago, IL to Miami, FL
On-Board Injuries   Fatal 63
Aircraft Damage   Destroyed

Description   Source: Aviation Safety Network[1]

Northwest Orient Airlines Flight 710, was a scheduled service from Minneapolis (MSP) to Chicago (MDW) and Miami (MIA).

The Lockheed L-188C *Electra* departed Minneapolis at 12:51 and arrived at Chicago-Midway Airport at 13:55. During the short time the aircraft was on the ground at Chicago, approximately 30 minutes, it was refueled and prepared for continuation of the flight to Miami. The flight took off from Midway Airport at 14:38.

At 15:13, Flight 710 reported over Scotland, Indiana, maintaining 18000 feet and estimating Bowling Green, Kentucky, at 15:35. This was the last radio contact with the flight.

At 15:25 the outboard engines and engine support structures, the complete right wing, and the outer portions of the left wing and ailerons separated in flight. The aircraft crashed out of control.

Probable Cause   Source: Accident Report[5]

The Board determines that the probable cause of this accident was the separation of the right wing in flight due to flutter induced by oscillations of the outboard nacelles. Contributing factors were a reduced stiffness of the structure and the entry of the aircraft into an area of severe clear air turbulence.

Regulatory Outcome

On March 20, 1960, the FAA issued, as a temporary measure, an emergency AD applicable to Lockheed Model 188 series airplanes: To reduce the maximum structural cruising speed (V_{NO}) from 324 knots calibrated air speed (CAS) to 275 knots CAS.

On March 25, 1960, the FAA issued a second emergency AD, AD-60-09-03, amendment 134, applicable to Lockheed Model 188 series airplanes: To limit V_{NO} to 225 knots CAS and establish a never-exceed speed (V_{NE}) of 245 knots CAS. Also included were requirements for immediate propeller feathering if the torque meter indicator registered zero or full scale; deactivation of the autopilot until appropriate modifications could be designed and installed; and adherence to Lockheed prescribed procedures in refueling operations.

During the original certification, the *Electra* was equipped with Allison engines and Aeroproducts propellers. Certification included a vibratory stress survey of the propellers. It was determined, based on experience, that the inboard propellers were more critical, and only the inboard propellers were instrumented. Later, a Hamilton Standard propeller was installed on the airplane and a supplemental type certificate was sought. At this time, it was decided that a vibratory stress survey on one outboard and one inboard propeller was to be conducted. As a result of this test, it was found that the outboard propellers were more highly stressed than the inboard propellers and that these stresses exceeded acceptable levels.

Following the *Electra* accidents, CAR 4b was revised, in October 1964, to add new aero elastic requirements to CAR §4b.308. The preamble to Amendment 16 of CAR 4b and the NPRM for CAR §4b.308 describe the changes to the regulation, specifically requiring the consideration of the effects of "forces associated with rotations and displacements of the plane of the propeller". The preamble material referred to the two *Electra* accidents, and required the consideration of the effects of failures of airplane structure, including engine mounts, on the aero elastic stability of an airplane design. The amended rule also required consideration of the failure of other structural elements for which fatigue and/or fail-safe analyses had been performed per §4b.270, and for elements of the flight control system.

When Part 25 was adopted in November, 1964, §25.629 adopted the requirements of §4b.308, and added a more specific speed envelope inside which an airplane was required to be free from flutter. In addition, §25.629 introduced fail-safe criteria, and required consideration of failures of structural elements for which fail-safe criteria/analyses had been applied. The rule further linked these criteria to fail-safe criteria in §25.571(c).

June 12, 1972: McDonnell Douglas DC-10, N-103AA

Location	Windsor, Ontario
Flight	AAL 96
Operation	American Airlines
Operation	Scheduled Airline
Route	Detroit, MI to Buffalo, NY
On-Board Injuries	Serious 11 Minor/None 56
Aircraft Damage	Substantial

Description   Source: Aviation Safety Network[1]

American Flight 96 had departed Detroit-Metropolitan Airport and was climbing through 11,750 feet at 260 kt IAS when the flightcrew felt a "thud". Simultaneously dust and dirt flew up into their faces, the rudder pedals moved to the full left-rudder position, the three thrust levers moved back to the near flight-idle position and the airplane yawed to the right. The crew managed to regain control of the plane, although a.o. elevator response was sluggish, and rudder control was not available. An emergency was declared and the crew returned to Detroit. A safe runway 03 long final approach and landing were carried out.

It appeared that a cargo door had separated, causing a rapid decompression, which, in turn, caused failure of the cabin floor over the bulk cargo compartment.

The separated door caused minor damage to the fuselage above the door and substantial damage to the leading edge and upper surface of the left horizontal stabilizer.

Probable Cause   Source: Accident Report[6]

The improper engagement of the latching mechanism for the aft bulk cargo compartment door during the preparation of the airplane for flight. The design characteristics of the door latching mechanism permitted the door to be apparently closed when, in fact, the latches were not fully engaged, and the latch lockpins were not in place.

Safety Recommendations

In its report, the NTSB made two recommendations to the FAA:

1. Require modification to the DC-10 cargo door locking system to make it physically impossible to position the external locking hand and vent door to their normal door locked position unless the locking pins are fully engaged.
2. Require the installation of relief vents between the cabin and aft cargo compartment to minimize the pressure loading on the cabin flooring in the event of a sudden depressurization of the cargo compartment.

Regulatory Outcome

[W]hen the FAA was about to issue an airworthiness directive (AD) making interim and long-term solutions mandatory for all US operators of the DC-10 (which foreign operators would have followed), discussions between the FAA administrator and the president of the Douglas division of McDonnell Douglas led to the senior FAA technical staff being overruled. Douglas and the FAA were no doubt being subjected to pleading from US airlines, who would not want to take their aircraft out of service in the peak summer season. So, the FAA did not issue that airworthiness directive. Instead, McDonnell Douglas almost immediately issued recommendations, in particular the installation of a "lock mechanism viewing window."

This gentleman's agreement between the FAA administrator and McDonnell Douglas's division president sufficed to prevent a repeat accident *in the United States*, but not overseas.[7]

March 3, 1974: McDonnell Douglas DC-10, TC-JAV

Location	Bois D'Ermeninville, France
Flight	THY 981*
Operation	Turkish Airlines
Operation	Scheduled Airline
Route	Paris-Orly, France to London, UK
On-Board Injuries	Fatal 346
Aircraft Damage	Destroyed

Description   Source: Aviation Safety Network[1]

On Sunday March 3, 1974 flight TK981 departed Istanbul for a flight to Paris and London. The DC-10 landed at Paris-Orly at 11:02 and taxied to stand A2. There were 167 passengers on board, of whom 50 disembarked. The aircraft was refueled and baggage was loaded onto the plane. The planned turnaround time of one hour was delayed by 30 minutes. An additional 216 passengers embarked. Most of the passengers were booked on this flight because of a strike at British Airways. The door of the aft cargo compartment on the left-hand side was closed at about 10:35. When all preparations were complete the flight received permission to taxi to runway 08 at 12:24. Four minutes later the crew were cleared to line up for departure and were cleared for departure route 181 and an initial climb to flight level 40. The aircraft took off at approximately 12:30 hours and was cleared by Orly Departure to climb to FL60, which was reached at 12:34. The North Area control centre then cleared TK981 further to FL230. Three or four seconds before 12:40:00 hours, the noise of decompression was heard and the co-pilot said: "the fuselage has burst" and the pressurization aural warning sounded. This was caused by the opening and separation of the aft left-hand cargo door. The pressure difference in the cargo bay and passenger cabin, the floor above the cargo door partly collapsed. Two occupied triple seat units were ejected from the aircraft. All the horizontal stabilizer and elevator control cables routed beneath the floor of the DC-10 and were thus also severely disrupted. Also the no. 2 engine power was lost almost completely. The aircraft turned 9 deg to the left and pitched nose down. The nose-down attitude increased rapidly to -20 deg. Although the no. 1 and 3 engines were throttled back the speed increased to 360 kts. The pitch attitude then progressively increased to -4 degrees and the speed became steady at 430 kts (800 km/h). At a left bank of 17 degrees the DC-10 crashed into the forest of Ermenonville, 37 km NE of Paris.

Probable Cause   Source: Accident Report[8]

The accident was the result of the ejection in flight of the aft cargo door on the left-hand side: the sudden depressurization which followed led to the disruption of the floor structure, causing six passengers and parts of the aircraft to be ejected, rendering No.2 engine inoperative and impairing the flight controls (tail surfaces) so that it was impossible for the crew to regain control of the aircraft.

The underlying factor in the sequence of events leading to the accident was the incorrect engagement of the door latching mechanism before takeoff. The characteristics of the design of the mechanism made it possible for the vent door to be apparently closed and the cargo door apparently locked when in fact the latches were not fully closed and the lock pins were not in place.

It should be noted, however that a view port was provided so that there could be a visual check of the engagement of the lock pins.

This defective closing of the door resulted from a combination of various factors:
- incomplete application of Service Bulletin 52-37;
- - incorrect modifications and adjustments which led, in particular, to insufficient protrusion of the lock pins and to the switching off of the flight deck visual warning light before the door was locked;
- - the circumstances of the closure of the door during the stop at Orly, and, in particular, the absence of any visual inspection, through the viewport to verify that the lock pins were effectively engaged, although at the time of the accident inspection was rendered difficult by the inadequate diameter of the view port.

Finally, although there was apparent redundancy of the flight control systems, the fact that the pressure relief vents between the cargo compartment and the passenger cabin were inadequate and that all the flight control cables were routed beneath the floor placed the aircraft in grave danger in the case of any sudden depressurization causing substantial damage to that part of the structure.

All these risks had already become evident, nineteen months earlier, at the time of the Windsor accident, but no efficacious corrective action had followed.

Certification Issues

A Federal official said today that he understood that a "gentleman's agreement" between the Government and the plane's builder had sidetracked the issuance of a mandatory requirement for a design change that might have prevented the March 3 crash of a DC - 10 jumbo jet airliner near Paris.

Instead, with Government approval, the company - recommended improvement was left as a voluntary measure both for the manufacturer itself, as new planes progressed down the assembly line, and for the airlines operating DC - 10's delivered earlier.

---

*   TK is the IATA code for Turkish Airlines and is used in the source material.

The change was intended as a "foolproof" safeguard against failure of a DC 10's rear cargo door. Most specialists are all but convinced that the Paris accident was caused by failure of the door, which was found nine miles from the main crash area.

The McDonnell Douglas corporation admitted yesterday that the change evidently had not been made on the Turkish Airlines plane involved in the Paris catastrophe even though the manufacturer's own records showed the change had in fact been made. The crash, in which all 346 persons aboard were killed, was the worst in aviation history[at the time].[9]

Five airworthiness directives were issued:[2]

- *AD 74-08-04 R4* mandated the incorporation of seven McDonnell-Douglas service bulletins to prevent possible in-flight depressurization of the airplane that might result from the opening of an improperly secured cargo door.
- *AD 74-12-07 R1* mandated the incorporation of five McDonnell-Douglas service bulletins to assure that in-flight depressurization will not occur as a result of the opening of a lower cargo door.
- *AD 75-15-05 R1* was probably the widest ranging result of the Paris crash. It became known as the "floors and doors" AD, and was applied to all wide bodies in service at the time. This AD mandated modification to improve the capability of the passenger and crew compartment floors to withstand, without collapse, an in-flight depressurization caused by the sudden opening of a large hole in the fuselage, and was applied to McDonnell-Douglas Model DC-10 Series, Lockheed Model L-1011 Series, Boeing Model 747 Series, and Airbus Industrie Model A-300 Series airplanes certificated in all categories.
- AD 74-08-04 was issued as a telegraphic AD on March 7, 1974, ninety-six hours after the crash to all operators of DC-10 aircraft. The AD directed them to perform the modifications on the cargo doors described in Douglas service bulletins. All of theses service bulletins were issued shortly after the American Airlines accident 19 months previously. (*SB 52-27, SB 52-37, SB 52-38*)

On March 22, 1974, the FAA amended AD 74-08-04 to require the installation of the so-called "closed loop" system for locking the cargo doors.

The FAA did not take any action to review the design of the cargo door systems on other transport airplanes until after the United Airlines B-747 and Evergreen DC-9 inadvertent opening of fuselage doors accidents on February 24, 1989 and March 18, 1989, respectively. After the accidents the *Transport Airplane Directorate* (TAD), Los Angeles and Seattle ACO's, conducted a Special certification Review (SCR) of the Type certificated cargo door systems on transport airplanes for compliance with Amendment 25-54 to 14 CFR 25.783 and AC 25.783-1. The SCR was conducted in compliance with Transport Airplane Directorate memorandum, "Modification of Outward Opening Doors on Existing Transport Airplanes," dated March 20, 1992. Numerous ADs were issued to ensure compliance with Amendment 25-54 to 14 CFR 25.783. For information on the SCR, see *Order 8110.4c*, paragraph 2-7.e.

February 24, 1989: Boeing 747, N-4713U

Location	Near Hawaii
Flight	UAL 811
Operation	United Airlines
Operation	Scheduled Airline
Route	Honolulu, HI to Aukland, New Zealand
On-Board Injuries	Fatal <u>9</u> Serious <u>5</u> Minor/None <u>337</u>
Aircraft Damage	Substantial

Description   Source: Aviation Safety Network[1]

United Airlines Flight 811, a Boeing 747-122, N4713U, took off from Honolulu (HNL), Hawaii at 01:33 local time, bound for Sydney, Australia, with an intermediate stop at Auckland, New Zealand. The initial climb passed through an area of thunderstorms, so the captain elected to keep the seat belt sign on. As the aircraft was climbing, between 22,000 and 23,000 feet, explosive decompression was experienced. An emergency was declared at approximately 02:20 HST. The captain initiated a 180-degree left turn to avoid a thunderstorm and proceeded toward HNL. The forward lower lobe cargo door had opened in flight, taking with it a large portion of the forward right side of the cabin fuselage. The starboard side engines (no.3 and 4) were damaged and had to be shut down. Parts of the leading and trailing edge flaps where also damaged resulting in the crew electing to use only 10-degrees trailing edge flaps for landing (a non-normal configuration). This resulted in the aircraft having to land at a higher speed than it would under normal conditions. The aircraft was cleared to land at HNL runway 8L. At 02:34 HST, Honolulu tower was notified by the flight crew that the airplane was stopped and an emergency evacuation had commenced on the runway. During the decompression, nine passengers had been ejected from the airplane and lost at sea.

Probable Cause   Source: Aviation Safety Network[1]

The sudden opening of the forward lower lobe cargo door in flight and the subsequent explosive decompression. The door opening was attributed to a faulty switch or wiring in the door control system which permitted electrical actuation of the door latches toward the unlatched position after initial door closure and before takeoff. Contributing to the cause of the accident was a deficiency in the design of the cargo door locking mechanisms, which made them susceptible to deformation, allowing the door to become unlatched after being properly latched and locked. Also contributing to the accident was a lack of timely corrective actions by Boeing and the FAA following a 1987 cargo door opening incident on a Pan Am B-747.

Birdstrikes/Bird Ingestion

October 4, 1960: Lockheed L-188A, N-5533
Location	Boston, Logan Airport, MA
Flight	EAL 375
Operator	Eastern Air Lines
Operation	Scheduled Airline
Route	Boston, MA to Philadelphia, PA
On-Board Injuries	Fatal 5 Minor/None 137
Aircraft Damage	Destroyed

Description   Source: Aviation Safety Network[1]

A few seconds after taking off from runway 05, the Electra struck a flock of starlings. A number of these birds were ingested in engine no.1, 2 and 4. Engine no. 1 was shut down and the prop feathered. Shortly after that the no. 2 and 4 engines experienced a substantial momentary loss of power. This caused the plane to yaw to the left and decelerate to stall speed. The left wing then dropped, the nose pitched up and the L-188 rolled left into a spin and fell almost vertically into the water.

Probable Cause   Source: Accident Report[10]

The unique and critical sequence of the loss and recovery of engine power following bird ingestion, resulting in loss of airspeed and control during takeoff.

January 15, 2009: Airbus A320, N-106US
Location	Hudson River, NJ, NY
Flight	USA 1549
Operator	US Airways
Operation	Scheduled Airline
Route	New York La Guardia, NY to Charlotte, NC
On-Board Injuries	Minor/None 155
Aircraft Damage	Destroyed*

Description   Source: Aviation Safety Network[1]

US Airways flight 1549, an Airbus A320-214, experienced an almost total loss of thrust in both engines after encountering a flock of birds and was subsequently ditched on the Hudson River near New York-LaGuardia Airport, USA.

The flight was en route to Charlotte Douglas International Airport (CLT), North Carolina, USA, and departed LaGuardia runway 04 at 15:24. At this time, the first officer was the pilot flying (PF), and the captain was the pilot monitoring (PM).

The takeoff and initial portion of the climb were uneventful. At 15:25:45, the LaGuardia ATCT local controller instructed the flight crew to contact the New York Terminal Radar Approach Control (TRACON) LGA departure controller. The captain contacted the departure controller at 15:25:51, advising him that the airplane was at 700 feet and climbing to 5,000 feet. The controller then instructed the flight to climb to and maintain 15,000 feet, and the captain acknowledged the instruction.

At 15:27:10, the captain stated, "birds." One second later, at an altitude of 2,818 feet above ground level, the crew heard thumps and thuds followed by a shuddering sound. The aircraft had struck several Canada geese.

Immediately after the bird encounter, both engines' fan and core (N1 and N2, respectively) speeds started to decelerate.

At 15:27:14, the first officer stated, "uh oh," followed by the captain stating, "we got one roll- both of 'em rolling back." The captain then stated he would start the APU and took over control of the airplane. At 15:27:28, the captain instructed the first officer to "get the QRH [quick reference handbook] loss of thrust on both engines", and reported the emergency situation to the LGA departure controller, stating, "mayday mayday mayday...this is...Cactus fifteen thirty nine hit birds, we've lost thrust in both engines, we're turning back towards LaGuardia."

The LGA departure controller acknowledged the captain's statement and then instructed him to turn left heading 220°.

The first officer began conducting Part 1 of the QRH ENG DUAL FAILURE checklist (Engine Dual Failure checklist), stating, "if fuel remaining, engine mode selector, ignition," and the captain responded, "ignition." The first officer

---

* The NTSB describes the damage as "Substantial." In fact, the airplane registration was cancelled following the accident.

then stated, "thrust levers confirm idle," and the captain responded, "idle." About 4 seconds later, the first officer stated, "airspeed optimum relight. three hundred knots. we don't have that," and the captain responded, "we don't."

At 15:28:05, the LGA departure controller asked the captain if he wanted to try to land on runway 13 at LGA if it was available, and the captain responded, "we're unable. we may end up in the Hudson [River]."

The LGA departure controller cleared the flight for a left-hand traffic pattern for runway 31, but the captain responded, "unable." The controller then stated that runway 4 at LGA was available, and the captain responded, "I'm not sure we can make any runway. Uh what's over to our right anything in New Jersey maybe Teterboro?" The controller replied, "ok yeah, off your right side is Teterboro Airport [TEB]."

Subsequently, the departure controller asked the captain if he wanted to try going to TEB, and the captain replied, "yes." At 15:29:11, the captain announced on the public address (PA) system, "this is the Captain, brace for impact." Meanwhile the pilots were working the checklist to restart the engines.

At 15:29:21, the LGA departure controller instructed the captain to turn right 280° for runway 1 at TEB. But the captain responded: "we can't do it."

The departure controller then asked the captain which runway at TEB he would like, and the captain responded, "we're gonna be in the Hudson." When it became clear that the engines would not restart, the captain requested to first officer to select the flaps.

At 15:30:43, the aircraft landed on the surface of the Hudson River. Within seconds after the ditching, the crewmembers and passengers initiated evacuation of the airplane. Subsequently, all of the occupants were evacuated from the airplane and rescued by area responders.

Probable Cause   Source: Aviation Safety Network[1]

"The ingestion of large birds into each engine, which resulted in an almost total loss of thrust in both engines and the subsequent ditching on the Hudson River. Contributing to the fuselage damage and resulting unavailability of the aft slide/rafts were (1) the Federal Aviation Administration's approval of ditching certification without determining whether pilots could attain the ditching parameters without engine thrust, (2) the lack of industry flight crew training and guidance on ditching techniques, and (3) the captain's resulting difficulty maintaining his intended airspeed on final approach due to the task saturation resulting from the emergency situation.

Contributing to the survivability of the accident was (1) the decision-making of the flight crewmembers and their crew resource management during the accident sequence; (2) the fortuitous use of an airplane that was equipped for an extended overwater flight, including the availability of the forward slide/rafts, even though it was not required to be so equipped; (3) the performance of the cabin crewmembers while expediting the evacuation of the airplane; and (4) the proximity of the emergency responders to the accident site and their immediate and appropriate response to the accident."

November 12, 1975: McDonnell Douglas DC-10, N-1032F

Location	JFK Airport, NY
Flight	ONA 032
Operator	Overseas National Airways
Operation	Non-Revenue Flight
Route	New York, NY to Frankfurt, Germany
On-Board Injuries	Minor/None 137
Aircraft Damage	Destroyed

Description   Source: Aviation Safety Network[1]

ONA Flight 032 left the gate at 12:56 EST for a ferry flight to Jeddah via Frankfurt. The passengers were all ONA employees. The aircraft taxied to runway 13R and commenced takeoff at 13:10. Shortly after accelerating through 100 knots, but before reaching the V1 speed, a flock of birds were seen to rise from the runway. The aircraft struck many birds and the takeoff was rejected. Bird strikes had damaged the no. 3 engine's fan blades, causing rotor imbalance. Fan-booster stage blades began rubbing on the epoxy micro balloon shroud material; pulverized material then entered into the engine's HPC area, ignited and caused the compressor case to separate. A fire erupted in the right wing and no. 3 engine pylon. The aircraft couldn't be stopped on the runway. The pilot-in-command steered the aircraft off the runway onto taxiway Z at a 40 knots speed. The main undercarriage collapsed and the aircraft came to rest against the shoulder of the taxiway. The successful evacuation may be partially attributed to the fact that nearly all passengers were trained crew members.

Probable Cause   Source: Accident Report[11]

The disintegration and subsequent fire in the No.3 engine when it ingested a large number of seagulls. Following the disintegration of the engine, the aircraft failed to decelerate effectively because: 1) The No.3 hydraulic system was inoperative, which caused the loss of the no.2 brake system and braking torque to be reduced 50 percent; 2) the no.3 engine thrust reversers were inoperative; 3) at least three tyres disintegrated; 4) the no.3 system spoiler panels on each wing could not deploy; and 5) the runway surface was wet. The following factors contributed to the accident: 1) The bird-control program at John F. Kennedy airport did not effectively control the bird hazard on the airport; and 2) the FAA and the General Electric Company failed to consider the effects of rotor imbalance on the abradable epoxy shroud material when the engine was tested for certification.

Regulatory Outcome

At the time of the L-188 accident, the Civil Air Regulations (CAR) were the effective regulations regarding certification and operation of airplanes and engines. By the time regulatory changes were issued, 14 CFR part 25 and 33 had replaced the CARs. As operational history was collected on turbine powered transport airplanes, a number of engine failure threats from foreign object ingestion had surfaced, bird threats being only one.

Advisory Circular 33-1A was published in June 1968 and provided the first guidance on bird ingestion and criteria for engine certification.[12]

While not a direct result of this accident, continued service experience with turbine engine bird ingestion, and a greater understanding of the bird threat, resulted in regulatory changes. These changes were subsequently adopted as:

- Foreign object ingestion, §33.77 was adopted in October 1974 to initially address the threat from bird ingestion.[13]

- The bird ingestion rules of §33.77 were moves to a new paragraph, §33.76 adopted in December 2000.[14] These changes were intended to improve the tolerance of turbine engines to bird ingestion.

January 4, 2009: Sikorsky S-76, N-748P

Location	Morgan City, LA
Identification	N-748P
Operator	Petroleum Helicopters
Operation	Off-Shore Personnel Transportation
Route	Amelia, LA to Tambelier 301B Oil Rig
On-Board Injuries	Fatal 8 Serious 1
Aircraft Damage	Destroyed

Description   Source: Aviation Safety Network[1]

A Sikorsky S-76C departed on an air taxi flight from PHI, Inc.'s heliport en route to an offshore oil platform with two pilots and seven passengers. Data from the helicopter's flight data recorder indicated that the helicopter established level cruise flight at 850 feet mean sea level and 135 knots indicated air speed. About 7 minutes after departure, the cockpit voice recorder recorded a loud bang, followed by sounds consistent with rushing wind and a power reduction on both engines and a decay of main rotor revolutions per minute. Due to the sudden power loss, the helicopter departed controlled flight and descended rapidly into marshy terrain.

Examination of the wreckage revealed that both the left and right sections of the cast acrylic windshield were shattered. Feathers and other bird remains were collected from the canopy and windshield at the initial point of impact and from other locations on the exterior of the helicopter. Laboratory analysis identified the remains as coming from a female red-tailed hawk; the females of that species have an average weight of 2.4 pounds. No defects in the materials, manufacturing, or construction were observed. There was no indication of any preexisting damage that caused the windshield to shatter. Thus, the fractures at the top of the right section of the windshield and damage to the canopy in that area were consistent with a bird impacting the canopy just above the top edge of the windshield. The fractures in the other areas of the windshields were caused by ground impact.

Probable Cause   Source: NTSB Accident File[15]

The National Transportation Safety Board determines the probable cause(s) of this accident to be
(1) the sudden loss of power to both engines that resulted from impact with a bird (red-tailed hawk), which fractured the windshield and interfered with engine fuel controls, and
(2) the subsequent disorientation of the flight crewmembers, which left them unable to recover from the loss of power.
Contributing to the accident were
  (1) the lack of Federal Aviation Administration regulations and guidance, at the time the helicopter was certificated, requiring helicopter windshields to be resistant to bird strikes;
  (2) the lack of protections that would prevent the T handles from inadvertently dislodging out of their detents; and
  (3) the lack of a master warning light and audible system to alert the flight crew of a low-rotor-speed condition.
Based on main rotor speed decay information provided by Sikorsky, the accident flight crew had, at most, about 6 seconds to react to the decaying rotor speed condition. Had they quickly recognized the cause of the power reduction and reacted very rapidly, they would likely have had enough time to restore power to the engines by moving the ECLs back into position. However, the flight crewmembers were likely disoriented from the bird strike and the rush of air through the fractured windshield; thus, they did not have time to identify the cause of the power reduction and take action to move the ECLs back into position.

The accident helicopter was not equipped with an audible alarm or a master warning light to alert the flight crew of a low-rotor-speed condition. An enhanced warning could have helped the accident flight crew quickly identify the decaying rotor speed condition and provided the flight crew with more opportunity to initiate the necessary corrective emergency actions before impact.

Certification Issues

In 1978, when the S-76 was certificated, there were no bird-strike requirements. Currently, 14 Code of Federal Regulations 29.631 (in effect since August 8, 1996) states that, at a minimum, a transport-category helicopter, such as the S-76C++, should be capable of safe landing after impact with a 2.2-pound bird at a specified velocity. This requirement includes windshields. Current FAA requirements for transport-category helicopter windshields also state that windshields and windows must be made of material that will not break into dangerous fragments.

The S-76C helicopter has an overhead engine control quadrant that houses, among other components, two engine fire extinguisher T-handles and two engine power control levers (ECL). The fire extinguisher T-handles, which are located about 4 inches aft of the captain's and first officer's windshields, are normally in the full-forward position during flight, and each is held in place by a spring-loaded pin that rests in a detent; aft pulling force is required to move the T-handles out of their detents. If the T handles are moved aft, a mechanical cam on each T-handle pushes the trigger on the associated ECL out of its wedge-shaped stop, allowing the ECL to move aft, reducing fuel to the engine that the ECL controls. (Flight crews are trained to move an engine's fire extinguisher T-handle full aft in the event of an in-flight fire so that the ECL can move aft and shut off the fuel flow to the affected engine.)

The impact of the bird on the canopy just above the windshield near the engine control quadrant likely jarred the fire extinguisher T-handles out of their detents and moved them aft, pushing both ECL triggers out of their stops and allowing them to move aft and into or near the flight-idle position, reducing fuel to both engines. A similar incident occurred on November 13, 1999, in West Palm Beach, Florida, when a bird struck the windshield of an S-76C helicopter, N276TH, operated by Palm Beach County. The bird did not penetrate the laminated glass windshield, but the impact force of the bird cracked the windshield and dislodged the fire extinguisher T-handles out of their detents; however, in that case, the force was not great enough to move the ECLs.

Regulatory Outcome

The design of the original ECL was modified by installing an improved throttle stop and a wider trigger. This redesign was mandated in Airworthiness Directive 2012-22-13. The investigation discovered that a bird-strike to the windshield area of the helicopter resulted in unintended movement of the engine control levers from the forward position and towards the flight-idle position, which reduced power on both engines. This redesign was intended to prevent inadvertent movement of the ECLs, which could result in main rotor speed decay and subsequent loss of control of the aircraft.

This recommended practice was disseminated to all helicopter operators. The Helicopter Safety Advisory Conference recommended operational means (primarily flight at higher altitudes) by which to reduce the probability of a bird strike.[16]

No other initiatives were reported.

Controlled Flight Into Terrain

May 6, 1935: Douglas DC-2, NC-13785

Location   Near Atlanta, MO
Flight   TWA 2
Operator   Transcontinental & Western Air
Operation   Scheduled Airline
Route   Albuquerque, NM to Kansas City, MO
On-Board Injuries   Fatal <u>6</u>   Serious <u>7</u>
Aircraft Damage   Destroyed

Description   Source: Aviation Safety Network[1]

A Transcontinental and Western Air (TWA) DC-2 crashed near Atlanta, MO., killing five of the eight persons aboard. Senator Bronson M. Cutting (R-NM) was among the fatalities. A Bureau of Air Commerce report cited the accident's causes as the U.S. Weather Bureau's failure to predict hazardous weather and misjudgments by the pilot and TWA ground personnel. In June 1936, however, a committee chaired by Sen. Royal S. Copeland (D-NY) issued a report alleging that the tragedy was caused by malfunctioning navigational aides and voicing other criticisms of the Bureau of Air Commerce. The controversy gave impetus to legislative efforts that eventuated in the Civil Aeronautics Act of 1938 which included establishing an independent Air Safety Board.

Probable Cause   Source: Accident Report[17] [18]

It is the opinion of the Accident Board that the probable direct cause of this accident was an unintentional collision with the ground while the airplane was being maneuvered at a very low altitude in fog and darkness.

     The probable contributory causes of this accident were:

(a) A forecast by the United States Weather Bureau which did not predict the hazardous weather that developed during the latter part of the forecast period.

(b) Improper clearance of the airplane from Albuquerque by the company's ground personnel because of their knowledge that the plane's two-way radio was not functioning on the western night frequency.

(c) Improper control by the company's ground personnel at Albuquerque for not calling the airplane back or ordering it to stop at an intermediate point when it was found that two-way radio communication could not be established.

(d) Error on the part of the pilot for proceeding in the flight after discovering that he was unable to effectively communicate with the ground.

(e) Failure of the company's ground personnel at Kansas City to expeditiously redispatch the airplane to a field where better weather existed when it became apparent that the ceiling at Kansas City was dropping to and below the authorized minimum for landing and while the airplane still had sufficient fuel to meet the Department of Commerce requirement of 45-minute fuel reserve after effecting a landing.

Regulatory Outcome

The controversy gave impetus to legislative efforts that resulted in the Civil Aeronautics Act of 1938 which established an independent Air Safety Board.

September 4, 1971: Boeing 727, N-2969G

Location	Near Juneau, AK
Flight	ASA* 66
Operator	Alaska Airlines
Operation	Scheduled Airline
Route	Yakutat, AK to Juneau, AK
On-Board Injuries	Fatal 111
Aircraft Damage	Destroyed

Description   Source: Aviation Safety Network[1]

Alaska Airlines, Flight 1866 (AS66) was a scheduled passenger flight from Anchorage (ANC), to Seattle (SEA), with intermediate stops at Cordova (CDV), Yakutat (YAK), Juneau (JNU), and Sitka (SIT). The IFR flight departed Anchorage at 09:13 and landed at Cordova at 09:42. AS66 departed Cordova at 10:34 after a delay, part of which was attributable to difficulty in securing a cargo compartment door. The flight landed at Yakutat at 11:07. While on the ground, AS66 received an air traffic control clearance to the Juneau Airport via Jet Route 507 to the Pleasant Intersection, direct to Juneau, to maintain 9,000 feet or below until 15 miles southeast of Yakutat on course, then to climb to and maintain FL230. The flight departed Yakutat at 11:35, with 104 passengers and seven crew members on board.

At 11:46, AS66 contacted the Anchorage ARTCC and reported level at FL230, 65 miles east of Yakutat. The flight was then cleared to descend at the pilot's discretion to maintain 10,000 ft so as to cross the Pleasant Intersection at 10,000 feet and was issued a clearance limit to the Howard Intersection.

The clearance was acknowledged correctly by the captain and the controller provided the Juneau altimeter setting of 29.46 inches and requested AS66 to report leaving 11,000 ft. At 11:51, AS66 reported leaving FL230. Following this report, the flight's clearance limit was changed to the Pleasant Intersection. At 11:54, the controller instructed AS66 to maintain 12,000 feet. Approximately 1 minute later, the flight reported level at 12,000 feet. The changes to the flight's original clearance to the Howard Intersection were explained to AS66 by the controller as follows: "I've got an airplane that's not following his clearance, I've got to find out where he is." The controller was referring to N799Y, a Piper Apache which had departed Juneau at 11:44 on an IFR clearance, destination Whitehorse, Canada.

On two separate occasions, AS66 acted as communications relay between the controller and N799Y.

At 11:58, AS66 reported that they were at the Pleasant Intersection, entering the holding pattern, whereupon the controller recleared the flight to Howard Intersection via the Juneau localizer. In response to the controller's query as to whether the flight was "on top" at 12,000 feet, the captain stated that the flight was "on instruments." At 12:00, the controller repeated the flight's clearance to hold at Howard Intersection and issued an expected approach time of 12:10. At 12:01, AS66 reported that they were at Howard, holding 12,000 feet. Six minutes later, AS66 was queried with respect to the flight's direction of holding and its position in the holding pattern. When the controller was advised that the flight had just completed its inbound turn and was on the localizer, inbound to Howard, he cleared AS66 for a straight-in LDA approach, to cross Howard at or below 9,000 feet inbound. The captain acknowledged the clearance and reported leaving 12,000 feet. At 12:08 the captain reported "leaving five thousand five... four thousand five hundred," whereupon the controller instructed AS66 to contact Juneau Tower. Contact with the tower was established shortly thereafter when the captain reported, "Alaska sixty-six Barlow inbound." (Barlow Intersection is located about 10 nautical miles west of the Juneau Airport). The Juneau Tower Controller responded, "Alaska 66, understand, ah, I didn't, ah, copy the intersection, landing runway 08, the wind 080° at 22 occasional gusts to 28, the altimeter now 29.47, time is 09 1/2, call us by Barlow". No further communication was heard from the flight.

The Boeing 727 impacted the easterly slope of a canyon in the Chilkat Range of the Tongass National Forest at the 2475-foot level. The aircraft disintegrated on impact.

Probable Cause   Source: Accident Report[19]

A display of misleading navigational information concerning the flight's progress along the localizer course which resulted in a premature descent below obstacle clearance altitude. The origin or nature of the misleading navigational information could not be determined. The Board further concludes that the crew did not use all available navigational aids to check the flight's progress along the localizer nor were these aids required to be used. The crew also did not perform the required audio identification of the pertinent navigational facilities.

---

*   The source material uses the AS (IATA code for Alaska Airlines). It is not clear why the flight number 1866 was truncated to 66.

October 30, 1974: Lockheed L-188, CF-PAB

Location	3 km S Rea Point Airfield, NT, Canada
Identification	CF-PAB*
Operator	Panarctic Oils
Operation	Business Transportation
Route	Edmonton, AB, Canada to Rea Point, NT, Canada
On-Board Injuries	Fatal <u>32</u> Serious <u>2</u>
Aircraft Damage	Destroyed

Description   Source: Aviation Safety Network[1]

Lockheed L-188 Aircraft CF-PAB operated as Panarctic flight 416 and departed Calgary International Airport at 18:05 hours 29 October 1974. The aircraft was on a routine positioning flight to Edmonton with a pilot-in-command, co-pilot and flight engineer on board. The 30-minute flight was uneventful with no unserviceabilities reported by the crew. The aircraft was prepared for the continuing flight north with the loading of 20000 lb of baggage and freight and 21000 lb of jet B fuel. The aircraft pilot-in-command and flight engineer were replaced by those scheduled for the Edmonton to Rea Point leg.

The pilot-in-command received a weather briefing; an IFR flight plan was filed at an initial cruising altitude of 18000 ft with Pedder Point as the alternate.

The estimated time en-route was 4 hours 12 minutes. After loading 30 passengers and a fourth crew man, the loadmaster/flight attendant, the aircraft departed the Edmonton International Airport at 20:04 hours. The flight proceeded uneventfully, cruising at 18 000ft to Fort Smith where it was cleared to flight level 210. The aircraft reported over Byron Bay at 23:04 hours with an estimated time of arrival at Rea Point of 00:16. About 100 miles north of Byron Bay the aircraft was cleared to flight level 250.

Radio contact was established with Rea Point about 150 miles out and a descent was started for a straight-in VOR/DME approach to runway 33. The descent was smooth except for some turbulence at 4000 ft. The aircraft levelled at 17 miles DME from Rea Point at 2000 ft for a period of 1 minute 45 seconds. The aircraft then slowly descended to about 875 ft ASL at 6 miles DME. A call was made to Rea Point advising them of the DME range on final. There was light turbulence. Fifteen hundred horsepower was selected on the engines; both the VHF navigation radios were selected to the Rea Point VOR frequency and both ADF's were selected to the Rea Point OX nondirectional beacon. Both cockpit barometric altimeters were set to 29.91 in of mercury, the latest Rea Point setting. The airspeed was indicating 150 kt which, with a 30 kt headwind component, resulted in a ground speed of 120 kt. The pre-landing check had been completed, 100 per cent flap selected and the landing gear was down. The landing lights were extended but were off, the wing leading edge lights as well as the alternate taxi lights were on. Glare had been experienced from external lights early in the descent from 10000 ft, but not thereafter. There was no pre-landing briefing conducted by the pilot-in-command.

The flight engineer was able to see what appeared to be open water below with ice. The co-pilot set his radio altimeter warning to 450 ft and the pilot-in-command set his to 300 ft. When the warning light came on the co-pilot's radio altimeter, he advised the pilot-in-command. As the descent continued through the minimum descent altitude of 450 ft, the co-pilot reset his radio altimeter to 300 ft and so advised the pilot-in-command. The aircraft was still in a shallow descent. At 300 ft radio altitude the co-pilot checked the DME reading as 3 miles, saw a dark area of open water and an ice line and reported to the pilot-in-command that they seemed to be approaching an ice ridge and that they had visual contact. The pilot-in-command reset his radio altimeter to about 150 ft. Also, close to this time the pilot-in-command said he believed they were on top of a layer of cloud, repeated the statement, following which he retarded the throttles and pushed forward on the control column with sufficient force to produce perceptible negative G.

The rate of descent increased rapidly to between 1700 and 2000 ft a minute. The co-pilot shouted at the pilot-in-command reporting their descent through 200 ft at 2 miles DME but there was no response. The flight engineer and the co-pilot both called through 50 ft without an observed reaction from the pilot-in-command. The co-pilot reached for the right side power levers and found the flight engineer's hands already on them. On impact, the cockpit area broke away from the remainder of the fuselage and with the cargo continued along the ice surface for 900 ft. After the cockpit came to rest, the flight engineer undid his seat belt and saw both the pilot-in-command and co-pilot in their seats. The co-pilot although injured was able to undo his seat belt and the flight engineer pulled him on to the ice before the cockpit section sank completely.

Probable Cause   Source: Accident Report[20]

In his perceptual state the Captain interpreted the visual information as requiring an immediate steep descent. Misinterpretations could include: the ice/water line; pitchup from the dark/light difference; the ice being a cloud layer; or variations in light intensity and/or image shifting.

After the steep descent was established the Captain did not respond to the warnings of the First Officer and Flight Engineer. He also failed to respond to the instrumentation that showed a hazardous rate of descent at low level three miles from the airstrip. This failure to respond indicates a degree of incapacitation.

Comments

It appears that this CFIT accident has been mis-interpreted as a tailplane stall accident.

---

*   Aircraft Registration used *vice* flight number

December 1, 1974: Boeing 727, N-54328

Location	Mount Weather (Berryville), VA
Flight	TWA 514
Operator	Trans World Airlines
Operation	Scheduled Airline
Route	Columbus, OH to Washington, DC
On-Board Injuries	Fatal 92
Aircraft Damage	Destroyed

Description

Source: Aviation Safety Network[1]

Trans World Airlines Flight 514 was a regularly scheduled flight from Indianapolis, IN (IND), to Washington-National Airport, DC (DCA), with an intermediate stop at Columbus-Port Columbus International Airport, OH (CMH).

Flight 514 departed Indianapolis at 08:53 EST and arrived in Columbus at 09:32. The Boeing 727 departed Columbus at 10:24, eleven minutes late. There were 85 passengers and 7 flight crew members aboard the aircraft when it departed Columbus.

At 10:36, the Cleveland Air Route Traffic Control Center (ARTCC) informed the crew of Flight 514 that no landings were being made at Washington National Airport because of high crosswinds, and that flights destined for that airport were either being held or being diverted to Dulles International Airport (IAD). At 10:38, the captain of Flight 514 communicated with the dispatcher in New York and advised him of the information he had received. The dispatcher, with the captain's concurrence, subsequently amended Flight 514's release to allow the flight to proceed to Dulles.

One minute later, the controller cleared the flight to descend to FL230 and to cross a point 40 miles west of Front Royal at that altitude. Control of the flight was then transferred to the Washington ARTCC and communications were established with that facility at 10:48. In the meantime, the flightcrew discussed the instrument approach to runway 12, the navigational aids, and the runways at Dulles, and the captain turned the flight controls over to the first officer. When radio communications were established with Washington ARTCC, the controller affirmed that he knew the flight was proceeding to Dulles. Following this contact, the crew discussed the various routings they might receive to conduct a VOR/DME approach to runway 12 at Dulles.

At 10:51, the Washington ARTCC controller requested the flight's heading. After being told that the flight was on a heading of 100 degrees, the controller cleared the crew to change to a heading of 090°, to intercept the 300° radial of the Armel VOR, to cross a point 25 miles northwest of Armel and to maintain 8,000 feet, "...and the 300° radial will be for a VOR approach to runway 12 at Dulles." He gave the crew an altimeter setting of 29.74 for Dulles. The crew acknowledged this clearance. The pilots again discussed the VOR/DME approach At 10:55, the landing preliminary checklist was read by the flight engineer and the other crewmembers responded to the calls. A reference speed of 127 kts was calculated and set on the airspeed indicator reference pointers. The altimeters were set at 29.74. The crew then again discussed items on the instrument approach chart including the Round Hill intersection, the final approach fix, the visual approach slope indicator and runway lights, and the airport diagram. At 10:59, the captain commented that the flight was descending from 11,000 feet to 8,000 feet. He then asked the controller if there were any weather obstructions between the flight and the airport. The controller replied that he did not see any significant weather along the route. The captain replied that the crew also did not see any weather on the aircraft weather radar.

At 11:01, the controller cleared the flight to descend to and maintain 7,000 feet and to contact Dulles approach control.

Twenty-six seconds later, the captain initiated a conversation with Dulles approach control and reported that the aircraft was descending from 10,000 feet to maintain 7,000 feet. He also reported having received the information "Charlie" transmitted on the ATIS broadcast. The controller replied with a clearance to proceed inbound to Armel and to expect a VOR/DME approach to runway 12. The controller then informed the crew that ATIS information Delta was current and read the data to them. The crew determined that the difference between information Charlie and Delta was the altimeter setting which was given in Delta as 29.70. There was no information on the CVR to indicate that the pilots reset their altimeters from 29.74. At 11:04, the flight reported it was level at 7,000 feet.

Five seconds after receiving that report, the controller said, "TWA 514, you're cleared for a VOR/DME approach to runway 12." This clearance was acknowledged by the captain. The CVR recorded the sound of the landing gear warning horn followed by a comment from the captain that "Eighteen hundred is the bottom." The first officer then said, "Start down." The flight engineer said, "We're out here quite a ways. I better turn the heat down." At 11:05:06, the captain reviewed the field elevation, the minimum descent altitude, and the final approach fix and discussed the reason that no time to the missed approach point was published. At 11:06:15, the first officer commented that, "I hate the altitude jumping around." Then he commented that the instrument panel was bouncing around. At 11:06:15, the captain said, "We have a discrepancy in our VOR's, a little but not much." He continued, "Fly yours, not mine." At 11:06:27, the captain discussed the last reported ceiling and minimum descent altitude. concluded, "...should break out."

At 11:06:42, the first officer said, "Gives you a headache after a while, watching this jumping around like that." At 11:07:27, he said, "...you can feel that wind down here now." A few seconds later, the captain said, "You know, according to this dumb sheet it says thirty-four hundred to Round Hill --- is our minimum altitude." The flight engineer then asked where the captain saw that and the captain replied, "Well, here. Round Hill is eleven and a half DME." The first officer said, "Well, but ---" and the captain replied, "When he clears you, that means you can go to your ---" An unidentified voice said, "Initial approach, and another unidentified voice said, "Yeah!" Then the captain said "Initial approach altitude." The flight engineer then said, "We're out a --- twenty-eight for eighteen." An unidentified voice said, "Right, and someone said, "One to go." At 11:08:14, the flight engineer said, "Dark in here," and the first officer stated, "And bumpy too." At 11:08:25, the sound of an altitude alert horn was recorded. The captain said, "I had ground contact a minute ago," and the first officer replied, "Yeah, I did too." At 11:08:29, the first officer said, "...power on this.... " The captain said "Yeah --- you got a high sink rate." "Yeah," the first officer replied. An unidentified voice said, "We're going uphill, " and the flight engineer replied, "We're right there, we're on course." Two voices responded, "Yeah!" The captain then said, "You ought to see ground outside in just a minute -- Hang in there boy." The flight engineer said, "We're getting seasick.

At 1108:57, the altitude alert sounded. Then the first officer said, "Boy, it was --- wanted to go right down through there, man," to which an unidentified voice replied, "Yeah!" Then the first officer said, "Must have had a of a downdraft."

At 1109:14, the radio altimeter warning horn sounded and stopped. The first officer said, "Boy!" At 11:09:20, the captain said, "Get some power on." The radio altimeter warning horn sounded again and stopped. At 11:09:22, the sound of impact was recorded.

After the aircraft left 7,000 feet, the descent was continuous with little rate variation until the indicated altitude was about 1,750 feet. [then] increased about 150 feet over a 15-second period and then decreased about 200 feet during a 20-second period. The recorded altitude remained about 1,750 feet until the airplane impacted the west slope of Mount Weather, Virginia, about 25 NM from Dulles, at an elevation of about 1,670 feet (509 m).

Probable Cause   Source: Accident Report[21]

The National Transportation Safety Board determines that the probable cause of the accident was the crew's decision to descend to 1,800 feet before the aircraft had reached the approach segment where that minimum altitude applied. The crew's decision to descend was a result of inadequacies and lack of clarity in the air traffic control procedures which led to a misunderstanding on the part of the pilots and of the controllers regarding each other's responsibilities during operations in terminal areas under instrument meteorological conditions. Nevertheless, the examination of the plan view of the approach chart should have disclosed to the captain that a minimum altitude of 1,800 feet was not a safe altitude.

Regulatory Outcome[22]

Following this accident, the FAA issued a fleet-wide requirement for installation of Ground Proximity Warning Systems in all air carrier aircraft by December 1975.

January 1, 1985: Boeing 727, N-819EA

Location	Near La Paz, Bolivia
Flight	EAL 980
Operator	Eastern Air Lines
Operation	Scheduled Airline
Route	Asunción, Paraguay to La Paz, Bolivia
On-Board Injuries	Fatal 29
Aircraft Damage	Destroyed

Description   Source: Aviation Safety Network[1]

Eastern flight 980 was a scheduled flight from Ascuncion (ASU), Paraguay to La Paz (LPB), Bolivia. The aircraft had reported crossing the DAKON intersection, 55 NM southeast of La Paz, at 25,000 ft. MSL. They were then cleared by La Paz ATC to descend to 18,000 feet, and the crew acknowledged this clearance. Although the Boeing 727 was supposed to be approaching La Paz along airway UA 320, on a 134° radial from the La Paz VOR, it veered significantly off course beyond DAKON; perhaps because the crew were maneuvering to avoid weather in the vicinity. The aircraft impacted the 19,600-ft. level of Mt. Illimani, a 21,000-ft. Andean peak, approximately 26 NM from the La Paz VOR/DME and 25 NM from La Paz runway 09R. Dark night, weather, and lack of visual references in the area all contributed to the crew's inability to see and avoid the high terrain in their path.

A climbing expedition was organized the following summer to retrieve the flight recorders. The expedition reached the crash site and was able to dig through accumulated snow and examine the wreckage. However, bad weather and altitude sickness forced the expedition to turn back without recovering the recorders. A June 2016 an expedition found parts of the aircraft's cockpit voice recorder and possibly flight data recorder as well.

Probable Cause*

:The aircraft had reported crossing the DAKON intersection, 55 NM southeast of La Paz, at 25,000 ft. MSL. They were then cleared by La Paz ATC to descend to 18,000 feet, and the crew acknowledged this clearance. Although the accident aircraft was supposed to be approaching La Paz along airway UA 320, on a 134° radial from the La Paz VOR, it veered significantly off course beyond DAKON; the impact location is along the 106° radial from La Paz. Investigators speculate the flight crew were maneuvering to avoid weather in the vicinity, and that impact occurred with the aircraft in cruise con-

---

*   The official Bolivian accident report was not located. A description of the circumstances was reported in a Department of Transportation Study.*

figuration, in a shallow descent. Dark night, weather, and lack of visual references in the area all contributed to the crew's inability to see and avoid the high terrain in their path.[24]

Due to the extreme high altitude and inaccessibility of the accident location, the FDR and CVR were never recovered. A climbing expedition was organized the following summer to retrieve these recorders. The expedition reached the crash site and was able to dig through accumulated snow and examine the wreckage. However, bad weather and altitude sickness forced the expedition to turn back without recovering the recorders.[25]

Regulatory Outcome

While the GPWS improved things greatly, it could not help in situations where the airplane was flying level towards a vertical cliff. As happened in this case. The US Department of Transportation studied the available data and published a profile of the accident trajectory as shown in **Figure 61** (page 144).

The La Paz accident pointed out the problem was the steep approach into near vertical terrain. GPWS would not have given enough warning to avoid the mountain. The accident gave impetus to the development of world-wide digital terrain databases. In 2000, NASA flew the Shuttle Radar Topography Mission aboard the Space Shuttle *Endeavour*. This mission used dual radar antennas to acquire interferometric data processed to digital topographic data at 1 arc-sec resolution.[26]

The original GPWS was improved by including digital terrain databases. provided no preliminary advisories and started with a loud warning: "Whoop, Whoop, Pull Up!" The second version provided advisories in some cases, alerting the crew that the system was approaching safe limits. They might hear a preliminary warning "Too Low – Gear", followed by the pull up command. This type of sequence provided confidence that the system was working by providing a rationale for the final alert.

This lead to the requirement for Terrain Awareness and Warning Systems (TAWS) which would be required in all US Airline turbine powered airplanes in 2005.[27]

January 20, 1992: Airbus A320, F-GGED

Location	19.5 km SW Strasbourg Airport, France
Flight	ITF 148
Operator	Air Inter
Operation	Scheduled Airline
Route	Lyon, France to Strasbourg, France
On-Board Injuries	Fatal <u>87</u> Serious <u>5</u> Minor/None <u>4</u>
Aircraft Damage	Destroyed

Description   Source: Aviation Safety Network[1]

Air Inter Flight ITF148, an Airbus A320, took off from Lyon (LYS) at 18:20 on a domestic service to Strasbourg-Entzheim Airport (SXB). Following an uneventful flight the crew prepared for a descent and approach to Strasbourg. At first the crew asked for an ILS approach to runway 26 followed by a visual circuit to land on runway 05. This was not possible because of departing traffic from runway 26. The Strasbourg controllers then gave flight 148 radar guidance to ANDLO at 11DME from the Strasbourg VORTAC. Altitude over ANDLO was 5000 feet. After ANDLO the VOR/DME approach profile calls for a 5.5% slope (3.3deg angle of descent) to the Strasbourg VORTAC. While trying to program the angle of descent, "-3.3", into the Flight Control Unit (FCU) the crew did not notice that it was in HDG/V/S (heading/vertical speed) mode. In vertical speed mode "-3.3" means a descent rate of 3300 feet/min. In TRK/FPA (track/flight path angle) mode this would have meant a (correct) -3.3deg descent angle. A -3.3deg descent angle corresponds with an 800 feet/min rate of descent. The Vosges mountains near Strasbourg were in clouds above 2000 feet, with tops of the layer reaching about 6400 feet when flight 148 started descending from ANDLO. At about 3nm from ANDLO the aircraft struck trees and impacted a 2710 feet high ridge at the 2620 feet level near Mt. Saint-Odile. Because the aircraft was not GPWS-equipped, the crew were not warned.

Probable Cause   Source: Accident Report[28]

After analysing the accident mechanisms, the commission reach the following conclusions:

1   The crew was late in modifying its approach strategy due to ambiguities in communication with air traffic control. They then let the controller guide them and relaxed their attention, particularly concerning their aircraft position awareness, and did not sufficiently anticipated preparing the aircraft configuration for landing.

2   In this situation, and because the controller's radar guidance did not place the aircraft in a position which allowed the pilot flying to align it before ANDLO, the crew was faced with a sudden workload peak in making necessary lateral corrections, preparing the aircraft configuration and initiating the descent.

3   The key event in the accident sequence was the start of aircraft descent at the distance required by the procedure but at an abnormally high vertical speed (3300 feet/min) instead of approx. 800 feet/min, and the crew failure to correct this abnormally high rate of descent.

4   The investigation did not determined, with certainty, the reason for this excessively high rate of descent. Of all the possible explanations it examined, the commission selected the following as seen most worthy of wider investigation and further preventative actions:

4.1 The rather probably assumptions of confusions in vertical modes (due either to the crew forgetting to change the trajectory reference or to incorrect execution of the change action) or of incorrect selection of the required value (for example, numerical value stipulated during briefing selected unintentionally).

4.2 The highly unlikely possibility of a FCU failure (failure of the mode selection button or corruption of the target value the pilot selected on the FCU ahead of its use by the auto-pilot computer).

5   Regardless of which of these possibilities short-listed by the commission is considered, the accident was made possible by the crew's lack of noticing that the resulting vertical trajectory was incorrect, this being indicated, in particular, by a vertical speed approximately four times higher than the correct value, an abnormal nose-down attitude and an increase in speed along the trajectory.

6   The commission attributes this lack of perception by the crew to the following factors, mentioned in an order which in no way indicates priority:

6.1 Below-average crew performance characterised by a significant lack of cross-checks and checks on the outputs of actions delegated to automated systems. This lack is particularly obvious by the failure to make a number of the announcements required by the operating manual and a lack of the height/range check called for as part of a VOR DME approach.

6.2 An ambiance in which there was only minimum communication between crew members;

6.3 The ergonomics of the vertical trajectory monitoring parameters display, adequate for normal situations but providing insufficient warning to a crew trapped in an erroneous mental representation.

6.4 A late change to the approach strategy caused by ambiguity in crew-ATC communication ;

6.5 A relaxation of the crew's attention during radar guidance followed by an instantaneous peak workload which led them to concentrate on the horizontal position and the preparation of the aircraft configuration, delegating the vertical control entirely to the aircraft automatic systems;

6.6 During the approach alignment phase, the focusing of both crew members attention on the horizontal navigation and their lack of monitoring of the auto-pilot controlled vertical trajectory.

Certification Issues

The major issue in this accident is the display of descent. Depending on the mode selected, the rate of descent would be shown as, for example, -3.3 for 3.3 degrees [-800 ft/min] or -33 for -3300 ft/min. This is clearly an example of poor ergonomics.

December 20, 1995: Boeing 757, N-651AA

Location	Near Cali, Columbia
Flight	AAL 965
Operator	American Airlines
Operation	Scheduled Airline
Route	Miami, FL, USA to Cali, Columbia
On-Board Injuries	Fatal 159 Serious 4
Aircraft Damage	Destroyed

Description   Source: Aviation Safety Network[1]

At about 18:34 EST, American Airlines Flight 965 took off from Miami for a flight to Cali. At 21:34, while descending to FL200, the crew contacted Cali Approach. The aircraft was 63nm out of Cali VOR (which is 8nm South of the airport)) at the time. Cali cleared the flight for a direct Cali VOR approach and report at Tulua VOR. Followed one minute later by a clearance for a straight in VOR DME approach to runway 19 (the Rozo 1 arrival).

The crew then tried to select the Rozo NDB (Non Directional Beacon) on the Flight Management Computer (FMC). Because their Jeppesen approach plates showed 'R' as the code for Rozo, the crew selected this option. But 'R' in the FMC database meant Romeo. Romeo is a navaid 150nm from Rozo, but has the same frequency. The aircraft had just passed Tulua VOR when it started a turn to the left (towards Romeo). This turn caused some confusion in the cockpit since Rozo 1 was to be a straight in approach. 87 Seconds after commencing the turn, the crew activated Heading Select (HDG SEL), which disengaged LNAV and started a right turn. The left turn brought the B757 over mountainous terrain, so a Ground Proximity (GPWS) warning sounded. With increased engine power and nose-up the crew tried to climb. The spoilers were still activated however. The stick shaker then activated and the aircraft crashed into a mountain at about 8900 feet (Cali field elevation being 3153 feet).

The accident report commented on the confusion: "Both of AA965's pilots were experienced in the airplane, and were described as proficient in the use of the FMS by their peers. Yet, most likely because of the self-induced time pressure and continued attempts to execute the approach without adequate preparation, the flightcrew committed a critical error by executing a change of course through the FMS without verifying its effect on the flightpath. The evidence indicates that either the captain or the first officer selected and executed a direct course to the identifier "R," in the mistaken belief that "R" was Rozo as it was identified on the approach chart. The pilots could not know with out verification with the EHSI display or considerable calculation that instead of selecting Rozo, they had selected the Rome beacon, located near Bogota, some 132 miles east-northeast of Cali. Both beacons had the same radio frequency, 274 kilohertz, and had the same identifier "R" provided in Morse code on that frequency. In executing a turn toward Romeo rather than Rozo, the flightcrew had the airplane turn away from Cali and towards mountainous terrain to the east of the approach course, while the descent continued. At this time, both pilots also attempted to determine the airplane's position in relation to ULQ, the initial approach fix. Neither flight crew member was able to determine why the navaid was not where they believed it should be, and neither noted nor commented on the continued descent. The CVR indicates that the flightcrew became confused and attempted to determine their position through the FMS. For example, at 2138:49 the first officer asked, "Uh, where are we?" and again, 9 seconds later asked, "Where [are] we headed? "The captain responded, "I don't know ... what happened here?"[29]

Probable Cause   Source: Accident Report[29]

Aeronautica Civil determines that the probable causes of this accident were:
- The flightcrew's failure to adequately plan and execute the approach to runway 19 at SKCL and their inadequate use of automation.
- Failure of the flightcrew to discontinue the approach into Cali, despite numerous cues alerting them of the inadvisability of continuing the approach.
- The lack of situational awareness of the flightcrew regarding vertical navigation, proximity to terrain, and the relative location of critical radio aids.
- Failure of the flightcrew to revert to basic radio navigation at the time when the FMS-assisted navigation became confusing and demanded an excessive workload in a critical phase of the flight.

Contributing to the cause of the accident were:
- The flightcrew's ongoing efforts to expedite their approach and landing in order to avoid potential delays.
- The flightcrew's execution of the GPWS escape maneuver while the speedbrakes remained deployed.
- FMS logic that dropped all intermediate fixes from the display(s) in the event of execution of a direct routing.
- FMS-generated navigational information that used a different naming convention from that published in navigational charts

Safety Recommendations
Aeronautica Civil made several recommendations related to aircraft certification:
- Evaluate all FMS-equipped aircraft and, where necessary, require manufacturers to modify the FMS logic to retain those fixes between the airplane's position and one the airplane is proceeding towards, following the execution of a command to the FMS to proceed direct to a fix.
- Encourage manufacturers to develop and validate methods to present accurate terrain information as part of a system of early ground proximity warning (Enhanced GPWS).
- Evaluate the dynamic and operational effects of automatically stowing the speedbrakes when high power is commanded and determine the desirability of incorporating on existing automatic speedbrake retraction during windshear and GPWS escape maneuvers, or other situations demanding maximum thrust and climb capability.
- Evaluate the possibility of requiring that flight crew generated inputs to the FMC be recorded as parameters in the FDR in order to permit investigators to reconstruct pilot-FMS interaction.

The NTSB made several recommendations related to aircraft certification to the FAA:
- Evaluate the effects of automatically stowing the speedbrakes on existing airplanes when high power is commanded and determine the desirability of incorporating automatic speedbrake retraction on these airplanes for windshear and terrain escape maneuvers, or other situations demanding maximum thrust and climb capability.(Class II, Priority Action) (A-96-90)
- Require that newly certified transport-category aircraft include automatic speedbrake retraction during windshear and ground proximity warning system escape maneuvers, or other situations demanding maximum thrust and climb capability. (Class II, Priority Action) (A-96-91)
- Evaluate the Boeing Commercial Airplane Group procedure for guarding the speedbrake handle during periods of deployment, and require airlines to implement the procedure if it increases the speed of stowage or decreases the likelihood of forgetting to stow the speedbrakes in an emergency situation. (Class II, Priority Action) (A-96-92)
- Evaluate the terrain avoidance procedures of air carriers operating transport-category aircraft to ensure that the procedures provide for the extraction of maximum escape performance and ensure that those procedures are placed in procedural sections of the approved operations manuals. (Class II, Priority Action) (A-96-93)
- Require that all transport-category aircraft present pilots with angle-of-attack information in a visual format, and that all air carriers train their pilots to use the information to obtain maximum possible airplane climb performance. (Class II, Priority Action) (A-96-94)
- Develop a controlled flight into terrain training program that includes realistic simulator exercises comparable to the successful windshear and rejected takeoff training programs and make training in such a program mandatory for all pilots operating under 14 CFR Part 121.(Class II, Priority Action) (A-96-95)
- Require all flight management system (FMS)-equipped aircraft, that are not already capable of so doing, to be modified so that those fixes between the airplane's position and the one towards which the airplane is proceeding are retained in the FMS control display unit and FMS-generated flightpath following the execution of a command to the FMS to proceed direct to a fix. (Class II, Priority Action) (A-96-96)
- Inform pilots of flight management system (FMS)-equipped aircraft of the hazards of selecting navigation stations with common identifiers when operating outside of the United States and that verification of the correct identity and coordinates of FMS-generated waypoints data is required at all times.(Class II, Priority Action) (A-96-97)

- Develop and implement standards to portray instrument approach criteria, including terminal environment information and navigational aids, on FMS-generated displays that match, as closely as possible, the corresponding information on instrument approach charts.(Class II, Priority Action) (A-96-98)

Certification Issues

The FAA's Lessons Learned website states "The accident was attributed almost exclusively to crew error. Since no unsafe condition of design flaw was implicated, there was no basic for AD action relative to the 757.

They cite three contributing errors: 1. Not reporting passing the Tulia VOR. (They never reached Tulia.); 2. Incorrectly an "R" for Rozo. The real issue -- A poor FMS interface; and 3. Failing to recognize they had lost situation awareness.

The FAA did implement a rule requiring the installation of Terrain Awareness Warning system (TAWS) in the US commercial fleet after the Shuttle Radar Topography Mission (SRTM).

Comments

We agree that crew error was involved; however the existence of high workload, rushed operations should have been evaluated during the design of the crew-system interfaces of the FMS. In addition, there should have been a means of verifying that the data in the radio database matched the data in the crew's charts.

March 14, 2017: Sikorsky S-92A, EI-ICR

Location	Atlantic Ocean, 6 Mi W Black Sod, County Mayo, Ireland
Flight	Rescue 116*
Operator	CHC Ireland
Operation	Search and Rescue on behalf of Irish Coast Guard
Route	Operating West of Black Sod
On-Board Injuries	Fatal 4
Aircraft Damage	Destroyed

Description   Source: Aviation Safety Network[1]

A Sikorsky S-92A helicopter, registration EI-ICR (call sign Rescue 116), which was being operated by a private operator on behalf of the Irish Coast Guard (IRCG), was en route from Dublin, on Ireland's east coast, to Blacksod, Co. Mayo, on Ireland's west coast. The flight crew's intention was to refuel at Blacksod before proceeding, as tasked, to provide Top Cover for another of the operator's helicopters, which had been tasked to airlift a casualty from a fishing vessel, situated approximately 140 nautical miles off the west coast of Ireland. At 00.46 hrs, on 14 March 2017, while positioning for an approach to Blacksod from the west, the helicopter, which was flying at 200 feet above the sea, collided with terrain at the western end of Black Rock, departed from controlled flight, and impacted with the sea.

During the immediate search and rescue response, the commander was found in the sea to the south-east of Black Rock and was later pronounced dead. Subsequently, the main wreckage of the helicopter was found close to the south-eastern tip of Black Rock, on the seabed at a depth of approximately 40 metres. The deceased co-pilot was located within the cockpit section of the wreckage and was recovered by naval service divers. Extensive surface and underwater searches were conducted; however, the two rear crew members were not located and remain lost at sea.

Probable Cause   Source: Accident Report[30]

The Helicopter was manoeuvring at 200 ft, 9 NM from the intended landing point, at night, in poor weather, while the Crew was unaware that a 282 ft obstacle was on the flight path to the initial route waypoint of one of the Operator's pre-programmed FMS routes.

Contributory Cause(s)

1.   The initial route waypoint, towards which the Helicopter was navigating, was almost coincident with the terrain at Black Rock.
2.   The activities of the Operator for the adoption, design and review of its Routes in the FMS Route Guide were capable of improvement in the interests of air safety.
3.   The extensive activity undertaken by the Operator in respect of the testing of routes in the FMS Route Guide was not formalised, standardised, controlled or periodic.
4.   The Training provided to flight crews on the use of the routes in the (paper) FMS Route Guide, in particular their interface with the electronic flight management systems on multifunction displays in the cockpit, was not formal, standardised and was insufficient to address inherent problems with the FMS Route Guide and the risk of automation bias.
5.   The FMS Route Guide did not generally specify minimum altitudes for route legs.

---

*   Operational call sign used *vice* flight number.

6.   The Flight crew probably believed, as they flew to join it, that the APBSS (waypoint BLKMO to BLKSD as described in legs 1 to 4 of narrative and on the map in FMS Route Guide in respect of APBSS) route by design provided adequate terrain separation from obstacles.
7.   Neither Flight Crew member had operated recently into Blacksod.
8.   EGPWS databases did not indicate the presence of Black Rock, and neither did some Toughbook and Euronav imagery.
9.   It was not possible for the Flight Crew to accurately assess horizontal visibility at night, under cloud, at 200 ft, 9 NM from shore, over the Atlantic Ocean.
10. The Flight Crew members' likely hours of wakefulness at the time of the accident were correlated with increased error rates and judgment lapses.
11. There were serious and important weaknesses with aspects of the Operator's SMS including in relation to safety reporting, safety meetings, its safety database SQID and the management of FMS Route Guide such that certain risks that could have been mitigated were not.
12. There was confusion at the State level regarding responsibility for oversight of SAR operations in Ireland.

Comments

This report was located during the final editing of the book. We thought it important enough to include, even though it will not receive the attention it deserves in the book. We plan to study it further.

The text of the report points out a major problem with "preprogrammed operator-developed routes". It appears that the route programmed in the FMS had not been properly validated and allowed the aircraft to fly into a large rock at night. This is troubling.

It is even more troubling that the rock was not included in the EGPWS database. According to the accident report, the EGPWS database was based on Russian military topographic maps and supplied to the EGPWS manufacturer in Digital Elevation Model (DEM) format

It is not clear if the rock was not included in the particular EGPWS database or if it was or wasn't present in the NASA database. This is an important question and must be determined.

In addition, the rock location was reported to the EGPWS manufacturer with ni resolution at the accident report date.

Display Blanking

November 6, 2001: Boeing 747, 9V-SPP

Location	B150 nm Northwest of Port Hedland, Australia
Identification	9V-SPP*
Operator	Singapore Airlines
Operation	Scheduled Airline
Route	Sydney, Australia to Singapore
On-Board Injuries	None Reported
Aircraft Damage	None

Description   Source: ATSB Safety Report[31]

On 6 November 2001, as the Boeing 747-400 aircraft was approaching flight level 360 en-route from Sydney to Singapore, the flight crew observed a CABIN ALT AUTO message and an increase in cabin altitude. The crew carried out the non-normal checklist, which included manually closing the outflow valves and turning off one of the air conditioning packs. However, the cabin altitude continued to rise and could not be controlled. The pilot in command (PIC) decided to conduct an emergency descent. The flight crew donned their oxygen masks and manually deployed the passenger oxygen masks.

As the aircraft commenced descent, one of the flight crew noticed the PIC's primary flight display screen go blank, with its data transferring to the PIC's navigation display screen. At that point a number of messages momentarily flashed onto the centre, Engine Indicating and Crew Alerting System (EICAS) screen, before all the display screens that are referred to as integrated display units (IDUs) blanked. The crew continued the emergency descent using standby instruments for reference, leveling at 10,000ft. The IDU system circuit breakers (CBs) were checked with all CBs found to be in their normal configuration. Manual selection between alternate system controllers was made but the IDUs remained blank.

After the post decompression drill announcement was made to the cabin crew, the PIC was advised that one passenger had sustained an injury in the form of bleeding from the ear. The PIC was also advised that there appeared to be no oxygen flowing through any of the passenger cabin oxygen masks, however this did not seem to have had any ill effect on the passengers. After the PIC cycled the passenger oxygen switch, a momentary surge of oxygen through the passenger masks was observed.

With Melbourne Air Traffic Control Centre providing vectoring assistance and using the standby flight instrumentation, the flight crew returned the aircraft to Sydney. Due to the loss of the display screens monitoring information, dumping of excess fuel was not possible. Information regarding the position of the landing gear or flaps was also not available. On approach to Sydney, the flight crew requested that the tower controller visually confirm that the landing gear was down and after receiving that confirmation, the crew conducted an uneventful, but overweight landing.

An inspection of the aircraft by engineers revealed that a cabin pressure relief valve had operated, indicating that the cabin had been over-pressurised at some stage during the flight. After a number of system CBs were cycled, the flight deck display screens returned to normal operation. An engine ground run and system check was then carried out, but the inflight faults could not be reproduced. The primary components of the aircraft's pressurisation and flight deck display systems were removed for further testing.

January 23, 2003: Boeing 747, Registration not available

Location	B150 nm Northwest of Port Hedland, Australia
Identification	None Reported
Operator	Singapore Airlines
Operation	Scheduled Airline
Route	Singapore to Sydney, Australia
On-Board Injuries	None Reported
Aircraft Damage	Unknown[32]

Description   Source: NTSB Safety Recommendation[33]

O n January 23, 2003, a Singapore Airlines (SIA) Boeing 747-400 experienced a complete loss of information on all six integrated display units (IDU) on the flight deck instrument panels while in cruise flight from Singapore to Sydney, Australia. The pilots flew the airplane for 45 minutes using standby flight instruments while they communicated with SIA maintenance personnel about the problem. SIA maintenance personnel advised the flight crew to pull out then push back in (or cycle) the circuit breakers for the EFIS/EICAS interface units (EIU), which returned the IDUs to normal operation. The flight continued to Sydney and landed without further incident.

---

* Registration used *vice* flight number.

The six IDUs of the airplane's integrated display system include the captain's primary flight display (PFD) and navigation display (ND), the first officer's PFD and ND, and the main and auxiliary engine indication and crew alerting system (EICAS) displays. The PFD and ND displays provide the pilots with attitude, altitude, airspeed, heading, and rate of climb and descent information. The EICAS displays provide the flight crew with the airplane's engine indicating information and annunciate advisories, cautions, and warnings. Without these displays, the flight crew is required to use standby flight instruments, which consist of an altimeter, airspeed indicator, and artificial horizon/attitude indicator; the Boeing 747-400 does not have standby engine instruments. The loss of the IDUs would also eliminate the flight crew's access to data from the traffic alert and collision avoidance system, enhanced ground proximity warning system, and weather radar

Boeing has provided a procedure to B747-400 operators that is similar to the method used to restore the displays in both SIA B747-400 display-blanking events. On February 25, 2003, Boeing issued Operations Manual Bulletin (OMB) #SIA-186, which states, "In the unlikely event all six display units blank, cycling the left and center EIU circuit breakers may recover the display units." The OMB also requests that the bulletin be placed in the OMB Record page of the airplane operations manual.

Based on findings to date in the investigations of both SIA IDU blanking events and a review of the B747-400 integrated display system, Boeing's recommendation to cycle the EIU circuit breakers to restore EIU functionality appears to be a reasonable temporary action until a permanent solution is determined. However, in the event that all six flight displays blank, flight crew workload would significantly increase, and a crew would need to be able to rapidly identify and execute the appropriate sequence of actions to address the problem. It is unrealistic to expect a B747-400 flight crew faced with a loss of cockpit displays and autothrottles to rapidly locate corrective actions in the operations manual. Instead, the procedure outlined in Boeing's OMB should be identified as a non-normal checklist procedure and incorporated in the quick-reference handbook to facilitate flight crews' access to this information

Comments

We were aware of the preliminary approach to restoring the failed display: resetting the display circuit breakers one at a time. However, the official position of our office was to never reset circuit breakers. As result the procedure was changed to removing all electrical power, then turning the displays one at a time. The argument was that C/B's were not designed to serve as switches and should not routinely be used as switches. One of us (RLN) strongly disagreed with prohibiting any resetting of C/B's, but was unable to convince our colleagues. Since the principal proponent of banning resetting wanted an annual pull and rest to ensure the breakers still worked, Perhaps the solution would be to simply require a logbook entry each time the breaker is reset.

The results of the investigations have not been forthcoming. The conduct of these two incidents are not typical of the usual open investigations by the various safety boards.

Ditchings

October 16, 1956: Boeing 377, N-90943

Location	1125 Mi SW San Francisco
Flight	PAA 6
Operator	Pan American World Airways
Operation	Scheduled Airline
Route	Honolulu, HI to San Francisco, CA
On-Board Injuries	Minor/None <u>31</u>
Aircraft Damage	Destroyed

Description
Source: Accident Report[34]

Trip 6 of October 13 was a regularly scheduled "around-the-world" flight eastbound from Philadelphia, Pennsylvania to San Francisco, California with en route stops in Europe, Asia and various Pacific Islands.

All prior segments had been routine and the flight departed Honolulu on the last leg of the trip on October 15. It was cleared to San Francisco Airport via Green Airway 9, then track to position 30 degrees N. 140 degrees W. at 13,000 feet then 21,000 feet to San Francisco. There were 24 passengers aboard, including 3 infants, and [7] crew.

The 8 hour, 54 minute flight was planned IFR and the aircraft carried sufficient fuel for 12 hours. 18 minutes. The gross takeoff weight of the aircraft was 138,903 pounds (maximum allowable 144,000) and the center of gravity was located within limits.

N 90943 departed Honolulu at 2026. The climb to initial altitude was normal and the flight proceeded in a routine manner. At 0102, the approximate midpoint of the flight, a request for VFR climb to its secondary altitude of 21,000 feet was approved by ATC.

After reaching 21,000 feet and simultaneously with the reduction of power, the No. 1 engine oversped. Airspeed was immediately reduced by the use of flaps and reduction of power. Attempts were also made to feather the No. 1 propeller. It was impossible to control the engine or to feather the propeller and the captain decided to freeze the engine by cutting off the oil supply. Shortly after this was done there was a momentary decrease in the r. p. m., followed by a heavy thud. The propeller continued to windmill. At this time airspeed had slowed to 150 knots and the aircraft was losing altitude at a rate of approximately 1.000 feet per minute.

The flight course was altered to "home in" on station "November" and climb power applied to engines Nos. 2, 3, and 4 to cheek the rate of descent. At this time it was noticed that No. 4 engine was only developing partial power at full throttle. At 0125 the flight notified "November" that ditching was imminent and received a ditching heading from the cutter. During the descent the crew found they could maintain altitude at an airspeed of 135 knots with rated power on engines Nos. 2 and 3 and the partial power on No. 4. About 0[137 the flight overheaded the cutter.

Prior to overheading the cutter, the maximum range with the fuel remaining had been computed and it was determined to be insufficient either to complete the flight to San Francisco or return to Honolulu. Mortar flares had been fired by the cutter and electric water lights laid to illuminate a ditching track for the aircraft. However, it was decided to postpone the ditching until daylight, if possible, meanwhile remaining close to the cutter.

About 0245 the No. 4 engine backfired and power dropped off. Its propeller was feathered normally. The flight was still able to maintain altitude and continued to orbit "November" to burn the fuel aboard down to a minimum while awaiting daylight.

At 0540 Captain Ogg notified the U. S. S. Pontchartrain he was preparing to ditch the aircraft. A foam path was laid along the ditching heading of 3150 by the cutter and the aircraft was ditched at 0615. Passengers and crew safely evacuated the aircraft, boarded liferafts, and were completely clear of the aircraft at 0632. The aircraft sank at 0635 at position 30 degrees 01.5' N. 140 degrees 09' W.

The most likely causes of the overspeed and inability to feather are that oil was being misdirected at the governor pilot valve or that there was insufficient oil pressure at the dome piston. Improper direction of the oil would involve governor malfunctions, caused either by a fault within the unit itself or by contaminated oil being supplied to the governor. Contaminated oil would indicate some failure with the engine which would most likely be of a progressive nature. No such failure was evident to the crew prior to the overspeed. Insufficient oil pressure at the dame piston is most generally due to excessive leakage. Leakage usually involves seals, passages, transfer tubes, or bearings in the propeller, propeller control, or the engine. The Board believes that a sing] e failure occurred which affected the portion of the system common to the constant speed and feathering portion of the propeller control system. Oil was being delivered to the system by 'the feathering pump and then dumped into the engine. A more specific reason for the overspeed cannot be determined

Probable Cause
Source: Accident Report[34]

An initial mechanical failure which precluded feathering the no. 1 propeller and a subsequent mechanical failure which resulted in a complete loss of power from the no. 4 engine, the effects of which necessitated a ditching.

The Board believes that a single failure occurred which affected the portion of the system common to the constant speed and feathering portion of the propeller control system. Oil was being delivered to the system by 'the feathering pump and then dumped into the engine. A more specific reason for the overspeed cannot be determined

Regulatory Outcome

Subsequent to this accident PAWA Pacific-Alaska Division experienced two uncontrollable engine overspeeds and inability to feather propellers due to failure of the propeller oil transfer bearing. A redesigned propeller oil transfer bearing has been provided by the manufacturer and its use was made mandatory by CAA Airworthiness Directive issued March 25, 1957

Comments

The crews of both the airplane and the Coast Guard cutter clearly deserve "attaboys". On his deathbed, Captain Off was said be thinking of the 3000 canaries in the cargo hold that drowned when the airplane sank.[35] This accident was the likely inspiration for the movie *The High and the Mighty*.

May 24, 1985: Cessna U206F, N-1803Q

Location	Off Piney Point, MD
Identification	N-1803Q*
Operator	Anti-Icing Research
Operation	Flight Test
Route	Springfield, OH to Patuxent River NAS, MD
On-Board Injuries	Minor/None 1
Aircraft Damage	Substantial

Description   Source: Aviation Safety Network[1]

While being vectored for GCA approach, engine seized at 4000 ft. Broke out of clouds at 800 ft and ditched in the bay. Airplane remained upright, but sank in less than 1 minute. Pilot picked up by Navy H-3 after 10 minutes in water and transported to Patuxent River NAS (KNHK).

Probable Cause   Source: Accident Report[36]

While descending in IFR conditions near the destination, the eng lost power at an alt of about 3800 TO 4000 ft. The plt requested vectors toward land & the nearest arpt; however, when it broke out of the clouds at about 1000 ft AGL, The acft was still over water. Unable to reach land, the plt ditched the acft in the mount of the Potomac River about 3/4 mi from shore. An exam of the eng revealed holes in each side of the top portion of the crankcase between the #1 & #2 cylinders. A tear down of the eng revealer the #2 connecting rod had failed. There was evidence of a lack of lubrication. The connecting rod bearings had evidence of extreme wear & heat, & the respective crankshaft journals showed signs of hear & wear. Also, the oil pump cavity & walls were scratched & galled.

Certification Issues

If a flaps down landing is made, egress through the right side door (cargo door) will be hindered by interference between the door and the flaps. During the landing, the pilot's right hand inadvertently hit the flap handle causing the flaps to deploy. When the master switch was opened following the landing, this caused the flaps to stop. They stopped at the point where the interference just began. This was not discussed in the Pilot Operating Handbook (POH).[37]

Regulatory Outcome

There have been a number of proposals to require new designs, procedures, or modification to correct this issue,[38][39][40] with no real action.

It seems that the POH is overly concerned about glassy water ditchings.† The instructions call for full flaps and 65 kts and no flare. With no power (the usual case in a ditching), the resulting steep approach such a landing could better be described as a crash. In our opinion, these POH instructions should be scrapped and rewritten. A recommendation to change the POH was made to the manufacturer with no action.

---

* Aircraft registration used *vice* flight number.

† When your senior author (RLN) took his check ride for a seaplane rating, the pilot examiner asked how he would perform a forced landing in glassy water. This was a trick question. According to the examiner, there is no safe way to accomplish a power-off, glassy water landing and it would be better to crash on land than attempt such a landing.

Engine (FADEC Failure)

June 5, 2008, Eclipse EA-500, N-612KB

Location   Chicago, IL
Identification   N-612KB*
Operator   Private Operator
Operation   Business Transportation
Route   Cleveland, OH to Chicago, IL
On-Board Injuries   Minor/None 4
Aircraft Damage   None

Description   Source: NTSB Incident Report[41]

The flying pilot of a very light jet airplane encountered windshear and a high sink rate on approach which he arrested by applying engine power. During landing he lightly applied brakes and observed that the airplane was accelerating. The pilot initiated a go-around. The engines were at maximum power and the crew alerting system illuminated left and right engine control failure messages. The automatic power reserve armed light was also illuminated. The crew maneuvered the airplane and troubleshot the engine control fail indication using the quick reference handbook while on downwind. The right engine was shutdown and the left engine would not respond to throttle movements during the remainder of the incident flight. The crew subsequently completed the emergency landing. On scene and subsequent throttle quadrant assembly examinations showed the assembly could produce an out of specified range signal when the throttles had forward force applied to them. Diagnostic storage unit data showed both engines' full authority digital engine control (FADEC)s encountered a fault signal from the engine controls about the time of the pilot's reported application of the throttles in response to the windshear during the incident approach. The airplane's FADEC fault logic was designed to "latch" or hold the last good value when the upper specified parameter limit was exceeded. The data showed the left engine went to flight idle when the right engine was shut down in accordance with the fault logic. The airplane's design was Federal Aviation Administration (FAA) certified with that fault logic.

Probable Cause   Source: NTSB Incident Report[41]

The National Transportation Safety Board determines the probable cause(s) of this incident to be: The airplane manufacturer's inadequate software design requirements of the engine's full authority digital electronic controls (FADEC) fault logic that resulted in a simultaneous unrecoverable loss of thrust control on both engines when the FADEC's input data values exceeded specified ranges during the approach. Contributing to the incident was the Federal Aviation Administration's failure to recognize and correct this condition during the certification of the airplane.

Recommendations   Source: NTSB Incident Report[41]

As a result of the investigation, the National Transportation Safety Board issued two urgent recommendations to the FAA.

A-08- 46 recommended an immediate inspection of all throttle quadrants to ensure that pushing the throttle levers against the maximum power stops will not result in an engine control failure, and further require that any units that fail the inspection be replaced and that the replacement parts be similarly inspected.

A-08-47 recommended an immediate emergency procedure for a dual engine control failure on the incident airplane model and incorporate the procedure into the airplane flight manual and quick reference handbook via an airworthiness directive.

Certification Issues   Source: DOT Inspector General Testimony[42]

A recent incident involving the EA-500 ha s heightened attention regarding the aircraft's design certification. On June 5, 2008, an EA-500 on approach to Chicago Midway airport experienced a throttle failure that resulted in an uncontrollable maximum power thrust from its engines. After consulting the emergency procedures, the pilots shut down one of the engines; however, this action caused the second engine to roll back to idle power and be unresponsive to the throttle. The pilots declared an emergency and were able to land the plane without injury to the two pilots or two passengers.

During its investigation into the incident, the National Transportation Safety Board (NTSB) expressed concern about the reliability of an assembly that failed after accumulating only 238 hours and 192 cycles. The NTSB also raised concerns that the problem could be due to flaws in the design logic for the software that controls the engines and issued two recommendations to FAA requiring (1) immediate inspection of all EA-500 engine throttles and (2) an emergency procedure to address dual engine control failure.

On June 12, 2008, FAA issued an Airworthiness Directive (AD) 2 that requires operators to examine throttle controls for identified faults and replace assemblies as necessary. Since awarding the design certificate to Eclipse, FAA has issued a total of six ADs for various components of the EA-500.

---

* Aircraft registration used *vice* flight number

As a result of this incident, FAA engineers re-examined the software that controls the engines and discovered software logic flaws that should have been resolved before design certification. At the end of June 2008, the local FAA certification manager sent a memorandum to the manufacturer requiring Eclipse to develop an approach to bring the aircraft design into certification compliance for that system. Eclipse is currently addressing FAA's requirement.

Engine (Thrust Reverser

May 26, 1991: Boeing 767, OE-LAV

Location	3.5 miles NNE Phu Toey, Thailand
Flight	LDA 004
Operator	Lauda Air
Operation	Scheduled Airline
Route	Bangkok, Thailand to Vienna, Austria
On-Board Injuries	Fatal 223
Aircraft Damage	Destroyed

Description   Source: Aviation Safety Network[1]

Lauda Air Flight 004 was a scheduled service from Hong Kong (HKG) back to Vienna (VIE), Austria. An intermediate stop was made in Bangkok (BKK), Thailand. The flight departed Bangkok at 23:02 hours. Some five minutes after takeoff the pilot-in-command stated "that keeps coming on," referring to a REV ISLN advisory warning. This indication appears when a fault has been detected in the thrust reverser system. The crew discussed the REV ISLN indication for about four and one-half minutes. The co-pilot read information from the Airplane Quick Reference Handbook as follows: "Additional systems failures may cause in- flight deployment" and "Expect normal reverser operation after landing." The pilot-in-command remarked "....its not just on, its coming on and off," he said, "...its just an advisory thing...," and shortly thereafter stated, "could be some moisture in there or something." At 23:12, the co-pilot advised the pilot-in-command that there was need for, "a little bit of rudder trim to the left." Fifteen minutes and one second into the flight the co-pilot exclaimed, "ah reverser's deployed," accompanied by sound similar to airframe shuddering, sounds of metallic snaps and the pilot-in-command stating "here wait a minute." With the deployment of the no. 1 engine thrust reverser, engine thrust was reduced to idle. Aerodynamic effects of the reverser plume in-flight during the engine run down to idle resulted in a 25 percent lift loss across the wing. The airplane stalled and entered an uncontrolled descent. Buffeting, maneuvering overload, and excessive speed caused pieces of the rudder and the left elevator to separate. This was followed by the down-and-aft separation of most of the right horizontal stabilizer from maneuvering overloads, as the crew attempted to control the airplane and arrest the high-speed descent. A torsional overload then caused the separation of the vertical and left horizontal stabilizers. The loss of the tail resulted in a sharp nose-over of the airplane, producing excessive negative loading of the wing. A downward wing failure was probably followed by the breakup of the fuselage. The complete breakup of the tail, wing, and fuselage occurred in a matter of seconds. The wreckage fell in mountainous jungle terrain.

Probable Cause   Source: Aviation Safety Network[1]

The Accident Investigation Committee of the Government of Thailand determines the probable cause of this accident to be uncommanded in-flight deployment of the left engine thrust reverser, which resulted in loss of flight path control. The specific cause of the thrust reverser deployment has not been positively identified.

Certification Issues[22]

In October 1991, the FAA chartered a joint FAA/AIA (Aerospace Industries Association) task force to examine the thrust reverser controllability and reliability aspects of the transport fleet. The *task force report** was released in March 1992 and resulted in the following determinations:

A subset of the transport fleet airplanes with twin, tail-mounted engines, for example, was established as "controllable" following an in-flight reverser deployment.

A much larger subset, including most large twin-engine transports, such as the 767, were determined to be "uncontrollable" following an in-flight reverser deployment, and in need of reverser system safety improvements to preclude in-flight deployments.

The task force further advised that airplanes deemed "controllable" should also require sufficient system safety to takeoff and attain cruise configuration without a reverser deployment. This would allow attainment of an altitude and speed combination that would be advantageous to recovery should a deployment occur relatively soon after takeoff. At the time the task force was making its recommendations, sufficient reverser system safety did not exist - additional failsafe features were deemed necessary.

As part of the fleet controllability assessment, the task force also arranged for the National Aeronautics and Space Administration (NASA) to develop a model of the effects of in-flight reverser deployment, and visualize the resulting upper wing surface flow disruption. A test flight on a NASA-owned DC-8 was conducted to validate NASA's modeling techniques. The flight consisted of several in-flight deployments of the inboard thrust reversers (the DC-8 is approved for in-flight use of reverse thrust) at various Mach numbers. The left wing upper surface was tufted in order to visualize the airflow effects over the upper wing surface with a reverser deployed.

The results of that flight test were published in a report dated June 1995.[43]

---

*   The "Lessons Learned" website included a URL for the transmittal document of the report, but not to the actual report.

Regulatory Outcome

[O]n February 24, 1994, the FAA issued AD 94-12-10 which required installation of the Boeing designed "sync" lock system (per Boeing service bulletin 767-78-0062) on PW400-equipped 767 airplanes. Later AD's levied similar requirements on GE and Rolls-Royce powered airplanes. Among the Boeing fleet, only the 707, 727, and 737-100/200 were assessed as controllable by the task force. Following the actions targeting the 767, all other Boeing models were required by AD to incorporate a third locking system.[22]

March 22, 1998: Airbus A320, RP-C3222

Location	Bacolod, Phillipines
Flight	PAL 137
Operator	Phillipines Air Lines
Operation	Scheduled Airline
Route	Manila, Phillipines to Bacolod, Phillipines
On-Board Injuries	Fatal 124
Ground Injuries	Fatal 3
Aircraft Damage	Destroyed

Description   Source: Aviation Safety Network[1]

Flight PR* 137 was a regular scheduled passenger flight and departed Manila for Bacolod at 18:40. The airplane departed with the thrust reverser of engine no. 1 inoperative.

At 19:20, PR137 called Bacolod Approach Control and reported passing FL260 and 55 DME to Bacolod . The crew then requested landing instructions and was instructed to descend to FL90 after passing Iloilo and descend to 3,000 ft for a VOR runway 04 approach. Wind was 030° at 08 kts, altimeter 1014 mbs, transition level at FL60 and temperature at 28°C. At 19:28, the flight requested to intercept the final approach to runway 04 and Approach Control replied "PR 137 visual approach on final". At 19:37, Bacolod Tower cleared the flight to land at runway 04 and the clearance was acknowledged by the pilot.

The approach was flown with the Autothrust system was engaged in SPEED mode. The thrust lever of engine no.1 was left in Climb detent. Upon touchdown the first officer called out "no spoilers, no reverse, no decel". Engine no.2 was set to full reverse thrust after touchdown, but the engine no .1 thrust lever was not retarded to idle and remained in the climb power position. Consequently, the spoilers did not deploy.

Because one engine was set to reverse, the autothrust system automatically disengaged. With the autothrust disengaged, no. 1 engine thrust increased to climb thrust. Due to the asymmetrical thrust condition, the A320 ran off the right side of the runway. At this speed, rudder and nosewheel steering are ineffective. Engine no.2 was moved out of reverse up to more than 70 percent N1 and the airplane swerved back onto the runway. The A320 continued past the runway end. The aircraft hit the airport perimeter fence and then jumped over a small river. It continued to slice through a hallow block fence where it went through several clusters of shanties and trees. No fire ensued after the crash.

Probable Cause   Source: Aviation Safety Network[1]

The probable cause of this accident was the inability of the pilot flying to assess properly the situational condition of the aircraft immediately upon touch down with No. 1 engine reverse inoperative, thereby causing an adverse flight condition of extreme differential power application during the landing roll resulting in runway excursion and finally an overshoot.

Contributory to this accident is the apparent lack of technical systems knowledge and lack of appreciation of the disastrous effects of misinterpreting provisions and requirements of a Minimum Equipment List (MEL).

---

*   PR is the IATA code for Phillipines Air Lines and is used in the source material

August 28, 2002: Airbus A320, N-635AW

Location   Phoenix, AZ
Flight   AWE 794
Operator   America West Airlines
Operation   Scheduled Airline
Route   Houston, TX to Phoenix, AZ
On-Board Injuries   Serious 1   Minor/None   158
Aircraft Damage   Substantial [ASN: Destroyed]

Description   Source: Aviation Safety Network[1]

Flight 794 departed Houston with a an inoperative no. 1 thrust reverser. On August 20, 2002, the number one thrust reverser had been deactivated by maintenance personnel. The airplane touched down at Phoenix on the centerline of runway 08 about 1,200 feet beyond its threshold. During rollout the captain positioned both thrust levers into reverse but then took the number one thrust lever out of the reverse position and inadvertently moved it to the Take-Off/Go-Around (TOGA) position, while leaving the #2 thrust lever in the full reverse position. Full left rudder and full left brake application did not compensate for the yaw. The airplane continued swerving to the right until exiting the right side of the runway. It crossed the apron east of intersection B8, and experienced the collapse and partial separation of its nose gear strut assembly upon traversing the dirt infield area south of the runway between intersections B9 and B10.

Probable Cause   Source: Aviation Safety Network[1]

The captain's failure to maintain directional control and his inadvertent application of asymmetrical engine thrust while attempting to move the #1 thrust lever out of reverse. A factor in the accident was the crew's inadequate coordination and crew resource management.

October 18, 2004: Airbus A320, B-22310

Location   Taipei, Taiwan
Flight   TNA 536
Operator   TransAsia Airways
Operation   Scheduled Airline
Route   Tainan, Taiwan to Taipei, Taiwan
On-Board Injuries   Minor/None 106
Aircraft Damage   Substantial

Description   Source: Aviation Safety Network[1]

The flight departed Tainan at 19:24. The pilot-in-commend (CM-1) was on the left seat as the pilot monitor (PM), the first officer (CM-2) was on the right seat as the pilot flying (PF). The crew executed an ILS approach to runway 10 at Taipei-Sung Shan. At 19:58:12, the flight received landing clearance. At 19:28, the ATIS broadcast for RCSS read "...expect ILS Approach. Runway one zero in use; wind variable at three; visibility four thousand five hundred meters;

light rain; cloud scattered eight hundred feet, broken one thousand eight hundred feet, overcast three thousand five hundred feet; temperature two three; dew point two two; QNH one zero zero eight hectopascals; wind shear on runway one zero; Low Level Wind Shear Advisory in effect, moderate to severe...".

The total landing weight was 55,140 kilogram, flaps were selected at 3 and the approach speed was 137 knots, ground spoilers were ARMED, autobrake selected MEDIUM, and antiskid selected ON at final approach.

At 19:59:04, radio altitude 282 feet, auto-pilot disengaged, autothrottle activated. The "RETARD" sound was heard four times between radio altitude 20 feet and the moment when the main landing gear touched the ground.

At 19:59:27, the main landing gear touched ground, airspeed 138 knots, ground speed 146 knots, heading 093 degrees, wind direction 297 degrees, wind speed 11 knots, the number 1/2 throttle level angle positioned at 19.7/22.5 degrees. Three seconds later the nose gear touched ground, the number 1/2 throttle level angle positioned at -22.5/22.5 degrees.

At 19:59:32, the autothrottle disconnected, number one thrust reverser deployed, but number two thrust reverser did not deploy. At 19:59:37 CM-1 called out "no brake", until 19:59:50, CM-1 called "no brake" five times. In the meantime, air speed 112 knots, ground speed 109 knots. The ground spoilers did not extend.

The aircraft touched down at 1,750 feet on runway 10, and rolled off the runway 321 feet from the end of runway 10, ending up with the nosegear in a ditch. According to the records in the Technical Log Book of the aircraft: The number 2 engine thrust reverser system was malfunctioned and was transferred to deferred defect (DD) item and the thrust reverser was deactivated in accordance with the procedures in the Minimum Equipment List.

Probable Cause   Source: Aviation Safety Network[1]

FINDINGS RELATED TO PROBABLE CAUSES:

1. When the aircraft was below 20 ft RA and Retard warnings were sounded, the pilot flying didn't pull thrust lever 2 to Idle detent which caused the ground spoilers were not deployed after touchdown though they were at Armed position, therefore the auto braking system was not triggered. Moreover, when the auto thrust was changed to manual operation mode automatically after touchdown, the thrust lever 2 was remained at 22.5 degrees which caused the Engine 2 still had an larger thrust output (EPR1.08) than idle position's. Thereupon, the aircraft was not able to complete deceleration within the residual length of the runway, and deviated from the runway before came to a full stop, even though the manual braking was actuated by the pilot 13 seconds after touchdown.

2. The pilot monitoring announced "spoiler" automatically when the aircraft touched down without checking the ECAM display first according to SOP before made the announcement, as such the retraction of ground spoilers was ignored.

FINDINGS RELATED TO RISK:

1. After touchdown, when the thrust lever 2 was not pulled back to Idle position and the Retard warning sounds have ceased, there were no other ways to remind pilots to pull back the thrust lever.

2. The diminution of Runway Safety Zone proclaimed by Sungshan airport, and the fixed objects of non auxiliary aviation facilities and uncovered drainage ditch within the area do not meet the requirements of Civil Airports Designing and Operating Regulations.

July 17, 2007: Airbus A320, PR-MBK

Location	São Paulo-Congonhas Airport, SP, Brazil
Flight	TAM-054
Operator	TAM Linhas Aéreas
Operation	Scheduled Airline
Route	Porto Alegre, RS, Brazil to São Paulo, SP, Brazil
On-Board Injuries	Fatal 187
Ground Injuries	Fatal 12
Aircraft Damage	Destroyed

Description   Source: Aviation Safety Network[1]

An Airbus A320-233 operated by TAM Linhas Aéreas of Brazil was destroyed when it suffered a runway excursion after landing at São Paulo-Congonhas Airport, Brazil.

TAM Flight 3054 was a regular passenger flight from Porto Alegre to São Paulo-Congonhas. The aircraft departed Porto Alegre at 17:19 with 181 passengers and 6 crew members on board.

The aircraft was operating with the number 2 engine reverser de-activated, in accordance with the Minimum Equipment List (MEL).

The weather prevailing along the route and at the destination was adverse, and the crew had to make a few deviations. Up to the moment of the landing, the flight occurred within the expected routine.

According to information provided to the Tower controller by crews that had landed earlier, the active runway at Congonhas (35L) was wet and slippery. Runway 35L is a 6365 x 147 feet (1940 x 45 meters) asphalt runway with a Landing Distance Available of 1880 m.

In the beginning of 2007, the runways at Congonhas were subjected to a restoration to correct surface irregularities and correct problems of gradient, so as to prevent water accumulation. With the new pavement, the coefficient of friction of the runway surface was improved. However, since more time was be needed until the Tarmac could be ready to receive the grooving, the airport administration decided to put the main runway in operation even without the grooves on 29 June 2007.

The flight touched down at 18:54 local time at a speed of 142 knots. Airbus had introduced a simplified procedure for landing with a deactivated thrust reverser in which crews had to select reversers for both engines after landing. Computer logic would [detect] which reverser was inoperative and thus block the increase in power. This simplified procedure however added an additional 55 meters in the calculations of the runway length required for landing if the runway was contaminated.

However, the flight crew failed to use this procedure and left the thrust lever of engine no.2 positioned at "CL".

This caused the autothrust to try to maintain the speed previously selected. Another consequence was the non-deflection of the ground spoilers, since, in accordance with their logic of operation, it is necessary that both thrust levers be at the "IDLE" position, or one of them be at "IDLE" and the other at "REV" (reverse), for the ground spoilers to be deflected.

The non-deflection of the ground spoilers significantly degraded the aircraft braking capability, increasing the distance necessary for a full stop of the airplane by about 50%.

Additionally the autobrake function, although armed, was not activated, because the opening of the ground spoilers is a prerequisite for such activation.

When the nose gear touched the runway, about 2.5 seconds after the left main gear, the number 1 engine thrust lever was moved to the "REV" position. With this action, the autothrust function of the aircraft was disconnected and the thrust lock function was activated, with the purpose of preventing the acceleration to reach the climb power level.

As a result, this function froze the number 2 engine power in the value it was at that moment (EPR2 = 1.18). The thrust lock function is disabled by the movement of the thrust lever, but since this movement did not occur, the number 2 engine remained with that power until the collision.

About six seconds after the main gear touched the runway, there was the first activation of the brakes by means of the pedals, which reached the maximum deflection five seconds later.

The FDR also recorded the use of the rudder and the differential braking by the pedals as the aircraft veered to the left.

Since the runway is at a higher elevation than the surrounding street and residential area, the A320 crossed over the Washington Luís Avenue, and collided with a concrete TAM Express building and a fuel service station at a ground speed of 96 knots.

All the persons onboard suffered fatal injuries. The accident also caused 12 fatalities on the ground among the people that were in the TAM Express building.

The accident caused severe damage to the convenience shop area of the service station and to some vehicles that were parked there. The TAM Express building sustained structural damages that determined its demolition.

Certification Issues

Typically, most landings are flown with autothrust set to maintain airspeed and the thrust levers (T/L) left in the Climb detent. During the landing flare, the T/Ls are retarded.* Pulling the levers back to reverse engages reverse thrust. Only if both engines have their T/Ls in idle or reverse will the ground spoilers deploy and auto-braking engages

All four of these accidents share the same fundamental issue: a confusing, frequently changed, procedure. Sometimes the procedure was to pull both levers into reverse and rely on the software to only activate the operative reverser. Sometimes a different procedure was in place: Only the T/L on the engine with the operative reverser was to be pulled into reverse.

It seems that many of the mishap pilots only retarded the thrust lever on the engine with the operative reverser to idle. The other engine' with the inoperative reverser would be left in the climb detent. When the airplane landed the ground spoilers did not deploy and the auto-brakes did not activate. Worse, autothrust was still active and the engine with the inoperative reverser would go to climb thrust.

As a result, we would now have one engine in reverse, one engine in forward climb thrust; No auto-braking and no ground spoilers!

The real issue, however, is wording in the manuals was difficult to read, particularly for a non-native English speaker. Following the Sao Paulo accident, we had the occasion to review the TAM manual. I found this language in this section very difficult to read.

Comments

This is an on-going problem for transport airplane manufacturers whether they write the original text in English or in French. The manuals should be written in clear English, but written for a technically well trained non-native English speaker.

This is important for flight crew manuals; it's even more true for maintenance manuals.

---

*   There is a voice prompt calling "Retard" if the thrust levers are not in idle during the flare. There have been a number of jokes about this prompt suggesting that the pilot is mentally retarded (Typical pilot humor).

Engine Failure (Uncontained Failure)

July 19, 1989: McDonnell Douglas DC-10, N-1819U

Location	Sioux City, IA
Flight	UAL 232
Operator	United Airlines
Operation	Scheduled Airline
Route	Denver, CO to Chicago, IL
On-Board Injuries	Fatal <u>111</u> Serious <u>47</u> Minor/None <u>138</u>
Aircraft Damage	Destroyed

Description   Source: Aviation Safety Network[1]

United Flight 232 departed Denver-Stapleton International Airport, Colorado, USA at 14:09 CDT for a domestic flight to Chicago-O'Hare, Illinois and Philadelphia, Pennsylvania. There were 285 passengers and 11 crewmembers on board.

The takeoff and the en route climb to the planned cruising altitude of FL370 were uneventful. The first officer was the flying pilot.

About 1 hour and 7 minutes after takeoff, at 15:16, the flightcrew heard a loud bang or an explosion, followed by vibration and a shuddering of the airframe. After checking the engine instruments, the flightcrew determined that the No. 2 aft (tail-mounted) engine had failed.

The captain called for the engine shutdown checklist. While performing the engine shutdown checklist, the flight engineer observed that the airplane's normal systems hydraulic pressure and quantity gauges indicated zero.

The first officer advised that he could not control the airplane as it entered a right descending turn. The captain took control of the airplane and confirmed that it did not respond to flight control inputs. The captain reduced thrust on the No. 1 engine, and the airplane began to roll to a wings-level attitude.

The flightcrew deployed the air driven generator (ADG), which powers the No. 1 auxiliary hydraulic pump, and the hydraulic pump was selected "on." This action did not restore hydraulic power.

At 15:20, the flightcrew radioed the Minneapolis Air Route Traffic Control Center (ARTCC) and requested emergency assistance and vectors to the nearest airport. Initially, Des Moines International Airport was suggested by ARTCC. At 15:22, the air traffic controller informed the flightcrew that they were proceeding in the direction of Sioux City; the controller asked the flightcrew if they would prefer to go to Sioux City. The flightcrew responded, "affirmative." They were then given vectors to the Sioux Gateway Airport (SUX) at Sioux City, Iowa.

A UAL DC-10 training check airman, who was off duty and seated in a first class passenger seat, volunteered his assistance and was invited to the cockpit at about 15:29.

At the request of the captain, the check airman entered the passenger cabin and performed a visual inspection of the airplane's wings. Upon his return, he reported that the inboard ailerons were slightly up, not damaged, and that the spoilers were locked down. There was no movement of the primary flight control surfaces. The captain then directed the check airman to take control of the throttles to free the captain and first officer to manipulate the flight controls.

The check airman attempted to use engine power to control pitch and roll. He said that the airplane had a continuous tendency to turn right, making it difficult to maintain a stable pitch attitude. He also advised that the No. 1 and No. 3 engine thrust levers could not be used symmetrically, so he used two hands to manipulate the two throttles.

About 15:42, the flight engineer was sent to the passenger cabin to inspect the empennage visually. Upon his return, he reported that he observed damage to the right and left horizontal stabilizers.

Fuel was jettisoned to the level of the automatic system cutoff, leaving 33,500 pounds. About 11 minutes before landing, the landing gear was extended by means of the alternate gear extension procedure.

The flightcrew said that they made visual contact with the airport about 9 miles out. ATC had intended for flight 232 to attempt to land on runway 31, which was 8,999 feet long. However, ATC advised that the airplane was on approach to runway 22, which was closed, and that the length of this runway was 6,600 feet. Given the airplane's position and the difficulty in making left turns, the captain elected to continue the approach to runway 22 rather than to attempt maneuvering to runway 31. The check airman said that he believed the airplane was lined up and on a normal glidepath to the field. The flaps and slats remained retracted.

During the final approach, the captain recalled getting a high sink rate alarm from the ground proximity warning system (GPWS). In the last 20 seconds before touchdown, the airspeed averaged 215 KIAS, and the sink rate was 1,620 feet per minute. Smooth oscillations in pitch and roll continued until just before touchdown when the right wing dropped rapidly.

The captain stated that about 100 feet above the ground the nose of the airplane began to pitch downward. He also felt the right wing drop down about the same time. Both the captain and the first officer called for reduced power on short final approach.

The check airman said that based on experience with no flap/no slat approaches he knew that power would have to be used to control the airplane's descent. He used the first officer's airspeed indicator and visual cues to determine the flightpath and the need for power changes. He thought that the airplane was fairly well aligned with the runway during the latter stages of the approach and that they would reach the runway. Soon thereafter, he observed that the airplane was positioned to the left of the desired landing area and descending at a high rate. He also observed that the right wing began to drop. He continued to manipulate the No. 1 and No. 3 engine throttles until the airplane contacted the ground. He said that no steady application of power was used on the approach and that the power was constantly changing. He believed that he added power just before contacting the ground.

The airplane touched down on the threshold slightly to the left of the centerline on runway 22 at 16:00. First ground contact was made by the right wing tip followed by the right main landing gear. The airplane skidded to the right of the runway and rolled to an inverted position. Witnesses observed the airplane ignite and cartwheel, coming to rest after crossing runway 17/35. Firefighting and rescue operations began immediately, but the airplane was destroyed by impact and fire.

The accident resulted in 111 fatal, 47 serious, and 125 minor injuries. The remaining 13 occupants were not injured.

Probable Cause   Source: Accident Report[44]

The National Transportation Safety Board determines that the probable cause of this accident was the inadequate consideration given to human factors limitations in the inspection and quality control procedures used by United Airlines' engine overhaul facility which resulted in the failure to detect a fatigue crack originating from a previously undetected metallurgical defect located in a critical area of the stage 1 fan disk that was manufactured by General Electric Aircraft Engines. The subsequent catastrophic disintegration of the disk resulted in the liberation of debris in a pattern of distribution and with energy levels that exceeded the level of protection provided by design features of the hydraulic systems that operate the DC-10's flight controls.

Safety Recommendations
The NTSB issued several recommendations to the FAA.
Regarding Occupant Restraints:
Current FAA regulations allow occupants who have not reached their second birthday to be held in the lap of an adult. The Safety Board believes that this regulation does not adequately protect occupants under age 2 and urged the FAA to require that infants and small children be restrained in child safety seats appropriate to their height and weight. The Safety Board believes that time consuming flight attendant duties, such as providing special brace-for-impact instructions for unrestrained infants, answering questions about those instructions, and distributing pillows in an effort to enhance the effectiveness of adult lap belts on small children, could be reduced if child restraint was mandatory. Thus, flight attendants could devote more time to other important duties while they prepare the cabin for an emergency landing. The Safety Board issued Recommendations A-90-78 andA-90-79 to address the child restraint issue .

A-90-78: Revise 14 CFR 91,121and 135 to require that all occupants be restrained during takeoff, landing, and turbulent conditions, and that all infants and small children below the weight of40 pounds and under the height of 40 inches to be restrained in an approved child restraint system appropriate to their height and weight.

A-90-79: Conduct research to determine the adequacy of aircraft seatbelts to restrain children too large to use child safety seats and to develop some suitable means of providing adequate restraint for such children.

Regulatory Outcome
Regarding Occupant Restraints: None

November 4, 2010: Airbus A380, VH-OQA

Location	Near Bantam Island, Indonesia
Flight	QFA 32
Operator	QANTAS
Operation	Scheduled Airline
Route	Singapore to Sydney, Australia
On-Board Injuries	Minor/None469
Aircraft Damage	Substantial

Description   Source: Aviation Safety Network[1]

An Airbus A380-842 passenger jet, registered VH-OQA, incurred substantial damage in an accident near Batam Island, Indonesia. There were no fatalities. The airplane operated on Qantas flight QF32 from Singapore-Changi International Airport (SIN) to Sydney-Kingsford Smith International Airport, NSW (SYD).

The airplane took off from runway 20C at 09.56. Following a normal takeoff, the crew retracted the landing gear and flaps. The crew reported that, while maintaining 250 kts in the climb and passing 7,000 ft above mean sea level, they heard two almost coincident 'loud bangs', followed shortly after by indications of a failure of the No 2 engine.

The aircraft levelled off and because of an overheat warning of engine No 2, thrust for this engine was moved to 'idle'. Meanwhile, at 10:02, when the airplane was flying over Batam Landmass, the crew radioed a PAN call to the Approach Controller citing a possible engine failure. At that time, the pilot of QFA 32 maintained height on 7,500 feet and requested to be on heading 150 degrees to investigate the problem, but did not request to return to Singapore immediately. Later on at 10.21, the crew reported that they had been gone through an extensive checklist and found that there was a hole in the side of engine number 2 and it had damaged a part of the wing. The pilot then requested to hold for half an hour before making an approach to Changi Airport.

A moment later, an air traffic controller from Batam Tower, who had received a report stating that parts of an aircraft had been found on Batam Center-Batam Island made a report to Singapore ATC Approach Sector about the finding.

The pilot informed Singapore ATC that other engines apart from engine number 2 appeared to be functioning normal; thus required an approach on Runway 20C at Changi Airport and a towing assistance when the aircraft stopped at the end of the runway.

While the aircraft was stopping abeam taxiway E10, Changi's Airport Emergency Service (AES) found that engine number 2 was damaged near the rear of the engine and fuel had leaked from the port side (left wing). Moreover, there was smoke from tyre number 7 and there were 4 tyres deflated, meanwhile, the pilot was not able to shut off engine number 1. Nevertheless, it was safe to disembark passengers. Exactly at 13:54, all passengers had been disembarked, and finally at 14:53, engine number 1 was finally able to be shutdown.

Analysis of the preliminary elements from the incident investigation shows that an oil fire in the HP/IP structure cavity may have caused the failure of the Intermediate Pressure Turbine (IPT) Disc.

Probable Cause   ## Source: Accident Report[45]

CONTRIBUTING SAFETY FACTORS:

1. Disc failure during the occurrence flight
- Over time, a fatigue crack had developed in the thin-wall section of the oil feed stub pipe in the No. 2 engine to the extent that, during the occurrence flight, opening of the crack through normal movement within the engine released oil into the HP/IP buffer space.
- Auto-ignition of the oil leaking from the oil feed stub pipe created an intense and sustained fire within the HP/IP buffer space that resulted in localised heat damage to the intermediate pressure (IP) turbine disc.
- The IP turbine disc separated from the drive arm and accelerated.
- Following the separation of the IP turbine disc from the drive arm, the engine behaved in a manner that differed from the engine manufacturer's modelling and experience with other engines in the Trent family, with the result that the IP turbine disc accelerated to a rotational speed in excess of its design capacity whereupon it burst in a hazardous manner. [Safety issue]

2. Manufacture and release into service of engine serial number 91045
- During the manufacture of the HP/IP bearing support assembly fitted to the No. 2 engine (serial number 91045), movement of the hub during the machining processes resulted in a critically reduced wall thickness within the counter bore region of the oil feed stub pipe.
- It was probable that a non-conformance in the location of the oil feed stub pipe interference bore was reported by the coordinate measuring machine during the manufacturing process, but that the non-conformance was either not detected or not declared by inspection personnel, resulting in the assembly being released into service with a reduced wall thickness in the oil feed stub pipe.

3. Opportunity to manage the non-conforming oil feed stub pipes in the Trent 900 fleet
- The statistical analysis used to estimate maximum likely oil feed stub pipe counter bore misalignment, and resulting thin wall section, did not adequately represent the population of actual misalignments in engines already released into service, nor did it implicitly provide a level of uncertainty in the results.
- The language used to define the size of the non-conformance on the retrospective concession form did not effectively communicate the uncertainty of the statistical analysis to those assessing and approving the concession.
- The engine manufacturer did not have a requirement for an expert review of statistical analyses used in retrospective concession applications. [Safety issue]
- The engine manufacturer's process for retrospective concessions did not specify when in the process the Chief Engineer and Business Quality Director approvals were to be obtained. Having them as the final approval in the process resulted in an increased probability that the fleet-wide risk assessment would not occur. [Safety issue]
- The retrospective concession was not approved by the Chief Engineer and Business Quality Director, as required by the group quality procedures relating to retrospective concessions, denying them the opportunity to assess the risk to the in-service fleet.

Regulatory Outcome[2]

As part of the process for cataloging findings and issuing safety recommendations, the ATSB identified a number of safety issues. Safety issues, for which resolution activity undertaken by the engine or airplane manufacturer were considered to be satisfactory, were closed by the ATSB and subsequent safety recommendations were not developed. Among the actions taken by Roll-Royce, Airbus, and EASA are:
Actions taken by Rolls-Royce:
1   Rolls-Royce incorporated a software change in the Engine Electronic Controller (EEC) to introduce an IP turbine overspeed protection system.

2   Rolls-Royce, after discovery in the A380 fleet of a number of oil feed stub pipes that were non-conforming, undertook a detailed analysis of those stub pipes, removing 40 engines from service and revising the minimum stub pipe wall thickness from .5 mm to .7 mm.

3   Rolls-Royce introduced new processes that required closer coordination between manufacturing engineers and design engineers to reduce manufacturing errors and maintain criteria for acceptance/rejection of non-conforming stub pipes.

4   Rolls-Royce reclassified the HP/IP bearing support structure to identify it as being "reliability sensitive," which invoked a higher level of process control during manufacture and installation.

Actions Taken by Airbus:

1   In response to Rolls-Royce's action to incorporate an EEC software revision which introduced an IP turbine overspeed protection system, Airbus issued a service bulletin to incorporate the software revision on all Trent-equipped A380 airplanes.

Actions taken by EASA:

1   EASA released an airworthiness directive 2010-0262 to require, within ten flights, an EEC software revision to incorporate an IP turbine overspeed protection per a Rolls-Royce service bulletin.

Comments

The crew on the flight deck was very likely the most experienced crew ever. In addition to theCaptain and First Officer, there was a Second Officer and two Check Pilots. One Check Pilot was giving a route check to the Captain;* the second was giving one to the first Check Pilot. In all, there were 72,411 hours of flying experience with 5,000 hours of A380 experience on the flight deck.. The reader might be interested in the Captain's book describing the experience.[46]

---

*   He didn't pass.[46]

Fire (In-Flight)

These accidents are discussed in Chapter 12 beginning on page 182

July 11, 1973: Boeing 707, PP-VJZ

Location	Paris, France
Flight	VRG 820
Operator	Varig (Viação Aérea RIo-Grandense)
Operation	Scheduled Airline
Route	Rio de Janeiro, Brazil to Paris, France
On-Board Injuries	Fatal 123 Serious 11
Aircraft Damage	Destroyed

Description   Source: Aviation Safety Network[1]

Varig Flight 820 departed Rio de Janeiro (GIG) at 03:03 for a flight to Paris-Orly (ORY). The en route part of the flight was uneventful. At 13:57 the aircraft had descended to FL80 and contacted Orly approach, who told the crew to maintain FL80 and head to the OLS VOR which would take the aircraft to the downwind leg of runway 26. At 13:58:20 the flight crew contacted Orly approach and reported a "problem with fire on board". An emergency descent was requested. At 13:59 clearance was given to descend to 3000 feet for a runway 07 landing, making a straight-in approach possible. While the situation on board was getting worse (smoke entering the cockpit and passengers becoming asphyxiated), a clearance to descend to 2000 feet was given at 14:01:10. The flight crew put on oxygen masks as smoke was making it impossible to read the instruments. At 14:03 the pilot decided to make an emergency landing 5 km short of the runway with gear down and flaps at 80deg. The Boeing approached with considerable nose-up attitude, in a slight left bank. The aircraft truncated some small trees and made a heavy landing on a field. Both main gears collapsed and the engines were torn off in the subsequent skid. The fuselage however, remained intact. Ten occupants (all crewmembers) evacuated the aircraft. By the time the firemen arrived (6-7 minutes later) the fire had burned through the roof and there was no sign of life. Of the four unconscious occupants the firemen could evacuate, only one survived.

Probable Cause   Source: Accident Report[47]

A fire which appears to have started in the washbasin unit of the aft right toilet. It was detected because smoke had entered the adjacent left toilet. The fire may have been started by an electrical fault or by the carelessness of a passenger. The difficulty in locating the fire made the actions of cabin personnel ineffective. The flight crew did not have the facilities to intervene usefully from the cockpit against the spread of the fire and the invasion of smoke.

The lack of visibility in the cockpit prompted the crew to decided on a forced landing. At the time of touch-down the fire was confined to the area of the aft toilets. The occupants of the passenger cabin were poisoned, to varying degrees by carbon monoxide and other combustion products.

Regulatory Outcome

In the aftermath of this accident, Varig Airlines sued the US Government claiming the CAA inspectors were negligent in not inspecting the heater installation. The case was finally decided by the US Supreme Court who found that such inspection was a discretionary function and was exempted from the Federal Tort Claims Act.[48] Review of this decision was part of the regulatory course for new FAA employees.

June 2, 1983: McDonnell Douglas DC-9, C-FTLU

Location	Cincinnati Airport, KY
Flight	ACA 797
Operator	Air Canada
Operation	Scheduled Airline
Route	Fort Worth, TX to Toronto, Ontario, Canada
On-Board Injuries	Fatal 23 Serious 3 Minor/None 20
Aircraft Damage	Destroyed

Description   Source: Aviation Safety Network[1]

Air Canada Flight 797, a McDonnell Douglas DC-9-32, was a scheduled flight from Dallas (DFW) to Montreal, Canada, with an en route stop at Toronto (YYZ).

At 16:25 CDT, Flight 797 left Dallas and climbed to FL330. Except for a deviation to the south of their filed flight plan route to avoid weather, the flight progressed without incident until it entered the Indianapolis Air Route Traffic Control Center's (ARTCC) airspace.

At 18:51:14 EDT, the three circuit breakers associated with the aft lavatory's flush motor and located on a panel on the cockpit wall behind the captain's seat, tripped in rapid succession. After identifying the circuit breakers, the captain immediately made one attempt to reset them; the circuit breakers would not reset. The captain assumed that the flush motor had probably seized and took no further action at this time. About 18:59:58, the captain again unsuccessfully tried to reset the three circuit breakers. He told the first officer that the circuit breaker "Pops as I push it". About 19:00, a passenger seated in the last row asked a flight attendant to identify a strange odor. The flight attendant thought the odor was coming from the aft lavatory. She took a CO_2 fire extinguisher from the cabin wall and opened the lavatory door a few inches. She saw that a light gray smoke had filled the lavatory from the floor to the ceiling, but she saw no flames. While she was inspecting the lavatory, she inhaled some smoke and closed the door. The No.3 flight attendant then saw the No.2 flight attendant nearby and asked her to tell the flight attendant in charge of the situation. Upon being advised there was a fire, the flight attendant in charge instructed the No. 2 flight attendant to inform the captain and then to assist the No. 3 flight attendant in moving the passengers forward and in opening the eyebrow air vents over the passenger seats to direct air to the rear of the cabin. The flight attendant in charge then took the CO_2 extinguisher and opened the lavatory door about three-quarters open. He also saw no flames, but he observed thick curls of black smoke coming out of the seams of the aft lavatory walls at the top of the wash basin behind the vanity and at the ceiling. He then proceeded "to saturate the washroom with CO_2" by spraying the paneling and the seam from which smoke was seeping and spraying the door of the trash bin. He then closed the lavatory door.

At 19:02:40, the No. 2 flight attendant reached the cockpit and told the captain, "Excuse me, there's a fire in the washroom in the back, they're just... went back to go to put it out". Upon being notified of the fire, the captain ordered the first officer to inspect the lavatory. The captain then donned his oxygen mask and selected the 100-percent oxygen position on his regulator. The first officer left the cockpit but did not take either smoke goggles or a portable oxygen bottle with him. (The airplane was not equipped with nor was it required to be equipped with self-contained breathing equipment or a full face smoke mask.) The first officer said that he could not get to the aft lavatory because the smoke, which had migrated over the last three to four rows of seats, was too thick. The flight attendant in charge told the first officer what he had seen when he opened the lavatory door, that he had discharged the CO2 extinguisher into the lavatory, and that he had not been able to see the source of the smoke before closing the door. He told the first officer, however, that he did not believe the fire was in the lavatory's trash container. The first officer told the flight attendant in charge that he was going forward to get smoke goggles. At 19:04:07, the first officer returned to the cockpit and told the captain that the smoke had prevented him from entering the aft lavatory and that he thought "we'd better go down". He did not tell the captain that the flight attendant in charge had told him that the fire was not in the trash bin. However, at 19:04:16, before the captain could respond, the flight attendant in charge came to the cockpit and told the captain that the passengers had been moved forward and that the captain didn't "have to worry, I think its gonna be easing up". The first officer looked back into the cabin and said that it was almost clear in the back. He told the captain, "it's starting to clear now," and that he would go aft again if the captain wanted him to do so. According to the captain, the first officer's smoke goggles were stored in a bin on the right side of the cockpit and were not easily accessible to the first officer while he was not in his seat. Since the first officer needed the goggles and since there was a hurry, the captain gave him his goggles and, at 19:04:46, directed him to go aft.

Two minutes later, while the first officer was out of the cockpit, the flight attendant in charge told the captain again that the smoke was clearing. The captain believed the fire was in the lavatory trash bin and decided not to descend at this time because he "expected it (the fire) to be put out". About 19:05:35, while the first officer was still aft to inspect the aft lavatory, the airplane experienced a series of electrical malfunctions. The master caution light illuminated, indicating that the airplane's left AC. and DC electrical systems had lost power.

In the meanwhile, the first officer proceeded to the aft lavatory and put on the smoke goggles. The lavatory door felt hot to the touch, so he decided not to open it and instructed the cabin crew to leave it closed. At that time, he noticed a flight attendant signaling him to hurry back to the cockpit. The first officer returned to the cockpit and got into his seat, and at 19:07:11, he told the captain, "I don't like what's happening, I think we better go down, okay?" Then the master warning light illuminated and the annunciator lights indicated that the emergency AC. and DC electrical buses had lost power. The captain's and first officer's attitude directional indicators tumbled. The captain ordered the first officer to activate the emergency power switch, thereby directing battery power to the emergency AC. and DC buses. The attitude directional indicators' gyros began erecting, however, because of the loss of AC. power, the stabilizer trim was inoperative and remained so during the rest of the flight. At 19:08:12, Flight 797 called the radar high sector controller at Indianapolis Center: "Mayday, Mayday, Mayday". The Louisville radar high sector controller acknowledged the call, and at 19:08:47, the flight told the controller that it had a fire and was going down. The controller told the flight that it was 25 nautical miles from Cincinnati and asked "can you possibly make Cincinnati". The flight answered that it could make Cincinnati and then requested clearance; it was then cleared to descend to 5,000 feet. At 19:09:05, Flight 797 reported that it was leaving FL330. The flight then told the controller that it needed to be vectored toward Cincinnati, that it was declaring an emergency, and that it had changed its transponder code to 7700 -- the emergency code. However, the transponder was inoperative due to the power loss.

When the captain sighted the runway, he extended the landing gear. Since the horizontal stabilizer was inoperative, the captain extended the flaps and slats incrementally through the 0, 5, 15, 25, and 40-degree positions. He allowed his indicated airspeed to stabilize at each flap position as he slowed to approach speed. He flew the final approach at 140 KIAS and completed the landing. After touchdown, he made a maximum effort stop using extended spoilers and full brakes. Because of the loss of the left and right AC. buses, the antiskid system was inoperative and the four main wheel tires blew out. The airplane was stopped just short of the intersection of taxiway J. After the captain completed the emergency engine shutdown checklist, both he and the first officer attempted to go back into the cabin and assist in the passenger evacuation, but were driven back by the smoke and heat. Thereafter, they exited the airplane through their respective cockpit sliding windows. After the airplane stopped, the left (L-1) and right (R-1) forward cabin doors, the left forward (L-2) overwing exit, and the right forward (R-2) and aft (R-3) overwing exits were opened, and the slides at the L-1 and R-1 doors were deployed and inflated. The 3 cabin attendants and 18 passengers used these 5 exits to evacuate the airplane. After the 18 passengers and 5 crewmembers left the airplane, the cabin interior burst into flames. Twenty-three passengers perished in the fire. Neither the passengers, crew, nor witnesses outside of the airplane saw flames inside the cabin before the survivors left the plane. The fuselage and passenger cabin were gutted before airport fire personnel could extinguish the fire.

Probable Cause   Source: Accident Report[49]

A fire of undetermined origin, an underestimate of fire severity, and conflicting fire progress information provided to the captain.

Contributing to the severity of the accident was the flight crew's delayed decision to institute an emergency descent.

Regulatory Outcome

Several regulatory changes were listed in the "Lessons Learned" website. These include changes to

Flammability Requirements For Seat Cushions (NPRM 83-14)

Floor Proximity Emergency Escape Path Marking (NPRM 83-15)

Lavatory Smoke Detection, Automatic Lavatory Trash Fire Extinguishers and Halon Fire Extinguisher Requirements (NPRM 84-05)

Lavatory Smoke Detection, Automatic Lavatory Trash Fire Extinguishers and Halon Fire Extinguisher Requirements (NPRM 89-01)

Advisory Circular: Smoke Detection, Penetration, and Evacuation Tests and Related Flight Manual Emergency Procedures (AC-25-9)

November 28, 1987: Boeing 747, ZS-SAS

Location	Indian Ocean
Flight	SAA 295*
Operator	South African Airways
Operation	Scheduled Airline
Route	Taipei, Taiwan to Plaisance Airport, Mauritius
On-Board Injuries	Fatal <u>159</u> Minor/None <u>4</u>
Aircraft Damage	Destroyed

Description   Source: Aviation Safety Network[1]

On November 27th 1987 flight SA295 was scheduled to depart from Taipei's Chiang Kai Shek Airport at 13:00 UTC for Mauritius' Plaisance Airport and Johannesburg, South Africa on a scheduled international air transport service. Due to adverse weather and the late arrival of a connecting flight the departure time was delayed and the airplane took off at 14:23 UTC with 149000 kg of fuel, 43225 kg of baggage and cargo, 140 passengers and a crew comprising 5 flight crew members and 14 cabin crew members. The calculated flight time was 10 hours 14 minutes.

The takeoff was normal. At 14:56 UTC the crew communicated with Hong Kong Radar and thereafter routine position reports were given to the flight information centers (FICs) at Hong Kong, Bangkok, Kuala Lumpur, Colombo, Cocos Islands and Mauritius. At 15:55 a routine report was made to the Operator's base at Johannesburg. The information given was that the airplane had taken off from Taipei at 14:23, was flying at FL310 and that the arrival time at Mauritius was estimated as 00:35 UTC.

At about 22:30 the pilot called Mauritius FIC, using HF radio, and advised that the aircraft had been at position 070° East at 22:29 at FL350 and that the time at position 065° East was estimated as 23:12. At 23:13 the position report of 065° East at FL350 was given to Mauritius FIC. The estimated time of arrival (ETA) over position 060° East was given as 23:58.

About 23:45 the master fire warning alarm sounded on the flight deck. Somebody, probably the pilot, inquired where the warning had come from and received the reply that it had come from the main deck cargo. The pilot then asked that the check list be read. Some 30 seconds later somebody on the flight deck uttered an oath.

The pilot called Mauritius Approach Control at 23:49 and said that they had a smoke problem and were doing an emergency descent to FL140.

The approach controller gave clearance for the descent and the pilot asked that the fire services be alerted. The controller asked if full emergency services were required to which the pilot replied in the affirmative.

At 23:51 the approach controller asked the pilot for his actual position. The pilot replied: "Now we have lost a lot of electrics, we haven't got anything on the aircraft now". At 23:52 the approach controller asked for an ETA at Plaisance and was given the time of 00:30. At 23:52:50 the pilot made an inadvertent transmission when he said to the senior flight engineer: "Hey Joe, shut down the oxygen left". From this time until 00:01:34 there was a period of silence lasting 8 minutes and 44 seconds. From 00:01:34 until 00:02:14 the pilot inadvertently transmitted instructions, apparently to the senior flight engineer, in an excited tone of voice. Most of the phrases are unintelligible. At 00:02:43 the pilot gave a distance

---

*   SA is the IATA code for South African Airways and is used in the source material

report as 65 nautical miles. This was understood by the approach controller to be the distance to the airport. In fact it was the distance to the next waypoint, Xagal. The distance to the airport at that point was approximately 145 nautical miles. At 00:02:50 the approach controller recleared the flight to FL50 and at 00:03:00 gave information on the actual weather conditions at Plaisance Airport, which the pilot acknowledged. When the approach controller asked the pilot at 00:03: 43 which runway he intended to use he replied one three but was corrected when the controller asked him to confirm one four. At 00:03:56 the controller cleared the flight for a direct approach to the Flic-en-Flac (FF) non-directional beacon and requested the pilot to report on approaching FL50. At 00:04:02 the pilot said: "Kay". From 00:08:00 to 00:30:00 the approach controller called the aircraft repeatedly but there was no reply.

The aircraft crashed into the Indian Ocean at a position determined to be about 134 nautical miles North-East of Plaisance Airport. The accident occurred at night, in darkness, at about 00:07 UTC. The local time was 04:07.

Within a few days drifting pieces of wreckage were found, but it took until January 28th, 1988 for the main wreckage field to be found on the Ocean floor, at a depth of 4400 meters.

The cockpit voice recorder was recovered on 6 Jan. 1989.

Probable Cause   Source: Board of Inquiry Report[50]

Fire of an unknown origin had possibly: 1) incapacitated the crew; 2) caused disorientation of the crew due to thick smoke; 3) caused crew distraction; 4) weakened the aircraft structure, causing an in-flight break-up.; 5) burned through several control cables; 6) caused loss of control due to deformation of the aircraft fuselage.

Regulatory Outcome

On November 28, 1987, South African Airways Flight 295 crashed into the Indian Ocean en route from Taipai's Chiang Kai Shek Airport to Mauritius's Plaisance Airport as the result of an uncontrolled fire in the airplane's main deck cargo compartment. All 159 people on board were lost. The crash occurred about ten hours into the flight, less than twenty minutes after the flight crew reported smoke to air traffic control in Mauritius. The airplane, a 747-200, was configured as a "combi," on which the main deck of the airplane is divided into two large sections, a passenger compartment and a cargo compartment. The method for controlling fires in main deck cargo compartments on combi airplanes relies heavily on manual firefighting using hand held fire extinguishers. Whether or not manual firefighting was attempted on Flight 295 is unclear, but considering the size of the 747's main deck cargo compartment and the apparent rapid growth of the fire that occurred, the chance that such an attempt would have been successful is questionable. After the accident, the FAA and other airworthiness authorities concluded that reliance on manual firefighting as the primary means to control fires in large Class B cargo compartments was inadequate, and regulatory action was taken to require design and operational changes.

May 11, 1996: McDonnell Douglas DC-9, N-904VJ

Location	Everglades, FL
Flight	VJC 592
Operator	ValuJet
Operation	Scheduled Airline
Route	Miami, FL to Atlanta, GA
On-Board Injuries	Fatal 100
Aircraft Damage	Destroyed

Description   Source: Accident Report[51]

The airplane crashed into the Everglades about 10 minutes after takeoff from Miami International Airport. Safety issues discussed in the Board's report include minimization of the hazards posed by fires in class D cargo compartments; equipment, training, and procedures for addressing in-flight smoke and fire aboard air carrier airplanes; guidance for handling of chemical oxygen generators and other hazardous aircraft components; SabreTech's and ValuJet's procedures for handling company materials and hazardous materials; ValuJet's oversight of its contract heavy maintenance facilities; FAA's oversight of ValuJet and ValuJet's contract maintenance facilities; FAA's and the Research and Special Programs Administration's (RSPA) hazardous materials program and undeclared hazardous materials in the U.S. mail; and ValuJet's procedures for boarding and accounting for lap children. Safety recommendations concerning these issues were made to the FAA, RSPA, the U.S. Postal Service, and the Air Transport Association.

Probable Cause   Source: Accident Report[51]

The National Transportation Safety Board determines the probable causes of this accident, which resulted from a fire in the airplane's class D cargo compartment that was initiated by the actuation of one or more oxygen generators being improperly carried as cargo, were

- The failure of SabreTech to properly prepare, package, and identify unexpended chemical oxygen generators before presenting them to ValuJet for carriage;
- The failure of ValuJet to properly oversee its contract maintenance program to ensure compliance with maintenance, maintenance training, and hazardous materials requirements and practices; and

- The failure of the Federal Aviation Administration (FAA) to require smoke detection and fire suppression systems in class D cargo compartments.

  Contributing to the accident was the failure of the FAA to adequately monitor ValuJet's heavy maintenance programs and responsibilities, including ValuJet's oversight of its contractors, and SabreTech's repair station certificate; the failure of the FAA to adequately respond to prior chemical oxygen generator fires with programs to address the potential hazards; and ValuJet's failure to ensure that both ValuJet and contract maintenance facility employees were aware of the carrier's 'no-carry' hazardous materials policy and had received appropriate hazardous materials training.

Regulatory Outcome

As a direct result of the Flight 592 accident and it's precursors, the Class D cargo compartment design methodology was abandoned and the Class D classification was removed from multiple Part 25 regulations. Legislation was enacted to increase protection from possible inflight fires. All existing Class D cargo compartments were required to be upgraded to Class C or Class E compartments in accordance with 121.314. In addition, all existing cargo smoke or fire detection systems were required to be updated to meet the more stringent requirements of §25.858 which required detection to occur within one minute.

Comments

The point that Class D cargo compartments was predicated on denying fires a source of oxygen seems to have been missed. The cargo that caught fire would self-generate oxygen. This would allow the fire to produce its own source of oxygen. Year later, with the advent of lithium batteries, a similar effect would result.

It is not clear to us that Class D cargo compartments were the problem; hazardous materials which provided a self-contained supply of oxygen were. A review of the NTSB accident records shows a single accident using the search term "Class D Cargo Compartment". ValuJet 592 was the only accident listed.

September 2, 1998: McDonnell Douglas MD-11, HB-IWF

Location	Near Peggy's Cove, NS, Canada
Flight	SWR 111
Operator	Swissair
Operation	Scheduled Airline
Route	New York, NY to Geneva, Switzerland
On-Board Injuries	Fatal 229
Aircraft Damage	Destroyed

Description   Source: Aviation Safety Network[1]

At 20:18 Swissair flight SR111,* departed New York-JFK Airport on a flight to Geneva, Switzerland. Forty minutes later the copilot contacted Moncton ACC, reporting FL330. At 21:10 the pilots detected an unusual odor in the cockpit and began to investigate. They determined that some smoke was present in the cockpit, but not in the passenger cabin. They assessed that the odor and smoke were related to the air conditioning system. Four minutes later a Pan Pan radio call was made to Moncton ACC. The aircraft was about 66 NM southwest of Halifax. The pilots reported that there was smoke in the cockpit and requested an immediate return to a convenient place. The pilots named Boston, which was about 300 NM behind them. The Moncton ACC controller immediately cleared SR111 to turn right toward Boston and to descend to FL310. Then the controller asked SR111 whether they preferred to go to Halifax. The pilots expressed a preference for Halifax. They immediately received an ATS clearance to fly directly to Halifax. At this time, the pilots donned their oxygen masks. At 21:16, the controller cleared SR111 to descend to 10000 feet. Two minutes later they were cleared down to 3000 feet. At 21:19, the controller instructed SR111 to turn left to a heading of 030 for a landing on runway 06 at the Halifax, and advised that the aircraft was 30 NM from the runway threshold. The aircraft was descending through approximately FL210 and the pilots indicated that they needed more than 30 NM. The controller instructed SR111 to turn to a heading of 360 to provide more track distance for the aircraft to lose altitude. The flight crew discussed internally the dumping of fuel based on the aircraft's gross weight, and on their perception of the cues regarding the aircraft condition, and agreed to dump fuel. The flight was vectored to the south to dump fuel. At 21:24, both pilots almost simultaneously declared an emergency. The co-pilot indicated to the controller that they were starting to dump fuel and that they had to land immediately. Last radio contact was one minute later when they again declared an emergency. By now the fire had propagated, causing severe disturbances of the electric system. In the last minutes of the flight, the electronic navigation equipment and communications radios stopped operating. The aircraft descended over the dark waters off the coast of Nova Scotia until it stuck the water in a 20 degrees nose down and 110 degrees right bank.

---

*   SR is the code for Swissair and is used in the source material.

Probable Cause

Findings as to Causes and Contributing Factors:

- Aircraft certification standards for material flammability were inadequate in that they allowed the use of materials that could be ignited and sustain or propagate fire. Consequently, flammable material propagated a fire that started above the ceiling on the right side of the cockpit near the cockpit rear wall. The fire spread and intensified rapidly to the extent that it degraded aircraft systems and the cockpit environment, and ultimately led to the loss of control of the aircraft.
- Metallized polyethylene terephthalate (MPET)-type cover material on the thermal acoustic insulation blankets used in the aircraft was flammable. The cover material was most likely the first material to ignite, and constituted the largest portion of the combustible materials that contributed to the propagation and intensity of the fire.
- Once ignited, other types of thermal acoustic insulation cover materials exhibit flame propagation characteristics similar to MPET-covered insulation blankets and do not meet the proposed revised flammability test criteria. Metallized polyvinyl fluoride–type cover material was installed in HB-IWF and was involved in the in-flight fire.
- Silicone elastomeric end caps, hook-and-loop fasteners, foams, adhesives, and thermal acoustic insulation splicing tapes contributed to the propagation and intensity of the fire.
- The type of circuit breakers (CB) used in the aircraft were similar to those in general aircraft use, and were not capable of protecting against all types of wire arcing events. The fire most likely started from a wire arcing event.
- A segment of in-flight entertainment network (IFEN) power supply unit cable (1-3791) exhibited a region of resolidified copper on one wire that was caused by an arcing event. This resolidified copper was determined to be located near manufacturing station 383, in the area where the fire most likely originated. This arc was likely associated with the fire initiation event; however, it could not be determined whether this arced wire was the lead event.
- There were no built-in smoke and fire detection and suppression devices in the area where the fire started and propagated, nor were they required by regulation. The lack of such devices delayed the identification of the existence of the fire, and allowed the fire to propagate unchecked until it became uncontrollable.
- There was a reliance on sight and smell to detect and differentiate between odour or smoke from different potential sources. This reliance resulted in the misidentification of the initial odour and smoke as originating from an air conditioning source.
- There was no integrated in-flight firefighting plan in place for the accident aircraft, nor was such a plan required by regulation. Therefore, the aircraft crew did not have procedures or training directing them to aggressively attempt to locate and eliminate the source of the smoke, and to expedite their preparations for a possible emergency landing. In the absence of such a firefighting plan, they concentrated on preparing the aircraft for the diversion and landing.
- There is no requirement that a fire-induced failure be considered when completing the system safety analysis required for certification. The fire-related failure of silicone elastomeric end caps installed on air conditioning ducts resulted in the addition of a continuous supply of conditioned air that contributed to the propagation and intensity of the fire.
- The loss of primary flight displays and lack of outside visual references forced the pilots to be reliant on the standby instruments for at least some portion of the last minutes of the flight. In the deteriorating cockpit environment, the positioning and small size of these instruments would have made it difficult for the pilots to transition to their use, and to continue to maintain the proper spatial orientation of the aircraft.

Certification Issues

The TSB found several certification irregularities:

- The inflight entertainment system STC project management structure did not ensure that the required elements were in place to design, install, and certify a system that included emergency electrical load-shedding procedures compatible with the MD-11 type certificate. No link was established between the manner in which the system was integrated with aircraft power and the initiation or propagation of the fire.
- The FAA STC approval process for the inflight entertainment system did not ensure that the DAS employed personnel with sufficient aircraft-specific knowledge to appropriately assess the integration of the system's power supply with aircraft power before granting certification.
- The FAA allowed a de facto delegation of a portion of their AEG function to the DAS even though no provision existed within the FAA's STC processes to allow for such a delegation.

§25.1309 requires that a system safety analysis be accomplished on every system installed in an aircraft; however, the requirements are not sufficiently stringent to ensure that all systems, regardless of their intended use, are integrated into the aircraft in a manner compliant with the aircraft's type certificate.

September 3, 2010: Boeing 747, N-571UP

Location   Dubai Airport
Flight   UPS 006
Operator   United Parcel Service
Operation   Cargo
Route   Dubai, UAE to Cologne, Germany
On-Board Injuries   Fatal 2
Aircraft Damage   Destroyed

Description   Source: Aviation Safety Network[1]

A Boeing 747-44AF(SCD) cargo plane, registered N571UP, was destroyed in an accident shortly after takeoff from Dubai Airport, UAE. Both crew members were killed in the crash.

On September 3rd 2010, Flight UPS6 arrived from Hong Kong on a scheduled cargo service flight into Dubai carrying among other items consignments of cargo that included lithium batteries. There were no declared shipments of hazardous materials onboard the airplane. However, at least three of the shipments contained lithium ion battery packs that met the Class 9 hazardous material criteria.

The aircraft was parked at the loading position at 15:35 local time. Several Unit Load Devices (ULD) were off-loaded and new ULDs were loaded onto the plane. The flight then departed from Dubai at 18:51. The First Officer was the Pilot Flying, the captain was the Pilot Non Flying (PNF) for the sector to Cologne.

The takeoff and climb out from Dubai was uneventful with the exception of a Pack 1 fault which was reset by the PNF at 18:55.

The flight transited from UAE airspace into Bahrain Airspace where, at 19:12, the fire bell alarm sounded on the flight deck. The airplane was approaching top of climb (FL320) at the time.

Following the fire bell annunciation, the captain assumed control of the aircraft as PF, and the First Officer reverted to PNF while managing the fire warnings and cockpit checklists. The captain advised Bahrain Air Traffic Control that there was a fire indication on the main deck of the aircraft. The crew informed Bahrain that they needed to land as soon as possible. Bahrain advised the crew that Doha International Airport was at the aircraft's 10 o'clock position at 100 NM. DOH was the nearest airport at the time the emergency was declared, Dubai was approximately 148 NM DME. The captain elected to return to Dubai, and following the request to land as soon as possible to Bahrain, the crew declared an emergency. Bahrain acknowledged the request, cleared the aircraft for a series of right hand heading changes back to Dubai onto a heading of 106°.

At approximately 19:14, the autopilot disconnected, followed at 19:15 by a second audible alarm similar to the fire bell. At about this time the flight crew put on the oxygen masks and goggles. The crew experienced difficulties communicating via the intercom with the masks on, which interfered with the Cockpit Resource Management (CRM).

Following the initiation of the turn back to Dubai, having been cleared to 27,000 ft, the crew requested an expedited, immediate descent to 10,000 feet. Following ATC clearance, the flight crew initiated a rapid descent to 10,000 ft. Bahrain advised the crew that the aircraft was on a direct heading to Dubai and cleared for landing on runway 12 Left at their discretion.

The Fire Main Deck checklist was activated. According to the system logic, the cabin began to depressurize, Packs 2 and 3 shut down automatically, and were then manually selected to off on the overhead panel in accordance with the checklist instructions.

At 19:15, PACK 1 shut down, with no corresponding discussion recorded on the CVR. A short interval after the A/P was disengaged, the captain informed the F/O that there was limited pitch control of the aircraft in the manual flying mode, the captain then requested the F/O to determine the cause of the pitch control anomaly.

During the turn back to Dubai, the autopilot was re-engaged, and the aircraft descent was stabilized at 19:17. The captain told the F/O to pull the smoke evacuation handle. This was not part of the Fire Main Deck Non-Normal checklist.

The captain informed Bahrain that the cockpit was 'full of smoke' and commented to the F/O about the inability to see the instruments. The captain instructed the F/O to input Dubai into the Flight Management System (FMS). The F/O acknowledged the request and commented about the increasing flight deck temperature. It was not clear from the CVR if the FMS was programmed for Dubai, although the FDR indicated that the ILS/VOR frequency was changed to 110.1 MHz which was the frequency for Dubai runway 12 Left. Based on the information available to date, it is likely that less than 5 minutes after the fire indication on the main deck, smoke had entered the flight deck and intermittently degraded the visibility to the extent that the flight instruments could not effectively be monitored by the crew.

At approximately 19:19, during the emergency descent, at approximately 20,000 ft cabin altitude, the Captain, as PF, declared a lack of oxygen supply. Following a brief exchange between the captain and F/O regarding the need for oxygen, the captain transferred control of the aircraft to the F/O as PF. Portable oxygen is located on the flight deck and in the supernumerary area, aft of the flight crew's positions when seated. At this point the recorded CVR is consistent with the captain leaving his seat, after which there is no further CVR information that indicates any further interaction from the captain for the remainder of the flight.

The normal procedural requirement of transiting into the Emirates FIR, inbound for Dubai was a radio frequency change from Bahrain to Emirates ATC. At 19:20, Bahrain advised the crew to contact Emirates. At approximately the same time, the PF transmitted 'mayday, mayday, mayday can you hear me?'.

The PF advised Bahrain that due to the smoke in the flight deck, the ability to view the cockpit instruments, the Flight Management System, Audio Control Panel, and radio frequency selection displays had been compromised. At 19:21, the PF advised Bahrain that they would stay on the Bahrain frequency as it was not possible to see the radios. The PF elected to remain on the Bahrain radio frequency for the duration of the flight. At approximately 19:22 the aircraft entered the Emirates FIR heading east, tracking direct to the Dubai RW12L intermediate approach fix. The aircraft was now out of effective VHF radio range with Bahrain. In order for the crew to communicate with BAH-C, Bahrain advised transiting aircraft that they would act as a communication relay between BAH-C and the emergency aircraft.

At 19:22, the F/O informed the relay aircraft that he was 'looking for some oxygen'.

Following the rapid descent to 10,000 ft the aircraft leveled off at the assigned altitude approximately 84NM from Dubai. At approximately 19:26, the PF requested immediate vectors to the nearest airport and advised he would need radar guidance due to difficulty viewing the instruments.

At around 19:33, approximately 26 NM from Dubai, the aircraft descended to 9000 ft, followed by a further gradual descent as the aircraft approached Dubai, inbound for RW12L. The speed of the aircraft was approximately 340 kts.

19:38, approximately 10NM from RW12L, Bahrain, through the relay aircraft, advised the crew the aircraft was too high and too fast and requested the PF to perform a 360° turn if able. The PF responded 'Negative'. At this time the FDR data indicated the gear lever was selected down, the speed brake lever moved toward extend and at approximately the same time there was a sound consistent with the flap handle movement; shortly afterward the PF indicated that the landing gear was not functioning.

The aircraft over flew the Dubai northern airport boundary on a heading of 117°, the aircraft speed and altitude, was 340 kts at an altitude of 4500 ft and descending. Following the over flight of Dubai, on passing the south eastern end of RW12L, the aircraft was cleared direct to Sharjah airport as an immediate alternate– Sharjah J was to the aircraft's left and the runway was a parallel vector.

The relay pilot asked the PF if it was possible to perform a left hand turn. The PF responded requesting the heading to Sharjah.

The PF was advised that Sharjah was at 095° from the current position at 10 NM and that this left hand turn would position the aircraft on final approach for Sharjah runway30. The PF acknowledged the heading change for Sharjah. The PF selected 195° degrees on the Mode Control Panel.

The A/P disconnected at 19:40, the aircraft then entered a descending right hand turn at an altitude of 4000 ft as the speed gradually reduced to 240 kts until the impact.

Several Ground Proximity Warning System caution messages were audible on the CVR indicating: Sink Rate, Too Low Terrain and Bank Angle warnings. Radar contact was lost at approximately 19:41. The aircraft crashed 9nm south of Dubai onto a military installation near Minhad Air Force Base.

Probable Cause
Source: Accident Report[53]

1. A large fire developed in palletized cargo on the main deck at or near pallet positions 4 or 5, in Fire Zone 3, consisting of consignments of mixed cargo including a significant number of lithium type batteries and other combustible materials. The fire escalated rapidly into a catastrophic uncontained fire.
2. The large, uncontained cargo fire, that originated in the main cargo deck caused the cargo compartment liners to fail under combined thermal and mechanical loads.
3. Heat from the fire resulted in the system/component failure or malfunction of the truss assemblies and control cables, directly affecting the control cable tension and elevator function required for the safe operation of the aircraft when in manual control.
4. The uncontained cargo fire directly affected the independent critical systems necessary for crew survivability. Heat from the fire exposed the supplementary oxygen system to extreme thermal loading, sufficient to generate a failure. This resulted in the oxygen supply disruption leading to the abrupt failure of the Captain's oxygen supply and the incapacitation of the captain.
5. The progressive failure of the cargo compartment liner increased the area available for the smoke and fire penetration into the fuselage crown area.
6. The rate and volume of the continuous toxic smoke, contiguous with the cockpit and supernumerary habitable area, resulted in inadequate visibility in the cockpit, obscuring the view of the primary flight displays, audio control panels and the view outside the cockpit which prevented all normal cockpit functioning.
7. The shutdown of Pack 1 for unknown reasons resulted in loss of conditioned airflow to the upper deck causing the Electronic Equipment Cooling system to reconfigure to "closed loop mode". The absence of a positive pressure differential contributed to the hazardous quantities of smoke and fumes entering the cockpit and upper deck, simultaneously obscuring the crew's view and creating a toxic environment.
8. The fire detection methodology of detecting smoke sampling as an indicator of a fire is inadequate as pallet smoke masking can delay the time it takes for a smoke detection system to detect a fire originating within a cargo container or a pallet with a rain cover.

Regulatory Outcome

An AD[54] was issued requiring design, equipment, and operational changes related to Class B cargo compartments in large transport category airplanes manufactured by Boeing and McDonnell Douglas. This AD provides four options:

- Convert the Class B cargo compartment to Class C, as defined in 14 CFR part 25.857(c), which includes upgrading the liner and installing a fire suppression system.
- Use containers that meet the requirements of Class C for all cargo carried in the Class B cargo compartment.
- Use fire resistant covers or containers for all cargo, along with incorporating other design, equipment, and operational modifications.

- Incorporate a long duration fire suppression system, along with other design, equipment, and operational modifications.

Comments

Again, the cargo created more oxygen during a fire as in the ValuJet fire (see page 291). Fighting a fire on the main deck in a freighter is normally dealt with by depressurizing the airplane to reduce the oxygen to support combustion. However, if the cargo generates its own oxygen, this won't work. Carrying flammable, oxygen-generating cargo requires new safety measures.

November 9, 2010: Boeing 787, N-787EX

Location	Laredo, TX
Identification	N-787EX*
Operator	Boeing Company
Operation	Flight Test
Route	Yuma, AZ to Harlingen, TX
On-Board Injuries	Minor/None 42
Aircraft Damage	Unknown

Description   Source: Aviation Safety Network[1]

The aircraft made an emergency landing in Laredo, Texas after smoke was seen in the cabin due to electrical problems in the aft electronics bay which disabled the primary flight displays in the cockpit. The aircraft was evacuated using slides. One of the 42 occupants sustained a minor injury.

Boeing Statement[55]

Boeing continues to investigate Monday's incident on ZA002.†[56] We have determined that a failure in the P100 panel led to a fire involving an insulation blanket. The insulation self-extinguished once the fault in the P100 panel cleared. The P100 panel on ZA002 has been removed and a replacement unit is being shipped to Laredo. The insulation material near the unit also has been removed.

Damage to the ZA002 P100 panel is significant. Initial inspections, however, do not show extensive damage to the surrounding structure or other systems. We have not completed our inspections of that area of the airplane.

The P100 panel is one of several power panels in the aft electronics bay. It receives power from the left engine and distributes it to an array of systems. In the event of a failure of the P100 panel, backup power sources – including power from the right engine, the Ram Air Turbine, the auxiliary power unit or the battery – are designed to automatically engage to ensure that those systems needed for continued safe operation of the airplane are powered. The backup systems engaged during the incident and the crew retained positive control of the airplane at all times and had the information it needed to perform a safe landing.

Molten metal has been observed near the P100 panel, which is not unexpected in the presence of high heat. The presence of this material does not reveal anything meaningful to the investigation.

Inspection of the surrounding area will take several days and is ongoing. It is too early to determine if there is significant damage to any structure or adjacent systems.

As part of our investigation, we will conduct a detailed inspection of the panel and insulation material to determine if they enhance our understanding of the incident.

We continue to evaluate data to understand this incident. At the same time, we are working through a repair plan. In addition, we are determining the appropriate steps required to return the rest of the flight test fleet to flying status.

Boeing will continue to provide updates as new understanding is gained.

Comment

This incident should have been reported to the NTSB, since it met several criteria of NTSB Regulation §830.5: These include Inflight failure of electrical system requiring the use of an air-drive generator and Evacuation of an aircraft using emergency egress system.[57] We have not seen any further reference to this incident.

---

*   Aircraft registration used *vice* flight number.
†   ZZ002 is Boeing's flight reference.

January 16, 2013: Boeing 787, JA-804A

Location   Takamatsu, Japan
Flight   ANA 692
Operation   All Nippon Airlines
Operation   Scheduled Airline
Route   Yamaguchi, Japan to Tokyo, Japan
On-Board Injuries   Minor/None 137
Aircraft Damage   Minor

Description   Source: Serious Incident Report[58]

The pilot landed at the airport to refuel the airplane and pick up cargo. The pilot spoke with three employees of the fixed base operator who stated that he seemed alert and awake but wanted to make a "quick turn." After the airplane was fueled and the cargo was loaded, the pilot departed; the airplane crashed 1 minute later. Night visual meteorological conditions prevailed at the time. An aircraft performance GPS and simulation study indicated that the airplane entered a right bank almost immediately after takeoff and then made a 42 degree right turn and that it was accelerating throughout the flight, from about 75 knots groundspeed shortly after liftoff to about 145 knots groundspeed at impact. The airplane was climbing about 500 to 700 feet per minute to a peak altitude of about 260 feet above the ground before descending. The simulation showed a gas generator speed of about 93 percent throughout the flight. The study indicated that the load factor vectors, which were the forces felt by the pilot, could have produced a somatogravic illusion of a climb, even while the airplane was descending. The postaccident examination of the airframe and engine revealed no evidence of mechanical malfunctions or failures that would have precluded normal operation. Based on the findings from the aircraft performance GPS and simulation study, the degraded visual reference conditions present about the time of the accident, and the forces felt by the pilot, it is likely that he experienced spatial disorientation, which led to his inadvertent controlled descent into terrain.

The National Transportation Safety Board's Office of Research and Engineering conducted an Aircraft Performance GPS and Simulation Study. This study presents the results of using data from a portable GPS unit carried aboard the airplane, crash site information, and a simulator model of the Cessna 208B as the basis for a simulation that provides a physics-based estimate of the position and orientation of the airplane throughout the accident flight. The performance observations noted here are based on the results of this simulation.

The first GPS point showing the accident airplane clearly airborne was recorded at 19:57:19 as the airplane was climbing at 700 feet per minute (fpm) through about 730 feet MSL (14 feet AGL), on a track of about 223 degrees, and accelerating through 91 knots. The airplane continued accelerating while climbing at about 500 to 700 fpm to an altitude of about 960 feet MSL (240 feet AGL). The rate of climb then decayed, and after reaching a peak altitude of about 980 feet MSL (260 feet AGL) at 19:57:45, the airplane started to descend, and ultimately impacted terrain about 1 mile west-southwest of the departure end of the runway. The exact time of the impact is not known, but the simulation model flight time from the last recorded GPS position to the location and elevation of the impact site was estimated at 15 seconds, putting the time of impact at 19:58:13. The simulation rate of descent from 19:57:52 to the time of impact is about 650 to 680 fpm. The elapsed time from when the airplane became airborne at 19:57:19 to impact is 54 seconds.

The simulation indicated that the airplane was accelerating throughout the flight, from about 75 knots groundspeed shortly after liftoff to about 145 knots at impact. In addition, the airplane entered a right bank almost immediately after liftoff, and during the flight made a 42 degree right turn from the runway heading of 225 degrees to 267 degrees. The peak simulation bank angle during this turn was 12.3 degrees. At impact, the simulation indicated an airspeed of 156 knots, a pitch angle of negative 2 degrees, and a bank angle of 4.5 degrees.

Throughout the simulation, a constant power lever angle (PLA) setting of 72 percent was maintained. At the 72 percent PLA setting, the simulator reading results in a gas generator speed (Ng) of about 93 percent throughout the flight. This throttle setting resulted in the best match of the GPS and impact site data.

The load factors output by the simulation were used to compute "apparent" pitch and roll angles, defined as the angles that make the load factor vector in an unaccelerated reference system parallel (in airplane body axes) to the load factor vector in the actual accelerated reference system. These angles represent the attitude a pilot would "feel" the airplane to be in, based on a vestibular / kinesthetic perception of the components of the load factor vector in their own body coordinate system. Throughout the flight, the apparent roll angle was close to zero, and the apparent pitch angle was always greater than zero – even when the real pitch angle was less than zero.

Probable Cause   Source: Serious Incident Report[58]

The emergency evacuation was executed on Takamatsu Airport taxiway in the serious incident, which was a consequence of emergency landing deriving from the main battery thermal runaway during the airplane's takeoff climb.

Internal heat generation in cell 6 very likely developed into venting, making it the initiating cell, resulting in cell-to-cell propagation and subsequent failure of the main battery. It is very likely that cell 6 internal heat generation and increased internal pressure caused it to swell, melt the surrounding insulation material and contact the brace bar creating a grounding path that allowed high currents to flow through the battery box. The currents generated arcing internal to the battery that contributed to cell-to-cell propagation consequently destroying the battery. Cell 6 heat generation was probably caused by internal short circuit; however the conclusive mechanism thereof was not identified.

In the serious incident, the internal short circuit of a cell developed into cell heat generation, thermal propagation to other cells, and consequently damaged the whole battery. The possible contributing factors to the thermal propagation are that the test conducted during the developmental phase did not appropriately simulate the on-board configuration, and the effects of internal short circuit were underestimated,

Fire (After Landing)

January 7, 2013: Boeing 787, JA-829J

Location	Boston, MA
Flight	JAL 008 (After Landing)
Operator	Japan Airlines
Operation	Parked after Flight from Tokyo, Japan
On-Board Injuries	None
Aircraft Damage	Substantial

Description   Source: Aircraft Incident Report[59]

On January 7, 2013, about 1021 eastern standard time, smoke was discovered by cleaning personnel in the aft cabin of a Japan Airlines (JAL) Boeing 787-8, JA829J,* which was parked at a gate at General Edward Lawrence Logan International Airport (BOS), Boston, Massachusetts. About the same time, a maintenance manager in the cockpit observed that the auxiliary power unit (APU) had automatically shut down. Shortly afterward, a mechanic opened the aft electronic equipment bay and found heavy smoke coming from the lid of the APU battery case and a fire with two distinct flames at the electrical connector on the front of the case. None of the 183 passengers and 11 crewmembers were aboard the airplane at the time, and none of the maintenance or cleaning personnel aboard the airplane was injured. Aircraft rescue and firefighting personnel responded, and one firefighter received minor injuries. The airplane had arrived from Narita International Airport, Narita, Japan, as a regularly scheduled passenger flight operated as JAL flight 008 and conducted under the provisions of 14 Code of Federal Regulations (CFR) Part 129.

The APU battery model is the same model used for the 787 main battery. On January 16, 2013, an incident involving the main battery occurred aboard a 787 airplane operated by All Nippon Airways during a flight from Yamaguchi to Tokyo, Japan. The airplane made an emergency landing at Takamatsu Airport (TAK), Takamatsu, Japan, shortly after takeoff. The Japan Transport Safety Board investigated this incident with support from the National Transportation Safety Board (NTSB).

The National Transportation Safety Board determines that the probable cause of this incident was an internal short circuit within a cell of the auxiliary power unit (APU) lithium-ion battery, which led to thermal runaway that cascaded to adjacent cells, resulting in the release of smoke and fire. The incident resulted from Boeing's failure to incorporate design requirements to mitigate the most severe effects of an internal short circuit within an APU battery cell and the Federal Aviation Administration's failure to identify this design deficiency during the type design certification process.

Probable Cause   Source: Aircraft Incident Report[59]

The National Transportation Safety Board determines that the probable cause of this incident was an internal short circuit within a cell of the auxiliary power unit (APU) lithium-ion battery, which led to thermal runaway that cascaded to adjacent cells, resulting in the release of smoke and fire. The incident resulted from Boeing's failure to incorporate design requirements to mitigate the most severe effects of an internal short circuit within an APU battery cell and the Federal Aviation Administration's failure to identify this design deficiency during the type design certification process.

Safety Recommendations[59]

As a result of this investigation, the National Transportation Safety Board makes the following new safety recommendations:

To the Federal Aviation Administration:

Develop or revise processes to establish more effective oversight of production approval holders and their suppliers (including subtier suppliers) to ensure that they adhere to established manufacturing industry standards. (A-14-113)

Work with aviation industry experts to develop or modify design safety standards for large-format lithium-ion batteries to require that sources of excessive heating, including electrical contact resistance from components and connections, be identified, minimized, and documented as part of the design. The standards should include measures for identifying and minimizing potential sources of heating that consider the range of operating temperatures and the most extreme electrical currents that the battery could be expected to experience during repeated charge and discharge cycles. (A-14-114)

Work with aviation industry experts to develop or modify existing safety standards related to the design of permanently installed lithium-ion batteries to require monitoring of individual cell temperature and voltage and recording of exceedances to prevent internal cell damage during operations under the most extreme operating temperatures and currents. (A-14-115)

Once the guidance requested in Safety Recommendation A-14-115 has been issued, require type certification applicants to demonstrate that the battery monitoring system maintains each individual cell within safe temperature limits at the most extreme battery operating temperatures and the heaviest electrical current loads approved for operation. (A-14-116)

---

*   JJ is the IATA Code for Japan Airlines.

Work with lithium-ion industry experts to (1) conduct research into battery monitoring system technologies that could improve the recognition of conditions leading to thermal runaway, (2) develop active mitigation of such conditions to minimize damage, and (3) update design and safety standards accordingly. (A-14-117)

Work with industry experts to develop appropriate test methods for determining the initial point of self-heating in a lithium-ion cell to establish objective margins of thermal safety for future battery designs. (A-14-118)

Provide your certification engineers with written guidance and training to ensure that (1) assumptions, data sources, and analytical techniques are fully identified and justified in applicants' safety assessments for designs incorporating new technology and (2) an appropriate level of conservatism is included in the analysis or design, consistent with the intent of Advisory Circular 25.1309 (Arsenal draft). (A-14-119)

During annual recurrent training for engineering designees, discuss the need for applicants to identify, validate, and justify key assumptions and supporting engineering rationale used in safety assessments addressing new technology. (A-14-120)

Develop written guidance for your certification engineers and engineering designees about the use of traceability principles to verify that the methods of compliance proposed by type certification applicants for special conditions involving new technology are correct and complete. (A-14-121)

Once the guidance requested in Safety Recommendation A-14-121 has been issued, provide training to your certification engineers and engineering designees on the subjects discussed in the guidance. (A-14-122)

Require applicants to discuss key assumptions related to safety-significant failure conditions, their validation, and their traceability to requirements and proposed methods of compliance during certification planning meetings for type designs involving special conditions. (A-14-123)

Require Boeing 787 operators to incorporate guidance about the enhanced airborne flight recorder stale data issue in their maintenance manuals to prevent stale data from being used for maintenance activities or flight recorder maintenance. (A-14-124) Evaluate whether the recording of stale data by the Boeing 787 enhanced airborne flight recorder, including whether the data are specifically identified as stale, impacts the certification of the recording system regarding the ranges, accuracies, and sampling intervals specified in 14 Code of Federal Regulations Part 121 Appendix M, and take appropriate measures to correct any problems found. (A-14-125)

Require Boeing to improve the quality of (1) the enhanced airborne flight recorder radio/hot microphone channels by using the maximum available dynamic range of the individual channels and (2) the cockpit area microphone airborne recordings by increasing the crew conversation signals over the ambient background noise. (A-14-126)

Either remove the current exception to European Organization for Civil Aviation Equipment ED-112A, "Minimum Operational Performance Specification for Crash Protected Airborne Recording Systems," chapter I-6 in Technical Standard Order 123B, "Cockpit Voice Recorder Equipment," or provide installers and certifiers with specific guidance to determine whether a cockpit voice recorder installation would be acceptable. (A-14-127)

July 12, 2013: Boeing 787, ET-AOP

Location	London-Heathrow, England
Flight	ETH-700 (After Landing)
Operator	Ethiopian Airlines
Operation	Parked after Scheduled Airline Flight
On-Board Injuries	None
Aircraft Damage	Substantial

Description   Source: Aviation Safety Network[1]

A Boeing 787-8 Dreamliner sustained substantial damage in a ground fire while parked at Stand 592 at London-Heathrow Airport (LHR), U.K.

The airplane had arrived from Addis Ababa (ADD), Ethiopia as flight ET700* at 06:27 hours local time in the morning and was scheduled to depart as ET701 back to ADD at 21:10.

After passenger and crew disembarkation, the aircraft was towed to Stand 592, next to a fire station along taxiway E. Ground power was switched off.

Approximately at 16:34 an employee in the air traffic control tower noticed smoke emanating from the aircraft and activated the crash alarm.

The Rescue and Fire Fighting Service (RFFS) arrived on scene at 16:35 and discharged water and foam onto the outside of the aircraft. One fire fighter removed the power umbilical cables from the aircraft as a precaution. Fire fighters equipped with breathing apparatus entered the aircraft at 15:37 via the L2 door and encountered thick smoke. As they moved to the rear of the aircraft the smoke became denser so they opened further cabin doors to clear the smoke. At the rear of the passenger cabin they observed indications of fire in a gap between two overhead luggage bins. They were unable to use a hose-reel as the gap was too small and discharged a handheld 'Halon' extinguisher through the gap, about 20 minutes after entering the cabin. This was ineffective, so they removed some ceiling panels to expose the area and to get better access. At this point a small amount of flame was visible. This was extinguished with several pulses of water spray from their hose-reel, about 25 minutes after entering the cabin. A thermal-imaging camera was used to identify affected areas requiring further cooling.

---

*   ET is the IATA code for Ethiopian Airlines.

Investigation showed that the fire was initiated by the uncontrolled release of stored energy from the lithium-metal battery in the ELT. The ELT battery wires, crossed and trapped under the battery compartment cover-plate, probably created a potential short-circuit current path which could allow a rapid discharge of the battery.

Neither the cell-level nor battery-level safety features were able to prevent this single-cell failure, which then propagated to adjacent cells, resulting in a cascading thermal runaway, rupture of the cells and consequent release of smoke, fire and flammable electrolyte.

The trapped battery wires compromised the environmental seal between the battery cover-plate and the ELT, providing a path for flames and battery decomposition products to escape from the ELT. The flames directly impinged on the surrounding thermo-acoustic insulation blankets and on the composite aircraft structure in the immediate vicinity of the ELT. This elevated the temperature in the fuselage crown to the point where the resin in the composite material began to decompose, providing further fuel for the fire. As a result of this a slow-burning fire became established in the fuselage crown, which continued to propagate from the ELT location at a slow-rate, even after the energy from the battery thermal runaway was exhausted.

Probable Cause   Source: Incident Report[60]

The following causal factors were identified in the ground fire:

a) A thermal runaway failure of the lithium manganese dioxide battery in the ELT resulted in the uncontrolled release of stored energy within the battery cells.

b) The location and orientation of the ELT, and the compromised seal on the battery cover-plate, allowed the resulting hot gas, flames and battery decomposition products to impinge directly on the aircraft's composite fuselage structure, providing sufficient thermal energy to initiate a fire in the rear fuselage crown.

c) The resin in the composite material provided fuel for the fire, allowing a slow-burning fire to become established in the fuselage crown, which continued to propagate from the ELT location even after the energy from the battery thermal runaway was exhausted.

d) The Navigation Radio System safety assessment conducted in support of the ELT certification, did not identify any ELT battery failure modes which could represent a hazard to the aircraft, and therefore these failure modes were not mitigated in the ELT design or the B787 ELT installation.

The following factors most likely contributed to the thermal runaway of the ELT battery:

a) The trapped ELT battery wires created a short-circuit condition, providing a current path for an unplanned discharge of the ELT battery.

b) The ELT battery may have exhibited an unbalanced discharge response, resulting in the early depletion of a single cell which experienced a voltage reversal, leading to a thermal runaway failure.

c) The Positive Temperature Coefficient (PTC) protective device in the battery did not provide the level of external short-circuit protection intended in the design.

d) There was no evidence that the reset behaviour, and the implications of the variable switching point of the PTC, had been fully taken into account during the design of the ELT battery.

e) The absence of cell segregation features in the battery or ELT design meant the single-cell thermal runaway failure was able to propagate rapidly to the remaining cells.

Flight Control System (Control Hardover)

March 3, 1991: Boeing 737, N-999UA

Location	4 mi S Colorado Springs, CO
Flight	UAL 5858
Operator	United Airlines
Operation	Scheduled Airline
Route	Denver, CO to Colorado Springs, CO
On-Board Injuries	Fatal <u>25</u>
Aircraft Damage	Substantial

Description   Source: Aviation Safety Network[1]

United Airlines flight 585 left Peoria for Colorado Springs, with intermediate stops at Moline, IL and Denver, CO. The aircraft took off from Denver at 09:23 for the last segment of the flight, estimating Colorado Springs at 09:42. The aircraft was cleared for a visual approach to runway 35. The aircraft then suddenly rolled to the right and started to pitch nose down. The crew tried to initiate a go-around by selecting 15-deg. flaps and an increase in thrust. The altitude decreased rapidly, acceleration increased to over 4G until the aircraft struck the ground of Widefield Park almost vertically.

After a 21-month investigation, the NTSB issued a report on the crash in December 1992. In that report, the NTSB said it 'could not identify conclusive evidence to explain the loss of' the aircraft, but indicated that the two most likely explanations were a malfunction of the airplane's directional control system or an encounter with an unusually severe atmospheric disturbance.

Investigation into a September 1994 crash of a USAir Boeing 737-300 and an loss of control incident on June 9, 1996 (Eastwind Airlines Boeing 737-200), cited a malfunction in the plane's rudder system as the most likely cause of all three events.

Probable Cause   Source: Accident Report[61]

The National Transportation Safety Board determines that the probable cause of the United Airlines flight 585 accident was a loss of control of the airplane resulting from the movement of the rudder surface to its blowdown limit. The rudder surface most likely deflected in a direction opposite to that commanded by the pilots as a result of a jam of the main rudder power control unit servo valve secondary slide to the servo valve housing offset from its neutral position and overtravel of the primary slide.

In this amended report, new or revised text that was added as a result of information developed during the USAir flight 427 accident investigation is shown with a change bar along the outside of the page. Although some additional corrections and revisions have been made, the remainder of the report is substantially the same as the December 1992 report.

September 8, 1994, Boeing 737, N-513AU

Location	Aliquippa, PA
Flight	USA 427
Operator	USAir
Operation	Scheduled Airline
Route	Chicago, IL to Pittsburgh, PA
On-Board Injuries	Fatal 132
Aircraft Damage	Destroyed

Description   Source: Aviation Safety Network[1]

USAir Flight 427, a Boeing 737-300, crashed following a loss of control during the approach to Pittsburgh International Airport, Pennsylvania, USA. All 132 on board were killed.

The flight departed Chicago-O'Hare International Airport, Illinois at 18:10 hours on a flight to Pittsburgh. The first officer was Pilot Flying (PF) and the captain was the pilot-not-flying (PNF) on this leg. The flight until the descent part of the flight was uneventful.

About 19:02:22 the Pittsburgh TRACON controller issued instructions to turn to a heading of 100 degrees and told the crew there was other traffic in the area.

About 19:02:53, flight 427 was rolling out of the left bank as it approached the ATC-assigned heading of 100° and was maintaining the ATC-assigned airspeed (190 knots) and altitude (6,000 feet msl). About four seconds later the aircraft suddenly entered the wake vortex of Delta Airlines flight 1083, a Boeing 727, that preceded it by 69 seconds (4,2 miles).

Over the next 3 seconds the aircraft rolled left to approx. 18 deg of bank. The autopilot attempted to initiate a roll back to the right as the aircraft went in and out of a wake vortex core, resulting in two loud "thumps". The first officer then manually overrode the autopilot without disengaging it by putting in a large right-wheel command at a rate of 150deg/second. The airplane started rolling back to the right at an acceleration that peaked 36deg/sec, but the aircraft never reached a wings level attitude.

At 19:03:01 the aircraft's heading slewed suddenly and dramatically to the left (full left rudder deflection). Within a second of the yaw onset the roll attitude suddenly began to increase to the left, reaching 30deg. The aircraft pitched down, continuing to roll through 55 degree left bank. At 19:03:07 the pitch attitude approached -20 degrees, the left bank increased to 70 degrees and the descent rate reached 3600 ft/min. At this point, the aircraft stalled. Left roll and yaw continued, and the aircraft rolled through inverted flight as the nose reached 90 degree down, approx. 3600 feet above the ground. The 737 continued to roll, but the nose began to rise. At 2000 feet above the ground the aircraft's attitude passed 40 degrees nose low and 15 degrees left bank. The left roll hesitated briefly, but continued and the nose again dropped.

The plane descended fast and impacted the ground nose first at 261 knots in an 80 degree nose down, 60 degree left bank attitude and with significant sideslip.

Probable Cause   Serious Incident Report[62]

The National Transportation Safety Board determines that the probable cause of the USAir flight 427 accident was a loss of control of the airplane resulting from the movement of the rudder surface to its blow down limit. The rudder surface most likely deflected in a direction opposite to that commanded by the pilots as a result of a jam of the main rudder power control unit servo valve secondary slide to the servo valve housing offset from its neutral position and over travel of the primary slide.

June 9, 1996: Boeing 737, N-221US

Location	Near Richmond, VA
Flight	SGR 517
Operator	Eastwind Airlines
Operation	Scheduled Airline
Route	Trenton, NJ to Richmond, VA
On-Board Injuries	Minor/None 53
Aircraft Damage	None

Description   Source: Aviation Safety Network[1]

Eastwind Airlines flight 517 experienced a reported loss of rudder control while on approach to Richmond, VA. There were no injuries or damage to the airplane as a result of the incident. At the time of the event the airplane's airspeed was about 250 knots and at 4,000 feet MSL. The captain reported that he was hand flying the airplane and he felt a slight rudder "bump" to the right. He asked the first officer if he had felt the bump, then the airplane suddenly rolled to the right. He reported that he applied opposite rudder but that the rudder felt stiff. He stated the he applied opposite aileron and used asymmetric power to keep the airplane upright. He stated that after he declared an emergency to the approach controller, he and the first officer performed the emergency checklist. The captain reported that as part of the checklist they turned off the yaw damper. He reported that the airplane became controllable, but was not certain if the problem when away at the same time that the yaw damper was turned off.

It is reported that the airplane has previously had problems with uncommanded rudder deflections. Previous reports have been of "rudder bumps" during departure and that the airplane would not trim properly.

Probable Cause   Source: Accident File[63]

The National Transportation Safety Board determines the probable cause(s) of this incident to be: The yaw/roll upset of the airplane resulting from the movement of the rudder surface to its blowdown limit. The rudder surface most likely deflected in a direction opposite to that commanded by the pilots as a result of a jam of the main rudder power control unit servo valve secondary slide to the servo valve housing offset from its neutral position and overtravel of the primary slide.

Comments

It worth noting that the crew reported using asymmetric thrust to maintain control.

Flight Control System (Jammed or Disconnection)

July 18, 1998: Cessna 152, C-GZLZ

Location	Lac Saint-François, QC, Canada
Identification	C-GZLZ*
Operator	Laurentide Aviation
Operation	Flight Instruction
Route	Local Flight
On-Board Injuries	Fatal 1 Serious 1
Aircraft Damage	Destroyed

Description   Source: Aviation Safety Network[1]

At about 0850 eastern daylight time, a flight instructor and a student took off on a local training flight from runway 25 at Montréal/Les Cèdres Aerodrome, Quebec. The student pilot was practicing spins and recoveries. The student initiated a spin to the left, his sixth of the day, at an altitude of 3 600 feet above sea level. The first five spins were to the right. The aircraft entered the spin normally. After one and a half turns, the flight instructor asked the student to recover. The student applied pressure on the right rudder pedal, as taught by the flight instructor, and the rotation did not stop. The flight instructor took over the controls and applied pressure on the right rudder pedal to stop the rotation, but the rotation did not stop. The aircraft, by then, was established in a stabilized spin, rotating to the left, and continuing its descent. The flight instructor applied full power for a moment, then full flaps, to no avail. Throughout the recovery attempt, the flight instructor continued in his efforts to avoid the crash. The aircraft struck the surface of Lac Saint-François. The student pilot sustained serious injuries but managed to evacuate the sinking aircraft through the right, rear window. He then tried to pull out the unconscious flight instructor, but without success. A fisherman close to the scene rescued the student and transported him ashore where emergency vehicles were standing by. The flight instructor did not exit the aircraft and died in the accident.

Probable Cause   Source: Accident Report[64]

The rudder was found jammed at 34° left, beyond the control stop setting of 23°. The TSB laboratory found that force required to break the rudder from the jam was 36 lb applied to the trailing edge of the rudder.[65] This equates to 180 lb if the force was applied to the rudder pedal. "However, given that the direction of cable pull tended to increase the jamming by closing the horn, it would not have been possible to break the rudder jam with the application of right rudder.

   To better understand whether and how the rudder could have over-traveled and jammed on the accident aircraft, additional tests were conducted on a similar aircraft. The test conditions included removing the right rudder bar return spring and disconnecting the right rudder cable. As with a loose rudder cable, this facilitated over-deflection of the rudder to the left. It was also determined that moving the elevator to a position more than two-thirds up--increasing the clearance between the rudder and the elevator--permitted further travel of the rudder. In that condition, a very hard left rudder pedal input permitted the rudder to over-travel and the stop plate locked below and behind the stop bolt, exactly as had been found in the accident aircraft. The rudder was locked irreversibly and had to be released by levering the rudder horn with a screwdriver. A second test, with the rudder cable reconnected, but slightly loose, and with other conditions the same, again led to a locked rudder. These tests showed that the design and condition of the stop bolt and rudder horn stop plate allowed the stop plate to over-travel the stop bolt and jam.†

Recommendations

The TSB made the following recommendations:

   The [Canadian] Department of Transport issue an Airworthiness Directive for Cessna 150 and 152 aircraft addressing a mandatory retrofit design change of the rudder horn stop bolt system to preclude over-travel and jamming of the rudder following a full rudder input.   (A00-09)

   The [Canadian] Department of Transport, in conjunction with the Federal Aviation Administration, take steps to have all operators of Cessna 150 and 152 aircraft notified about the circumstances and findings of this accident investigation and the need to restrict spin operations until airworthiness action is taken to prevent rudder jamming.   (A-00-10)

   The National Transportation Safety Board review the circumstances and findings of this investigation and evaluate the need for mandatory airworthiness action by the Federal Aviation Administration.   (A00-12)

Regulatory Outcome

The following actions were taken in response to these recommendations;

---

*   Aircraft registration used *vice* flight number.

†   Direct quote from the accident report.

- The manufacturer, Cessna issued a Service Bulletin, SEB01-1,[66] to provide an enhanced rudder stop and other hardware to prevent the possibility of the rudder overriding the stop bold during full rudder deflection;* and
- The Canadian Department of Transport issued an AD[67] applying to all Canadian registered Cessna 150 and 152 aircraft.

These two actions addressed Recommendation A00-09. There was no apparent action on responses to Safety Recommendations A00-10 or A00-12 by US authorities.

April 11, 2005: Cessna 152, N-24779

Location	Williamsburg, OH
Flight	N-24779†
Operator	Sporty's Academy
Operation	Flight Instruction
Route	Local Flight
On-Board Injuries	Fatal 2
Aircraft Damage	Destroyed‡

Description   Source: Aviation Safety Network[1]
The certificated flight instructor and student pilot were practicing spins about 3,000 feet gal, and did not recover. The airplane subsequently descended in a nose down spiral, and impacted a field. Examination of the wreckage revealed that the rudder was jammed beyond its left travel limit. Further examination revealed that the two rudder bumpers had been installed inverted, and the right side rudder bumper had traveled over and beyond the rudder stop, and locked behind it. Review of the maintenance records revealed no specific mention of the rudder bumpers during the 28-year history of the airplane. However, work had been performed near the rudder bumpers on several occasions. Additionally, the paint on the inverted rudder bumpers was consistent with a paint job completed about 8 years prior to the accident. The investigation could not determine if the rudder bumpers were inverted at the time of production, or if they had been inverted during the maintenance history of the airplane. The airplane manufacturer issued a service bulletin about 3 1/2 years prior to the accident. The purpose of the service bulletin was to provide an enhanced rudder stop, bumper, doublers and attachment hardware designed to assist in preventing the possibility of the rudder overriding the stop bolt during full left or right operation of the rudder. Specifically, the new rudder stop was much larger than the original rudder stop. The service bulletin had not been complied with on the accident airplane, and under 14 CFR Part 91, was not required.

Probable Cause   Source: NTSB Accident File[68]
An improperly installed rudder bumper, which resulted in a rudder jam during spin training and subsequent uncontrolled descent into terrain.

Regulatory Outcome
On March 21, 2007, the NTSB issued Safety Recommendation A-07-33 recommending that the FAA issue an airworthiness directive requiring that Cessna 150 and 152 models comply with Cessna's Service Bulletin No. SEB10-1 and undergo a one-time inspection at the next 100-hour or annual inspection to verify that the rudder bumpers are correctly installed on the rudder horn assembly.[69]

On January 22, 2001, Cessna Aircraft Company issued Service Bulletin SEB01-1, and designated it mandatory.§ The purpose of the service bulletin was to provide an enhanced rudder stop, bumper, doubler and attachment hardware designed to assist in preventing the possibility of the rudder overriding the stop bolt during full left or right operation of the rudder. Specifically, the new rudder stop was much larger than the original rudder stop. The service bulletin had not been complied with on the accident airplane, and under 14 CFR Part 91, compliance was not mandatory.

On October 10, 2003, Transport Canada issued Airworthiness Directive (AD) CF-2000-20R2, which made the service bulletin mandatory for all applicable Canadian registered airplanes. It should be observed that issuing an airworthiness directive unilaterally to an aircraft model originally certified by another country is extremely rare. It is customary to "work with the home country".

---

* Cessna designated the service bulletin as "mandatory". Current US regulations do not recognize such designations as requiring compliance.

† Aircraft registration used *vice* flight number.

‡ The NTSB accident description describes the damage as "substantial." However the damage description describes an airplane damaged beyond economic repair. A review of the FAA records show that the airplane's registration is now longer in force and there is no record of the serial number being assigned a new registration.

§ "Mandatory" designation by the manufacturer have no legal effect.

At the time of the accident, the FAA had no similar AD. During the course of the accident investigation, the FAA began procedures to issue a similar AD. The AD would make the service bulletin mandatory and require a check of the correct orientation of the rudder bumpers.* The final AD provided the alternative of adding a placard prohibiting intentional spins without installing the service bulletin.

Comment

We became aware of this issue shortly after Recommendation A-07-33 was issued. We routinely review NSTB and other investigative agencies' recommendations. As owners of an airplane covered by SEB01-1,† this immediately aroused interest. We contacted the responsible Aircraft Certification Office and obtained a copy of SEB01-1. We had the service bulletin performed within a month at a cost of about $250.00. (Parenthetically, the final rule of the AD estimated the cost as $430.00.)

January 31, 2000: McDonnell-Douglas MD-83, N-963AS

Location	Off Point Mugu, CA
Flight	ASA 261
Operator	Alaska Airlines
Operation	Scheduled Airline
Route	Puerto Vallarta, Mexico to San Francisco, CA, USA
On-Board Injuries	Fatal <u>88</u>
Aircraft Damage	Destroyed

Description   Source: Aviation Safety Network[1]

Alaska Airlines Flight 261 departed Puerto Vallarta at about 14:30 PST for a flight to San Francisco and Seattle. En route to San Francisco a FL310 a problem arose with the stabilizer trim. At 16:10 the crew radioed Los Angeles ARTCC that they were having control problems and that they were descending through FL260. At 16:11 Los Angeles ARTCC asked the condition of the flight and were told that they were troubleshooting a jammed stabilizer. The crew requested, and were granted, a FL200-FL250 block altitude clearance. At 16:15 the crew were handed off to Los Angeles sector control. The Alaska Airlines crew reported problems maintaining their altitude and told their intentions to divert to Los Angeles International Airport. They were cleared to do so at 16:16. The crew then requested permission to descend to FL100 over water to change their aircraft configuration. Los Angeles cleared them to FL170. Last message from Flight 261 was when they requested another block altitude. The request was granted at 16:17, without a readback from the crew. During the descent the crew was also talking to Alaska Airlines maintenance personnel in Seattle and Los Angeles to troubleshoot their stabilizer trim problems. As the crew attempted to diagnose or correct the problem the out-of-trim condition became worse, causing a tendency for the plane to pitch nose-down. When preparing the plane for landing control was lost and the MD-83 was seen 'tumbling, spinning, nose down, continuous roll, corkscrewing and inverted'. The aircraft crashed off Point Mugu in 650 feet deep water.

Probable Cause   Source: Accident Report[70]

A loss of airplane pitch control resulting from the in-flight failure of the horizontal stabilizer trim system jackscrew assembly's acme nut threads. The thread failure was caused by excessive wear resulting from Alaska Airlines' insufficient lubrication of the jackscrew assembly.

Contributing to the accident were Alaska Airlines' extended lubrication interval and the Federal Aviation Administration's (FAA) approval of that extension, which increased the likelihood that a missed or inadequate lubrication would result in excessive wear of the acme nut threads, and Alaska Airlines' extended end play check interval and the FAA's approval of that extension, which allowed the excessive wear of the acme nut threads to progress to failure without the opportunity for detection. Also contributing to the accident was the absence on the McDonnell Douglas MD-80 of a fail-safe mechanism to prevent the catastrophic effects of total acme nut thread loss.

---

* When the AD was first published, the pilots' association issued strongly worded statements against this AD. As a member, we complained about the tone of their statements and expressed support for the AD. They told us there were only two fatal accidents. They were silent when asked how many fatal accidents were needed.

† A Cessna 150F, the first swept-tail model. The problem is common to all swept-tail Cessna 150's or 152's.

Regulatory Outcome

The FAA Commercial Airplane Certification Process Study (CPS), issued in March, 2002,[71] was a direct result of this accident and the TWA B-747 accident. The report focused on numerous process categories, including human factors in airplane design, operation and maintenance, flight critical systems and structures, safety data management, maintenance/operations coordination, major repair and modification, and safety oversight. The findings and observations directly relate to all lessons learned from this accident. The CPS findings and observations directly relate to all lessons learned from this accident. The findings are contained in Appendix B, beginning on page 102 of the report.

February 16, 2000: Douglas DC-8, N-8079U

Location	Rancho Cordova, CA
Flight	EWW 17
Operator	Emery Worldwide
Operation	Cargo
Flight Route	Sacramento, CA to Dayton, OH
On-Board Injuries	Fatal _3_ Serious _1_
Aircraft Damage	Destroyed

Description   Source: Aviation Safety Network[1]

Emery Flight 17 had just departed Sacramento-Mather Airport runway 22L when the crew reported balance problems. A little later the aircraft was seen to crash in a left wing low, nose low attitude, into the Insurance Auto Auctions salvage yard, setting fire to 100-200 cars. Debris cut a swath about 250 yards wide and a 600yds long. The plane's cargo included clothing, transmission fluid and a small amount of 9 grams fuses used to activate automobile air bags. The NTSB is focusing on the possibility that a push rod in an elevator control tab may have been separated from the control system of the plane before impact.

Probable Cause   Source: Accident Report[72]

A loss of pitch control resulting from the disconnection of the right elevator control tab. The disconnection was caused by the failure to properly secure and inspect the attachment bolt.

Certification Issues

CAR 4b standards allowed single disconnection to have catastrophic results.

Flight Control System (Misrigging)

September 17, 1961: Lockheed L-188, N-137US

Location	O'Hare Airport, Chicago, IL
Flight	NWA 706
Operator	Northwest Orient Airlines
Operation	Scheduled Airline
Route	Chicago, IL to Tampa, FL
On-Board Injuries	Fatal 37
Aircraft Damage	Destroyed

Description   Source: Aviation Safety Network[1]

Lockheed Electra N137US operated on Northwest flight 706 from Milwaukee to Miami (MIA) with intermediate stops at Chicago, Tampa and Fort Lauderdale. After a crew change at Chicago the plane taxied to runway 14R at 08:55 and was cleared for takeoff. Between the 8,000 and the 9,000-foot runway marker the aircraft was observed to commence an apparently coordinated right turn with a slowly increasing rate of bank. When the bank angle was 30 to 45 degrees, the crew made a short, garbled transmission. Immediately thereafter, at a bank angle of 50 to 60 degrees, the aircraft began to lose altitude. The maximum altitude attained in the entire turn was 200 to 300 feet. The right wing struck powerlines adjacent to the Chicago Northwestern Railroad tracks, severing the lines at an angle of about 70 degrees from the horizontal. It then continued in a direction of about 271 degrees magnetic and, when in a bank of about 85 degrees and a nose-down attitude of about 10 degrees, the right wing of the aircraft struck the railroad embankment.

Continuing to roll about its longitudinal axis, the aircraft cartwheeled, the nose crashing into the ground 380 feet beyond the point of first impact, and landed right side up. It then slid tail first another 820 feet. The aircraft disintegrated throughout its path.

Probable Cause   Accident Report[73]

The Board determines that the probable cause of this accident was a mechanical failure in the aileron primary control system due to an improper replacement of the aileron boost assembly, resulting in a loss of lateral control of the aircraft at an altitude too low to effect recovery.

March 20, 2001: Airbus A320, D-AIPW

Location	Frankfurt/Main, Germany
Flight	DLH D-AIPW*
Operator	Lufthansa
Operation	Scheduled Airline
Route	Frankfurt, Germany to Paris, France
On-Board Injuries	Minor/None 121
Aircraft Damage	None

Description   Source: Aviation Safety Network[1]

The Airbus 320 hit turbulence just after rotation from runway 18 and the left wing dipped. The captain responded with a slight sidestick input to the right but the aircraft banked further left. Another attempt to correct the attitude of the plane resulted in a left bank reaching ca 22deg. The first officer then said "I have control", and switched his sidestick to priority and recovered the aircraft. The left wingtip was reportedly just 0.5m off the ground. The aircraft climbed to FL120 where the crew tried to troubleshoot the problem. When they found out that the captain's sidestick was reversed in roll, they returned to Frankfurt. Investigation revealed that maintenance had been performed on the Elevator Aileron Computer no. 1 (ELAC). Two pairs of pins inside the connector had accidentally been crossed during the repair.

Probable Cause†   Serious Incident Report[74]

The BFU comes to the conclusion that the serious incident occurred because:
- during maintenance of the ELAC Nr. 1 connector, two cable pairs were crossed
- the error remained undetected
- the error with the "FLIGHT CONTROL CHECK" was not recognized by the crew

Contributing to the causes were:

---

*   Flight number not available. Identification uses aircraft registration *vice* flight number.

†   Unofficial translation

- an unclear, with difficulty manageable documentation because of which a wrong "Wiring Diagram" was used - a deviation from the manufacturers regulations by "maintenance support"
- unclear formulated manufacturers manual
- an incorrect execution of the functioning check
- an insufficiently functioning quality assurance
- a failing monitoring of the maintenance organization by the airline.
- a quantitatively and consequently qualitatively insufficient monitoring of the maintenance organization and the airline by the authorities
- a shortcoming in the "AFTER START CHECK LIST" for the execution of the "FLIGHT CONTROL CHECK"

Comment

This incident convinces us that non-backdriven controls are not necessarily bad. In this case, they probably prevented a catastrophic crash. Compare with the similar accident to the Northwest *Electra* on the previous page.

Flight Control System (Pilot Induced Oscillation)

March 17, 2001: Airbus A320, N-357NW

Location	Detroit, MI
Flight	NWA 985
Operator	Northwest Airlines
Operation	Scheduled Airline
Route	Detroit, MI to Miami, FL
On-Board Injuries	Minor/None 153
Aircraft Damage	Substantial

Description   Source: Aviation Safety Network[1]

Takeoff was initiated on runway 3C (8,500 feet by 200 feet, wet). The flight crew reported that at an airspeed of approximately 110 knots, the nose of the airplane began to lift off. The captain attempted to lower the nose, however, the airplane was unresponsive. The airplane became airborne and climbed to an altitude of 20 to 30 feet above the runway. The captain reduced the power on both engines and the airplane settled to the runway, striking the tail. The airplane traveled approximately 700 feet off the end of the runway where it came to rest in the muddy terrain. An emergency evacuation was performed during which the L2 evacuation slide did not deploy. During the investigation, NTSB determined that the airplane was loaded so that its center of gravity (CG), although within limits, was in the aft region of the permissible range. Further, the flight crew had incorrectly set the trim for the trimmable horizontal stabilizer (THS) at -1.7°UP (airplane nose up). This setting resulted in a pitch-up trim condition. The proper trim setting, 1.7°DN (airplane nose down), would have resulted in a correct trim condition for the way the airplane was loaded. The improperly set trim caused the nose of the airplane to lift off the runway prematurely.

Probable Cause   Source: Accident Report[75]

The pilot induced oscillations and the delay in aborting the takeoff. Factors associated with the accident were the first officer used an improper trim setting and the captain did not identify and correct the setting during the taxi checklist, and the wet runway conditions.

December 7, 2002: Airbus A321, C-GJVX and C-GIUF

Location	Toronto Airport, Canada	
Flight	ACA 457	ACA 1130
Operator	Air Canada	Air Canada
Operation	Scheduled Airline	Scheduled Airline
Route	Not Reported	Not Reported
On-Board Injuries	No Injuries 129	No Injuries 172
Aircraft Damage	None	None

Description   Source: Aviation Investigation Report[76]

At approximately 1607 Eastern standard time (EST), Air Canada Flight 457 (ACA457), an Airbus A321-211 aircraft (registration C-GJVX, serial number 1726) was on approach to Toronto/Lester B. Pearson International Airport (LBPIA), Ontario, with 123 passengers and 6 crew on board. At approximately 140 feet above ground level (AGL), on final approach to Runway 24R with full flaps selected, the aircraft experienced roll oscillations. The flight crew leveled the wings and the aircraft touched down firmly. During the approach, the aircraft had accumulated mixed ice on areas of the wing and the leading edge of the horizontal stabilizer that are not protected by anti-ice systems.

Approximately three hours later on the same day, Air Canada Flight 1130 (ACA1130), an Airbus A321-211 aircraft (registration C-GIUF, serial number 1638), with 165 passengers and 7 crew onboard was on approach to Runway 24R at LBPIA. At 1859 EST and approximately 50 feet AGL, the aircraft experienced roll oscillations. The flight crew conducted a go-around, changed flap settings, and returned for an uneventful approach and landing. At the gate, it was noted that the aircraft had accumulated ice on areas of the wing and the leading edge of the horizontal stabilizer that are not protected by anti-ice systems. There was no damage to either aircraft nor injury to the crew or passengers.

The pilots of both aircraft were aware that moderate icing conditions were forecast for the region during the period the aircraft were scheduled to arrive at Toronto/Lester B. Pearson International Airport (LBPIA), Ontario.

As the aircraft were being vectored for landing, ° to Ω inch of ice accumulated on the visual ice indicators. Both aircraft used engine and wing anti-ice while flying in the icing conditions. In addition, as both aircraft were landing with full flaps (CONFIG FULL), their approach speeds(V_{APP}) were increased, in accordance with Air Canada's A321 Aircraft Operating Manual, to VLS(lowest selectable airspeed) plus five knots to accommodate an approach conducted with CONFIG FULL in icing conditions. The autopilot systems were selected OFF between 500 and1000 feet above ground level (AGL) in preparation for landing.

The two aircraft performed as expected until the roll oscillations began at approximately 140 feet AGL during the approach of ACA457 and at approximately 50 feet AGL for ACA1130. At that point the pilot at the control of each aircraft attempted to bring the wings level by initiating side-to-side stick movement up to the stops at a frequency of approximately 0.5 hertz. Although the pilots attempted to dampen these oscillations by applying right and left stick inputs up to the stops, the magnitude of the oscillations actually increased. At this point, the pilot flying the first aircraft (ACA457) decreased power, and the aircraft touched down. In the second aircraft(ACA1130), the pilot flying decided to advance the throttles and conduct a go-around. Both decisions resulted in successful landings. In preparation for their second landing attempt, the pilots of ACA1130 decided to use CONFIG 3 and adjusted the approach speed accordingly. The second approach and landing were uneventful. No other landing anomalies were reported at LBPIA during this time frame.

Prior to the occurrences of 07 December 2002, data regarding the A321 aircraft performance in icing were based on its certification and in-service history. As a derivative of the A320, the extent of the A321 icing trials was based in part on the A320 certification results. Specifically, although the A321 has a double-slotted flap system rather than the single-slotted flap arrangement in theA320, both aircraft were considered similar, and the effect of ice accretion on the leading edge of the flaps was not examined during A321 certification. Similarly, any A321 in-service icing anomalies were viewed in the context of the in-service performance of the entire A320 fleet, which revealed no systemic problem. Indeed, for all its certification programmes Airbus considers that ice is accreted in clean configuration.

Analysis of the post-occurrence icing trials concluded that ice accretion on the leading edge of the flaps of both the A320 and A321, and on the flap tabs of the A321, contributed to unusual roll behaviour by modifying the aerodynamic characteristics. Flights performed in roll direct laws on both aircraft showed that both A320 and A321 ice accretion with flaps extended induces an increase in roll spoiler efficiency in both CONFIG 3 and CONFIG FULL. This increase was higher in CONFIG FULL than in CONFIG 3

Findings   Aviation Investigation Report[76]

Findings as to Causes and Contributing Factors
1. The A321 normal, lateral flight control laws programmed into the elevator aileron computer provided higher roll efficiency in CONFIG FULL than in CONFIG 3, which resulted in a reduced stability margin in icing conditions.
2. The external influence of the ice on the leading edge of the flaps changed the aircraft aerodynamics, which, when combined with pilot input and the normal, lateral flight control law, resulted in airplane-pilot coupling, which produced an unstable aircraft.

   Findings as to Risk
1. Flight tests in natural icing conditions were not accomplished in any configuration in the A321 to determine if an acceptable level of safety existed in the handling characteristics.
2. It is likely that the icing conditions encountered by both aircraft were outside the Federal Aviation Regulation 25, Appendix C envelopes used for certification of theA321.
3. After these occurrences, flight tests in natural icing confirmed that with flaps extended, ice accretion on the flap leading edges increased roll sensitivity in the normal, lateral flight control law on the A321.

Flight Control System (Sudden Gain-Change)

June 21, 1996: Airbus A340, D-AIBE

Location	East of Dallas, TX
Flight	DLH D-AIBE*
Operator	Lufthansa
Operation	Scheduled Airline
Route	Frankfurt, Germany to Dallas, TX, USA to Houston, TX, USA
On-Board Injuries	None reported
Aircraft Damage	None

Description   Source: Aviation Safety Network[1]

During climb, the TCAS accrued a TA then a RA down advisory 'don't climb greater than 500 fpm.' The crew simultaneously operated their respective side stick controls when the captain responded with an abrupt maneuver without making a verbal announcement that he was taking command. Input from both side sticks changed the aircraft nose-up pitch from 0.7 degrees to 4.22 degrees in 1.1 second and an aft galley g load from -0.36 g's to 2.27 g's. Subsequently, the Captain Side Stick Position gave an aircraft pitch change to 2.11 degrees nose-down and an aft galley CG of -0.76 g's in a span of one second. Both Side Stick positions cumulatively gave 1.8 degrees nose-up and 2.09 g's at the aft galley in 1.2 seconds. The VFR traffic had visual contact on the Airbus and both aircraft were advised by ATC that the traffic was not a factor. The seat belt sign had been turned 'OFF' and the FA's were preparing for passenger service. FA's reported strong consecutive jolts, without warning, that threw them and items against the ceiling. The captain had the TCAS CBT training. He had not received the simulator TCAS training. The Captain had not participated in CRM.

Probable Cause   Source: Aviation Safety Network[1]

The National Transportation Safety Board determines the probable cause(s) of this accident to be: The flightcrew's abrupt maneuver/descent in response to a TCAS RA. Factors were the company assignment of the captain to the crew position with a lack of captain training.

Flight Control System: Uncommanded Control Inputs

July 6, 1969: Beechcraft 99, N-844NS

Location	Near Monroe, GA
Flight	KKB 168
Operator	Air South
Operation	Scheduled Airline
Route	Atlanta, GA to Greenville, SC
On-Board Injuries	Fatal 14
Aircraft Damage	Destroyed

Description   Source: Aviation Safety Network[1]

Air South Flight 168 departed Atlanta at 21:07. At 21:13 the flight reported level at its assigned cruising altitude of 7,000 feet. The Beech had been cruising for eleven minutes when it attained a gradual nose down attitude due to a change in the longitudinal trim. The pilots noticed the change after about six seconds and initiated a recovery action. The horizontal stabilizer continued to move to a full nose down position. Excessive pulling force on the control column was necessary to recover from the high speed dive. The necessary stick forces for such an out-of-trim condition can exceed the capability of one pilot, and in some cases two pilots, to control. The Beech continued to descend until both wings failed at high speed, just before the airplane crashed into the ground in a near vertical attitude.

---

* Identification uses the aircraft registration *vice* a flight number.

Probable Cause   Source: Accident Report[77]

An unwanted change in longitudinal trim which resulted in a nose down high-speed flight condition that was beyond the physical capability of the pilots to overcome. The initiating element in the accident sequence could not be specifically determined. However, the design of the aircraft flight control system was conducive to malfunctions which, if undetected by the crew, could lead to a loss of control.

Regulatory Outcome

The testimony at the [NTSB] hearing indicated that the FAA policy regarding the Delegated Option certification procedure was to accept certification data from the manufacturer and to review the data in the areas the FAA felt were necessary. The FAA also indicated that they participated in flight tests only when a new regulation was being applied to an aircraft, or when the manufacturer produced a new design feature that had not previously been certificated by them. The trimmable stabilizer in the E99 was such a new design feature, but the FAA did not participate in the flight testing of this item

This type of stabilizer has been in use for a long period of time on various commercial and military aircraft, and the problems that were associated with it should have been well known throughout the industry. These problems have included excess stabilizer-up angle, might not be capable of overcoming the stabilizer power. Since this runaway trim potential, and flight conditions where the elevator power type of stabilizer has been in use, various devices have been incorporated in the systems to provide more information to the crew and to eliminate some of the known hazards that could evolve from its use. These devices have included audible warning of trim motion, stabilizer position indicators, restrictions to stabilizer-up angles, and published emergency procedures developed to deal with the results of various malfunctions in the system.

The Board notes that the modifications applied to the trim system of the B99, since the accidents, are similar to those which have been previously applied to large aircraft.

The fault analysis used by the manufacturer and the FAA to certificate the longitudinal trim system of the E99 and the Board concludes it was inadequate. As stated in this report, a fault analysis that did not consider the total operating environment was not complete.

The Board recommends that the FAA take action to

(1)   require direct participation of FAA personnel in the certification of all newly designed aircraft components;

(2)   review its aircraft certification system for possible procedural changes which would ensure that lessons learned in investigation of large aircraft accidents and incidents would be applied, when appropriate, to certification of small aircraft.

(3)   bring recommendation (2) above, to the attention of those units within the FAA that are charged with the certification of small aircraft.

October 2, 2000: Airbus A340, TC-JDN and Airbus A330, C-GGWD

Location	Atlantic Ocean (58°N28'/16°W46')	
Aircraft	A340	A330
Flight	TC-JDN*	C-GGWD†
Operator	Turkish Airlines	Air Canada
Operation	Scheduled Airline	Scheduled Airline
Route	Istanbul, Turkey to New York, USA	London, UK to Ottawa, Canada
On-Board Injuries	None Reported	None Reported
Aircraft Damage	None Reported	None Reported

Description   Source: Incident Report[78]

Both aircraft were westbound on North Atlantic Track E with an entry point of 58°/10°W. Both aircraft were cleared by air traffic control (ATC) to cruise at 0.82 M. The A-340 at FL360 and the A-330 at FL370. The vertical separation of 1000 ft was in accordance with Reduced Vertical Separation Minima (RVSM).

The post-incident meteorological analysis described a low pressure area centered to the west and southwest of Iceland with an occluded from lying through 60°N/16°W. A change in the height of the tropopause was very likely at or near the incident the incident position and an associated change in air temperature at the frontal boundary. Track E lay some 150 NM to the north of the core of a westerly jetstream with a of 170 kts at FL310. The 1200 hrs significant weather chart showed the position of the jetstream and the occluded front. The incident location was within a area marked as likely to contain clear air turbulence between FL200 and FL400.

Both aircraft were in clear air as the A-330 was slowly overtaking the A-340 below it. The A-330 pilot stated his aircraft was slightly to the right of the A-340 and also abeam it when he saw the A-340's wings flex. At the same time, he felt a bump similar to entering a mountain wave. He (the A-330 pilot) received a TCAS resolution advisory to climb. The A-340 zoom climbed past the A-330's altitude before he (the A-330 pilot) could react.

---

*   Flight numbers were not reported.

†   Flight numbers were not reported.

At 14:20:40 hrs, the A-340 entered a region of successive and increasing variations in wind and air temperature. This caused fluctuations in pitch attitude, normal load factor, altitude, airspeed, engine rpm, and Mach. At 14:21:40 hrs, the Mach briefly increased to 0.87 which triggered a Master Warning and disconnected (probably by the handling pilot). At 14:21:50 hrs, the flight controls switched to Angle-of-Attack (AOA) Protection law.

When AOA Protection was invoked, the flight controls immediately commanded *Alpha-prot*. As this was an increase in AOA a zoom climb resulted.

Probable Cause   Source: Incident Report[78]

Ten seconds after the autopilot disengaged, the corrected or phase-advanced angle-of-attack (a computed parameter which is not recorded but can be corrected by Airbus Industrie from the DFDR data) reached the *alpha-prot* value. This ... caused a change in the pitch flight control law from normal law (N_z law) to angle-of-attack law (AOA law). If both sidesticks are at neutral the AOA protection law seeks to hold the angle at *alpha-prot* until a sidestick pitch command is made. If the stick is pulled full aft then the angle-of-attack increases to *alpha-max*. AOA protection remains active until a nose-down command greater than half forward travel is made or until a nose-down sidestick input has been applied for more than one second. The first recorded sidestick input was made at 14:22:08.

Regulatory Outcome

The only regulatory topic raised was the lateral deviation of the overtaking A-330 to ensure safe distance from the "zooming" A-340. There was no interest in dealing with the design of the AOA law software

Comments

At the time we became aware of this incident, we felt this was a deficiency in the underlying assumptions of the AOA law. Software is certified (in the civil community) by reference to DO-178.[79][80] DO-178 involves a tedious verification of the behavior of the software. It has been said that most of the software problems in aviation involve the basic assumptions in the software, not in the behavior of the actual coding.

As we view the problem in this near miss was that the basic assumptions in the design of the software is that the stall protection would involve inadvertent slowing of the aircraft during approach and landing. In such a case, the software activation would prevent increasing the AOA beyond a safe value (*alpha-prot*). Unfortunately, in the A340 incident, the activation actually increased the AOA; yes, increased the AOA. *Alpha-prot* varies from 8° at low Mach to 4.5° at 0.7 Mach. The cruise AOA was reported to be about 2.5°. This would translate to an immediate load factor of about 1.8g.

In our opinion, stall protection routines should never increase angle-of-attack when engaging.

February 7, 2001: Airbus A320, EC-HKJ

Location	Bilbao, Spain
Flight	IBE 1456
Operator	Iberia
Operation	Scheduled Airline
Route	Barcelona, Spain to Bilbao, Spain
On-Board Injuries	Minor/None 142
Aircraft Damage	Destroyed

Description   Source: Aviation Safety Network[1]

Following a nighttime flight from Barcelona to Bilbao, the crew positioned the plane for a runway 30 approach and landing. During their final ILS approach, the aircraft encountered heavy turbulence at about 200 feet AGL. with gusts up to 65 mph. The aircraft encountered windshear with 1.25G updraft, downdraft and a tailwind gust at just 70 feet AGL. When the Ground Proximity Warning System (GPWS) sounded, the captain called for a go-around while pulling on the sidestick, reportedly without pressing his priority control button. The combination of dynamic winds and the crew actions created a situation that triggered the airplane's alpha protection system. As the crew applied TOGA power for a go-around, with both pilots pulling back on their sidesticks, the alpha protection law reduced the elevator nose-up command. Instead of a go-around, the aircraft struck the runway with a vertical speed of approx. 1,200 fpm. The nosegear collapsed and the aircraft skidded 3,280 feet (about 1000 m) down the runway before coming to a stop.

Probable Cause   Source: Accident Report[81]

The cause of the accident was the activation of the angle of attack protection system which, under a particular combination of vertical gusts and windshear and the simultaneous actions of both crew members on the sidesticks, not considered in the design, prevented the aeroplane from pitching up and flaring during the landing.

Regulatory Outcome

Alpha protection logic holds *Alpha-prot* with no stick input. This would have exacerbated any deviation.

October 7, 2008: Airbus A330, VH-QPA

Location   Indian Ocean
Flight   QFA 72
Operator   Qantas
Operation   Scheduled Airline
Route   Singapore to Perth, Australia
On-Board Injuries   Serious<u>12</u> Minor/None <u>303</u>
Aircraft Damage   Minor

Description   Source: Aviation Safety Network[1]

At 09:32 local time (01:32 UTC) on 7 October 2008, an Airbus A330-303 aircraft, registered VH-QPA, departed Singapore (SIN) on a scheduled passenger transport service to Perth (PER), Australia. On board flight QF72* were 303 passengers, nine cabin crew and three flight crew. At 12:40:28, while the aircraft was cruising at 37,000 ft, the autopilot disconnected. That was accompanied by various aircraft system failure indications. At 12:42:27, while the crew was evaluating the situation, the aircraft abruptly pitched nose-down. The aircraft reached a maximum pitch angle of about 8.4 degrees nose-down, and descended 650 ft during the event. After returning the aircraft to 37,000 ft, the crew commenced actions to deal with multiple failure messages. At 12:45:08, the aircraft commenced a second uncommanded pitch-down event. The aircraft reached a maximum pitch angle of about 3.5 degrees nose-down, and descended about 400 ft during this second event.

At 12:49, the crew made a PAN emergency broadcast to air traffic control, and requested a clearance to divert to and track direct to Learmonth. At 12:54, after receiving advice from the cabin crew of several serious injuries, the crew declared a MAYDAY. The aircraft subsequently landed at Learmonth Airport, WA (LEA) at 13:50.

At least 110 of the 303 passengers and nine of the 12 crew members were injured; 12 of the occupants were seriously injured and another 39 received hospital medical treatment. Most of the injuries involved passengers who were seated without their seatbelts fastened.

Probable Cause   Source: Accident Report[82]

There was a limitation in the algorithm used by the A330/A340 flight control primary computers for processing angle of attack (AOA) data. This limitation meant that, in a very specific situation, multiple AOA spikes from only one of the three air data inertial reference units could result in a nose-down elevator command. [Significant safety issue]

When developing the A330/A340 flight control primary computer software in the early 1990s, the aircraft manufacturer's system safety assessment and other development processes did not fully consider the potential effects of frequent spikes in the data from an air data inertial reference unit. [Minor safety issue]

One of the aircraft's three air data inertial reference units (ADIRU 1) exhibited a data-spike failure mode, during which it transmitted a significant amount of incorrect data on air data parameters to other aircraft systems, without flagging that this data was invalid. The invalid data included frequent spikes in angle of attack data. Including the 7 October 2008 occurrence, there have been three occurrences of the same failure mode on LTN-101 ADIRUs, all on A330 aircraft. [Minor safety issue]

The LTN-101 air data inertial reference unit involved in the occurrence (serial number 4167) also had a previous instance of the data-spike failure mode, indicating that it probably contained a marginal weakness in its hardware, which reduced the resilience of the unit to some form of triggering event.

For the data-spike failure mode, the built-in test equipment of the LTN-101 air data inertial reference unit was not effective, for air data parameters, in detecting the problem, communicating appropriate fault information, and flagging affected data as invalid. [Minor safety issue]

The air data inertial reference unit manufacturer's failure mode effects analysis and other development processes for the LTN-101 ADIRU did not identify the data-spike failure mode.

Comments

The ATSB considered this to be a minor safety issue! Actually, the safety level of an event should be based on the worst credible outcome. It should be obvious that for an uncommanded pitch down, the worst credible outcome would the airplane colliding with the ground – a catastrophic outcome. Fortunately in this case, the airplane was at a safe altitude although several occupants were injured, including a crew member. This rises well beyond the "minor safety issue."

The behaviour of the sensor error checking in this accident is discussed beginning on page 150.

---

*   QF is the IATA code for Qantas used in the source material.

December 27, 2008: Airbus A330, VH-QPG*

Location	Indian Ocean
Flight	QFA 71
Operator	Qantas
Operation	Scheduled Airline
Route	Perth, Australia to Singapore
On-Board Injuries	None reported
Aircraft Damage	None reported

Probable Cause   Source: QFA 72 Accident Report[82]

On 27 December 2008, an Airbus A330-303 aircraft, registered VH-QPG (QPG) and being operated as Qantas flight 71, was on a scheduled passenger transport¶service from Perth to Singapore. The aircraft departed Perth at 0750 UTC (1550 local time) and reached its cruise altitude of FL360 at 0814. At 0829 (1729 local time), ADIRU 1 failed.
    ¶The flight crew reported that:
- The autopilot (autopilot 1) disconnected and the ECAM started providing a¶series of caution messages, including a NAV IR 1 FAULT
- They actioned the relevant operational procedure204 by selecting the IR 1 pushbutton to OFF and the ADR 1 pushbutton to OFF, and both OFF lights¶illuminated.
- They continued to receive multiple ECAM messages, and those messages were¶constantly scrolling on the display. One of the ECAM messages was a NAV IR¶1 FAULT, and the recommended crew action was to switch the ATT HDG¶switch to the CAPT ON 3 position. They completed this action, but it did not¶stop the ECAM messages.
- They returned to Perth and conducted an uneventful landing. At no stage was there any effect on the aircraft's flight controls.

    At the time that the autopilot disconnected, the aircraft was about 480 km (260 NM) north-west of Perth and about 650 km (350 NM) south of Learmonth (Figure 26).
    ¶The PFR for the 27 December 2008 flight contained a series of messages associated with ADIRU 1 that were very similar to those for the PFR from the 7 October 2008 occurrence (Table D1). Consistent with there being no in-flight upset, there were no PRIM FAULTS or PRIM PITCH FAULTS.

October 29, 2018: Boeing 737-8(MAX), PK-LQP

Location	15 km N off Tanjung Bungin, Indonesia
Flight	LNI 610
Operator	Lion Air
Operation	Scheduled Airline
Route	Jakarta, Indonesia to Pangkai Pinan, Indonesia
On-Board Injuries	Fatal 189
Aircraft Damage	Destroyed

Description   Source: Aviation Safety Network[1]

Lion Air flight 610, a Boeing 737 MAX 8, crashed into the sea shortly after takeoff from Jakarta-Soekarno-Hatta International Airport, Indonesia, killing all 189 on board.
    The aircraft, registration PK-LQP, had entered service with Lion Air on August 18, 2018. The Aircraft Flight Maintenance Log (AFML) recorded that since October 26, 2018 until the occurrence date several problems occurred related to airspeed and altitude flags that appeared on the captain's (left) Primary Flight Display (PFD) on three occasions, SPEED TRIM FAIL light illumination and MACH TRIM FAIL light illumination that occurred twice.
    Several attempts were made by engineers to rectify these issues. The day before the accident the Angle of Attack (AOA) sensor was replaced by engineers at Denpasar Airport.
    The flight from Denpasar to Jakarta (JT43) was the flight prior to JT610. During rotation of flight JT43,† the stick shaker activated and an IAS DISAGREE warning showed on the captain's PFD at 400 feet. The flight was handled by the copilot as it was determined that the captain's PFD was unreliable. The flight crew moved the STAB TRIM (stabilizer trim) switch to CUT OUT due to three automatic nosedown trim occurrences. The crew worked checklists and continued the flight to CGK.

---

*   This mishap was reported shortly after the previous upset to another QANTAS aircraft. The investigation was included as an appendix in the QFA Accident report.[82]

†   JT is the IATA code for Lion Air used in the source material.

Based on the crew's entry in the AFML, the engineer at Jakarta flushed the left Pitot Air Data Module (ADM) and static ADM to rectify the reported IAS and ALT disagree and cleaned the electrical connector plug of the elevator feel computer. The aircraft was subsequently released to carry out flight JT610.

During takeoff from Jakarta, the DFDR recorded a difference between left and right Angle of Attack of about 20° which continued until the end of the recording. During rotation of the aircraft, the left control column stick shaker activated and continued for most of the flight.

After the flaps were retracted, the FDR recorded automatic aircraft nose down (AND) trim for 10 seconds followed by flight crew commanded aircraft nose up (ANU) trim. Automatic AND trim briefly stopped when the flaps were temporarily extended to 5.

Last radio contact was at 06:31 local time when the captain requested the arrival controller to block altitude 3,000 feet above and below for traffic avoidance. The controller asked what altitude the pilot wanted, to which the captain responded "five thou". The controller approved the pilot request. The FDR stopped recording within twenty seconds of the pilot's response.

The aircraft impacted the sea some 15 km north off Tanjung Bungin. All 189 persons on board died in the accident.

Search and rescue personnel recovered the flight data recorder (FDR) and other debris on November 1, at 30-35 m below the water surface.

Probable Cause   Source: Accident Report[83]

1. During the design and certification of the Boeing 737-8 (MAX), assumptions were made about flight crew response to malfunctions which, even though consistent with current industry guidelines, turned out to be incorrect.

2. Based on the incorrect assumptions about flight crew response and an incomplete review of associated multiple flight deck effects, MCAS's reliance on a single sensor was deemed appropriate and met all certification requirements.

3. MCAS was designed to rely on a single AOA sensor, making it vulnerable to erroneous input from that sensor.

4. The absence of guidance on MCAS or more detailed use of trim in the flight manuals and in flight crew training, made it more difficult for flight crews to properly respond to uncommanded MCAS.

5. The AOA DISAGREE alert was not correctly enabled during Boeing 737-8 (MAX) development. As a result, it did not appear during flight with the mis-calibrated AOA sensor, could not be documented by the flight crew and was therefore not available to help maintenance identify the mis-calibrated AOA sensor.

6. The replacement AOA sensor that was installed on the accident aircraft had been mis-calibrated during an earlier repair. This mis-calibration was not detected during the repair.

7. The investigation could not determine that the installation test of the AOA sensor was performed properly. The mis-calibration was not detected.

8. Lack of documentation in the aircraft flight and maintenance log about the continuous stick shaker and use of the Runaway Stabilizer NNC meant that information was not available to the maintenance crew in Jakarta nor was it available to the accident crew, making it more difficult for each to take the appropriate actions.

9. The multiple alerts, repetitive MCAS activations, and distractions related to numerous ATC communications were not able to be effectively managed. This was caused by the difficulty of the situation and performance in manual handling, NNC execution, and flight crew communication, leading to ineffective CRM application and workload management. These performances had previously been identified during training and reappeared during the accident flight.

March 10, 2019: Boeing 737-8(MAX), ET-AVJ

Location	50 km ESE Addis Ababa Airport
Flight	ETH 302
Operator	Ethiopian Airlines
Operation	Scheduled Airline
Route	Addis Ababa, Ethiopia to Nairobi, Kenya
On-Board Injuries	Fatal157
Aircraft Damage	Destroyed

Description   Source: Aviation Safety Network[1]

Ethiopian Airlines flight ET302, * a Boeing 737 MAX 8, crashed shortly after takeoff from Addis Ababa-Bole Airport, Ethiopia. There were no survivors among the 157 occupants.

---

* ET is the IATA code for Ethiopian Airlines and is used in the source material.

Takeoff roll began from runway 07R at a field elevation of 2333 m at 08:38 hours local time, with a flap setting of 5 degrees and a stabilizer setting of 5.6 units. The takeoff roll appeared normal, including normal values of left and right angle-of-attack (AOA). During takeoff roll, the engines stabilized at about 94% N1. At 08:38:44, shortly after liftoff, the left and right recorded AOA values deviated. Left AOA decreased to 11.1° then increased to 35.7° while value of right AOA indicated 14.94°. Then the left AOA value reached 74.5° in less than a second while the right AOA reached a maximum value of 15.3°. At this time, the left stick shaker activated and remained active until near the end of the flight. Also, the airspeed, altitude and flight director pitch bar values from the left side noted deviating from the corresponding right side values. The left side values were lower than the right side values until near the end of the flight.

The captain attempted to engage the autopilot twice, but this resulted in two autopilot warnings.

At 08:40:00 shortly after the autopilot disengaged, the FDR recorded an automatic aircraft nose down (AND) activated for 9.0 seconds and pitch trim moved from 4.60 to 2.1 units. The climb was arrested and the aircraft descended slightly.

At 08:40:05, the First-Officer reported to ATC that they were unable to maintain SHALA 1A and requested runway heading which was approved by ATC.

The column moved aft and a positive climb was re-established during the automatic AND motion.

At 08:40:12, approximately three seconds after AND stabilizer motion ends, electric trim (from pilot activated switches on the yoke) in the Aircraft nose up (ANU) direction is recorded on the DFDR and the stabilizer moved in the ANU direction to 2.4 units. The Aircraft pitch attitude remained about the same as the back pressure on the column increased.

At 08:40:20, approximately five seconds after the end of the ANU stabilizer motion, a second instance of automatic AND stabilizer trim occurred and the stabilizer moved down and reached 0.4 units.

At 08:40:27, the Captain advised the First-Officer to trim up with him.

At 08:40:28 Manual electric trim in the ANU direction was recorded and the stabilizer reversed moving in the ANU direction and then the trim reached 2.3 units.

At 08:40:35, the First-Officer called out "stab trim cut-out" two times. Captain agreed and First-Officer confirmed stab trim cut-out.

At 08:40:41, approximately five seconds after the end of the ANU stabilizer motion, a third instance of AND automatic trim command occurred without any corresponding motion of the stabilizer, which is consistent with the stabilizer trim cutout switches were in the "cutout" position.

At 08:40:44, the Captain called out three times "Pull-up" and the First-Officer acknowledged.

At 08:40:50, the Captain instructed the First Officer to advise ATC that they would like to maintain 14,000 ft and they have flight control problem.

At 08:40:56, the First-Officer requested ATC to maintain 14,000 ft and reported that they are having flight control problem. ATC approved.

From 08:40:42 to 08:43:11 (about two and a half minutes), the stabilizer position gradually moved in the AND direction from 2.3 units to 2.1 units. During this time, aft force was applied to the control columns which remained aft of neutral position. The left indicated airspeed increased from approximately 305 kt to approximately 340 kt (VMO). The right indicated airspeed was approximately 20-25 kt higher than the left.

The data indicates that aft force was applied to both columns simultaneously several times throughout the remainder of the recording.

08:41:20 -   The right overspeed clacker was recorded on CVR. It remained active until the end of the recording.

08:41:21 -   The selected altitude was changed from 32000 ft to 14000 ft.

08:41:30 -   The Captain requested the First-Officer to pitch up with him and the First-Officer acknowledged.

08:41:32 -   The left overspeed warning activated and was active intermittently until the end of the recording.

08:41:46 -   The Captain asked the First-Officer if the trim is functional. The First-Officer replied that the trim was not working and asked if he could try it manually. The Captain told him to try.

08:41:54 -   The First-Officer replied that it is not working.

08:42:10 -   The Captain asked and the First-Officer requested radar control a vector to return and ATC approved.

08:42:30 -   ATC instructed ET302 to turn right heading 260 degrees and the First-Officer acknowledged.

08:42:43 -   The selected heading was changed to 262 degrees.

08:42:51 -   The First-Officer mentioned Master Caution Anti-Ice. The Master Caution is recorded on DFDR.

08:42:54 -   Both pilots called out "left alpha vane".

08:43:04 -   The Captain asked the First Officer to pitch up together and said that pitch is not enough.

08:43:11 -   About 32 seconds before the end of the recording, at approximately 13,400 ft, two momentary manual electric trim inputs are recorded in the ANU direction. The stabilizer moved in the ANU direction from 2.1 units to 2.3 units.

08:43:20- Approximately five seconds after the last manual electric trim input, an AND automatic trim command occurred and the stabilizer moved in the AND direction from 2.3 to 1.0 unit in approximately 5 seconds. The aircraft began pitching nose down. Additional simultaneous aft column force was applied, but the nose down pitch continues, eventually reaching 40° nose down. The stabilizer position varied between 1.1 and 0.8 units for the remainder of the recording.

The left Indicated Airspeed increased, eventually reaching approximately 458 kts and the right Indicated Airspeed reached 500 kts at the end of the recording. The last recorded pressure altitude was 5,419 ft on the left and 8,399 ft on the right.

Findings   ## Source: Interim Accident Report[84]

The takeoff roll and lift-off was normal, including normal values of left and right angle-of-attack (AOA). During takeoff roll, the engines stabilized at about 94% N1. From this point for most of the flight, the N1 Reference remained about 94%.

Shortly after lift-off, the left and right recorded AOA values deviated. The left AOA values were erroneous and reached 74.5° while the right AOA reached a maximum value of 15.3°.The difference between the left and the right AOA values was 59° and remained as such until near the end of the recording.

Right after the deviation of the AOA the left stickshaker activated and remained active until the near end of the recording. The pitch Flight Director (F/D) bars disappeared on both left hand and right hand Primary Flight Displays (PFD). As the aircraft crossed 400 ft Radio Altitude the right and left pitch F/D bars appeared again.

Immediately after the LH AOA sensor failure, the left AOA erroneous values affected the LH FD pitch command, and the RH and LH Flight Director (FD) pitch bars started to display different guidance.

The Stall Management Yaw Damper Computer -1 (SMYDC 1) computed LH minimum operational speed and LH stick shaker speed greater than VMO (340kt) without any alert or invalidity detection. Thus; the indicated LH airspeed was inside the minimum speed (red and black) band.

Approximately five seconds after the end of the crew manual electrical trim up inputs, a third automatic nose-down trim(MCAS) triggered. There was no corresponding motion of the stabilizer, which is consistent with the stabilizer trim cutout switches being in the "cutout" position

The right hand overspeed clacker sounded and it remained active until the end of the recording. The RH speed values varied between 360 kt and 375 kt (RH values). On the LH PFD, the LH computed airspeed oscillated between 335 kt and 350 kt.

Approximately five seconds after the last manual electric trim up input, a fourth automatic trim nose-down (MCAS) triggered. The stabilizer moved from 2.3 to 1 unit. The vertical speed decreased and became negative 3 s after the MCAS activation.

The difference training from B737NG to B737MAX provided by the manufacturer was found to be inadequate.

The AOA Disagree message did not appear on the accident aircraft as per the design described on the flight crew operation manual.

AOA failure detection feature of the ADIRU did not detect the erroneous AOA from the left AOA sensor because it only considers the value to be erroneous when the AOA value is outside the physical range. Thus; SPD and ALT flag never appeared on the PFD.

MCAS design on single AOA inputs made it vulnerable to undesired activation.

The specific failure modes that could lead to uncommanded MCAS activation, such as an erroneous high AOA input to the MCAS, were not simulated as part of the functional hazard assessment validation tests. As a result, additional flight deck effects (such as IAS DISAGREE and ALT DISAGREE alerts and stick shaker activation) resulting from the same underlying failure (for example, erroneous AOA) were not simulated and were not documented in the stabilizer trim and auto flight safety assessment.

Safety Recommendations

The NTSB identified several safety issues:

- The assumptions in Federal Aviation Administration (FAA) guidance for pilot response to flight control failure conditions, such as unintended Maneuvering Characteristics Augmentation System operation, do not adequately consider and account for the impact that multiple flight deck alerts and indications could have on pilots' responses to the hazard ;
- The need for a standardized methodology and/or tools for manufacturers use in evaluating and validating assumptions about pilot recognition and response to failure condition(s); and
- The need for aircraft systems that can more clearly and concisely inform pilots of the highest priority actions when multiple flight deck alerts and indications are present.

Regulatory Outcome

Worldwide grounding beginning with China on March 11. These were followed by Australia, Canada, EASA, and others on March 12. The USA followed on March 13.

Flight Test Accidents

August 27, 1992: de Havilland DHC-4A Prototype Conversion, N-400NC

Location   Gimli, Manitoba, Canada
Identification   N-400NC
Operator   NewCal Aviation
Operation   Development Test Flight
Route   Local flight
On-Board Injuries   Fatal   3
Aircraft Damage   Destroyed

Description   Source: TSB Accident Report[85]

The aircraft had just taken off on an experimental flight when it entered a gradually steepening climb. During the climb the aircraft rolled slowly to the right and, at approximately 200 feet above ground level (agl), it entered a steep nose-down, right-wing-low attitude and crashed. Upon impact, the on-board fuel ignited and the majority of the aircraft wreckage was destroyed by fire. The three crew members aboard the aircraft were fatally injured.

The occurrence flight was intended to be the first of several trips designed to flight-check the fuel and hydraulic systems. On the morning of the accident, the crew attended a preflight briefing, which included a thorough review of the flight test plan. The aircraft was lightly loaded at a mid centre of gravity (C of G) position. In-flight checks were scheduled to include simulated failures of both the wing fuel pumps and the in-line pumps; records were to be maintained regarding the resulting fuel pressures. A company engineer who had been involved in the design of the fuel and hydraulic modifications was included on this flight to record flight test results and to evaluate in-flight performance of the two systems

Both flight crew members were licensed and qualified to conduct this flight. Experience of either crew member on the turbo-conversion aircraft was limited in that it was a newly modified, "one-of-a-kind" aircraft. Neither pilot was an experienced flight test crew. Autopsy and toxicology examinations did not reveal any physiological, toxicological, or pathological factors that would have had a bearing on this accident.

It was reported that the scheduled copilot, a very experienced piston-Caribou captain, was replaced by another pilot with considerably less experience on type. He was, in fact, the aircraft owner's son.[86]

Probable Cause   Source: TSB Accident Report[85]

The Transportation Safety Board of Canada determined that the gust lock system was not fully disengaged prior to flight and one or more of the gust locking pins became re-engaged for undetermined reasons after lift-off. It is unlikely that a control check had been completed prior to take-off and, once airborne, the crew were unable to disengage the gust lock mechanism before losing control of the aircraft

Comments

The ultimate cause was likely inadequate preflight testing to verify gust lock operation and the absence of jamming issues. Since no control checks were observed, cockpit procedures were probably inadequate.

February 3, 1993: Lockheed L-382-HTTB, N-130X

Location	Marietta, GA
Identification	LAC N-130X*
Operator	Lockheed Corporation
Operation	Research Test Flight
Route	Local flight
On-Board Injuries	Fatal 7
Aircraft Damage	Destroyed

Description   Source: NTSB Accident Report[87]

The aircraft was designed and used as the company's engineering test bed. An evaluation of the fly-by-wire rudder actuator and ground minimum control speed (V_{MCG}) was being conducted. During the final high-speed ground test run the aircraft abruptly veered left and became airborne. It entered a left turn, climbed to about 250 ft, departed controlled flight and impacted the ground. Investigation revealed a design feature in the rudder actuator that removes hydraulic pressure within the actuator if the rudder position commanded by the pilot exceeded the actual rudder actuator position for a specified time, and the rudder aerodynamically trails. The actuator previously disengaged in flight. The company did not conduct a system safety review of the rudder bypass feature and its consequences to all flight regimes, nor of the V_{MCG} test. The flight test plan specified that engine power be retarded if the rudder became ineffective. Neither pilot had received training as an experimental test pilot. The company allowed experimental flight tests at a confined, metropolitan airport.

Probable Cause   Source: NTSB Accident Report[87]

Disengagement of the rudder fly-by-wire flight control system resulting in a total loss of rudder control capability while conducting ground minimum control speed tests. The disengagement was a result of the inadequate design of the rudder's integrated actuator package by its manufacturer; the operator's insufficient system safety review failed to consider the consequences of the inadequate design to all operating regimes. A factor which contributed to the accident was the flight crew's lack of engineering flight test training.

Comments

On the flight the preceded this accident, an inflight determination of V_{MCA} was conducted with the project test pilot in command. This type of test involves a sudden shut-down of the critical engine. When this was done, the pilot immediately applied full pedal. During this test, the rudder hydraulic power was immediately shut-down causing the rudder to trail and the aircraft to depart. Since this test was conducted at a safe altitude, the crew was able to recover. They tried a second time with the same result.

The problem was written up, but maintenance found no discrepancy. The system complied with the design documents and shop tests could not duplicate the sudden application of rudder inputs.

The ground test (V_{MCG}) was scheduled next. Since the project pilot was ill, a substitute was in command. The substitute pilot had not been trained as an experimental test pilot. According to company procedures, the flight crew was only briefed during the preflight briefing. The crew elected to start the test series below the predicted V_{MCG} speed. It must be noted that starting faster than the predicted speed and working down is much safer than starting below. It should also be noted that the pilot should abort the test by reducing power as this will reduce that adverse control problem and keep the airplane on the ground.

During the test, the airplane turned sharply to the side, departing the runway. The crew tried to reset the rudder power while the pilot tried to continue the takeoff. The result was a catastrophic crash.

Aboard the airplane were four engineers who were observing the test! Why are we carrying passengers on test flight?

June 30, 1994: Airbus A330, F-WWKH

Location	Toulouse, France
Identification	AIB 129
Operator	Airbus Industrie
Operation	Development Test Flight
Route	Local flight
On-Board Injuries	Fatal 7
Aircraft Damage	Destroyed

Description   Source: Aviation Safety Network[1]

An Airbus A330-300 aircraft crashed during a test flight at the Toulouse-Blagnac Airport in France, killing all seven on board.

---

*   Aircraft registration used *vice* flight number

The test flight was part of the preparation required to certify the autopilot, on this Airbus A330 equipped with Pratt and Whitney engines, to category III standards. The first part of the test flight was completed successfully when the aircraft landed on runway 15L. A 180 degree turn was made for takeoff from runway 33R.

The second takeoff was to be performed under conditions similar to those of the first takeoff. For this test however, the autopilot would incorporate the modification under study. The takeoff was performed by the co-pilot with TOGA (Takeoff Go Around) power instead of Flex 49, a lower power setting which was specified in the test procedure. Rotation was positive and pitch input was stopped when the attitude changed from 12 to 18 degrees nose-up. Within 5 seconds after takeoff several attempts to engage the autopilot were unsuccessful. After it was engaged, activation was delayed by two seconds because the first officer was exerting a slight nose down input on the side stick. The aircraft, still trimmed at 2.2 degrees nose-up, pitched up to reach 29 degrees and the speed had decreased to 145 knots. The captain meanwhile reduced thrust on the no. 1 engine to idle and cut off the hydraulic system in accordance with the flight test order. Immediately after it activated, the autopilot switched to altitude acquisition mode. The altitude had been set at 2000 feet on the previous flight phase. This caused the pitch attitude to increase to 32 degrees in an attempt to reach 2000 feet. The speed decreased further to 100 knots, whereas the minimum control speed is 118 kts. Roll control was lost and the captain reduced no. 2 engine thrust to idle to recover symmetry on the roll axis. Bank and pitch attitudes had reached 112 degrees left and -43 degrees respectively, before the pilot managed to regain control. It was however too late to avoid ground impact at a pitch attitude of around -15 degrees.

Probable Cause   Source: Aviation Safety Network[1]

At the present stage of its work, the commission estimates that the accident can be explained by a combination of several factors none of which, taken separately, would have led to an accident.

The initial causes are primarily related to the type of the test and its execution by the crew during the last takeoff:

1) Choice of maximum power (TOGA) instead of Flex 49;
2) Very aft CG for the last takeoff;
3) Trim set in the takeoff range, but in too high a nose-up position;
4) Selected altitude of 2000 feet;
5) Imprecise and late definition of the test to be conducted and the tasks to be performed by the captain and first officer, respectively;
6) Positive and very rapid rotation executed by the first officer;
7) The captain was busy with the test operations to be performed immediately after takeoff (engagement of the autopilot, reduce thrust on the engine and cut off the blue hydraulic system) which temporarily placed him outside the control loop;
8) In addition the absence of pitch attitude protection in the autopilot altitude acquisition mode played a significant role.

The following is also contributed to the accident:

1) The inability of the crew to identify the mode in which the autopilot was placed;
2) The confidence of the crew in the expected reactions of the aircraft;
3) The late reaction from the flight test engineer when faced with a potentially hazardous change in parameters (speed in particular);
4) The time taken by the captain to react to an abnormal situation.

The following is also contributed to the accident:[88]

As the pitch attitude was greater than 25 deg. After the autopilot was engages, the information concerning the autopilot and flight director activation modes was no longer displayed to the pilot and was not transcribed in the SSDFR. The recorder only stores the data presented to the crew on the primary flight display. This information was, however, transmitted by telemetry and is therefore available in this case.

The autopilot has no pitch attitude authority in the ALT mode. Consequently, at relatively low speeds, when a high change in thrust occurs after entry into this mode, the autopilot may command abnormal pitch attitudes as it attempts to achieve an attitude [to] capture [a] flight path that [has become] impossible [to attain].

Comments

Why are we carrying passengers on test flight?

November 27, 2008: Airbus A320, D-AXLA

Location	7 km S of Saint Cyprien, France
Identification:	D-AXLA
Operator	XL Airways Germany
Operation	Maintenance Test Flight
Route	Local flight followed by Positioning Flight
On-Board Injuries	Fatal 7
Aircraft Damage	Destroyed

Description   Source: Aviation Safety Network[1]

Airbus A320 D-AXLA had been leased by XL Airways Germany since May 2006. The airplane was due to be returned to its owner, Air New Zealand, on December 1, 2008. The Airbus was ferried to Perpignan Airport, France, where it underwent maintenance at EAS Industries. It was also repainted in full Air New Zealand livery, but retained the registration D-AXLA.

The leasing agreement specified a programme of in-flight checks to ensure the airplane was fit for purpose.

The crew was made up of a captain and a co-pilot from XL Airways Germany. A pilot and three engineers from Air New Zealand, as well as a representative of the New Zealand Civil Aviation authority were also on board. The pilot and one of the engineers had taken seats in the cockpit.

The takeoff took place at 15:44 hours local time. A few minutes after takeoff, the crew requested to perform a "360", but this not authorised by the controller. The en route controller explained to the crew that they could not undertake tests in general air traffic and that the flight plan filed was not compatible with the manoeuvres requested.

The crew announced that they wanted to continue on the route planned in the flight plan and asked to climb to FL310 before turning back towards Perpignan. At around 16:12, the crew turned back. Some checks planned in the flight programme were performed. The maximum flight level reached was FL390.

At 16:33, in descent towards the FL130, the crew contacted Perpignan Approach. They were then cleared to descend to FL120 towards the PPG VOR. The approach controller asked them to reduce speed to 250 kt and to plan a hold at the PPG VOR. They were number two on approach.

At 16:34, the crew requested radar vectoring. The approach controller asked the crew to turn left onto heading 090 and to reduce the speed to 200 kt. The approach controller asked the crew to reduce speed to 180 kt and to descend to FL 80 then to FL 60.

From 16:38:03 and for about forty seconds, the pilot from Air New Zealand described the actions to take to perform a check at low speed planned in the programme. At around 16:40, the approach controller asked the crew to turn right on heading 190 and to maintain 180 kt. The airplane speed was 215 kt. About one minute later, the approach controller cleared the crew to the LANET ILS approach for runway 33 and to descend towards 5,000 ft altitude. At the request of the crew, the approach controller repeated the message. While the co-pilot was reading back, the captain indicated to the Air New Zealand pilot that the low speed flight should be made later, during the flight towards Frankfurt. He even considered not performing it.

At 16:42:14, the approach controller asked for the speed of the airplane. The co-pilot answered that the speed was falling then at 16:42:25 that it was 180 kt. The approach controller then asked them to maintain 180 kt and to descend to 2,000 ft. The slat and flap controls selector was put in position 2.

At 16:42:46, the captain stated that the approach was not included in the database. Thirty-six seconds later, the co-pilot carried out the approach briefing.

At 16:43:37, the captain announced that he was passing under the cloud layer. He disengaged the autopilot and asked the Air New Zealand pilot what he wanted. The latter answered that it was necessary to go slowly and described to him the necessary actions to activate the alpha floor protection. During these exchanges, the captain called for gear extension and put the thrust levers in the IDLE position. At the same time, the approach controller asked the crew its intentions twice. The co-pilot answered that they wanted to make a go-around and continue towards Frankfurt.

At 16:44:30, the captain stabilised the airplane at an altitude of 3,000 ft. The airplane was in FULL configuration. In thirty-five seconds, the speed went from 136 to 99 kt and the horizontal stabilizer went to the pitch-up stop. The stall warning sounded. The pitch angle was then slightly below 19 degrees. The thrust levers were advanced towards the TO/GA position in the following second. While the thrust on the engines increased in a symmetrical manner, the speed continued to drop to 92.5 kt, then began to increase. The airplane started to roll slightly to the left, then to the right. The captain countered these movements.

At 16:45:15, the flight control laws changed to "direct" law. The bank angle was 50 degrees to the right. At 16:45:19, the stall warning stopped. The bank angle was 40 degrees to the left.

One second later, the pitch angle was 7 degrees, the wings were close to horizontal and the speed was 138 kt. The airplane's pitch and altitude then began to increase. During this climb, the stall warning sounded a second time. The crew retracted the landing gear and the flight control law in pitch changed to "alternate".

At 16:45:44 the maximum recorded values were: pitch 57 degrees, altitude 3,800 ft. The speed was below 40 kt.

At 16:45:47 the stall warning stopped. It sounded again five seconds later. From 16:45:55, the airplane banked to the right up to 97 degrees and its pitch reached 42 degrees nose-down. At 16:45:58, the stall warning stopped. The slat and flap controls selector was placed in position 1, then 0 two seconds later. The captain used the flight controls and the thrust levers to try to recover the airplane. At 16:46:00, the stall warning stopped. At 16:46:06.8, the last recorded values were a pitch of 14° nose down, a bank angle of 15° to the right, a speed of 263 kt and an altitude of 340 ft. Less than a second later, the airplane crashed into the sea.

Probable Cause

Source: Accident Report[89]

The loss of control of the aeroplane by the crew following the improvised demonstration of the functioning of the angle of attack protections, while the blockage of the angle of attack sensors made it impossible for these protections to trigger.

The crew was not aware of the blockage of the angle of attack sensors. They did not take into account the speeds mentioned in the programme of checks available to them and consequently did not stop the demonstration before the stall.

The following factors contributed to the accident:
* The decision to carry out the demonstration at a low height;
* The crew's management, during the thrust increase, of the strong increase in the longitudinal pitch, the crew not having identified the pitch-up stop position of the horizontal stabiliser nor acted on the trim wheel to correct it, nor reduced engine thrust;
* The crew having to manage the conduct of the flight, follow the programme of in-flight checks, adapted during the flight, and the preparation of the following stage, which greatly increased the work load and led the crew to improvise according to the constraints encountered;
* The decision to use a flight programme developed for crews trained for test flights, which led the crew to undertake checks without knowing their aim;
* The absence of a regulatory framework in relation to non-revenue flights in the areas of air traffic management, of operations and of operational aspects;
* The absence of consistency in the rinsing task in the aeroplane cleaning procedure, and in particular the absence of protection of the AOA sensors, during rinsing with water of the aeroplane three days before the flight. This led to the blockage of the AOA sensors through freezing of the water that was able to penetrate inside the sensor bodies.

The following factors also probably contributed to the accident:
* Inadequate coordination between an atypical team composed of three airline pilots in the cockpit;
* The fatigue that may have reduced the crew's awareness of the various items of information relating to the state of the systems.

Comments

Again. Why are we carrying passengers on test flight?

November 9, 2010: Boeing 787, N-787EX

Location	Laredo, TX
Flight	BOE N-787EX*
Operator	Boeing Company
Operation	Development Test Flight
Route	Yuma, AZ to Harlingen, TX
On-Board Injuries	Minor/None 42
Aircraft Damage	Unknown

Details of this accident can be found on page 295

Comments

At least there were no passengers on this test flight. All occupants had inflight duties.

---

* Aircraft registration used *vice* flight number.

April 2, 2011: Gulfstream G650, N-652GD

Location   Roswell, NM
Flight   GLF N-652GD*
Operator   Gulfstream Aerospace
Operation   Development Test Flight
Route   Local Flight
On-Board Injuries   Fatal 4
Aircraft Damage   Destroyed

Description   Source: Aviation Safety Network[1]

A Gulfstream G650 jet was damaged beyond repair in a takeoff accident at Roswell International Air Center Airport, NM (ROW). The two pilots and the two flight test engineers were fatally injured. The accident occurred during a planned one-engine-inoperative (OEI) takeoff when a stall on the right outboard wing produced a rolling moment that the flight crew was not able to control, which led to the right wingtip contacting the runway and the airplane departing the runway from the right side. After departing the runway, the airplane impacted a concrete structure and an airport weather station, resulting in extensive structural damage and a postcrash fire that completely consumed the fuselage and cabin interior.

The National Transportation Safety Board's (NTSB) investigation of this accident found that the airplane stalled while lifting off the ground. As a result, the NTSB examined the role of "ground effect" on the airplane's performance. Ground effect refers to changes in the airflow over the airplane resulting from the proximity of the airplane to the ground. Ground effect results in increased lift and reduced drag at a given angle of attack (AOA) as well as a reduction in the stall AOA. In preparing for the G650 field performance flight tests, Gulfstream considered ground effect when predicting the airplane's takeoff performance capability but overestimated the in ground effect stall AOA. Consequently, the airplane's AOA threshold for stick shaker (stall warning) activation and the corresponding pitch limit indicator (on the primary flight display) were set too high, and the flight crew received no tactile or visual warning before the actual stall occurred.

The accident flight was the third time that a right outboard wing stall occurred during G650 flight testing. Gulfstream did not determine (until after the accident) that the cause of two previous uncommanded roll events was a stall of the right outboard wing at a lower-than-expected AOA. (Similar to the accident circumstances, the two previous events occurred during liftoff; however, the right wingtip did not contact the runway during either of these events.) If Gulfstream had performed an in-depth aerodynamic analysis of these events shortly after they occurred, the company could have recognized before the accident that the actual in-ground-effect stall AOA was lower than predicted.

During field performance testing before the accident, the G650 consistently exceeded target takeoff safety speeds (V2). V2 is the speed that an airplane attains at or before a height above the ground of 35 feet with one engine inoperative. Gulfstream needed to resolve these V2 exceedances because achieving the planned V2 speeds was necessary to maintain the airplane's 6,000-foot takeoff performance guarantee (at standard sea level conditions). If the G650 did not meet this takeoff performance guarantee, then the airplane could only operate on longer runways. However, a key assumption that Gulfstream used to develop takeoff speeds was flawed and resulted in V2 speeds that were too low and takeoff distances that were longer than anticipated.

Rather than determining the root cause for the V2 exceedance problem, Gulfstream attempted to reduce the V2 speeds and the takeoff distances by modifying the piloting technique used to rotate the airplane for takeoff. Further, Gulfstream did not validate the speeds using a simulation or physics-based dynamic analysis before or during field performance testing. If the company had done so, then it could have recognized that the target V2 speeds could not be achieved even with the modified piloting technique. In addition, the difficulties in achieving the target V2 speeds were exacerbated in late March 2011 when the company reduced the target pitch angle for some takeoff tests without an accompanying increase in the takeoff speeds.

Gulfstream maintained an aggressive schedule for the G650 flight test program so that the company could obtain Federal Aviation Administration (FAA) type certification by the third quarter of 2011. The schedule pressure, combined with inadequately developed organizational processes for technical oversight and safety management, led to a strong focus on keeping the program moving and a reluctance to challenge key assumptions and highlight anomalous airplane behavior during tests that could slow the pace of the program. These factors likely contributed to key errors, including the development of unachievable takeoff speeds, as well as the superficial review of the two previous uncommanded roll events, which allowed the company's overestimation of the in-ground-effect stall AOA to remain undetected.

---

*   Aircraft registration used *vice* flight number.

Probable Cause

Source: Accident Report[90]

"The National Transportation Safety Board determines that the cause of this accident was an aerodynamic stall and subsequent uncommanded roll during a one engine-inoperative takeoff flight test, which were the result of (1) Gulfstream's failure to properly develop and validate takeoff speeds for the flight tests and recognize and correct the takeoff safety speed (V2) error during previous G650 flight tests, (2) the G650 flight test team's persistent and increasingly aggressive attempts to achieve V2 speeds that were erroneously low, and (3) Gulfstream's inadequate investigation of previous G650 uncommanded roll events, which indicated that the company's estimated stall angle of attack while the airplane was in ground effect was too high. Contributing to the accident was Gulfstream's failure to effectively manage the G650 flight test program by pursuing an aggressive program schedule without ensuring that the roles and responsibilities of team members had been appropriately defined and implemented, engineering processes had received sufficient technical planning and oversight, potential hazards had been fully identified, and appropriate risk controls had been implemented and were functioning as intended."

Comments

This appears to be another example where organizational policy outweighed professional judgment.

Fuel Exhaustion/Starvation

August 24, 2001: Airbus A330, C-GITS

Location	Terceira-Lajes, Azores
Flight	TSC 236
Operator	Air Transat
Operation	Non-Scheduled Airline
Route	Toronto, Canada to Lisbon, Portugal
On-Board Injuries	Serious 2 Minor/None 304
Aircraft Damage	Substantial

Description   Source: Aviation Safety Network[1]

Air Transat Flight TS236,* was en route at FL390 when at 05:36 UTC, the crew became aware of a fuel imbalance be-tween the left and right-wing main fuel tanks. Five minutes later the crew concerned about the lower-that-expected fuel quantity indication, decided to divert to Lajes Airport in the Azores. At 05:48 UTC, when the crew ascertained that a fuel leak could be the reason for the possible fuel loss, an emergency was declared to Santa Maria Oceanic Control. At 06:13, at a calculated distance of 135 miles from Lajes, the right engine (Rolls-Royce Trent 772B) flamed out. At 06:26, when the aircraft was about 85 nm from Lajes and at an altitude of about FL345, the left engine flamed out. At 06:39 the aircraft was at 13,000 feet and 8 miles from the threshold of runway 33. An engines-out visual approach was carried out and the air-craft landed on runway 33. Eight of the plane's ten tyres burst during the landing.

Investigation has determined that a low-pressure fuel line on the right engine, had failed probably as the result of its coming into contact with an adjacent hydraulic line.

Probable Cause   Source: Accident Report[91]

FINDINGS AS TO CAUSES AND CONTRIBUTING FACTORS:

1   The replacement engine was received in an unexpected pre-SB configuration to which the operator had not previ-ously been exposed.
2   Neither the engine-receipt nor the engine-change planning process identified the differences in configuration be-tween the engine being removed and the engine being installed, leaving complete reliance for detecting the differ-ences upon the technicians doing the engine change.
3   The lead technician relied on verbal advice during the engine change procedure rather than acquiring access to the relevant SB, which was necessary to properly complete the installation of the post-mod hydraulic pump.
4   The installation of the post-mod hydraulic pump and the post-mod fuel tube with the pre-mod hydraulic tube assem-bly resulted in a mismatch between the fuel and hydraulic tubes.
5   The mismatched installation of the pre-mod hydraulic tube and the post-mod fuel tube resulted in the tubes coming into contact with each other, which resulted in the fracture of the fuel tube and the fuel leak, the initiating event that led to fuel exhaustion.
6   Although the existence of the optional Rolls-Royce SB RB.211-29-C625 became known during the engine change, the SB was not reviewed during or following the installation of the hydraulic pump, which negated a safety defence that should have prevented the mismatched installation.
7   Although a clearance between the fuel tube and hydraulic tube was achieved during installation by applying some force, the pressurization of the hydraulic line forced the hydraulic tube back to its natural position and eliminated the clearance.
8   The flight crew did not detect that a fuel problem existed until the Fuel ADV advisory was displayed and the fuel imbalance was noted on the Fuel ECAM page.
9   The crew did not correctly evaluate the situation before taking action.
10   The flight crew did not recognize that a fuel leak situation existed and carried out the fuel imbalance procedure from memory, which resulted in the fuel from the left tanks being fed to the leak in the right engine.
11. Conducting the FUEL IMBALANCE procedure by memory negated the defence of the Caution note in the FUEL IM-BALANCE checklist that may have caused the crew to consider timely actioning of the FUEL LEAK procedure.
12. Although there were a number of other indications that a significant fuel loss was occurring, the crew did not conclude that a fuel leak situation existed - not actioning the FUEL LEAK procedure was the key factor that led to the fuel exhaustion.

---

*   TS is the IATA code for Air Transat used in the source material.

July 23, 1983: Boeing 767, C-GAUN

Location	Gimli, Manitoba, Canada
Flight	ACA 143
Operator	Air Canada
Operation	Scheduled Airline
Route	Montreal, Canada to Edmonton, Canada
On-Board Injuries	Minor/None 69
Aircraft Damage	Destroyed

Description   Source: Aviation Safety Network[1]

Boeing 767 C-GAUN was one of four brand new 767's delivered to Air Canada at the time of the accident. On July 22, 1983 C-GAUN underwent a routine service check in Edmonton (YEG). During this check the three fuel quantity indicators, situated on an overhead panel between the two pilots, were found to be blank. The technician found that he could obtain fuel indication by pulling and deactivating the channel 2 circuit breaker. He marked the circuit breaker 'inoperative' and made an entry in the log book. Because the fuel quantity system was not redundant anymore, the fuel load had to be confirmed by the use of the fuel measuring sticks located under the wings of the aircraft.

The following day the 767 was flown to Montreal via Ottawa (YOW). A crew change took place for the return flight, AC143 to Edmonton. Before the new flight crew arrived on board, a technician entered the cockpit. He noted the entry made in the log book and saw the circuit breaker which had been pulled and tagged. He was confused by the entry in the log book which did not appear to coincide with what he had been taught about the processor in recent training. Because of his confusion, he attempted a self-test of the system and reset the number 2 channel circuit breaker. This caused the fuel gauges in the cockpit to go blank. Not being satisfied with the test, he decided that the ´Fuel Quantity Information System Processor´ had to be replaced. However, none were available in Montreal. On returning to the flight deck, the technician was distracted by the arrival of the fueller and forgot to pull the number 2 circuit breaker again. When the captain for the return flight arrived on board, he saw the blank fuel gauges. This did not surprise him. In a brief conversation with the arriving crew, he was told that the fuel gauges were inoperative and that a fuel drip had to be done to ascertain the amount of fuel on the aircraft. Similarly, the log book entry further confirmed his false assumption about the fuel gauges.

The captain then consulted the Minimum Equipment List (MEL), which clearly indicates that the aircraft was not legal to go with blank fuel gauges. Nevertheless, because of the mistaken assumption already in his mind, the captain formed the opinion that he could safely take and fly the aircraft, provided the fuel quantity on board the aircraft was confirmed by use of the fuel quantity measuring sticks in the fuel tanks. Because of the problem with the gauges, it was decided to load enough fuel to go right through to Edmonton with a drip check to be made both in Montreal and in Ottawa.

Maintenance crew at Montreal calculated the 767's fuel load by hand. They dripped the tanks and the flightcrew calculated the total amount of fuel by using 1.77 pounds/litre as the specific gravity factor. This was the factor written on the refueler's slip and used on all of the other planes in Air Canada's fleet. On the all-metric Boeing 767 however they should have used 0.8 kg/litre of kerosene. The aircraft departed Montreal and landed at Ottawa, a scheduled stop on its way to Edmonton. At Ottawa the plane was re-dripped and the crew were told 11,430 liters of fuel were on board. The flightcrew then thought they had 20,400 kilos of fuel (instead of only 9144 kilos !). This amount was entered in the FMS. En route to Edmonton, at FL410, the EICAS warned low fuel pressure in the left fuel pump. The captain at once decided to divert the flight to Winnipeg, then 120 miles (192 km) away, and commenced a descent. Within seconds, warning lights appeared indicating loss of pressure in the right main fuel tank. Within minutes, the left engine failed, followed by failure of the right engine. The aircraft was then at 35,000 feet, 65 miles (104 km) from Winnipeg and 45 miles (72 km) from Gimli. Without power to generate electricity all the electronic gauges in the cockpit became blank, leaving only stand-by instruments, consisting of a magnetic compass, an artificial horizon, an airspeed indicator and an altimeter. Vectors were given to Gimli. The captain, who had flying experience on a glider, used gliding techniques to manoeuver the airplane for the approach. The landing gear was lowered, but the nose gear could not be lowered and locked. The 767 touched down on runway 32L within 800 feet of the threshold. The nose contacted the runway and the airplane came to rest short of a part of the runway which was at the time being used as a drag racing strip.

Probable Cause   Source: Aviation Safety Network[1]

Since the investigation was not done in accordance with ICAO Annex 13, no Probable Cause was identified.

Comments

The accident crews deserve "attaboys" for using their flying skills. However, the flight crews also deserve some criticism for poor preflight planning in the case of the "Gimli Glider" and for poor trouble shooting in the case of the Air Transat flight.

Fuel Tank Damage

July 25, 2000: Aerospatiale/BAC *Concorde*, F-BTSC

Location	Paris, France
Flight	AFR 4590
Operator	Air France
Operation	Non-scheduled Charter
Route	Paris, France to New York, USA
On-Board Injuries	Fatal 109
Ground Injuries	Fatal 4
Aircraft Damage	Destroyed

Description   Source: Aviation Safety Network[1]

The Air France *Concorde*, registered F-BTSC, was to depart Paris-Charles de Gaulle for a flight to New York-JFK. Departure was delayed by about one hour because the crew had requested a replacement of the thrust reverser pneumatic motor of the no. 2 engine. Also, the rear bogie truck of the left hand main undercarriage was replaced. When all 100 passengers had boarded, the plane taxied to runway 26R (4217 m long). Takeoff weight was calculated to be 186.9 tons, including 95 tons of fuel, which was one tone over the maximum takeoff weight. At 14:42:17 the crew were cleared for takeoff.

At 14:42:31, the captain commenced takeoff. At 14:42:54.6, the co-pilot called one hundred knots, then V1 nine seconds later. A few seconds after that, tyre No 2 (right front) on the left main landing gear was destroyed after having run over a strip of metal lost by a Continental Airlines DC-10-30, registered N-13067 which departed Paris as flight 055 to Newark five minutes before. The destruction of the tyre in all probability resulted in large pieces of rubber being thrown against the underside of the left wing and the rupture of a part of tank 5. A severe fire broke out under the left wing and around the same time engines 1 and 2 suffered a loss of thrust, severe for engine 2, slight for engine 1. By 14:43:13, as the captain commenced the rotation, the controller informed the crew the presence of flames behind the aircraft. The co-pilot acknowledged this transmission and the flight engineer announced the failure of engine no. 2. Nine seconds later the engine fire alarm sounded and the flight engineer announced "shut down engine 2" then the captain called for the "engine fire" procedure. A few seconds later, the engine 2 fire handle was pulled and the fire alarm stopped. The co-pilot drew the captain's attention to the airspeed, which was 200 kt. At 14:43:30, the captain called for landing gear retraction. The controller confirmed the presence of large flames behind the aircraft. Twelve seconds later the engine fire alarm sounded again for around 12 seconds. It sounded for the third time at about 14:43:58 and continued until the end of the flight. At 14:43:56, the co-pilot commented that the landing gear had not retracted and made several callouts in relation to the airspeed. Three seconds later, the GPWS alarm sounded several times. The co-pilot informed ATC that they were trying for Le Bourget aerodrome. Then the number 1 engine lost power as well. The aircraft entered a left turn until control was lost, crashing into hotel 'Hotellisimo' and bursting into flames.

Probable Cause   Source: Accident Report[92]

- High-speed passage of a tyre over a part lost by an aircraft that had taken off five minutes earlier and the destruction of the tyre.
- The ripping out of a large piece of tank in a complex process of transmission of the energy produced by the impact of a piece of tyre at another point on the tank, this transmission associating deformation of the tank skin and the movement of the fuel, with perhaps the contributory effect of other more minor shocks and /or a hydrodynamic pressure surge.
- Ignition of the leaking fuel by an electric arc in the landing gear bay or through contact with the hot parts of the engine with forward propagation of the flame causing a very large fire under the aircraft's wing and severe loss of thrust on engine 2 then engine 1.
- In addition, the impossibility of retracting the landing gear probably contributed to the retention and stabilisation of the flame throughout the flight.

Regulatory Outcome

Type certificates for all *Concordes* revoked.

EASA issued amendment 14 to CS 25 to include: §25.734 (new) "Protection against Wheel and Tyre Failures" This requires that The safe operation of the aeroplane must be preserved in case of damaging effects on systems or structures from:

- tyre debris;
- tyre burst pressure;
- flailing tyre strip; and
- wheel flange debris.

The amendment also revised §25.963(e) "Fuel Tank Protection" which sets forth a means of compliance with the provisions of CS 25 dealing with the certification requirements for fuel tanks (including skin and fuel tank access covers) on large aeroplanes. Guidance information is provided for showing compliance with the impact and fire resistance requirements of CS §25.963(e).

This revised rule deleted reference to Fuel Tank Access Covers, which had been the focus of the original rule (reference the 1985 B737-200 accident in Manchester, England), and expanded the area to be protected from expected debris impact to include the entire tank surface.

The FAA did not add §25.734 to Part 25. They did issue special conditions for the Boeing Model 787, and the Airbus Model A350, both incorporating unique composite wing structure forming the fuel tank surfaces in the areas near the wheels, tires, and engines.

High Intensity Radiated Field (HIRF)

April 15, 1990: Airship 600, N-602SK

Location	Creswell, NC
Identification	N-602SK
Operator	Airship Industries
Operation	Business Transportation
Route	Elizabeth City, NC to Florence, SC
On-Board Injuries	Minor/None 4
Aircraft Damage	Substantial

Description   Source: NTSB Accident File[93]

While cruising at 700 ft on a cross-country flight, the airship had a complete loss of power in both engines. Subsequently, the airship collided with trees & the terrain during a forced landing in 20 kt winds. An investigation revealed that the airship had traversed the beam of a highly directional radio frequency broadcast from a Voice of America antenna. The signal radiating from the antenna had been demonstrated to damage electronic ignition control units of the same type used on the airship engines. Subsequent tests of the failed ignition units showed that the failure mode in this accident replicated the failures of the previously tested ignition units. At the time of certification, Airship requirements for High Intensity Radiated Electromagnetic Fields (HIRF) testing had not been issued.

Probable Cause   Source: NTSB Accident File[93]

The failure of the electronic ignition units as the airship traversed the directional antenna beam of a high intensity radiated electromagnetic field (HIRF). Factors related to the accident were: the unfavorable wind and the lack of HIRF certification standards for airships at the time of FAA certification.

Certification Issues

The airship was a Skyship 600 (S/N 1215/07) certified in the Normal Category [94] based on a British Airworthiness Certificate. The airship was powered by 2 Porsche 930/67/A1/3 engines.

Comments

To the best of our knowledge, this is the only reported civil aircraft accident attributed to HIRF.

Hypoxia

October 25, 1999: Learjet 35A, N-47BA

Location	Aberdeen, SD Identification	N-47BA*
Operator	Sunjet Aviation	
Operation	Non-scheduled Air Taxi	
Route	Portland, FL to Dallas, TX	
On-Board Injuries	Fatal 6	
Aircraft Damage	Destroyed	

Description   Source: Aviation Safety Network[1]

On October 25, 1999, about 1213 central daylight time (CDT), a Learjet Model 35, N47BA, operated by Sunjet Aviation, Inc., of Sanford, Florida, crashed near Aberdeen, South Dakota. The airplane departed Orlando, Florida, for Dallas, Texas, about 0920 eastern daylight time (EDT). Radio contact with the flight was lost north of Gainesville, Florida, after air traffic control (ATC) cleared the airplane to flight level (FL) 390. The airplane was intercepted by several U.S. Air Force (USAF) and Air National Guard (ANG) aircraft as it proceeded northwest bound. The military pilots in a position to observe the accident airplane at close range stated (in interviews or via radio transmissions) that the forward windshields of the Learjet seemed to be frosted or covered with condensation. The military pilots could not see into the cabin. They did not observe any structural anomaly or other unusual condition. The military pilots observed the airplane depart controlled flight and spiral to the ground, impacting an open field. All occupants on board the airplane (the captain, first officer, and four passengers) were killed, and the airplane was destroyed.

Probable Cause   Source: Accident Report[95]

The National Transportation Safety Board determines the probable cause of this accident to be: Incapacitation of the flight crewmembers as a result of their failure to receive supplemental oxygen following a loss of cabin pressurization, for undetermined reasons.

Comment

Professional golfer Payne Stewart was a passenger on the flight.

August 14, 2005: Boeing 737, 5B-DBY

Location	Grammatikos, Greece
Flight	HCY 522
Operator	Helios Airways
Operation	Scheduled Airline
Route	Larnica, Cyprus to Prague, Czech Republic
On-Board Injuries	Fatal 121
Aircraft Damage	Destroyed

Description   Source: Aviation Safety Network[1]

Helios' Boeing 737-300 5B-DBY underwent maintenance on the night prior to the accident. The pressurization system was checked, but after completion of the tests the Pressurization Mode Selector (PMS) was reportedly left in the "Manual" position instead of the "Auto" mode. In manual mode the crew had to manually open or close the outflow valves in order to control the cabin pressure. The outflow valves were one-third in the open position which meant that the cabin would not pressurize after takeoff. The PMS mode was apparently not noted during the pre-departure checks by the crew.

In the morning the 737 was to operate Flight 522 from Larnaca to Prague, Czech Republic with an intermediate stop at Athens, Greece. The flight departed Larnaca at 09:07 for the leg to Athens with a planned flying time of 1 hour and 23 minutes. As the airplane climbed over the Mediterranean the cabin altitude alert horn sounded. This occurred as the 737 passed through an altitude of 10,000 feet. Cabin altitude is usually held around 8,000 feet. The crew possibly thought it was an erroneous takeoff configuration warning because the sound is identical. Then, at 14,000 feet, the oxygen masks automatically deployed and a master caution light illuminated in the cockpit. Because of a lack of cooling air another alarm activated, indicating a temperature warning for the avionics bay.

The German captain and the Cypriot co-pilot tried to solve the problem but encountered some problems communicating with each other. They contacted the Helios´ maintenance base to seek advice. The engineer told that they needed to pull the circuit breaker to turn off the alarm. The radio contact ended as the aircraft climbed through 28 900 ft.

---

*   Aircraft registration used *vice* flight number.

The circuit breaker was located in a cabinet behind the captain. The captain got up from his seat to look for the circuit breaker. The crew were not wearing their oxygen masks as their mindset and actions were determined by the preconception that the problems were not related to the lack of cabin pressure.

As the airplane was still climbing the lack of oxygen seriously impaired the flight crew. The captain probably became unconscious when he was trying to find the circuit breaker. The first officer was still in his seat when he also became unconscious. Because the plane's autopilot was programmed for FL340 the Boeing continued to climb until leveling out at that altitude some 19 minutes after takeoff. At 09:37 the 737 entered the Athens FIR but not contact was established with the flight. Over Rodos at about 09:52 the airplane entered the UL995 airway. At 10:21 the airplane passed the KEA VOR, which is located about 28 NM south of the Athens airport. The airplane then passed the Athens Airport and subsequently entered the KEA VOR holding pattern at 10:38. All efforts by Greek air traffic controllers to contact the pilots were futile. Around 11:00 two Greek F-16 fighter planes were scrambled from the Néa Anghialos air base. At 11:24, during the sixth holding pattern, the F-16's intercepted the airliner. The F-16 pilots reported that they were not able to observe the captain, while the first officer seemed to be unconscious and slumped over the controls.

At 11:49, the F-16's reported a person not wearing an oxygen mask entering the cockpit and occupying the captain's seat. The F-16 pilot tried to attract his attention without success. At 11:50, the left engine flamed out due to fuel depletion and the aircraft started descending. At 11:54, two Mayday messages were recorded on the CVR.

At 12:00, the right engine also flamed out at an altitude of approximately 7100 feet. The aircraft continued descending rapidly and impacted hilly terrain.

The same Boeing 737, 5B-DBY, suffered a loss of cabin pressure on December 20, 2004 during a flight from Warsaw to Larnaca. Three passengers needed medical treatment after landing in Larnaca. This incident was caused by a leaking door seal of the right hand rear door.

Probable Cause   Source: Accident Report[96]

1.   Non-recognition that the cabin pressurization mode selector was in the MAN (manual) position during the performance of the
      a)   Preflight procedure
      b)   Before Start checklist; and
      c)   After Takeoff checklist.
2.   Non-identification of the warnings and the reasons for the activation of the warnings (cabin altitude warning horn, passenger oxygen masks deployment indication, Master Caution), and continuation of the climb.
3.   Incapacitation of the flight crew due to hypoxia, resulting in continuation of the flight via the flight management computer and the autopilot, depletion of the fuel and engine flameout, and impact of the aircraft with the ground.
The accident board also found fault with
      Operator's safety culture,
      Regulatory Authority's oversight,
      Crew Resource Management by the flight crew, and
      Ineffectiveness and inadequacy of measures taken by the manufacturer.

Certification Issues

For both the LearJet 35 (N-47BA) and the B-737 (5B-DBY) accidents, the design of the cabin altitude warning system incorporated a warning horn that served two purposes. On the ground the warning served as a takeoff configuration (e.g., flap setting) warning and in flight, warned of excessive cabin altitude. A combination of pre-flight maintenance activities and multiple checklist errors resulted in an improperly configured airplane pressurization system. Following takeoff, as the cabin altitude exceeded safe limits, the warning system functioned properly to alert the crew. Due to the dual function of the warning, it was misinterpreted by the flightcrew who ultimately failed to take appropriate corrective actions.

The intermittent warning horn is a carryover from designs in the 1950s. The airplanes of the era did not have modern warning systems and only had a few attention getting methods, warning horns, fire bells, "clackers, long with warning lights. It seems strange for these simplistic warning systems should exist in modern airplanes.

Regulatory Outcome

Amendment 25-132 to §25.1322 made some general changes to the flightcrew alerts.[97]

September 5, 2014: Socata TBM-700, N-900KN

Location	Caribbean Sea
Flight	N-900KN*
Operator	Individual
Operation	Personal Transportation
Route	Rochester, NY to Naples, FL
On-Board Injuries	Fatal 2
Aircraft Damage	Destroyed

Description   Source: Aviation Safety Network[1]

The TBM-900 airplane was en route from Rochester, NY to Naples, FL when it failed to change course to Naples. On a recording made by Live ATC - a website that monitors air traffic control recordings - and quoted by the Associated Press, the pilot can be heard saying: "We need to descend to about (18,000 feet). We have an indication that's not correct in the plane." A controller replies: "Stand by." After a pause, the controller tells the pilot to fly at 25,000ft. "We need to get lower," the pilot responded. "Working on that," the controller says. Controllers then clear the plane to descend to 20,000 feet and the pilot acknowledges the command but does not descend.

The airplane continued off the east coast of Florida, over the Bahamas and Cuba before it lost altitude north of Jamaica. The airplane reportedly crashed near Jamaica after running out of fuel. According to a review of preliminary radar data received from the FAA, the airplane entered a high rate of descent from FL250 prior to impacting the water. The last radar target was recorded over open water about 10,000 feet at 18.3547N, 76.44049W. The two occupants (pilot and one passenger) are died in the crash. The remains of the deceased were recovered on Monday 19 January 2015.

In U.S. airspace the airplane was tracked by two F-15 jets who reported that the cockpit windows were frosted up.

Jamaica delegated the investigation to NTSB.

Probable Cause   Source: Accident Report[98]

The design of the cabin pressurization system, which made it prone to unnecessary shutdown, combined with a checklist design that prioritized troubleshooting over ensuring that the pilot was sufficiently protected from hypoxia. This resulted in a loss of cabin pressure that rendered the pilot and passenger unconscious during cruise flight and eventually led to an in-flight loss of power due to fuel exhaustion over the open ocean.

Regulatory Outcome

Daher-Socata has also issued Service Information 2014-008, "Review of High-altitude Operations" in response to this incident.[99]

Comments

The **Emergency Descent Mode** is the obvious solution for these types of accidents.(See page 157)

---

* Aircraft registration used *vice* flight number.

Icing (Airframe)

October 31, 1994: ATR 72, N-401AM

Location	Near Roselawn, IN
Flight	EGF 4184
Operator	Simmons Airlines dba American Eagle
Operation	Scheduled Airline
Route	Indianapolis, IN to Chicago, IL
On-Board Injuries	Fatal <u>68</u>
Aircraft Damage	Destroyed

Description
Source: <u>Aviation Safety Network</u>[1]

American Eagle Flight 4184 was scheduled to depart the gate in Indianapolis at 14:10; however, due to a change in the traffic flow because of deteriorating weather conditions at destination Chicago-O'Hare, the flight left the gate at 14:14 and was held on the ground for 42 minutes before receiving an IFR clearance to O'Hare. At 14:55, the controller cleared flight 4184 for takeoff. The aircraft climbed to an enroute altitude of 16,300 feet. At 15:13, flight 4184 began the descent to 10,000 feet. During the descent, the FDR recorded the activation of the Level III airframe de-icing system. At 15:18, shortly after flight 4184 leveled off at 10,000 feet, the crew received a clearance to enter a holding pattern near the LUCIT intersection and they were told to expect further clearance at 15:45, which was revised to 16:00 at 15:38. Three minutes later the Level III airframe de-icing system activated again. At 15:56, the controller contacted flight 4184 and instructed the flight crew to descend to 8,000 feet. The engine power was reduced to the flight idle position, the propeller speed was 86 percent, and the autopilot remained engaged in the vertical speed (VS) and heading select (HDG SEL) modes. At 15:57:21, as the airplane was descending in a 15-degree right-wing-down attitude at 186 KIAS, the sound of the flap overspeed warning was recorded on the CVR. The crew selected flaps from 15 to zero degrees and the AOA and pitch attitude began to increase. At 15:57:33, as the airplane was descending through 9,130 feet, the AOA increased through 5 degrees, and the ailerons began deflecting to a right-wing-down position. About 1/2 second later, the ailerons rapidly deflected to 13:43 degrees right-wing-down, the autopilot disconnected. The airplane rolled rapidly to the right, and the pitch attitude and AOA began to decrease. Within several seconds of the initial aileron and roll excursion, the AOA decreased through 3.5 degrees, the ailerons moved to a nearly neutral position, and the airplane stopped rolling at 77 degrees right-wing-down. The airplane then began to roll to the left toward a wings-level attitude, the elevator began moving in a nose-up direction, the AOA began increasing, and the pitch attitude stopped at approximately 15 degrees nose down. At 15:57:38, as the airplane rolled back to the left through 59 degrees right-wing-down (towards wings level), the AOA increased again through 5 degrees and the ailerons again deflected rapidly to a right-wing-down position. The captain's nose-up control column force exceeded 22 pounds, and the airplane rolled rapidly to the right, at a rate in excess of 50 degrees per second. The captain's nose-up control column force decreased below 22 pounds as the airplane rolled through 120 degrees, and the first officer's nose-up control column force exceeded 22 pounds just after the airplane rolled through the inverted position (180 degrees). Nose-up elevator inputs were indicated on the FDR throughout the roll, and the AOA increased when nose-up elevator increased. At 15:57:45 the airplane rolled through the wings-level attitude (completion of first full roll). The nose-up elevator and AOA then decreased rapidly, the ailerons immediately deflected to 6 degrees left-wing-down and then stabilized at about 1 degree right-wing-down, and the airplane stopped rolling at 144 degrees right wing down. At 15:57:48, as the airplane began rolling left, back towards wings level, the airspeed increased through 260 knots, the pitch attitude decreased through 60 degrees nose down, normal acceleration fluctuated between 2.0 and 2.5 g, and the altitude decreased through 6,000 feet. At 15:57:51, as the roll attitude passed through 90 degrees, continuing towards wings level, the captain applied more than 22 pounds of nose-up control column force, the elevator position increased to about 3 degrees nose up, pitch attitude stopped decreasing at 73 degrees nose down, the airspeed increased through 300 KIAS, normal acceleration remained above 2 g, and the altitude decreased through 4,900 feet. At 15:57:53, as the captain's nose-up control column force decreased below 22 pounds, the first officer's nose-up control column force again exceeded 22 pounds and the captain made the statement "nice and easy." At 15:57:55, the normal acceleration increased to over 3.0 g. Approximately 1.7 seconds later, as the altitude decreased through 1,700 feet, the elevator position and vertical acceleration began to increase rapidly. The last recorded data on the FDR occurred at an altitude of 1,682 feet (vertical speed of approximately 500 feet per second), and indicated that the airplane was at an airspeed of 375 KIAS, a pitch attitude of 38 degrees nose down with 5 degrees of nose-up elevator, and was experiencing a vertical acceleration of 3.6 g. The airplane impacted a wet soybean field partially inverted, in a nose down, left-wing-low attitude.

Probable Cause*   ## Source: Accident Report[100]

The loss of control, attributed to a sudden and unexpected aileron hinge moment reversal, that occurred after a ridge of ice accreted beyond the deice boots while the airplane was in a holding pattern during which it intermittently encountered supercooled cloud and drizzle/rain drops, the size and water content of which exceeded those described in the icing certification envelope. The airplane was susceptible to this loss of control, and the crew was unable to recover. Contributing to the accident were 1) the French Directorate General for Civil Aviation's (DGAC's) inadequate oversight of the ATR 42 and 72, and its failure to take the necessary corrective action to ensure continued airworthiness in icing conditions; 2) the DGAC's failure to provide the FAA with timely airworthiness information developed from previous ATR incidents and accidents in icing conditions, 3) the Federal Aviation Administration's (FAA's) failure to ensure that aircraft icing certification requirements, operational requirements for flight into icing conditions, and FAA published aircraft icing information adequately accounted for the hazards that can result from flight in freezing rain, 4) the FAA's inadequate oversight of the ATR 42 and 72 to ensure continued airworthiness in icing conditions; and 5) ATR's inadequate response to the continued occurrence of ATR 42 icing/roll upsets which, in conjunction with information learned about aileron control difficulties during the certification and development of the ATR 42 and 72, should have prompted additional research, and the creation of updated airplane flight manuals, flightcrew operating manuals and training programs related to operation of the ATR 42 and 72 in such icing conditions.

The French BEA submitted a rebuttal probable cause:[101] "The Probable Cause of this accident is the loss of control of the aircraft by the flight crew, caused by the accretion of a ridge of ice aft of the de-icing boots, upstream of the ailerons, due to a prolonged operation of Flight 4184 in a freezing drizzle environment, well beyond the aircraft's certification envelope, close to V_{FE}, and utilizing a 15 degree flap holding configuration not provided for by the Aircraft Operating Manuals, which led to a sudden roll upset following an unexpected Aileron Hinge Moment Reversal when the crew retracted the flaps during the descent."

The also cited contributing factors:

The failure of the flight crew to comply with basic procedures, to exercise proper situational awareness, cockpit resource management, and sterile cockpit procedures, in a known icing environment, which prevented them from exiting these conditions prior to the ice-induced roll event, and their lack of appropriate control inputs to recover the aircraft when the event occurred;

The insufficient recognition, by Airworthiness Authorities and the aviation industry worldwide, of freezing drizzle characteristics and their potential effect on aircraft performance and controllability;

The failure of Western Airworthiness Authorities to ensure that aircraft icing certification conditions adequately account for the hazards that can result from flight in conditions outside of 14 CFR Part 25, Appendix C, and to adequately account for such hazards in their published aircraft icing information; [and]

The lack of anticipation by the Manufacturer was well as by Airworthiness and Investigative Authorities in Europe and in the USA, prior to the Post Accident Edwards AFB testing program that the ice-induced Aileron Hinge moment reversal phenomenon could occur.

The ATC's improper release, control, and monitoring of Flight 4184.

The Post Accident Edwards AFB [testing] was a program to measure the ice accretion on an ATR-72 using the US Air Force icing tanker.

Comments

There were two findings that immediately drew our attention when this report was released. First, the flight was operating in freezing drizzle. The existing standard for ice protection, Part 25, Appendix C was intended to cover flight in clouds, not in freezing precipitation.

Second, the crew was flying in a holding pattern with the flaps extended to make the cabin floor more level for the flight attendants. After this accident, we immediately wondered what pilots of the late 1940's never would be flying (1) in icing conditions in a holding pattern, and (2) extending the flaps except to land (and possibly not then with an accumulation of ice. Holding with flaps extended has the effect of moving the impingement area further back on the upper wing surface. In this aircraft with it's light aileron forces, it certainly contributed to the roll upset.

We certainly agree with BEA's comments.

The Appendix C envelope was based on data from the late 1940s.[102][103] The maximum altitude of the data is 22,000 feet, because that was the maximum capabilities of the airplanes used to sample the data. Put simply, civil airplanes of the era had to fly in the icing conditions. Today, most fly above and only encounter icing in the terminal area. In the 1940's airplanes didn't remain in icing weather very long. The horizontal extend of Continuous Maximum conditions was 17.4 NM. An airplane would be through that in a few minutes. With present-day operations, an airplane might be in a holding pattern for some time.

Regulatory Outcome

The FAA issued an updated an Advisory Circular for pilots regarding the hazards of aircraft icing and the use of airplane ice protection systems.[104]

---

* Based on petitions filed for reconsideration of the probable cause, the NTSB updated it's findings on September 2002.

January 9, 1997: Embraer EMB-120, N-265CA

Location   Monroe, MI
Flight   CCM 3272
Operator   Comair
Operation   Scheduled Airline
Route   Cincinnati, OH to Detroit, MI
On-Board Injuries   Fatal 29
Aircraft Damage   Destroyed

Description   Source: Aviation Safety Network[1]

The Embraer aircraft had departed Cincinnati at 14:53 for a flight to Detroit. Thirty-eight seconds before impact the aircraft was flying at 4000 feet with a speed of 164 knots, flaps and gear retracted and engines at flight idle. After making a turn to a 090deg. heading, the Brasilia then started a slight roll to the left. When the angle was about 20deg. left wing down, the crew tried to counteract the roll (t-33 sec). The bank angle increased through 30deg. at t-29 sec, followed by an increase in engine power when the crew tried to stop the pitch from decreasing at t-26 sec. At t-17 sec airspeed had dropped to 145 knots when a rapid roll from 40deg. to 146deg. left bank was experienced. At t-10 sec the pitch decreased from 3deg. nose-up to 50deg. nose-down. The aircraft descended out of control, experiencing some steep roll and pitch oscillations and crashed nose down into a snowy field 18 miles short of the Detroit-Metropolitan Airport runway 03. There were indications that the pilots had tried to activate the engine-fire extinguishing system, though no traces of fire on the no. 2 engine have been found by the NTSB.

Probable Cause   Source: Accident Report[105]

The National Transportation Safety Board determines that the probable cause of this accident was the FAA's failure to establish adequate aircraft certification standards for flight in icing conditions, the FAA's failure to ensure that an FAA/CTA-approved procedure for the accident airplane's deice system operation was implemented by U.S.-based air carriers, and the FAA's failure to require the establishment of adequate minimum airspeeds for icing conditions, which led to the loss of control when the airplane accumulated a thin, rough accretion of ice on its lifting surfaces.

Contributing to the accident were the flight crew's decision to operate in icing conditions near the lower margin of the operating airspeed envelope (with flaps retracted), and Comair's failure to establish and adequately disseminate unambiguous minimum airspeed values for flap configurations and for flight in icing conditions.

Comments

We can't comment on this mishap since we were involved with the certification of the Embraer EMB-120.

Regulatory Outcome

In 2007, the FAA amended §25.1419 to require defined icing detection and activation of airframe ice protection systems. All procedures must be established and documented in the Airplane Flight Manual.[22]

In July 2007, The FAA issued an updated an Advisory Circular for pilots regarding the hazards of aircraft icing the use of airplane ice protection systems.[106]

In December 2007 the FAA, in an action coordinated with the NTSB, revised Advisory Circular 91-74 to clarify ambiguities related to operations in severe icing conditions.[107] This was in response to NTSB recommendation A-96-060 from the accident to the ATR-72 in October 1994.

In 2015, The FAA amended Part 25 by adding §25.1420 which applies to airplanes with a MTOW less than 60,000 lb or which has reversible flight controls. In addition, the NPRM added a new Appendix O which defines Super-cooled Large Drop (SLD) icing conditions. §25.1420 requires that the airplane must be capable of one of the following:
  (1)   Operating safely after encountering icing conditions defined in Appendix O.
        (i)   Must have a means of detecting the conditions of Appendix O, and
        (ii)   Must be able to exit these conditions safely after detections.
  (2)   Operating safely within part of Appendix O (as selected by the applicant).
        (i)   Must have a means of detecting conditions that exceed the selected
                 portion of Appendix O, and
        (ii)   Must be able to exit these conditions safely after detections.
  (3)   Operating safely within Appendix O.

February 12, 2009: DeHavilland DHC-8, N-200WQ

Location	Clarence Center, NY
Flight	CJC 3407
Operator	Colgan Flight dba Continental Connection
Operation	Scheduled Airline
Route	Newark, NJ to Buffalo, NY
On-Board Injuries	Fatal <u>49</u>
Ground Injuries	Fatal <u>1</u>
Aircraft Damage	Destroyed

Description   Source: Aviation Safety Network[1]

A Colgan Air DHC-8-400, N200WQ, operating as Continental Connection flight 3407, crashed during an instrument approach to runway 23 at the Buffalo-Niagara International Airport (BUF), Buffalo, New York. The crash site was approximately 5 nautical miles northeast of the airport in Clarence Center, New York, and mostly confined to one residential house (6038 Long St, Clarence). The four flight crew and 45 passengers were fatally injured and the aircraft was destroyed by impact forces and post crash fire. There was one ground fatality. Night visual meteorological conditions prevailed at the time of the accident.

Flight CJC3407 departed Newark-Liberty International Airport, NJ (EWR) at 21:20 on a domestic flight to Buffalo Airport, NY (BUF). At 22:15:14 the Buffalo Approach controller cleared the flight for an ILS approach to runway 23: "Colgan thirty four zero seven three miles from KLUMP turn left heading two six zero maintain two thousand three hundred till established localizer clear ILS approach runway two three." The flight acknowledged that clearance.

At 22:16:02, the engine power levers were reduced to flight idle. At that time Buffalo Approach instructed the crew to contact Buffalo Tower. The crew extended the landing gear and the auto flight system captured the ILS 23 localizer. The captain then moved the engine conditions levers forward to the maximum RPM position as the copilot acknowledged the instructions to Buffalo Tower.

At 22:16:28 the crew moved the flaps to 10°, and two seconds later the stall warning stick shaker activated. The autopilot disconnected at about the same time that the stick shaker activated. The crew added power to approximately 75% torque. The airplane began a sharp pitch up motion, accompanied by a left roll, followed by a right roll, during which the stick pusher activated. During this time, the indicated airspeed continued to decrease to less than 100 kts. Eight seconds after the flaps had been selected to 10°, and at an airspeed of less than 110 kts, the crew retracted the flaps. Sixteen seconds later the flaps were fully retracted.

Following further pitch and roll excursions the airplane pitched down and entered a steep descent from which it did not recover. The airplane impacted a residential house and was destroyed.

Probable Cause   Source: Accident Report[108]

The National Transportation Safety Board determines that the probable cause of this accident was the captain's inappropriate response to the activation of the stick shaker, which led to an aerodynamic stall from which the airplane did not recover. Contributing to the accident were (1) the flight crew's failure to monitor airspeed in relation to the rising position of the low-speed cue, (2) the flight crew's failure to adhere to sterile cockpit procedures, (3) the captain's failure to effectively manage the flight, and (4) Colgan Air's inadequate procedures for airspeed selection and management during approaches in icing conditions.

Comments

The crew responded quickly and properly to a tailplane stall emergency. Unfortunately the tailplane stall recovery exacerbates a normal wing stall. We have extensive experience flying icing test flights and have never experienced a tailplane stall. Furthermore, we don't know any pilot who has. We asked a colleague, perhaps the most experienced icing test pilot in the country about this. He has never encountered one, but, he knows someone who has experienced a tailplane stall. Unfortunately, the FAA misinterpreted a controlled flight into terrain (CFIT) mishap in the arctic.[109]

Icing (Fuel System)

January 17, 2008: Boeing 777, G-YMMM

Location	London Heathrow Airport, UK
Flight	BAW 38*
Operator	British Airways
Operation	Scheduled Airline
Route	Beijing, China to London, UK
On-Board Injuries	Minor/None 152
Aircraft Damage	Substantial

Description   Source: Aviation Safety Network[1]

Following an uneventful flight (BA38) from Beijing, China, the aircraft was established on an ILS approach to runway 27L at London Heathrow. Initially the approach progressed normally, with the Autopilot and Autothrottle engaged, until the aircraft was at a height of approximately 600 ft and 2 miles from touch down. The aircraft then descended rapidly and struck the ground, some 1,000 ft short of the paved runway surface, just inside the airfield boundary fence. The aircraft stopped on the very beginning of the paved surface of runway 27L. During the short ground roll the right main landing gear separated from the wing and the left main landing gear was pushed up through the wing root. A significant amount of fuel leaked from the aircraft but there was no fire. An emergency evacuation via the slides was supervised by the cabin crew and all occupants left the aircraft, some receiving minor injuries.

Initial indications from the interviews and Flight Recorder analyses show the flight and approach to have progressed normally until the aircraft was established on late finals for runway 27L. At approximately 600 ft and 2 miles from touch down, the Autothrottle demanded an increase in thrust from the two engines but the engines did not respond. Following further demands for increased thrust from the Autothrottle, and subsequently the flight crew moving the throttle levers, the engines similarly failed to respond. The aircraft speed reduced and the aircraft descended onto the grass short of the paved runway surface.

Probable Cause   Source: Accident Report[110]

Whilst on approach to London (Heathrow) from Beijing, China, at 720 feet agl, the right engine of G-YMMM ceased responding to autothrottle commands for increased power and instead the power reduced to 1.03 Engine Pressure Ratio (EPR). Seven seconds later the left engine power reduced to 1.02 EPR. This reduction led to a loss of airspeed and the aircraft touching down some 330 m short of the paved surface of Runway 27L at London Heathrow. The investigation identified that the reduction in thrust was due to restricted fuel flow to both engines.

It was determined that this restriction occurred on the right engine at its Fuel Oil Heat Exchanger (FOHE). For the left engine, the investigation concluded that the restriction most likely occurred at its FOHE. However, due to limitations in available recorded data, it was not possible totally to eliminate the possibility of a restriction elsewhere in the fuel system, although the testing and data mining activity carried out for this investigation suggested that this was very unlikely. Further, the likelihood of a separate restriction mechanism occurring within seven seconds of that for the right engine was determined to be very low.

The investigation identified the following probable causal factors that led to the fuel flow restrictions:

1) Accreted ice from within the fuel system released, causing a restriction to the engine fuel flow at the face of the FOHE, on both of the engines.
2) Ice had formed within the fuel system, from water that occurred naturally in the fuel, whilst the aircraft operated with low fuel flows over a long period and the localised fuel temperatures were in an area described as the 'sticky range'.
3) The FOHE, although compliant with the applicable certification requirements, was shown to be susceptible to restriction when presented with soft ice in a high concentration, with a fuel temperature that is below -10°C and a fuel flow above flight idle.
4) certification requirements, with which the aircraft and engine fuel systems had to comply, did not take account of this phenomenon as the risk was unrecognised at that time.

Recommendations

AAIB Recommendations:

2009-030 It is recommended that the Federal Aviation Administration and the European Aviation Safety Agency conduct a study into the feasibility of expanding the use of anti-ice additives in aviation turbine fuel on civil aircraft.

2009-031 It is recommended that the Federal Aviation Administration and the European Aviation Safety Agency jointly conduct research into ice formation in aviation turbine fuels.

2009-032 It is recommended that the Federal Aviation Administration and the European Aviation Safety Agency jointly conduct research into ice accumulation and subsequent release mechanisms within aircraft and engine fuel systems.

---

* BA is the IATA code for British Airways and is used in the source material.

2009-049 It is recommended that the Federal Aviation Administration and the European Aviation Safety Agency review the current certification requirements to ensure that aircraft and engine fuel systems are tolerant to the potential build up and sudden release of ice in the fuel feed system.

Regulatory Outcome

The FAA issued two interim and one final Airworthiness Directive (AD-2010-07-01) to improve the FOHE to improve ice removal of ice buildup.[111]

Comments

At this point, the AAIB's recommendations appear to have fallen on deaf ears. While the FAA did issue one AD to address the symptoms of the problem, the fact remains, the industry and certification authorities didn't fully understand the environment when these engines were certified, and we still don't now.

November 26, 2008: Boeing 777, N-862DA

Location	Bozeman, MT
Flight	DAL 18
Operator	Delta Air Lines
Operation	Scheduled Airline
Route	Shanghai, China to Atlanta, GA
On-Board Injuries	Minor/None 263
Aircraft Damage	None

Description   Source: : NTSB Incident Report[112]

About flight level 390 (about 39,000 feet), the airplane experienced an uncommanded thrust reduction (or "rollback") of the right engine. The flight crew actions did not initiate or exacerbate the engine rollback. Although the flight crew expressed some questions about the applicable procedures, they reduced thrust as required and the right engine recovered and responded normally for the remainder of the flight. (Following the Delta incident, Boeing revised its engine response non normal procedure.) The flight did not encounter any severe weather or abnormally low temperatures aloft, and the fuel did not reach a low enough temperature to freeze or solidify. All relevant systems on the airplane were operating as designed and certified. The water content and chemical makeup of the fuel met all international test standards.

Probable Cause   Source: NTSB Incident Report[112]

The National Transportation Safety Board determines the probable cause(s) of this incident to be: An accumulation of ice in the fuel system, which formed from the water normally present in jet fuel during commonly encountered flight conditions, which accreted and released, restricting the fuel flow at the right engine fuel-oil heat exchanger inlet face.

Contributing to the incident were certification requirements (with which the aircraft and engine fuel systems were in compliance), which did not account for the possibility of ice accumulating and subsequently releasing in the aircraft and engine fuel feed system upstream of the fuel-oil heat exchanger

March 22, 2009: Pilatus PC-12/45, N-128CM

Location	Near Butte, MT
Flight	N-128CM*
Operator	Eagle Cap Leasing
Operation	Personal Flight
Route	Orovill, CA to Bozeman, MT
On-Board Injuries	Fatal 14
Aircraft Damage	Destroyed

Description   Source: Accident Report[113]

In this accident, the airplane was flying from Oroville, CA to Bozeman, MT with thirteen passengers and a pilot. The airplane was equipped with eight passenger and two pilot seats. Seven of the passengers were children under the age of 9 years.

---

*   Aircraft registration used *vice* flight number.

While enroute, the fuel flow from the left tank became restricted as a result of ice accumulation at the fuel filter. The PC-12 AFM states that Fuel System Icing Inhibitor (FSII) must be mixed into the fuel tanks for all operations conducted in ambient temperatures below 0°C. Had such an FSII been added to the fuel, the accumulation of ice would have been avoided.

At about 1:21 into the flight, fuel supplied to the engine was being drawn solely from the right tank. A left-wing heavy condition resulted which continued to increase. The pilot passed possible diversion airports, but finally decided to divert to Bert Mooney Airport, Butte, Montana.

The airplane was controllable in static flight with the left-wing heavy fuel imbalance that existed at the time of the accident, but the pilot lost control of the airplane with the dynamic maneuvers during the final moments of the flight.

At least four of the seven children were not restrained or were improperly restrained.

Except for the pilot and the adult passenger in the right cockpit seat, the NTSB was unable to determine the seating position for the airplane occupants. The airplane owner who organized the trip stated that adults could hold the children on their laps. However, only one of the seven children was under the age of 2 years and was permitted by [the regulations] to be held on the lap of an adult. The other six children ranged in age from 3 to 9 years. Because the bodies of at least four of these children were found farthest from the impact site, the NTSB concluded that at least four of the seven children were not restrained or were improperly restrained.

Probable Cause   Source: Accident Report[113]

The National Transportation Safety Board determines that the probable cause of this accident was (1) the pilot's failure to ensure that a fuel system icing inhibitor was added to the fuel before the flights on the day of the accident; (2) his failure to take appropriate remedial actions after a low fuel pressure state (resulting from icing within the fuel system) and a lateral fuel imbalance developed, including diverting to a suitable airport before the fuel imbalance became extreme; and (3) a loss of control while the pilot was maneuvering the left wing-heavy airplane near the approach end of the runway.

Recommendations

The NTSB issued several recommendations to the FAA:

A-11-70   Amend certification requirements for aircraft requiring fuel additives, including fuel system icing inhibitors, so that those limitations are highlighted by a warning in the limitations section of the airplane flight manual.

A-11-71   Require all existing certificated aircraft (both newly manufactured and in-service aircraft) that require fuel additives, including fuel system icing inhibitors, to have those limitations highlighted by a warning in the limitations section of the airplane flight manual.

A-11-72   Amend aircraft certification fuel placarding requirements so that aircraft requiring fuel additives, including fuel system icing inhibitors, have a fuel filler placard that notes this limitation and refers to the airplane flight manual for specific information about the limitation.

A-11-73   Require all existing certificated aircraft (both newly manufactured and in-service aircraft) that require fuel additives, including fuel system icing inhibitors, to have a fuel filler placard that notes this limitation and refers to the airplane flight manual for specific information about the limitation.

A-11-74   Issue guidance on fuel system icing prevention that (1) includes pilot precautions and procedures to avoid fuel system icing problems aboard turbine engine-powered aircraft and (2) describes the possible consequences of failing to use a fuel system icing inhibitor, if required by the airplane flight manual, especially during operations at high altitudes and in cold temperatures.

The NTSB made similar recommendations to EASA.
on page 370.

Regulatory Outcome

With respect to fuel icing, none to date.

Latent Failures

February 6, 1975: Mooney M-20F, N-6371Q

Location	Near Syracuse, NY
Flight	N-6371Q
Operator	Private Individual
Operation	Personal Transportation
Route	Ottawa, Canada to Dayton, OH, USA
On-Board Injuries	Minor/None 1
Aircraft Damage	None

Description   Source: Incident Report[114]

A Mooney M-20F, N6371Q, lost all attitude and directional instruments shortly after departing the Syracuse, New York Airport on the evening of February 6, 1975. The incident occurred in instrument meteorological conditions (IMC) during the hours of darkness. The airplane entered a spiral dive and regained visual conditions in time for recovery. There were no injuries or property damage.

The shear failure of the vacuum pump driver shaft. Contributing to the seriousness of the malfunction was a previous, undetected failure of the turn rate gyro's electric drive. This electrical failure was masked by the backup vacuum drive of the turn rate gyro. In addition, the vacuum failure warning light did not show.

Additional Research

The NTSB reviewed incident data and found several similar incidents. Fortunately, all occurred in visual conditions.

Recommendation

All pilots and operators of Mooney M20-F's or aircraft with similar dual power supplies for one or more gyros should be made aware of the potential hazard. If possible, a simple preflight test of the adequacy of the separate power systems should be developed.

Part 23 should be amended to preclude all gyro instruments from using a common power source if the aircraft It to be operated in IMC.

This recommendation was rejected by the FAA citing "serious economic consequences".

August 1, 2005: Boeing 777, 9M-MRG

Location	240 km NW of Perth, Australia
Flight	MAS 124
Operator	Malaysia Airlines
Operation	Scheduled Airline
Route	Perth, Australia to Kuala Lumpur, Malaysia
On-Board Injuries	None reported
Aircraft Damage	None reported

Description   Source: Aviation Occurrence Report[115]

At approximately 1703 Western Standard Time, on 1 August 2005, a Boeing Company 777-200 aircraft, (B777) registered 9M-MRG, was being operated on a scheduled international passenger service from Perth to Kuala Lumpur, Malaysia. The crew reported that, during climb out, they observed a LOW AIRSPEED advisory on the aircraft's Engine Indication and crew Alerting System (EICAS), when climbing through flight level (FL) 380. At the same time, the aircraft's slip/skid indication deflected to the full right position on the Primary Flight Display (PFD). The PFD airspeed display then indicated that the aircraft was approaching the overspeed limit and the stall speed limit simultaneously. The aircraft pitched up and climbed to approximately FL410 and the indicated airspeed decreased from 270 kts to 158 kts. The stall warning and stick shaker devices also activated. The aircraft returned to Perth where an uneventful landing was completed.

The aircraft's flight data recorder (FDR), cockpit voice recorder and the air data inertial reference unit (ADIRU) were removed for examination. The FDR data indicated that, at the time of the occurrence, unusual acceleration values were recorded in all three planes of movement. The acceleration values were provided by the aircraft's ADIRU to the aircraft's primary flight computer, autopilot and other aircraft systems during manual and automatic flight.

Subsequent examination of the ADIRU revealed that one of several accelerometers had failed at the time of the occurrence, and that another accelerometer had failed in June 2001.

Contributing safety factors – An anomaly existed in the component software hierarchy that allowed inputs from a known faulty accelerometer to be processed by the air data inertial reference unit (ADIRU) and used by the primary flight computer, autopilot and other aircraft systems.

Other safety factors – The software anomaly was not detected in the original testing and certification of the ADIRU. – The aircraft documentation did not provide the flight crew with specific information and action items to assess and respond to the aircraft upset event.

Regulatory Outcome

On 29 August 2005, the US Federal Aviation Administration (FAA) issued Emergency Airworthiness Directive (AD) AD 2005-18-51 which required all B777 operators to install operational program software (OPS) part number 3470-HNc-100-03 (version -03) in the air data inertial reference unit (ADIRU) in accordance with the accomplishment instructions of Boeing Alert Service Bulletin 777-34A0137 dated 26 August 2005. In addition, the Limitations section of the Airplane Flight Manual was to be amended by inserting Boeing operations manual bulletins cS3-3093 and cS3-3155 dated 26 August 2005.

On 9 August 2005, the aircraft manufacturer issued a Multi Operators Message, to all B777 operators that recommended that they do not despatch an aircraft with an inoperative secondary attitude air data reference unit (SAARU), which was previously permitted under the conditions of the Master Minimum Equipment List.

Comments

These two incidents point out one of the unforeseen hazards of multiple redundancies in which operators continue to operate, in this instance, for years with inoperative components.

There is additional discussion on page 154.

Lightning Strike on Composite Structure

January 19, 1995: Aerospatiale AS-332L, G-TIGK

Location   North Sea
Flight   BHL 56C
Operator   Bristow Helicopters
Operation   Offshore Transport
Route   Aberdeen Airport, Scotland to Brae Alpha Platform
On-Board Injuries   Minor/None 18
Aircraft Damage   Destroyed

Description   Source: Aviation Safety Network[1]

Struck by lightning while descending to the Brae A platform in the North Sea. Part of a tail rotor blade was lost. The resulting heavy vibrations resulted in a fatigue failure of the Tail Gear Box mounting. This resulted in loss of tail rotor drive. The TGB however remained held in place by its hydraulic pipes, avoiding a shift in CG and a catastrophic pitch down. The helicopter successfully ditched in the rough sea (~Sea State 5), all POB (2 crew and 16 passengers) evacuated and were rescued. It sank some time after the after evacuation.

Probable Cause   Source: Accident Report[116]

1. One of the carbon composite tail rotor blades suffered a lightning strike which exceeded its lightning protection provisions, causing significant damage and mass loss.
2. The dynamic response of the gearbox/pylon boom assembly to the tail rotor system imbalance induced rapid cyclic overstressing of the gearbox attachment which accelerated by the early failure of the upper mounting bolt locking wire, allowing consequent loosening and fatigue failure of this bolt.
3. complete loss of the yaw control system and a momentary pitch-down as a result of detachment of the tail rotor, gearbox, and pitch servo assembly.
4. The lightning strike protection provisions on this design of carbon composite tail rotor blade were inadequate due to it having been developed from an earlier fibreglass blade which had been certificated to lighting test criteria which may have since become obsolete.

Recommendations

Review aircraft lightning certification requirements, and the Advisory Circular AC20-53A, to introduce the following more stringent requirements for rotary wing aircraft with composite rotor blades:
(1) Increase the specified Zone 1A action integral from $2 \times 10^6 A^2$ to a level compatible with the highest energy positive polarity lightning discharges likely to be encountered in service.
(2) Replace the existing 98% probability assurance with 100% probability target.
(3) Addition of specified arc attachment points to be used in the lightning certification tests on rotor blades, to include: leading edge tip; tip weight bolt(s) if used; trailing edge tip; trailing edge up to 0.5 metres inboard of the tip.

     In addition, the CAA and appropriate committees should review lightning certification requirements with regard to any corresponding, or other, improvements which may be deemed necessary for fixed wing aircraft with significant composite material structural elements.

     comment: The cited Advisory Circular[117] applies to lightning strikes on aircraft fuel systems. It has been revised twice, but still applies to aircraft fuel system protection.

Regulatory Outcome

There have been no regulatory changes to address lightning strikes on composite rotor blades resulting from this accident. The only lightning rule changes have dealt with effect on electronic equipment.

Maintenance Procedures

May 25, 1979: McDonnell Douglas DC-10, N-110AA

Location	Chicago O'Hare Airport, IL
Flight	AAL 191
Operator	American Airlines
Operation	Scheduled Airline
Route	Chicago, IL to Los Angeles, CA
On-Board Injuries	Fatal 271
Ground Injuries	Fatal 2
Aircraft Damage	Destroyed

Description   Source: Aviation Safety Network[1]

American Airlines Flight 191, a McDonnell-Douglas DC-10-10, crashed on takeoff from Chicago-O'Hare International Airport, Illinois, USA. The aircraft was destroyed and all 271 occupants were killed. Additionally, two persons on the ground sustained fatal injuries.

At 14:59 hours local time Flight 191 taxied from the gate at O'Hare Airport. The flight was bound for Los Angeles, California, with 258 passengers and 13 crewmembers on board. Maintenance personnel who monitored the flight's engine start, push-back, and start of taxi did not observe anything out of the ordinary.

The weather at the time of departure was clear, and the reported surface wind was 020° at 22 kts. Flight 191 was cleared to taxi to runway 32R for takeoff. The company's Takeoff Data card showed that the stabilizer trim setting was 5° aircraft noseup, the takeoff flap setting was 10°, and the takeoff gross weight was 379,000 lbs. The target low-pressure compressor (N1) rpm setting was 99.4 percent, critical engine failure speed (V1) was 139 kts indicated airspeed (KIAS), rotation speed (VR) was 145 KIAS, and takeoff safety speed (V2) was 153 KIAS.

Flight 191 was cleared to taxi into position on runway 32R and hold. At 15:02:38, the flight was cleared for takeoff, and at 15:02:46 the captain acknowledged, "American one ninety-one under way." The takeoff roll was normal until just before rotation at which time sections of the No.1 (left) engine pylon structure came off the aircraft. Witnesses saw white smoke or vapor coming from the vicinity of the No. 1 engine pylon. During rotation the entire No. 1 engine and pylon separated from the aircraft, went over the top of the wing, and fell to the runway.

Flight 191 lifted off about 6,000 ft down runway 32R, climbed out in a wings-level attitude.

About nine seconds after liftoff, the airplane had accelerated to 172 knots and reached 140 feet of altitude. As the climb continued, the airplane began to decelerate at a rate of about one knot per second, and at 20 seconds after liftoff, and an altitude of 325 feet, airspeed had been reduced to 159 knots. At this point, the airplane began to roll to the left, countered by rudder and aileron inputs. The airplane continued to roll until impact, 31 seconds after liftoff, and in a 112-degree left roll, and 21-degree nose down pitch attitude.

At 15:04 Flight 191 crashed in an open field and trailer park about 4,600 ft northwest of the departure end of runway 32R. The aircraft was demolished during the impact, explosion, and ground fire.

The No.1 engine pylon failure during takeoff was determined to have been caused by unintended structural damage which occurred during engine/pylon reinstallation using a forklift. The engine/pylon removal and reinstallation were being conducted to implement two DC-10 Service Bulletins. Both required that the pylons be removed, and recommended that this be accomplished with the engines removed. The Service Bulletin instructions assumed that engines and pylons would be removed separately, and did not provide instructions to remove the engine and pylon as a unit. Additionally, removal of the engines and pylons as a unit was not an approved Maintenance Manual procedure.

The lack of precision associated with the use of the forklift, essentially an inability to perform the fine manipulations necessary to accomplish reinstallation of the engine/strut assembly, in combination with the tight clearances between the pylon flange and the wing clevis resulted in damage to the same part that had just been inspected.

Inspections of other DC-10 pylon mounts following the accident resulted in nine additional cracked mounts being identified.

Probable Cause   Source: Accident Report[118]

"The asymmetrical stall and the ensuing roll of the aircraft because of the uncommanded retraction of the left wing outboard leading edge slats and the loss of stall warning and slat disagreement indication systems resulting from maintenance-induced damage leading to the separation of the no.1 engine and pylon assembly procedures which led to failure of the pylon structure.

"Contributing to the cause of the accident were the vulnerability of the design of the pylon attach points to maintenance damage; the vulnerability of the design of the leading edge slat system to the damage which produced asymmetry; deficiencies in FAA surveillance and reporting systems which failed to detect and prevent the use of improper maintenance procedures; deficiencies in the practices and communications among the operators, the manufacturer, and the FAA which failed to determine and disseminate the particulars regarding previous maintenance damage incidents; and the intolerance of prescribed operational procedures to this unique emergency."

Regulatory Outcome[22]

Unclear guidance concerning proper techniques for changing engines or executing certain critical maintenance tasks, and a lack of regulatory oversight led to the adoption of 14 CFR 25.1529, Instructions for Continued Airworthiness, and companion requirements in 14 CFR parts 23, 27, and 29.

The actual requirements of 14 CFR 25.1529 are contained in Appendix H.

In association with the creation of 14 CFR 25.1529, 14 CFR 43.16 and 14 CFR 91.403(c) were added at amendment 20 and amendment 267, respectively, as companion requirements to the part 25 requirements.

September 11, 1991: Embraer EMB-120, *Brasilia*, N-33701

Location	Eagle Lake, TX
Flight	BTA 2574
Operator	Britt Airways operating for Continental Express
Operation	Scheduled Commuter
Route	Laredo, TX to Houston, TX
On-Board Injuries	Fatal 14
Aircraft Damage	Destroyed

Description   Source: Aviation Safety Network[1]

Embraer 120 N33701 was pulled into the continental Express hangar at Houston around 21:30 for scheduled maintenance which included removal and replacement of both left and right hand horizontal stabilizer deice boots. The second shift mechanics started working on the right hand deice boot. Although planned for the third shift, the 47 screws from the top of the left leading edge assembly for the horizontal stabilizer were already removed by the second shift. The third shift mechanics finished the replacement of the right hand deice boot but did not have time to replace the left hand boot as well.

The first flight was a 07:00 scheduled flight from Houston to Laredo.

Flight 2574 departed Laredo for the return leg to Houston at 09:09. The cruise portion of the flight was uneventful and at 10:03 the aircraft was descending through 11800 feet to 9000 feet when the air loads caused the left horizontal stabilizer leading edge to bend downward and separate. A sudden severe nose down pitchover occurred and the wings stalled negatively. A negative g of 3.5 was recorded by the FDR. Eyewitnesses reported a bright flash and saw the aircraft breaking up while descending in a flat left spin until impact.

Probable Cause   Source: Accident Report[119]

"The failure of continental Express maintenance and inspection personnel to adhere to proper maintenance and quality assurance procedures for the airplane's horizontal stabilizer deice boots that led to the sudden in-flight loss of the partially secured left horizontal stabilizer leading edge and the immediate severe nose-down pitchover and breakup of the airplane. Contributing to the cause of the accident was the failure of the continental Express management to ensure compliance with the approved maintenance procedures, and the failure of FAA surveillance to detect and verify compliance with approved procedures."

NTSB board member John K. Lauber filed a dissenting statement on the investigation report, believing the probable cause should read as follows:

"1)   The failure of continental Express management to establish a corporate culture which encouraged and enforced adherence to approved maintenance and quality assurance procedures, and

"2)   The consequent string of failures by continental Express maintenance and inspection personnel to follow approved procedures for the replacement of the horizontal stabilizer deice boots. Contributing to the accident was the inadequate surveillance by the FAA of the continental Express maintenance and quality assurance programs."

Regulatory Outcome[22]

None

Metal Fatigue

January 10, 1954: De Havilland DH106 (*Comet*), G-ALYP

Location	Mediterranean, 10 mi S of Elba
Flight	G-ALYP*
Operator	British Overseas Airways Corporation
Operation	Scheduled Airline
Route	Rome, Italy to London, UK
On-Board Injuries	Fatal <u>35</u>
Aircraft Damage	Destroyed

<u>Description</u>   <u>Source: Aviation Safety Network</u>[1]

Comet G-ALYP left Rome-Ciampino Airport at 10:31 on a flight to London. After taking off the aircraft was in touch with Ciampino control tower by radio telephone and from time to time reported its position. These reports indicated that the flight was proceeding according to the BOAC flight plan and the last of them, which was received at 10:50, said that the aircraft was over the Orbetello Beacon. The captain of another BOAC aircraft, Argonaut G-ALHJ, gave evidence of communications which passed between him and G-ALYP. The last such message received by the Argonaut began " George How Jig from George Yoke Peter did you get my " and then broke off. At that time, approximately 10:51, the aircraft was probably approaching a height of 27,000 feet. The *Comet* descended and crashed into the sea off the Island of Elba.

Initial examination and reconstruction of the wreckage of G-ALYP revealed several signs of inflight break-up. Shreds of cabin carpet were found trapped in the remains of the *Comet*'s tail section; The imprint of a coin was found on a fuselage panel from the rear of the aircraft; and Smears and scoring on the rear fuselage were tested and found to be consistent to the paint applied to the passenger seats of the *Comet*.

When most of the wreckage was recovered, investigators found that fractures started on the roof, a window then smashed into the back elevators, the back fuselage then tore away, the outer wing structure fell, then the outer wing tips and finally the cockpit broke away and fuel from the wings set the debris on fire.

<u>Probable Cause</u>   <u>Source: Report of the Court of Inquiry</u>[120]

We have formed the opinion that the accident at Elba was caused by structural failure of the pressure cabin, brought about by fatigue. We reach this opinion for the following reasons:

(a)   The low fatigue resistance of the cabin has been demonstrated by the test described in Part 3, and the test result is interpretable as meaning that there was, at the age of the Elba aeroplane a definite risk of fatigue failure occurring.

(b)   The cabin was the first part of the aeroplane to fail in the Elba accident.

(c)   The wreckage indicates that the failure in the cabin was of the same basic type as that produced in the fatigue test.

(d)   This explanation seems to us to be consistent with all the circumstantial evidence.

(e)   The only other defects found in the aeroplane were not concerned at Elba, as demonstrated by the wreckage.

April 8, 1954: De Havilland DH106 (*Comet*), G-ALYY

Location	Mediterranean, near San Lucido
Flight	G-ALYY†
Operator	South African Airways
Operation	Scheduled Airline
Route	Rome, Italy to Cairo, Egypt
On-Board Injuries	Fatal <u>21</u>
Aircraft Damage	Destroyed

<u>Description</u>   <u>Source: Aviation Safety Network</u>[1]

DH106 *Comet* G-ALYY was leased to South African Airways by BOAC. It was flown by SAA crew members between Johannesburg and London with several intermediate stops. The aircraft was grounded in January 1954 following the in-flight structural breakup of sistership G-ALYP. Special checks were carried out and a number of modifications were made affecting the airframe, the controls and the fire detection and protection at the engines. On the 15th February, 1954, the fuselage was subjected to a proving test to 11 lb/sq. in. The aircraft was returned available for service on the 24th February, 1954.

---

*   Aircraft registration used *vice* flight number in the source material. The flight number was BOA 781.

†   Aircraft registration used *vice* flight number in the source material. The flight number was SAA-201.

Passengers services with *Comet* aircraft were permitted to resume on the 23rd March, 1954.

SAA Flight 201 departed Rome at 18:32. G-ALYY climbed through three moderately thick layers of cloud. After taking off the aircraft from time to time gave its position by radio telephone to Rome Air control at Ciampino and on the last such occasion at about 18:57 reported that it was abeam Naples and climbing to 35,000 ft. This position and those given earlier indicated that the flight was proceeding according to the BOAC flight plan. At 19:05 Cairo received a signal from the aircraft reporting its departure from Rome and giving its estimated time of arrival at Cairo. Thereafter no message was received from G-ALYY.

Based on the findings of the accident to G-ALYP, it was concluded that G-ALYY must also have suffered structural failure of the pressure cabin, due to fatigue.

Probable Cause   Source: Report of the court of Inquiry[120]

Owing to the absence of wreckage, we are unable to form a definite opinion on the cause of the accident near Naples, but we draw attention to the fact that the explanation offered for the accident at Elba appears to be applicable to that at Naples.

Regulatory Outcome

The Airworthiness certificate was revoked for the *Comet*.

As part of the investigation, another *Comet* airframe (G-ALYU) was put in a huge water tank, the tank was filled, and water was pumped into the plane to simulate flight conditions. After the equivalent of only 3,000 flights investigators at the Royal Aircraft Establishment (RAE) were able to conclude that the crash had been due to failure of the pressure cabin at the forward ADF window in the roof. This 'window' was in fact one of two apertures for the aerials of an electronic navigation system in which opaque fiberglass panels took the place of the window 'glass.' The failure was a result of metal fatigue caused by the repeated pressurization and de-pressurization of the aircraft cabin.

The *Comet*'s pressure cabin had been designed to a safety factor comfortably in excess of that required by British civil Airworthiness Requirements, 2.5×P as opposed to the requirement of 1.33×P and an ultimate load of 2x P, P being the cabin 'Proof' pressure) and the accident caused a revision in the estimates of the safe loading strength requirements of airliner pressure cabins.

The *Comet* design employed a general structural design philosophy referred to as "safe-life". critical parts are assigned a safe amount of time, cycles, or flight-hours in which they will function properly. This "life" is determined from extensive designing, testing, and evaluating. The *comet*s' safe-life was targeted at 10,000 cycles. After this, significant repairs or replacements would be required in order that the aircraft could be safe for another calculated number of cycles.

Other contemporary pressurized aircraft used a "fail-safe" design philosophy of "fail-safe". In fail-safe design, a critical part has redundancies, alternate load paths, and provisions for cracks to be stopped and turned, should they occur. Often, this may result in more complex, or heavier designs. Today, both the fail-safe and safe-life philosophies have evolved into "damage tolerant" standards, which integrates attributes of each approach into one objective standard. Damage tolerance involves a structural capability that provides safe operation with specific failure or damage conditions which may exist between prescribed inspection periods. Further information on damage tolerance can be found in and in §25.571 and in AC-25-571.[121]

April 28, 1988: Boeing 737, N-73711

Location	Near Maui, HI
Flight	AAH 243
Operator	Aloha Airlines
Operation	Scheduled Airline
Route	Hilo, HI to Honolulu, HI
On-Board Injuries	Fatal 1 Serious 65 Minor/None 29
Aircraft Damage	Destroyed

Description   Source: Aviation Safety Network[1]

On April 28, 1988, an Aloha Airline Boeing 737, N73711, was scheduled for a series of inter-island flights in Hawaii. The crew flew three uneventful roundtrip flights, one each from Honolulu to Hilo (ITO), Kahului Airport, HI (OGG) on the island of Maui, and Kauai Island Airport (LIH). At 11:00, a scheduled first officer change took place for the remainder of the day. The crew flew from Honolulu to Maui and then from Maui to Hilo. At 13:25, flight 243 departed Hilo Airport en route to Honolulu. The first officer conducted the takeoff and en route climb to FL240 in VMC. As the airplane leveled at 24,000 feet, both pilots heard a loud "clap" or "whooshing" sound followed by a wind noise behind them. The first officer's head was jerked backward, and she stated that debris, including pieces of gray insulation, was floating in the cockpit. The captain observed that the cockpit entry door was missing and that "there was blue sky where the first-class ceiling had been". The captain immediately took over the controls of the airplane. He described the airplane attitude as rolling slightly left and right and that the flight controls felt "loose". Because of the decompression, both pilots and the air traffic controller in the observer seat donned their oxygen masks. The captain began an emergency descent. He stated that he extended the speed brakes and descended at an indicated airspeed (IAS) of 280 to 290 knots. Because of ambient noise, the pilots initially used hand signals to communicate. The first officer stated that she observed a rate of descent of 4,100 feet per minute at some point during the emergency descent. The captain also stated that he actuated the passenger oxygen switch. The passenger oxygen manual tee handle was not actuated. When the decompression occurred, all the passengers were seated and the seat belt sign was illuminated. The No. 1 flight attendant reportedly was standing at seat row 5. According to passenger observations, the flight attendant was immediately swept out of the cabin through a hole in the left side of the fuselage. The No. 2 flight attendant, standing by row 15/16, was thrown to the floor and sustained minor bruises. She was subsequently able to crawl up and down the aisle to render assistance and calm the passengers. The No. 3 flight attendant, standing at row 2, was struck in the head by debris and thrown to the floor. She suffered serious injuries. The first officer tuned the transponder to emergency code 7700 and attempted to notify Honolulu Air Route Traffic control center (ARTCC) that the flight was diverting to Maui. Because of the cockpit noise level, she could not hear any radio transmissions, and she was not sure if the Honolulu ARTCC heard the communication. Although Honolulu ARTCC did not receive the first officer's initial communication, the controller working flight 243 observed an emergency code 7700 transponder return about 23 nautical miles south-southeast of the Kahalui Airport, Maui. Starting at 13:48:15, the controller attempted to communicate with the flight several times without success.

When the airplane descended through 14,000 feet, the first officer switched the radio to the Maui Tower frequency. At 13:48:35, she informed the tower of the rapid decompression, declared an emergency, and stated the need for emergency equipment. The local controller instructed flight 243 to change to the Maui Sector transponder code to identify the flight and indicate to surrounding air traffic control (ATC) facilities that the flight was being handled by the Maui ATC facility. The first officer changed the transponder as requested. At 13:50:58, the local controller requested the flight to switch frequencies to approach control because the flight was outside radar coverage for the local controller. Although the request was acknowledged, Flight 243 continued to transmit on the local controller frequency. At 13:53:44, the first officer informed the local controller, "We're going to need assistance. We cannot communicate with the flight attendants. We'll need assistance for the passengers when we land". An ambulance request was not initiated as a result of this radio call. The captain stated that he began slowing the airplane as the flight approached 10,000 feet MSL. He retracted the speed brakes, removed his oxygen mask, and began a gradual turn toward Maui's runway 02. At 210 knots IAS, the flightcrew could communicate verbally. Initially flaps 1 were selected, then flaps 5. When attempting to extend beyond flaps 5, the airplane became less controllable, and the captain decided to return to flaps 5 for the landing. Because the captain found the airplane becoming less controllable below 170 knots IAS, he elected to use 170 knots IAS for the approach and landing. Using the public address (PA) system and on-board interphone, the first officer attempted to communicate with the flight attendants; however, there was no response. At the command of the captain, the first officer lowered the landing gear at the normal point in the approach pattern. The main gear indicated down and locked; however, the nose gear position indicator light did not illuminate. Manual nose gear extension was selected and still the green indicator light did not illuminate; however, the red landing gear unsafe indicator light was not illuminated. After another manual attempt, the handle was placed down to complete the manual gear extension procedure. The captain said no attempt was made to use the nose gear downlock viewer because the center jumpseat was occupied and the captain believed it was urgent to land the airplane immediately. At 13:55:05, the first officer advised the tower, "We won't have a nose gear," and at 13:56:14, the crew advised the tower, "We'll need all the equipment you've got". While advancing the power levers to maneuver for the approach, the captain sensed a yawing motion and determined that the No.1 (left) engine had failed. At 170 to 200 knots IAS, he placed the No. 1 engine start switch to the "flight" position in an attempt to start the engine; there was no response. A normal descent profile was established 4 miles out on the final approach. The captain said that the airplane was "shaking a little, rocking slightly and felt springy". Flight 243 landed on runway 02 at Maui's Kahului Airport at 13:58:45. The captain said that he was able to make a normal touchdown and landing rollout. He used the No. 2 engine thrust reverser and brakes to stop the airplane. During the latter part of the rollout, the flaps were extended to 40° as required for an evacuation. An emergency evacuation was then accomplished on the runway.

Probable Cause   Source: Accident Report[122]

The failure of the Aloha Airlines maintenance program to detect the presence of significant disbonding and fatigue damage, which ultimately led to failure of the lap joint at S-10L and the separation of the fuselage upper lobe. Contributing to the accident were the failure of Aloha Airlines management to supervise properly its maintenance force as well as the failure of the FAA to evaluate properly the Aloha Airlines maintenance program and to assess the airline's inspection and quality control deficiencies. Also contributing to the accident were the failure of the FAA to require Airworthiness Directive 87-21-08 inspection of all the lap joints proposed by Boeing Alert Service Bulletin SB 737-53A1039 and the lack of a complete terminating action (neither generated by Boeing nor required by the FAA) after the discovery of early production difficulties in the 737 cold bond lap joint, which resulted in low bond durability, corrosion and premature fatigue cracking.

Regulatory Outcome

As a result of this accident, the following items were issued:

- Mandatory corrosion control programs which require that all operators have prevention and inspection systems sufficient to ensure that hazardous corrosion never occurs.
- ADs which mandated removal and replacement of certain areas of the skin lap splice.
- Amendment 25-96 to §25.571 which requires special consideration for WFD in the structure. This amendment, in part, says, "It must be demonstrated with sufficient full scale fatigue test evidence that WFD will not occur within the design service goal of the airplane."
- Several amendments to "Repair Assessment for Pressurized Fuselages Rule": §§91.410, 121.370, 125.248, and 129.32.

October 4, 1992: Boeing 747, 4X-AXG

Location   Bijlmermeer, Amsterdam, Netherlands
Flight   ELY 1862
Operator   El Al Israel Airlines
Operation   Scheduled Airline
Route   Amsterdam, Netherlands to Tel Aviv, Israel
On-Board Injuries   Fatal   4
Ground Injuries   Fatal 39
Aircraft Damage   Destroyed

Description   Source: Aviation Safety Network[1]

El Al flight 1862 departed New York-JFK Airport for a cargo flight to Tel Aviv, Israel via Amsterdam, the Netherlands. The aircraft, a Boeing 747-258F, arrived at Amsterdam-Schiphol Airport at 14:40 hours local time for a crew change, cargo processing and refueling. The total amount of cargo was 114.7 tons, gross weight of the aircraft 338.3 tons which was 21 tons below the maximum allowable.

The aircraft taxied out to runway 01L at 18:14 and started the takeoff roll at 18:21. At 18:28:30, as the aircraft was climbing through 6500 feet, the no. 3 engine and pylon separated from the wing in an outward and rearward movement, colliding with the no. 4 engine causing this engine and pylon to separate as well. An emergency was declared and the crew acknowledged their intention to return to Schiphol Airport and reported that they had a no. 3 engine failure and a loss of engine thrust of both no. 3 and 4 engine. At 18:28:57 the Amsterdam Radar controller informed the crew that runway 06 was in use with wind from 40 degrees at 21 knots. The crew however requested runway 27 for landing. A straight in approach to runway 27 was not possible because of airplane altitude (5000 feet) and distance to the runway (7 miles). The Amsterdam Arrival controller then instructed the crew to turn right heading 360 degrees and descend to 2000 feet. During this descending turn the flight crew reported that the no. 3 and 4 engine were out and that they were having flap problems. Final clearance was given to turn right heading 270 to intercept the final approach course. When it became apparent that the aircraft was going to overshoot the localizer, the controller informed the crew accordingly and directed them to turn to heading 290 to try and intercept the final approach path again. A further instruction was given for a 310 degree heading change and descent clearance for 1500 feet. These instructions were acknowledged and the crew added that they were experiencing control problems. While reducing speed in preparation for the final approach, control was lost and the aircraft crashed into an eleven-floor apartment building the Bijlmermeer suburb of Amsterdam.

Probable Cause   Source: Accident Report[123]

The design and certification of the B747 pylon was found to be inadequate to provide the required level of safety. Furthermore the system to ensure structural integrity by inspection failed. This ultimately caused - probably initiated by fatigue in the inboard midspar fuse-pin - the no. 3 pylon and engine to separate from the wing in such a way that the no. 4 pylon and engine were torn off, part of the leading edge of the wing was damaged and the use of several systems was lost or limited. This subsequently left the flight crew with very limited control of the airplane. Because of the marginal controllability a safe landing became highly improbable, if not virtually impossible.

Resulting Safety Initiatives[22]

No specific regulatory changes were made as a result of the El Al accident.

The significant regulatory requirement governing structural design and analysis, 14 CFR 25.571, had been updated to require that new designs incorporate the damage tolerance philosophy, rather than applying failsafe principles. This change came about as Amendment 45, issued in 1978.

The redesigned 747 strut, which was mandated for retrofit on the 747 fleet, was designed using the latest damage tolerance principles and technology.

The actual requirements of 14 CFR 25.1529 are contained in Appendix H.

In association with the creation of 14 CFR 25.1529, 14 CFR 43.16 and 14 CFR 91.403(c) were added at amendment 20 and amendment 267, respectively, as companion requirements to the part 25 requirements.

Mid-Air Collisions

June 30, 1956: Lockheed L-1049, N-6902C and Douglas DC-7, N-6324C

Location	Grand Canyon, AZ	
Aircraft	L-1049A	DC-7
Flight	TWA 2	UAL 718
Operator	Trans World Airlines	United Airlines
Operation	Scheduled Airline	Scheduled Airline
Route	Los Angeles, CA to Kansas City	Los Angeles, CA to Chicago, IL
On-Board Injuries	Fatal: <u>70</u>	Fatal: <u>58</u>
Aircraft Damage	Destroyed	Destroyed

Description   Source: Aviation Safety Network[1]

TWA 2 took off from Los Angeles at 09:01 PST for an IFR flight to Kansas City. After reporting "on top" of the clouds at 2400 feet the crew contacted Los Angeles ARTCC. clearance was given to climb to 19000 feet. Immediately thereafter TWA 2 asked for a routing change to Daggett via Victor Airway 210, This was approved. At 09:21 Flight 2 reported that it was approaching Daggett and requested a change in flight plan altitude assignment to 21,000 ft. The request was not approved because of traffic at 21,000 ft (United Flight 718). A request for 1000 on top was granted. At 09:59 Flight 2 reported its position through company radio at Las Vegas. It reported that it had passed Lake Mohave at 09:55, was 1,000 on top at 21,000 feet, and estimated it would reach the Painted Desert fix at 10:31. This was the last radio communication with the flight.

United 718 departed from Los Angeles at 09:04 for an IFR flight to Chicago cruising at FL210. At 09:58 United 718 made a position report to a CAA communications station that they were over Needles, CA at 09:58, at FL210 feet, and estimated the Painted Desert fix at 10:31.

Both aircraft were at the same altitude on an intersecting course over Grand canyon. United 718 was heading 046 degrees and TWA 2 was heading 59 degrees. The aircraft collided in mid-air. First contact involved the center fin leading edge of the constellation and the left aileron tip of the DC-7. The lower surface of the DC-7 left wing struck the upper aft fuselage of the L-1049 with disintegrating force. The collision ripped open the fuselage of the constellation from just forward of its tail to near the main cabin door. The empennage of the L-1049 separated almost immediately. The plane pitched down and fell to the ground. Most of the left outer wing of the DC-7 had separated and aileron control was restricted, causing the plane fell to the ground out of control.

Probable Cause   Source: Accident Report[124]

The pilots did not see each other in time to avoid the collision. It is not possible to determine why the pilots did not see ach other, but the evidence suggests that it resulted from any one or a combination of the following factors: 1) Intervening clouds reducing time for visual separation; 2) Visual limitations due to cockpit visibility, and; 3) Preoccupation with normal cockpit duties; 4) Preoccupation with matters unrelated to cockpit duties such as attempting to provide the passengers with a more scenic view of the Grand canyon area; 5) Physiological limits to human vision reducing the time opportunity to see and avoid the other aircraft, or; 6) Insufficiency of en-route air traffic advisory information due to inadequacy of facilities and lack of personnel in air traffic control.

Regulatory Outcome

This accident ultimately led to a wholesale reorganization of the US aviation oversight. The main effect was the replacement of the CAA by the new Federal Aviation Agency.

September 29, 2006: Boeing 737, PR-GTD & Embraer EMB-135, N-600XL

Location	Near Cachimbo Air Base, Brazil	
Flight	XLS N-600XL	GLO 1907
Operator	ExcellAire Services	Gol Transportes Aeros
Operation	Non Revenue (Positioning)	Scheduled Airline
Route	Sao Jose dos Campos to Manaus	Manaus to Rio de Janeiro
On-Board Injuries	Minor/None 7	Fatal 154
Aircraft Damage	Substantial	Destroyed

Description   Source: Aviation Safety Network[1]

On September 29, an Embraer Legacy 600 executive jet (N600XL) was scheduled to be delivered from the Embraer factory at São José dos Campos Airport (SJK) to the United States. An intermediate stop was planned at Manaus (MAO). N600XL took off at 14:51 with a routing via the OREN departure procedure to Pocos beacon, then airway UW2 to Brasilia VOR (BRS), airway UZ6 to Manaus. The cruise altitude was filed as FL370, with a planned change to FL360 at BRS, and to FL380 at the TERES navigational fix, approximately 282 miles north of BRS.

At 15:35, GOL Flight 1907, a Boeing 737-800, departed Manaus (MAO) on a scheduled flight to Brasilia (BSB) and Rio de Janeiro (GIG). The flight was also routed via UZ6 to BRS. cruising altitude was FL370, which was reached at 15:58. At that time, Legacy N600XL had just passed BRS, level at FL370. When N600XL was about 30 miles north-northwest of BRS, at 16:02, the transponder of N600XL was no longer received by ATC radar. At 16:26 the CINDACTA 1 controller made a "blind call" to N600XL. Subsequently until 16:53, the controller made an additional 6 radio calls attempting to establish contact. The 16:53 call instructed the crew to change to frequencies 123.32 or 126.45. No replies were received. Beginning at 16:48, the crew of N600XL made a series of 19 radio calls to ATC attempting to make contact.

Both the GOL Boeing 737 and the Legacy were now on a head on collision course on airway UZ6 at the same altitude. Because the transponder of N600XL was not functioning properly -for as of yet undetermined reasons- the TCAS equipment on both planes did not alert the crews. At FL370, over the remote Amazon jungle, both aircraft collided. The left winglet of the Legacy (which includes a metal spar) contacted the left wing leading edge of the Boeing 737. The impact resulted in damage to a major portion of the left wing structure and lower skin, ultimately rendering the 737 uncontrollable. The Boeing 737 was destroyed by in-flight breakup and impact forces.

The Embraer's winglet was sheared off and damage was sustained to the vertical stabilizer tip. The crew made numerous further calls to ATC declaring an emergency and their intent to make a landing at the Cachimbo air base. At 17:02, the transponder returns from N600XL were received by ATC. At 17:13, an uninvolved flight crew assisted in relaying communications between N600XL and ATC until the airplane established communication with Cachimbo tower.

A safe emergency landing was made [by the Embraer].

Probable Cause   Source: Accident Report[125]

Relatively to the crew of the N600XL, the following active failures were identified: lack of an adequate planning of the flight, and insufficient knowledge of the flight plan prepared by the Embraer operator; non-execution of a briefing prior to departure; unintentional change of the transponder setting, failure in prioritizing attention; failure in perceiving that the transponder was not transmitting; delay in recognizing the problem of communication with the air traffic control unit; and non-compliance with the procedures prescribed for communications failure.

Regulatory Outcome

During the investigation, the reason for the inoperative transponder was identified as the location of the transponder power switch just beneath the pilot's footrest. The switch was a push-on/push off switch. A simulator evaluation confirmed this. An Airworthiness Directive was proposed, but never implemented.

The two Embraer pilots were charged with 154 counts of manslaughter. Since they had already returned to the USA, they are unlikely to be tried.

Minimum Equipment List Events

January 18, 1969: Boeing 727, N-7434U

Location   Santa Monica Bay, CA
Flight   UAL 266
Operator   United Airlines
Operation   Scheduled Airline
Route   Los Angeles, CA to Denver, CO
On-Board Injuries   Fatal 38
Aircraft Damage   Destroyed

Description   Source: Aviation Safety Network[1]

United Airlines flight 266, a Boeing 727-22c impacted Santa Monica Bay approximately 11.3 miles west of the Los Angeles International Airport, California, USA. The aircraft was destroyed and the six crewmembers and 32 passengers on board were all fatally injured.

The aircraft, N7434U, performed a scheduled service from Los Angeles to Denver, Colorado and Milwaukee, Wisconsin. It had been operating since January 15, 1969, with the no. 3 generator inoperative. This was allowed because according to the Minimum Equipment List, the aircraft is airworthy with only two generators operable provided certain procedures are followed and electrical loads are monitored during flight.

Flight 266 was scheduled to depart the gate at 17:55, but was delayed until 18:07 because of the inclement weather and loading problems. The flight commenced its takeoff roll on runway 24 at approximately 18:17. At 18:18:30 the sound of an engine fire warning bell was heard in the cockpit. The crew reported a no. 1 engine fire warning and stated that they wanted to return to the airport. Shortly after shutdown of the no. 1 engine, electrical power from the remaining generator (no. 2) was lost. Following loss of all generator power, the standby electrical system either was not activated or failed to function. Electrical power at a voltage level of approximately 50 volts was restored approximately a minute and a half after loss of the no. 2 generator. The duration of this power restoration was just 9 to 15 seconds. The Boeing descended until it struck the sea. The ocean depth at this point was approximately 950 feet.

Probable Cause   Source: Aviation Safety Network[1]

The loss of attitude orientation during a night, instrument departure in which all attitude instruments were disabled by loss of electrical power. The Board has been unable to determine (a) why all generator power was lost or (b) why the standby electrical power system either was not activated or failed to function.

Regulatory Outcome

Amendment 25-23, May 8, 1970 (35 FR 5665, April 8, 1970), to §25.1333, "Instrument systems," of the regulations. Amendment 25-23, was adopted, and required that flight instruments needed for safe flight and landing of the airplane be available to the flightcrew after any failure or combination of failures not shown to be extremely improbable, without the need for crewmember action.[126]

In addition, the operating rules (FAR Part 121).were amended to require a third attitude indicator whose operation is independent of the electrical generating system.[127]

comment

A non-regulatory change has been the design of auxiliary power units (APUs) which could be operated in flight to provide redundant electrical power and bleed air. The B-727's APU can only be operated on the ground.

June 25, 2005: Airbus A320, I-BIKE

Location	Heathrow Airport, London, UK
Flight	AZA I-BIKE*
Operator	Alitalia
Operation	Scheduled Airline
Route	Milan, Italy to London, UK
On-Board Injuries	Minor/None <u>104</u>
Aircraft Damage	None

Description   Source: Mishap Report[128]

The flight departed from Milan Airport at 0547 hrs on a scheduled flight to London Heathrow Airport (LHR) with the commander as the Pilot Flying (PF). The previous day, the No 3 ADIRU was found to be unserviceable and had been turned off; the Minimum Equipment List (MEL) allowed the aircraft to depart in this condition, as both the Nos and 2 ADIRUs were serviceable. During the flight, as a precautionary measure, the commander and co-pilot reviewed the Flight Manual Abnormal Procedures for the actions to be taken in the event of a second ADIRU becoming unserviceable.

Following an uneventful transit, the aircraft was given radar vectors and became fully established on the ILS approach to Runway 09L at LHR. Two stages of flap were selected and, at, 820 ft (QNH), the landing gear was lowered. Some 16 seconds later, just as the landing gear locked down, the Inertial Reference (IR) part of the No 1 ADIRU failed and a 'NAV IR 1 FAULT' message appeared on the aircraft's Electronic Centralised Aircraft Monitor (ECAM). The autopilot and autothrottle both disconnected and much of the flight instrument information on the commander's Primary Flight Display (PFD) and Navigation Display (ND) was lost, with only the ILS localiser and glideslope, airspeed and altitude indications remaining on his PFD. In addition, the aircraft's flight control laws changed from NORMAL to DIRECT law and both flight directors and the No 1 yaw damper became unavailable. Some 14 seconds after the landing gear locked down, the Enhanced Proximity Warning System (EGPWS) indicated that the terrain warning function was no longer available. The commander handed over control of the aircraft to the co-pilot, whose PFD and ND were functioning normally, and the ILS approach was continued.

As the crew continued their approach, ATC advised that they would receive a late clearance to land. When the aircraft was at about 250 ft radio altitude an EGPWS "too low flap" warning was recorded on the CVR. The commander then decided to go-around in order to attempt to restore the NAV IR fault condition but, before he could do so, ATC instructed the aircraft to go-around as the preceding aircraft had not yet cleared the runway. The commander acknowledged this instruction and called

"GOING AROUND, REQUEST A HOLDING PATTERN OVERHEAD CHILTERN OR OCKAM TO RESOLVE A LITTLE FAILURE" but ATC were not advised of the specific nature of the failure. The thrust levers were set to the take-off/go-around (TOGA) detent and, having descended to a minimum radio altitude of 159 ft, the aircraft then started to climb, The landing gear lever was selected up, and the landing gear retracted normally. At this point, the EGPWS warning ceased. The controller became concerned that the aircraft was drifting south of the runway extended centerline and advised the crew of the missed approach procedure, but did not acknowledge the commander's request to enter a hold. He then transferred the aircraft to the Intermediate Approach controller. Following the frequency change, the commander again requested radar vectors and said "WE REQUIRE A FEW MINUTES TO RESOLVE A LITTLE...NAVIGATION FAILURE...". The controller asked for the message to be repeated, possibly due to the commander's heavily accented English, and subsequently acknowledged the request.

At about 0731 hrs, ATC requested if the aircraft had a problem. The commander reported that the aircraft had had "a double inertial reference failure" but the controller replied that the implications of this were not understood. The commander of I-BIKE then stated that they were able to perform a CAT 1 ILS approach only. At about 0734 hrs, he transmitted a PAN call requesting assistance for a radar vectored approach to Runway 09L, explaining the aircraft had suffered a navigation problem. ATC did not respond initially, due to a double transmission, but another aircraft brought it to their attention. Following this, the requested vectors were provided to position the aircraft at the agreed distance of 23 NM (track miles) to touchdown.

In attempting to address the problem with the No 1 ADIRU, the flight crew turned the No 1 ADIRU rotary switch to the OFF position. The ECAM actions did not call for this action in the event of the IR part of an ADIRU failing, but the crew recalled from their review of abnormal procedures in the Flight Manual during the transit from Milan, that there were circumstances when this was required. The commander attempted to find the relevant text in the Flight Manual but was unable to do so before ATC instructed the aircraft to turn onto base leg.

The crew's decision to deviate from the ECAM procedure, by switching off the No ADIRU (with the No 3 ADIRU unavailable) caused the loss of further information from the commander's instrument displays. The landing gear normal extension system was also rendered inoperative, but it was successfully lowered using the emergency gravity (free fall) extension system.

Another consequence of this was that the nosewheel steering system became inoperative. Accordingly, the commander advised ATC that he was not sure if the aircraft would be able to clear the runway after landing. As the aircraft was radar vectored onto an intercept heading for the localiser, the commander upgraded his PAN to a MAYDAY, transmitting "ON FINAL, MAYDAY FROM THIS MOMENT, WE CANNOTCPERFORM A GO-AROUND, AH FINALS 09L", in order to ensure priority. ATC switched traffic ahead of I-BIKE onto Runway 09R to proved a clear approach and, due to his reduced airspeed, also radar vectored a following aircraft to the north. At 0739 hrs, the crew advised ATC that the

---

*   Aircraft registration used *vice* flight number.

aircraft was fully established. control of the aircraft was transferred to the tower controller who advised that there was traffic on the runway to vacate. The crew responded by advising that "WE HAVE AN EMERGENCY", which the controller acknowledged. Landing clearance was given for Runway 09L a short time later.

Although the tower controller was aware that I-BIKE had a navigation problem and that it may not be able to clear the runway after landing, he was not made aware that the commander had declared a MAYDAY and so did not bring the airport Rescue and Fire Fighting Service (RFFS) to a Local Standby state.

The aircraft touched down at 0742:05 hrs at an airspeed of about 134 kt and began to decelerate. Some 50 seconds later, when the ground speed was about 50 kt, the aircraft made a right turn, using rudder and asymmetrical braking, onto the adjacent taxiway. The aircraft came to a stop and the parking brake was applied; the crew then requested a tug to tow the aircraft to the stand.

Aircraft History

On 24 June 2005, the aircraft suffered an IR fault in the No 3 ADIRU. An attempt was made to reset the unit, but this proved unsuccessful. The aircraft was released for service with this ADIRU selected OFF, in accordance with procedure 0-34-3-0-0 a), (562) of the operator's MEL and, accordingly, an Acceptable Deferred Defect was raised in the Aircraft Technical Log. This is a category C item under the JAA MMEL/MEL.040 definition, and such items must be rectified within ten calendar days, excluding the day of discovery of the defect. The Technical Log entry reflected that the defect must be rectified by 4 July 2005.

The ADIRS is controlled via the ADIRS control panel on the overhead panel, Figure 2. In normal operation, the rotary selector mode switches are set to the NAV position. In this configuration, the No 1 and 2 ADIRUs supply data to the commander's and co-pilot's EFIS displays respectively, with No 3 ADIRU available as a standby. Following loss of the ADR and/or IR function of ether the No or 2 ADIRU, rotary selector switches on the SWITCHING panel on the centre pedestal, Figure 3, enable air data and/or inertial data from the No 3 ADIRU to be selected to replace the data from the failed unit.

An IR fault in ADIRU No or 2 will cause a loss of attitude and navigation information on their associated PFD and ND screens. An ADR fault will cause the loss of airspeed and altitude information on the affected display. In either case the information is restored by selecting the No 3 ADIRU.

Conclusions[128]

During this investigation, it was apparent that the operator's training organisation train their flight crews to a high standard and that nothing in the tang of the I-BIKE crew should have led them to deviate from the checklist displayed on the ECAM. The operator's training organisation took the view that the commander had correctly elected to carry out a go-around and deal with the failure of the navigation equipment in a holding pattern. However, the reducing cloudbase, combined with being limited to a CAT 1 ILS approach, then became the main consideration of the crew to land the aircraft without unnecessary delay. The incorrect action by the crew of selecting the No ADIRU to OFF, rather than following the ECAM checklist, was carried out from memory at a time of relatively high workload, and led to further loss of aircraft systems.

By not adopting the usual protocol for declaring a MAYDAY, the commander may have contributed to ATC not being fully aware that the crew had declared an emergency situation. His heavy accent may also been a factor. This resulted in the airport RFFS not being brought to a Local Standby state of readiness for the landing.

Comments

Unfortunately, a detailed analysis of the failure was not reported in this incident report. The failure of two ADIRU units installed in the number 1 position is consistent with a defect in the wiring or mounting hardware in that position. If this was the case, moving the former ADIRU-1 to the number three position would not have corrected the failure which might result in loss of the new ADIRU-1. Additional discussion can be found on page 139.

In any event, the language difficulties during the approach into Heathrow points out the difficulties in international aviation.

The crew deserves an "attaboy".

Missing Aircraft

October 16, 1972: Cessna 310C, N-1812H

Location	Unknown Location
Flight	N-1812H
Operator	Pan Alaska Airways, Ltd.
Operation	Non-scheduled Air Taxi
Route	Anchorage, AK to Juneau, AK
Presumed Injuries	Fatal 4
Aircraft Damage	Missing (Presumed Destroyed)

Description   Source: Aviation Safety Network[1]

A Cessna 310C, N1812H, operated by the chief pilot of Pan Alaska Airways, departed from Anchorage International Airport, Alaska, at 08:59. Three passengers, including two United States Congressmen, were aboard.

At 0909, the pilot of N1812H filed a Visual Flight Rules flight plan with the Anchorage Flight Service Station.

He stated that he had departed Anchorage at 0900 and that his intended route of flight was via the V-317 airway to Yakutat, thence direct to his destination, Juneau International Airport, Alaska. He estimated his flying time en route at 3 hours and 30 minutes.

At 1315, the U.S. Air Force Rescue Coordination Center at Elmendorf Air Force Base, Alaska, received notification that N1812H was overdue at Juneau. An intensive search of areas along the proposed route and all probable alternate routes was conducted during the following 5-1/2 weeks. Nothing was found that could be identified with either the aircraft or its occupants.

The weather conditions along the proposed route from Anchorage to Juneau were not conducive to flight under Visual Flight Rules. NI812H, a Cessna 310C operated by the chief pilot of Pan Alaska Airways, Ltd., disappeared on a flight from Anchorage to Juneau, .Alaska, on October16,1972. In addition to the pilot, three passengers, including two U. S. Congressman, were aboard the aircraft.

Probable Cause   Source: Accident Report[129]

Conclusions:
1. The pilot of N1812H was certificated and qualified in accordance with the applicable regulations
2. The aircraft was certificat3d in accordance with the applicable regulations; it was not equipped with an emergency locator transmitter,
3   The pilot of N1812H was not in possession of a portable emergency locator transmitter nor survival equipment when he departed Anchorage.
4   The weather conditions along the proposed route of flight from Anchorage to Juneau were not conducive to flight under Visual Flight Rules.

The Safety Board is unable to determine the probable cause of this accident from the evidence presently avialable. If the aircraft is found, the Safety Board will continue the investigation and make a determination as to the probable cause of the accident.

Recommendation   Source: Accident Report[129]

Recommendation A-73-1: The National Transportation Safety Board recomments that the Federal Aviation Adinistration, through its accident prevention staff, make wide dissemination of the details of this accident to the general aviaiton community, particularly to those pilots and operators involved in operations in remote and environmentally hostile areas.

Regulatory Outcome[130]

ELTs [were] are required to be installed in almost all U.S.-registered civil aircraft, including general aviation aircraft, as a result of a congressional mandate. The mandate resulted from the 1972 loss of U.S. Representative Hale Boggs and Nick Begich in Alaska after their aircraft crashed and was never found.

Comments[131]

The pilot had written a controversial article on flight in icing conditions which was published in Flying the same month of the accident. In this article, he advises readers not to believe "99 per cent of the bullshit about icing".[132]

March 8, 2014: Boeing 777, 9M-MRO

Location	Unknown Location
Flight	MAS-370*
Operator	Malaysia Airlines
Operation	Scheduled Airline
Route	Kuala Lumpur, Malaysia to Beijing, China
Presumed Injuries	Fatal <u>239</u>
Aircraft Damage	Missing (Presumed Destroyed)

Description   Source: Aviation Safety Network[1]

Malaysia Airlines Flight MH370 from Kuala Lumpur, Malaysia to Beijing, China was reported missing on March 8, 2014. There were 227 passengers and 12 crew members on board.

The Boeing 777-2H6ER took off from Kuala Lumpur International Airport's runway 32R at 00:41. At 00:42 the flight was cleared to climb to FL180 and was issued a direct track by Lumpur Approach to the IGARI waypoint. MH370 was then transferred to Lumpur Radar and was cleared to climb to FL250. At 00:50 the flight was further cleared to the planned cruising altitude of FL350. MH370 reported maintaining FL350 at 01:07. Last radio contact was at 01:19 when the Kuala Lumpur Radar controller instructed the flight to contact the radio frequency of Ho chi Minh Air Traffic control centre, Vietnam: "Malaysian Three Seven Zero contact Ho chi Minh 120 decimal 9 Good Night". One of the flight crew members replied: "Good night, Malaysian Three Seven Zero."

At 01:21 MH370 was observed on the radar screen of the Kuala Lumpur Radar controller as it passed over waypoint IGARI. Nine seconds later the radar label for MH370 disappeared from the radar screen. The transponder was switched off.

At 01:38 Ho chi Minh ATCC contacted Kuala Lumpur ATCC on the whereabouts of MH370. Kuala Lumpur ATCC contacted the airline's operations centre, Singapore ACC, Hong Kong ACC, and Phnom Penh ACC, failing to establish the location of MH370.

Meanwhile, the airplane flew in a westerly direction back over peninsular Malaysia before turning northwest. Primary radar data showed that the aircraft tracked along the Malacca Strait. During this time the aircraft passed close to waypoints VAMPI, MEKAR, NILAM and possibly IGOGU along a section of airway N571. The final primary radar fix occurred at 02:22.

From then on seven handshakes between the aircraft's SATCOM system and the Inmarsat ground station were recorded. Last satellite data was recorded at 08:11 Malaysian time.

Initially search efforts focused on the South China Sea area. On 24 March 2014 further analysis of the Inmarsat satellite data indicated that MH370 flew south and ended its flight in the southern part of the Indian Ocean.

A surface search was conducted of probable impact areas along an arc, identified by calculations based on Inmarsat data. The search was carried out from 18 March - 28 April 2014. This search effort was undertaken by an international fleet of aircraft and ships with the search areas over this time progressing generally from an initial southwest location along the arc in a north-easterly direction. No debris associated with MH370 was identified either from the surface search, acoustic search or from the ocean floor search in the vicinity of acoustic detections, which were initially believed to have been from the pingers on the flight recorders. The ocean floor search was completed on 28 May 2014.

On June 26 the ATSB published a new search area based on refinements to the analysis of both the flight and satellite data. The priority area of approximately 60,000 km2 extends along the arc for 650 km in a northeast direction from Broken Ridge, an underwater ridge. The width of the priority search area is 93 km.

On July 29, 2015 a flapperon washed ashore on the French island of Réunion in the Indian Ocean. On August 5 it was established to have been from MH370.

Probable Cause   Source: Accident Report[133]

The Team is unable to determine the real cause for the disappearance of MH370.

Safety Recommendations

It is recommended that the International civil Aviation Organization examine the safety benefits of introducing a standard for real-time tracking of commercial air transport aircraft.[134]

To review the effectiveness of current ELTs fitted to passenger aircraft and consider ways to more effectively determine the location of an aircraft that enters water.(SR #19)

Regulatory Outcome

Based on the above recommendation,[134] the ICAO council has adopted Amendments 40 and 42 on 02 March 2016 and 27 February 2017, respectively, to the International Standards and Recommended Practices, Operation of Aircraft - International commercial Air Transport - Aeroplanes (Annex 6, Part I to the convention on International civil Aviation). Excerpts of the amendments are listed below.

---

* MH is the IATA code for Malaysia Airlines and is used in the source material.

Pilot Attempt to Override Autopilot

April 26, 1994: Airbus A300B4, B-1816

Location	Nagoya Airport, Japan
Flight	CAL 140
Operator	China Airlines
Operation	Scheduled Airline
Route	Taipei, Taiwan to Nagoya, Japan
On-Board Injuries	Fatal <u>264</u> Serious <u>7</u>
Aircraft Damage	Destroyed

Description   Source: Aviation Safety Network[1]

China Airlines Flight 140 departed from Taipei International Airport, Taiwan, bound for Nagoya Airport, Japan. After initial descent and contact with Nagoya Approach control, the flight was cleared for the Instrument Landing System (ILS) approach to runway 34 (ILS 34 approach) and was switched to the Nagoya tower frequency at approximately 2007 local time. It was nighttime and Nagoya airport weather at the time was reported as winds from 280 degrees at 8 knots, visibility of 20 kilometers, cumulus clouds at 3,000 feet and a temperature of 20°C. During the initial phase of the approach, both autopilot systems (AP1 and AP2) were engaged as well as the auto throttles. After passing the ILS outer marker and receiving landing clearance, the first officer, who was the pilot flying, disengaged the autopilot system and continued the ILS approach manually.

When passing through approximately 1,000 feet on the approach glidepath, the first officer inadvertently triggered the GO levers placing the auto throttles into go-around mode, which led to an increase in thrust. This increase in thrust caused the aircraft to level off at approximately 1,040 feet for 15 seconds and resulted in the flight path becoming high relative to the ILS glideslope. The captain recognized that the GO lever had been triggered and instructed the first officer to disengage it and correct the flight path down to the desired glide slope. While manually trying to correct the glide path with forward yoke, the first officer engaged the autopilot, causing it to be engaged in the go-around mode as well. As he manually attempted to recapture the glide slope from above by reducing thrust and pushing the yoke forward, he was providing pitch inputs to the elevator that were opposite the autopilot commands to the Trimmable Horizontal Stabilizer (THS), which was attempting to command pitch up for a go around.

The THS progressively moved from -5.3 degrees to its maximum nose-up limit of approximately -12.3 degrees as the aircraft passed through approximately 880 feet. During this period the first officer continued to apply increasing manual nose-down command through forward yoke control which resulted in increasing nose-down elevator movement, opposite the THS movement, masking the out-of-trim condition. The first officer attempted to use the pitch trim control switch to reduce the control force required on the yoke. However, because pitch trim control of the THS is inhibited during autopilot operation, it had no effect. In a normal, trimmed condition the THS and elevator should remain closely aligned. However, because of the opposing autopilot (nose up) commanded THS and manually commanded elevator (nose down) for approximately 30 seconds, the THS and elevator became "mistrimmed".

Passing through approximately 700 feet, the autopilot was disengaged but the THS remained at its last commanded position of -12.3 degrees. Also at this time, due to the thrust reduction commanded by the first officer, the airspeed decreased to a low level, resulting in an increasing angle-of-attack (also termed alpha, or AOA). As a result, the automatic alpha floor function of the aircraft was activated, causing an increase in thrust and a further pitch-up. The alpha floor function of the A300 is an AOA protection feature intended to prevent excessive angles of attack during normal operations. Because of the greater size of the THS relative to the elevator (approximately three times greater in terms of surface area), the available elevator control power or authority was overcome as the aircraft neared 570 feet on the approach. Upon hearing the first officer report that he could not push the nose further down and that the throttles had latched (alpha floor function engaged), the captain took over the controls unaware of the THS position.

Upon assuming control, the captain initially attempted to continue the approach but was surprised by the strong resistive force to his full nose-down control inputs. He retarded the throttles in an attempt to recapture glide slope. Unable to control the increasing nose-up pitch, which had reached 22 degrees, he called for the GO-lever shortly thereafter in attempt to execute a go around. The increasing thrust added additional nose-up pitch moment and resulted in and uncontrolled steep climb as airspeed continued to decrease and AOA continued to rise. During the attempted go-around, the captain only operated the pitch trim briefly, indicating he was unaware of the mistrimmed position (extreme nose-up) of the THS. Furthermore, flaps/slats had been retracted two positions (30/40 to 15/15) to the go-around setting, which increased the airplane pitch up and reduced the stall margin. The aircraft continued to climb steeply up to 1,730 feet with AOA rapidly increasing and airspeed decreasing, reaching a maximum pitch angle of approximately 53 degrees until the stall warning and subsequent stall.

Once stalled the aircraft nose lowered to a steep dive and the captain applied full aft yoke in an attempt to recover from the dive; however, the aircraft remained stalled until impacting the ground tail-first, 300 feet to the right of the runway and burst into flames.

Probable Cause

While the aircraft was making an ILS approach to runway 34 of Nagoya Airport, under manual control by the first officer, the first officer inadvertently activated the GO lever, which changed the FD (Flight Director) to GO AROUND mode and caused a thrust increase. This made the aircraft deviate above its normal glide path. The APs were subsequently engaged, with GO AROUND mode still engaged. Under these conditions the first officer continued pushing the control wheel in accordance with the captain's instructions. As a result of this, the THS (Horizontal Stabilizer) moved to its full nose-up position and caused an abnormal out-of-trim situation. The crew continued approach, unaware of the abnormal situation. The AOA increased the Alpha Floor function was activated and the pitch angle increased.

It is considered that, at this time, the captain (who had now taken the controls), judged that landing would be difficult and opted for go-around. The aircraft began to climb steeply with a high pitch angle attitude. The captain and the first officer did not carry out an effective recovery operation, and the aircraft stalled and crashed.

The AAIC determined that the following factors, as a chain or a combination thereof, caused the accident:

- The first officer inadvertently triggered the Go lever. It is considered that the design of the GO lever contributed to it: normal operation of the thrust lever allows the possibility of an inadvertent triggering of the GO lever.
- The crew engaged the APs while GO AROUND mode was still engaged, and continued approach.
- The first officer continued pushing the control wheel in accordance with the captain's instructions, despite its strong resistive force, in order to continue the approach.
- The movement of the THS conflicted with that of the elevators, causing an abnormal out-of-trim situation.
- There was no warning and recognition function to alert the crew directly and actively to the onset of the abnormal out-of-trim condition.
- The captain and first officer did not sufficiently understand the FD mode change and the A/P override function.
- It is considered that unclear descriptions of the AFS (Automatic Flight System) in the FCOM (Flight crew Operating Manual) prepared by the aircraft manufacturer contributed to this.
- The captain's judgment of the flight situation while continuing approach was inadequate, control take-over was delayed, and appropriate actions were not taken.
- The Alpha-Floor function was activated; this was incompatible with the abnormal out-of-trim situation, and generated a large pitch-up moment. This narrowed the range of selection for recovery operations and reduced the time allowance for such operations.
- The captain's and first officer's awareness of the flight conditions, after the Plc took over the controls and during their recovery operation, was inadequate respectively.
- crew coordination between the captain and the first officer was inadequate.
- The modification prescribed in Service Bulletin SB A300-22-602 1 had not been incorporated into the aircraft.
- The aircraft manufacturer did not categorise the SB A300-22-6021 as "Mandatory", which would have given it the highest priority. The airworthiness authority of the nation of design and manufacture did not issue promptly an airworthiness directive pertaining to implementation of the above SB.

July 13, 1996: McDonnell-Douglas MD-11, N-1768D

Location	Westerly RI
Flight	AAL 140
Operator	American Airlines
Operation	Scheduled Airline
Route	London, UK to New York, USA
On-Board Injuries	Serious 1 Minor/None 179
Aircraft Damage	None

Description

The first officer (F/O) was making a descent from FL350 to FL240 in smooth air in VMC conditions with the autopilot (A/P) engaged. As the flight neared FL250, the captain became concerned that the airplane would not level off at FL240. He instructed the F/O to slow the rate of descent. The F/O attempted this by using the pitch thumbwheel on the A/P control panel. The captain then took control, attempted to overpower the A/P, and pulled back on the control yoke. With back pressure on the control yoke, he disengaged the A/P. With the reduced control column resistance after the A/P was disengaged, there was further excursion of the elevators to the up position. According to the flight data recorder, the airplane was subjected to a +2.28 G-load. A passenger in the aft lavatory suffered a fractured ankle. A 2nd passenger and 2 flight attendants received minor injuries. The MD-11 Flight crew Operating Manual (FCOM) advised against attempting to overpower the A/P; however, this was contained under the title, 'SEVERE TURBULENCE AND/OR HEAVY RAIN INGESTION.' Each use of pitch thumbwheel interrupted the automatic level off, and the system would wait for 2 seconds after release of the thumbwheel before initiating a level off sequence again. continued use of the thumbwheel precluded the A/P from performing the level off at FL240. The manufacturer has issued FCOM changes to warn pilots about the hazards of applying force to the control wheel or column while the A/P is engaged and adjusting the pitch thumbwheel during a level off.

Probable Cause   NTSB Accident File[136]

The National Transportation Safety Board determines the probable cause(s) of this accident to be: insufficient information from the manufacturer in the airplane flight manual and flightcrew operating manual regarding the hazards of applying force to the control wheel or column while the autopilot is engaged and adjusting the pitch thumbwheel during a level off. Also causal was the flightcrew's lack of understanding of these items and the captain's improper decisions to overpower the engaged autopilot and then to disconnect the autopilot while holding back-pressure on the control yoke.

Regulatory Outcome

This and several other accidents involving human error sparked increased focus on consideration of human factors and automation in the design of transport aircraft.

In 1994, the FAA launched a study to evaluate all flightcrew/flight deck automation interfaces of current generation transport category airplanes. The FAA chartered a Human Factors Team to conduct the study. Team members included experts from the FAA, the European Joint Airworthiness Authorities (JAA), and academia. The objective of the study was to look beyond the label of "flightcrew error," and examine the contributing factors from the perspective of design; flightcrew training and qualifications; operations; and regulatory processes. The FAA also tasked the team to develop recommendations to address any problems identified.[137]

With regard to autopilot issues, the Team identified several specific problematic issues, including:

- Pilot/autopilot interactions that create hazardous out-of-trim conditions
- Autopilots that can produce hazardous speed conditions and may attempt maneuvers that would not normally be expected by a pilot; and
- Insufficient wording in the Airplane Flight Manual regarding the capabilities and limitations of the autopilot.

In the ILS approach to runway 34 of Nagoya Airport, under manual control by the first officer, the first officer inadvertently activated the GO lever, which changed the FD (Flight Director) to GO AROUND mode and caused a thrust increase. This made the aircraft deviate above its normal glide path. The APs were subsequently engaged, with GO AROUND mode still engaged. Under these conditions the first officer continued pushing the control wheel in accordance with the captain's instructions. As a result of this, the THS (Horizontal Stabilizer) moved to its full nose-up position and caused an abnormal out-of-trim situation. The crew continued approach, unaware of the abnormal situation. The AOA increased the Alpha Floor function was activated and the pitch angle increased.

It is considered that, at this time, the captain (who had now taken the controls), judged that landing would be difficult and opted for go-around. The aircraft began to climb steeply with a high pitch angle attitude. The captain and the first officer did not carry out an effective recovery operation, and the aircraft stalled and crashed.

The AAIC determined that the following factors, as a chain or a combination thereof, caused the accident:

- The first officer inadvertently triggered the Go lever. It is considered that the design of the GO lever contributed to it: normal operation of the thrust lever allows the possibility of an inadvertent triggering of the GO lever.
- The crew engaged the APs while GO AROUND mode was still engaged, and continued approach.
- The first officer continued pushing the control wheel in accordance with the captain's instructions, despite its strong resistive force, in order to continue the approach.
- The movement of the THS conflicted with that of the elevators, causing an abnormal out-of-trim situation.
- There was no warning and recognition function to alert the crew directly and actively to the onset of the abnormal out-of-trim condition.
- The captain and first officer did not sufficiently understand the FD mode change and the A/P override function.
- It is considered that unclear descriptions of the AFS (Automatic Flight System) in the FCOM (Flight crew Operating Manual) prepared by the aircraft manufacturer contributed to this.
- The captain's judgment of the flight situation while continuing approach was inadequate, control take-over was delayed, and appropriate actions were not taken.
- The Alpha-Floor function was activated; this was incompatible with the abnormal out-of-trim situation, and generated a large pitch-up moment. This narrowed the range of selection for recovery operations and reduced the time allowance for such operations.
- The captain's and first officer's awareness of the flight conditions, after the PIC took over the controls and during their recovery operation, was inadequate respectively.
- crew coordination between the captain and the first officer was inadequate.
- The modification prescribed in Service Bulletin SB A300-22-602 1 had not been incorporated into the aircraft.
- The aircraft manufacturer did not categorise the SB A300-22-6021 as "Mandatory", which would have given it the highest priority. The airworthiness authority of the nation of design and manufacture did not issue promptly an airworthiness directive pertaining to implementation of the above SB.

Pitot-Static Accidents

December 1, 1974: Boeing 727, N-274US

Location	Theils, NY
Flight	NWA 6231
Operator	Northwest Airlines
Operation	Non-Revenue Flight
Route	New York, NY to Buffalo, NY
On-Board Injuries	Fatal 3
Aircraft Damage	Destroyed

Description   Source: Aviation Safety Network[1]

Flight NW6231 departed New York-JFK at 19:14 for a ferry flight to Buffalo and was cleared by departure control to climb to FL140. At 19:21 the flight was cleared to climb to FL310. The aircraft began to climb 2500fpm at an airspeed of 305 knots. As the aircraft climbed through FL160, both the airspeed and the rate of climb began to increase. Reaching FL230, the airspeed had reached 405 knots and the rate of climb had exceeded 6500fpm. The overspeed warning horn sounded a little later, followed 10 seconds later by a stickshaker stall warning. The aircraft then leveled at 24800 feet with a speed of 420 knots until it turned rapidly to the right, 13 seconds later. The airplane started to descend out of control, reaching a vertical acceleration of +5g until it struck the ground in a slightly nose down and right wing-down attitude. The aircraft had descended from 24000 feet to 1090 feet in 83 seconds. The loss of control of the aircraft because the flight crew failed to recognize and correct the aircraft's high-angle-of-attack, low-speed stall and its descending spiral. The stall was precipitated by the flight crew's improper reaction to erroneous airspeed and Mach indications which had resulted from a blockage of the pitot heads by atmospheric icing. contrary to standard operational procedures, the flight crew had not activated the pitot heat.

Probable Cause   Source: Accident Report[138]

The loss of control of the aircraft because the flight crew failed to recognize and correct the aircraft's high-angle-of-attack, low-speed stall and its descending spiral. The stall was precipitated by the flight crew's improper reaction to erroneous airspeed and Mach indications which had resulted from a blockage of the pitot heads by atmospheric icing. contrary to standard operational procedures, the flight crew had not activated the pitot head heaters.

Regulatory Outcome

Certification rules were changed to automatically turn pitot heat on.

Comments

The NASA Loss-of-control Program studied a number of P-S mishaps.[139] Many of these were caused by blocked pitot tubes, a few by blocked static ports. One characteristics of most P-S mishaps is crew confusion. It is evident in the mishap reports there is confusion as the airplane climbs higher. In the stress of the situation, pilots seem to forget that overspeed warnings are triggered by airspeed (or Mach) indications, while low speed warnings are generally triggered by angle-of-attack.

Many airplanes do not show an indication of airspeed or altitude disagreement. This seems strange to us in view of their importance. Many of the airplanes cited in the P-S report detected the airspeed discrepancy. Unfortunately, they indicated this by simultaneous MACH/SPD TRIM and RUDDER RATIO. This arcane warning does not appear to adequately warn the flight crew.

February 6, 1996: Boeing 757, TC-GEN

Location   Off Puerto Plata, Dominican Republic
Flight   ALW 301
Operator   Birginair Airlines, operating for Alas Nacionales*
Operation   Non-Scheduled Passenger Flight
Route   Puerto Plata, Dominican Republic to Frankfurt, Germany
On-Board Injuries   Fatal 189
Aircraft Damage   Destroyed

Description   Source: Aviation Safety Network[1]

Alas Nacionales flight 301 departed Puerto Plata for a charter flight to Frankfurt via Gander and Berlin at 23:42 LT. At 80 knots on takeoff the captain found out that his air speed indicator (ASI) wasn't working properly. The co-pilot's indicator seemed to work fine. While climbing through 4700 feet the captain's ASI read 350 knots (real speed was about 220 kts); this resulted in an autopilot/autothrottle reaction to increase the pitch-up attitude and a power reduction in order to lower the airspeed. At that time the crew got 'Rudder ratio' and 'Mach airspeed' advisory warnings. Both pilots got confused when the co-pilot stated that his ASI read 200 knots decreasing while getting an excessive speed-warning, followed by a stick shaker warning. This led the pilots to believe that both ASIs were unreliable.

    Finally realizing that they were losing speed and altitude they disconnected the autopilot. The autopilot, fed by the captain's faulty ASI, had reduced the speed close to the stall speed. Full thrust was then applied. At 23:47:17 an aural GPWS warning sounded. Eight seconds later the aircraft struck the ocean.

    The incorrect ASI readings were probably caused by the obstruction of the pitot system by mud and/or debris from a small insect that was introduced in the pitot tube during the time the aircraft was on the ground. The aircraft was not flown for 20 days before the crash and was returned for service without a verification of the pitot-static system as recommended by Boeing.

Probable Cause   Source: Accident Report†[140]

The probable cause of the accident was the failure of the flight crew to recognize the activation of the stick-shaker as an imminent warning of an aerodynamic stall entry and the failure of the crew to execute the recovery procedures. of the entry in loss. Prior to the stick-shaker warning, there was a flight crew confusion due to erroneous relative speed increase indications and an overspeed warning. [141]

October 2, 1996: Boeing 757, N-52AW

Location   55 nm W Lima, Peru
Flight   PLI 603
Operator   Aero Peru
Operation   Scheduled Airline
Route   Lima, Peru to Santiago, Chile
On-Board Injuries   Fatal 70
Aircraft Damage   Destroyed

Description   Source: Aviation Safety Network[1]

    AeroPeru Flight 603, a Boeing 757-200, crashed into the sea off Lima, Peru, killing all 70 occupants.

    The aircraft performed a regular passenger service from Lima, Peru to Santiago, Chile. The flight took off from runway 15 at Lima Airport at 00:42 hours local time (05:42 C). While climbing through an altitude of approximately 200-300 ft, the pilots noted that the airspeed and altitude indications were too low. In calm winds, the windshear warning suddenly sounded three times. The flight crew then declared an emergency.

    The crew also started to receive rudder ratio and mach speed trim warnings, which were repeated throughout the flight, distracting their attention and adding to the problem of multiple alarms and warnings which saturated and bewildered them, creating confusion and chaos.

    At 00:53 the flight contacted the Departure controller again, eight minutes after the initial call, reporting they "request vectors from now on". By then the aircraft had climbed over sea and the flight crew were attempting to turn back, while trying to manage all conflicting warnings.

---

*   Birgenair Airlines (Turkey, ICAO code BHT) was operating flight 301 for Alas Nacionales (Dominican Republic, ICAO code ALW) on February 6, 1996. The accident report and most references cite ALW 301 as the flight number. Both airlines ceased operations in 1996.

†   Translation.

At 00:55:07, the crew radioed: "You're going to have to help us with altitudes and speed if that's possible." And from that moment until the end, the stick shaker, overspeed and "too low – terrain" alarms began to sound.

This caused confusion and the copilot said to the captain: "...right now we're stalling." The captain disagreed: "we're not stalling. it's fictitious, it's fictitious". Referring to the stick shaker, the copilot replies "... how can we not be stalling?"

At 01:04 the flight crew was attempting to maintain 4000 feet, however one minute later they start a measured and continuous descent. The aircraft kept descending and impacted the water with the left wing and no. 1 engine at a 10 degrees angle, at a speed of 260 knots. The aircraft pulled up to about 300 feet and flew for another 17 seconds. It then rolled inverted and crashed. The captain's airspeed indicated 450 knots and altitude 9500 feet.

Investigation results showed that the aircraft's three static ports on the left side were obstructed by masking tape. The tape had been applied before washing and polishing of the aircraft prior to the accident flight.

Probable Cause   Source: Aviation Safety Network[1]

It can be deduced from the investigation carried out that the maintenance staff did not remove the protective adhesive tape from the static ports. This tape was not detected during the various phases of the aircraft's release to the line mechanic, its transfer to the passenger boarding apron and, lastly, the inspection by the crew responsible for the flight (the walk-around or pre-flight check), which was carried out by the pilot-in-command, [name], according to the mechanic responsible for the aircraft on the day of the accident.
CONTRIBUTING CAUSES

The pilot-in-command, Mr [name], made a personal error by not complying with the procedure for GPWS alarms and not noticing the readings of the radio altimeters in order to discard everything which he believed to be fictitious.

The co-pilot, Mr [name], made a personal error by not being more insistent, assertive and convincing in alerting the pilot-in-command much more emphatically to the ground proximity alarms.

January 28, 2009: Boeing 757, G-STRZ

Location	Accra, Ghana
Flight	AEU G-STRZ*
Operator	Astraeus Airways for Ghana Airways
Operation	Scheduled Passenger Flight
Route	Accra, Ghana to Unknown Destination
On-Board Injuries	Minor/None <u>105</u>
Aircraft Damage	None

Description   Source: Serious Incident Report[142]

The aircraft had a blocked pitot tube, causing an airspeed discrepancy, which was detected early during the takeoff roll. The commander decided to continue the takeoff and deal with the problem whilst airborne. After passing FL180 the crew selected the left Air Data switch to ALTN, believing this isolated the left Air Data computer (ADC) from the Autopilot & Flight Director System (AFDS). Passing FL316, the VNAV mode became active and the Flight Management computers (FMCs), which use the left ADC as their input of aircraft speed, sensed an overspeed condition and provided a pitch-up command to slow the aircraft. The co-pilot was concerned about the aircraft's behaviour and, after several verbal prompts to the commander, pushed the control column forward. The commander, uncertain as to what was failing, believed that a stick-pusher had activated. He disengaged the automatics and lowered the aircraft's nose, then handed over control to the co-pilot. A MAYDAY was declared and the aircraft returned to Accra. The operator's subsequent engineering investigation discovered the remains of a beetle-like creature in the left pitot system.

The aircraft commenced its takeoff roll from Accra, Ghana at 2334 hrs with the commander as PF. Before the '80 kt' call was made, the commander noticed that his ASI was not functioning. He elected to continue the takeoff using the co-pilot's and standby ASIs, which appeared to be functioning normally, as he believed the weather conditions were suited to resolving the problem when airborne. The Engine Indication and crew Alerting System (EICAS) messages, AIRSPEED UNRELIABLE, MACH/SPEED TRIM and RUDDER RATIO illuminated during the initial climb below 400 ft. The commander handed over control to the co-pilot who, at 1000 ft, called for the Vertical Navigation (VNAV) mode of the AFDS and the right autopilot (AP), to be engaged. During flap retraction the crew considered that the aircraft was not accelerating normally so the autopilot was disconnected and the aircraft flown manually.

The crew asked a company engineer on board the aircraft to assist them with diagnosing the airspeed problem. The right AP was re-engaged, and the QRH consulted. The crew considered that the pitch attitude and thrust were relatively normal for the stage of flight so left the AP, Auto-Throttle (AT) and Flight Directors (FD) engaged. The engineer advised the crew that the EICAS messages were displayed because the left Air Data computer (ADC) was unserviceable; he had experienced the same defect on another company aircraft several months earlier when a bug had blocked the left pitot tube. On that occasion, the aircraft was flown without incident using the right AP until rectification could take place on the ground. In accordance with the QRH, the commander selected ALTN on the air data switch and believed that he had isolated the problem with the left ADC. He retook control of the aircraft and continued the climb with the right AP engaged using the Lateral Navigation (LNAV) and Flight Level change (FLCH) modes.

---

* Aircraft registration used *vice* flight number.

The commander recalled selecting VNAV at about FL250. At approximately FL320, the co-pilot became aware that the aircraft's rate of climb had started to increase, and that the indicated airspeed was decreasing. He called "climb rate" and the commander attempted to select vertical speed (VS) mode and reduce the rate of climb to 1,500 ft per minute. The commander recalled that the Mode control Panel (MCP) alternated between VS mode with a 4,000 fpm climb and altitude hold (alt hold) modes, but the aircraft's pitch attitude seemed normal. The co-pilot was now concerned with the situation and he urgently expressed concerns about the aircraft's deviations from the normal flight profile. As the AP did not appear to be following the MCP selections the co-pilot disconnected the AP and pushed forward on the control column to "increase the speed and prevent an increasing ROC (rate of climb)". He recalled calling out "I have it", but the commander had no recollection of this.

As the IAS reduced to approximately 250 kt, the commander noticed the control column move forward and he considered that a stick pusher must have activated1. He disconnected the AP and AT, moved the thrust levers forward, and pitched the aircraft to 3-5 degrees nose-down. Even with the AP and AT out, and the speed increasing through 245 kt, the commander could feel the control column was being pushed forward. He became aware that the co-pilot was on the controls and handed control to him while he transmitted a MAYDAY. A nearby aircraft observed from its Traffic collision Avoidance System (TCAS) that G-STRZ's indicated level was FL310. The FD's were disengaged and the aircraft returned to Accra with the co-pilot flying. As the aircraft neared Accra, and appeared to be operating normally, the MAYDAY was downgraded to a PAN and the commander flew an uneventful approach and overweight landing.

Probable Cause   Source: Accident Report[142]

The operator's engineers [mechanics] performed the investigation and they found the remains of a "beetle-like creature" in the left-hand pitot system. No faults were found with the ADC, the autopilots, or any of the relevant systems.

June 1, 2009: Airbus A330, F-GZCP

Location	South Atlantic Ocean
Flight	AFR 447
Operator	Air France
Operation	Scheduled Airline
Route	Rio de Janeiro, Brazil to Paris, France
On-Board Injuries	Fatal 228
Aircraft Damage	Destroyed

Description   Source: Aviation Safety Network[1]

An Air France Airbus A330-200 was destroyed when it crashed into the sea while on transatlantic flight from Rio de Janeiro-Galeao International Airport, RJ (GIG) to Paris-Charles de Gaulle Airport (CDG). All occupants, 12 crew members an 216 passengers, were killed.

Air France flight 447 departed at 19:29 hours local time (May 31) from Rio de Janeiro. The takeoff weight was 232.8t (for a MTOW of 233 t), including 70.4 tonnes of fuel. One of the co-pilots was Pilot Flying (PF).

The flight progressed as planned with the crew contacting several air traffic control centres along the way. cruising altitude was FL350.

At 23:35 local time (01:35 C), the crew informed the Atlantic Area control centre (CINDACTA III) controller that they had passed the INTOL waypoint. INTOL is an RNAV waypoint located in the Atlantic Ocean, 565 km from Natal, Brazil.

At 01:48 C the aircraft went out of the radar coverage of CINDACTA III, Fernando de Noronha. The meteorological situation in the area of AF447's flight path over the Atlantic was typical of that encountered in the month of June in the inter-tropical convergence zone. There were powerful cumulonimbus clusters on the route of AF447. Some of them could have been the centre of some notable turbulence.

At 01:55, the captain woke the second co-pilot and said "[...] he's going to take my place". The captain then attended the briefing between the two co-pilots, during which the Pilot Flying said, in particular "the little bit of turbulence that you just saw [...] we should find the same ahead [...] we're in the cloud layer unfortunately we can't climb much for the moment because the temperature is falling more slowly than forecast" and that "the logon with Dakar failed". The captain left the cockpit at 02:01:46 C.

The airplane was flying at FL350 and at Mach 0.82 and the pitch attitude was about 2.5 degrees. Autopilot 2 and auto-thrust were engaged. At 02:06:04, the PF called the cabin crew, telling them that "in two minutes we should enter an area where it'll move about a bit more than at the moment, you should watch out" and he added "I'll call you back as soon as we're out of it".

At 02:08:07, the PNF said "you can maybe go a little to the left [...]". The airplane began a slight turn to the left, the change in relation to the initial route being about 12 degrees. The level of turbulence increased slightly and the crew decided to reduce the speed to about Mach 0.8.

From 02:10:05, the autopilot then auto-thrust disengaged and the PF said "I have the controls". The airplane began to roll to the right and the PF made a left nose-up input. The stall warning sounded twice in a row. The recorded parameters show a sharp fall from about 275 kt to 60 kt in the speed displayed on the left primary flight display (PFD), then a few moments later in the speed displayed on the integrated standby instrument system (ISIS).

At 02:10:16, the PNF said "so, we've lost the speeds" then "alternate law [...]".

The airplane's pitch attitude increased progressively beyond 10 degrees and the plane started to climb. The PF made nose-down control inputs and alternately left and right roll inputs. The vertical speed, which had reached 7,000 ft/min, dropped to 700 ft/min and the roll varied between 12 degrees right and 10 degrees left. The speed displayed on the left side increased sharply to 215 kt (Mach 0.68). The airplane was then at an altitude of about 37,500 ft and the recorded angle of attack was around 4 degrees. From 02:10:50, the PNF tried several times to call the captain back.

At 02:10:51, the stall warning was triggered again. The thrust levers were positioned in the TO/GA detent and the PF maintained nose-up inputs. The recorded angle of attack, of around 6 degrees at the triggering of the stall warning, continued to increase. The trimmable horizontal stabilizer (THS) passed from 3 to 13 degrees nose-up in about 1 minute and remained in the latter position until the end of the flight.

Around fifteen seconds later, the speed displayed on the ISIS increased sharply towards 185 kt; it was then consistent with the other recorded speed. The PF continued to make nose-up inputs. The airplane's altitude reached its maximum of about 38,000 ft, its pitch attitude and angle of attack being 16 degrees.

At around 02:11:40, the captain re-entered the cockpit. During the following seconds, all of the recorded speeds became invalid and the stall warning stopped. The altitude was then about 35,000 ft, the angle of attack exceeded 40 degrees and the vertical speed was about -10,000 ft/min. The airplane's pitch attitude did not exceed 15 degrees and the engines' N1's were close to 100%. The airplane was subject to roll oscillations that sometimes reached 40 degrees. The PF made an input on the sidestick to the left and nose-up stops, which lasted about 30 seconds.

At 02:12:02, the PF said "I don't have any more indications", and the PNF said "we have no valid indications". At that moment, the thrust levers were in the IDLE detent and the engines' N1's were at 55%. Around fifteen seconds later, the PF made pitch-down inputs. In the following moments, the angle of attack decreased, the speeds became valid again and the stall warning sounded again.

At 02:13:32, the PF said "we're going to arrive at level one hundred". About fifteen seconds later, simultaneous inputs by both pilots on the sidesticks were recorded and the PF said "go ahead you have the controls".

The angle of attack, when it was valid, always remained above 35 degrees. The recordings stopped at 02:14:28. The last recorded values were a vertical speed of -10,912 ft/min, a ground speed of 107 kt, pitch attitude of 16.2 degrees nose-up, roll angle of 5.3 degrees left and a magnetic heading of 270 degrees. The airplane struck the surface of the sea.

Several attempts were made to locate the wreckage of the airplane. Finally on April 2, 2011, a search vessel using unmanned submarines located pieces of wreckage including an engine, landing gear and fuselage and wing parts on the Ocean floor. The flight recorders were recovered on May 2, 2011.

Probable Cause   Source: Accident Report[143]

The obstruction of the Pitot probes by ice crystals during cruise was a phenomenon that was known but misunderstood by the aviation community at the time of the accident. From an operational perspective, the total loss of airspeed information that resulted from this was a failure that was classified in the safety model. After initial reactions that depend upon basic airmanship, it was expected that it would be rapidly diagnosed by pilots and managed where necessary by precautionary measures on the pitch attitude and the thrust, as indicated in the associated procedure.

The occurrence of the failure in the context of flight in cruise completely surprised the pilots of flight AF 447. The apparent difficulties with aeroplane handling at high altitude in turbulence led to excessive handling inputs in roll and a sharp nose-up input by the PF. The destabilisation that resulted from the climbing flight path and the evolution in the pitch attitude and vertical speed was added to the erroneous airspeed indications and ECAM messages, which did not help with the diagnosis.

The crew, progressively becoming de-structured, likely never understood that it was faced with a 'simple' loss of three sources of airspeed information.

In the minute that followed the autopilot disconnection, the failure of the attempts to understand the situation and the de-structuring of crew cooperation fed on each other until the total loss of cognitive control of the situation. The underlying behavioural hypotheses in classifying the loss of airspeed information as 'major' were not validated in the context of this accident. confirmation of this classification thus supposes additional work on operational feedback that would enable improvements, where required, in crew training, the ergonomics of information supplied to them and the design of procedures.

The aeroplane went into a sustained stall, signaled by the stall warning and strong buffet. Despite these persistent symptoms, the crew never understood that they were stalling and consequently never applied a recovery manoeuvre. The combination of the ergonomics of the warning design, the conditions in which airline pilots are trained and exposed to stalls during their professional training and the process of recurrent training does not generate the expected behaviour in any acceptable reliable way.

In its current form, recognizing the stall warning, even associated with buffet, supposes that the crew accords a minimum level of 'legitimacy' to it. This then supposes sufficient previous experience of stalls, a minimum of cognitive availability and understanding of the situation, knowledge of the aeroplane (and its protection modes) and its flight physics. An examination of the current training for airline pilots does not, in general, provide convincing indications of the building and maintenance of the associated skills.

More generally, the double failure of the planned procedural responses shows the limits of the current safety model. When crew action is expected, it is always supposed that they will be capable of initial control of the flight path and of a rapid diagnosis that will allow them to identify the correct entry in the dictionary of procedures. A crew can be faced with an unexpected situation leading to a momentary but profound loss of comprehension. If, in this case, the supposed capacity for initial mastery and then diagnosis is lost, the safety model is then in 'common failure mode'. During this event, the initial inability to master the flight path also made it impossible to understand the situation and to access the planned solution.

Thus, the accident resulted from the following succession of events:

- Temporary inconsistency between the airspeed measurements, likely following the obstruction of the Pitot probes by ice crystals that, in particular, caused the autopilot disconnection and the reconfiguration to alternate law;
- Inappropriate control inputs that destabilized the flight path;
- The lack of any link by the crew between the loss of indicated speeds called out and the appropriate procedure;
- The late identification by the PNF of the deviation from the flight path and the insufficient correction applied by the PF;
- The crew not identifying the approach to stall, their lack of immediate response and the exit from the flight envelope;
- The crew's failure to diagnose the stall situation and consequently a lack of inputs that would have made it possible to recover from it.

    These events can be explained by a combination of the following factors:
- The feedback mechanisms on the part of all those involved that made it impossible:
* To identify the repeated non-application of the loss of airspeed information procedure and to remedy this,
* To ensure that the risk model for crews in cruise included icing of the Pitot probes and its consequences;
- The absence of any training, at high altitude, in manual aeroplane handling and in the procedure for 'Vol avec IAS douteuse';
   - Task-sharing that was weakened by:
    * Incomprehension of the situation when the autopilot disconnection occurred,
    * Poor management of the startle effect that generated a highly charged emotional factor for the two copilots;
- The lack of a clear display in the cockpit of the airspeed inconsistencies identified by the computers;
- The crew not taking into account the stall warning, which could have been due to:
    * A failure to identify the aural warning, due to low exposure time in training to stall phenomena, stall warnings and buffet,
    * The appearance at the beginning of the event of transient warnings that could be considered as spurious,
    * The absence of any visual information to confirm the approach-to-stall after the loss of the limit speeds,
    * The possible confusion with an overspeed situation in which buffet is also considered as a symptom,
    * Flight Director indications that may led the crew to believe that their actions were appropriate, even though they were not,
    * The difficulty in recognizing and understanding the implications of a reconfiguration in alternate law with no angle of attack protection.

Recommendations

The initiating event was the loss of airspeed data because of the icing of the pitot tubes. This a result of the high liquid water content in some tropical thunderstorms. Indeed, the accident report makes the recommendation that EASA and other regulatory agencies undertake studies to better define the composition of such cloud masses at high altitudes and modify the certification criteria based on these results.[143]

    Since it was very likely that the PF had limited experience in hand-flying the airplane at high altitude, that he was ill-prepared for the situation of hand-flying the airplane at high altitude and severe turbulence with degraded handing (in alternate law). The accident report also recommends additional training "dedicated to manual aircraft handling of approach to stall and stall recovery, including at high altitude."

February 3, 2013: Airbus A340, A6-EHF

Location	Indian Ocean Between Waypoints ELATI and PIPOV
Flight	EDT 460
Operator	Etihad Airways
Operation	Scheduled Airline
Route	Abu Dhabi to Melbourne, Australia
On-Board Injuries	Minor/None <u>295</u>
Aircraft Damage	None

Description
Source: Serious Incident Report[144]

The Airbus A340-600 aircraft, registration A6-EHF, operating a scheduled passenger flight to Melbourne International Airport, Australia, departed Abu Dhabi International Airport at approximately 19:35 C (February 2). There were a total of 295 persons on board: 17 crew members and 278 passengers. The captain was the pilot flying and the first officer was the pilot monitoring.

While cruising at FL350, just leaving the Colombo FIR and entering the Melbourne FIR, the aircraft encountered moderate to heavy turbulence, and experienced significant airspeed oscillations on both the captain's and the standby airspeed indicators. The autopilot, autothrust, and flight directors disconnected automatically. The flight control law changed from "Normal" to "Alternate" Law, leading to the loss of some flight mode and flight envelope protections. changes from Normal to Alternate Law occurred twice; thereafter the aircraft remained in Alternate Law until the end of the flight. The autothrust system and the flight directors were successfully re-engaged, however, neither autopilot (autopilots 1 or 2) could be re-engaged, thus the aircraft was flown manually until landing. In addition to the system anomalies, the aircraft experienced high N1 vibration on the No. 2 engine. As the aircraft had lost capability to maintain Reduced Vertical Separation Minima (RVSM) the flight crew decided to divert to Singapore-Changi International Airport. The diversion required the flight crew to dump fuel in order to land the aircraft below its maximum landing weight.

The landing was uneventful and none of persons on board were injured.

Findings

- The pitot probes on the left side of the Aircraft, most probably, were being intermittently obstructed by ice crystals.

- Three occasions of unreliable airspeed indication on CAS1 and CAS3 were experienced, which resulted in reversion of the flight control law to Alternate Law, and prevented the re-engagement of the autopilot during the remainder of the flight.

- The logics of the FMGES and FcS related to unreliable indicated airspeed functioned correctly as designed.

- No deficiencies in the systems functionality related to airspeed, altitude, AOA, or temperature were found.

- The ambient temperature and the Aircraft altitude were beyond the icing envelope of the JAR specification and the manufacturer requirements.

- On the No. 2 engine, outer surface the Omega Seal was in disband condition, which made the omega seal loose on the Nose cone and allowed water ingress

- Water or ice crystals entered and passed through the spinner fairing and accreted under the annulus fillers creating an ice out-of-balance situation, which led to the increase of N1 vibration beyond the maximum over a period of 1 hour and 20 minutes.

- The Aircraft lost the capability to operate in RVSM airspace.

- There was no link between the unreliable airspeed indication and the No. 2 engine N1 high vibration events.

- The Aircraft experienced icing conditions in which the altitude and static ambient temperature (SAT) conditions were outside the JAR specification and the Aircraft Manufacturer requirements.

Probable Cause   Source: Serious Incident Report[144]

- The cause of the Unreliable Airspeed Indication Serious Incident was the intermittent obstruction of the Aircraft left side pitot probes due to, most probably, accumulations of ice crystals.

- The cause of the No. 2 engine N1 high vibration was the ingress of water through a gap created after the Omega Seal disbanded. The water froze to ice, which entered and passed through the spinner fairing and accreted under the annulus fillers.

Contributing Factors

- An incorrect weather radar tilt setting was selected. Accordingly, there was no predictive detection of the cumulonimbus cloud that may have enabled the crew to take avoidance maneuvers.

- The ambient temperature and the Aircraft altitude were outside the icing envelope parameters of the JAR specification and the manufacturer's design requirements for pitot probes.

Certification Issues:

The certification issue is the inadequate specifications for the pitot tubes.

Comments

The initiating cause of this Serious Incident was essentially the same as for Air France 447: Flying in a tropical convective cloud with pitot icing beyond the capacity of the pitot heat. The difference was the pilots' reaction. Simply put, they flew the airplane.

Procedural Deviations

July 5, 1970: McDonnell Douglas DC-8, CF-TIW

Location	Toronto, ON, Canada
Flight	ACA 621
Operator	Air Canada
Operation	Scheduled Airline
Route	Montreal, QC, Canada to Toronto, ON, Canada
On-Board Injuries	Fatal 109
Aircraft Damage	Destroyed

Description
Source: Aviation Safety Network[1]

Air Canada flight 621 was a routine early morning flight originating from Montreal-Dorval International Airport, Qc (YUL), with destination Toronto International Airport, ON (YYZ). The DC-8-63 plane, a relatively new airplane which had been delivered just over two months ago, departed at 07:17 for a flight which was to take just over 50 minutes. The captain was pilot flying. The enroute and descent portion of the flight were uneventful. At a distance of 8 miles from Toronto Airport, about 08:02, the "Before-Landing check" was made. This included the lowering of the undercarriage and according to Air Canada procedures should include arming the spoilers. This item however was intentionally omitted. On previous flights were the captain and first officer had flown together they had disagreed on when to arm the spoilers. Both men did not like to arm the spoilers at the beginning of the final approach, fearing that this increased the chance of inadvertent spoiler activation. The captain preferred to arm the spoilers on the flare, while the first officer preferred to arm and extend them on the ground. Although both procedures where contrary to company policy, it was agreed between them that when the captain was flying the aircraft, the first officer would cause the spoilers to be extended on the ground, and when the first officer was flying the captain would arm the spoilers on the flare.

However on this particular occasion, the captain and first officer had a discussion about when the spoilers should be armed. The captain finally ordered: "All right, give them to me on the flare", which was contrary to their personal agreement on previous flights.

Power was reduced then on the aircraft for the purpose of the flare and the captain gave the order to the first officer by saying "O.K."; and immediately thereafter the ground spoilers were deployed. The aircraft was about 60 feet above runway 32 at that time and began to sink rapidly. The captain immediately noticed what had happened, applied full throttle to all four engines and pulled back the control column. The nose came up as the aircraft continued to sink. Realizing what he had done, the first officer apologized to the captain. The plane than struck the runway heavily, causing the number 4 engine and pylon to separate from the wing. It fell on the runway along with a piece of the lower wing plating, allowing fuel to escape and subsequently ignite. The DC-8 rose back into the air, at which time the ground spoilers retracted, and climbed to an altitude of 3100 feet. During this climb, fire and smoke were seen trailing behind the aircraft intermittently. The crew wanted to circle for an emergency landing on runway 32. This was not possible because of debris on the runway, so the controller suggested a landing on runway 23. About two and a half minutes after the initial touchdown an explosion occurred in the right wing outboard of the number 4 engine location causing parts of the outer wing structure to fall free to the ground. Six seconds later, a second explosion occurred in the vicinity of number 3 engine and the engine with its pylon ripped free of the wing and fell to the ground in flames. Six and one half seconds later, a third explosion occurred which caused the loss of a large section of the right wing, including the wing tip. The airplane then went into a violent manoeuvre, lost height rapidly and at the same time more wing plating tore free following which the DC-8 struck the ground at a high velocity, about 220 knots in a left wing high and nose low attitude.

Regulatory Outcome

As a result of this accident, the Federal Aviation Administration issued Airworthiness Directive AD 70-25-02 cautioning pilots against in-flight operation of ground spoilers by requiring the installation of a warning placard in the cockpit and the insertion of an additional Operating Limitation in the Flight Manual.

Comments

It appears that the crew were uncomfortable with arming the ground spoilers because they felt that this exposed them to inadvertent deployment on final approach. They had agreed to either arm them during the flare or once the airplane was on the ground. In this case, the spoilers were armed during the flare and immediately deployed triggering the hard landing. The most likely time for such an event is during the act of arming them. This argues for arming them when the gear is extended (as part of the SOP) or once the airplane was on the ground.

This is exacerbated when act of arming the spoilers cannot be quickly reversed, so called bang-bang controls.

This also points out the need to ensure that flight crews understand the implications of well meaning, but incorrect procedural deviations

Propeller and Rotor Blade Failures

May 22, 1968: Sikorsky S-61L, N-303Y

Location	Paramount, CA
Flight	LAA 841
Operator	Los Angeles Airways
Operation	Scheduled Airline
Route	Anaheim, CA to Los Angeles, CA
On-Board Injuries	Fatal 23
Aircraft Damage	Destroyed

Description   Source: Aviation Safety Network[1]

Los Angeles Airways Flight 841, a Sikorsky S-61LY N303Y named "Megapolis IV", crashed and burned at Paramount, California.

The accident sequence probably began while the aircraft was in cruising flight about 2,000 feet above the ground and about 2 miles east of the accident site. The black, yellow, and blue main rotor blades, followed by the red and white blades, underwent a series of extreme overtravel excursions in their lead/lag axis.

During the extreme excursions, the yellow main rotor blade overtraveled in the lead direction and, as a result, its pitch change control road was subjected to downward and rearward loading many times its design operating strength. Under these forces, the rod became detached at its lower trunnion end where it is normally secured to the attachment lugs of the main rotor rotating swashplate. With this separation, the blade went out of control and struck the right side of the aircraft diagonally across the baggage loading door. The other four main rotor blades then struck and penetrated both the aft and forward portions of the aircraft. The blade strikes destroyed the main rotor blades and separated major portions of the fuselage, including the tailrotor pylon and tail rotor assemblies. The aircraft, completely uncontrollable, crashed in a near-vertical descent.

The initial malfunction, failure , or condition which precipitated the accident sequence was probably a loss of main rotor blade damper integrity caused by either failure of the black main rotor blade damper or a loss of effective damper action by the white main rotor blade damper. An important portion of the black damper and a portion of the black blade horizontal hinge pin to which the damper attaches were not recovered.

Probable Cause   Source: Accident Report[145]

The Safety Board determines that the probable cause of this accident was the loss of main rotor blade damper integrity due to either a failure of the black blade damper or a loss of effective damping action by the white blade damper. This resulted in uncontrolled excursions of the main rotor blades in their leadllag axis, an overload detachment of the yellow main rotor blade pitch change control rod and destruction of the structural integrity of the aircraft by blade strikes. The precise reason for either of the possibilities for the loss of damper integrity is undetermined.

April 5, 1991: Embraer EMB-120, N-270AS

Location	Brunswick, GA
Flight	ASQ 2311
Operator	Atlantic Southeast Airlines
Operation	Scheduled Airline
Route	Atlanta, GA to Brunswick, GA
On-Board Injuries	Fatal 23
Aircraft Damage	Destroyed

Description   Source: Aviation Safety Network[1]

Flight 2311 was scheduled initially for airplane N228AS to depart at 13:24 EST. Because of mechanical problems an airplane change was made to N270AS. The flight departed Atlanta at 13:47 and arrived in the Brunswick area about 14:44. At 14:48 the flight was cleared for a visual approach to runway 07. The Embraer had just turned from base leg to final approach when the aircraft was seen to pitch up about 5deg and roll to the left until the wings were vertical. The airplane then nosed down into the ground, 9975 feet short of the runway.

Probable Cause   Source: Accident Report[146]

The loss of control in flight as a result of a malfunction of the left engine propeller control unit which allowed the propeller blade angles to go below the flight idle position. Contributing to the accident was the deficient design of the propeller control unit by Hamilton Standard and the approval of the design by the Federal Aviation Administration. The design did not correctly evaluate the failure mode that occurred during this flight, which resulted in an uncommanded and uncorrectable movement of the blades of the airplane's left propeller below the flight idle position.

Regulatory Outcome

In spite of the forgoing, no regulatory changes resulted.[22]

Radar Altimeter Failure

February 25, 2009: Boeing 737, TC-JGE

Location	Amsterdam Schiphol Airport, Netherlands
Flight	THY 1591
Operator	THY Turkish Airlines
Operation	Scheduled Airline
Route	Istanbul, Turkey to Amsterdam, Netherlands
On-Board Injuries	Fatal 9 Serious 27 Minor/None 99
Aircraft Damage	Destroyed

Description   Source: Aviation Safety Network[1]

Turkish Airlines Flight TK1951, a Boeing 737-800, departed Istanbul-Atatürk International Airport (IST) for a flight to Amsterdam-Schiphol International Airport (AMS), The Netherlands. The flight crew consisted of three pilots: a line training captain who occupied the left seat, a first officer under line training in the right seat and an additional first officer who occupied the flight deck jump seat. The first officer under line training was the pilot flying. The en route part of the flight was uneventful.

The flight was descending for Schiphol and passed overhead Flevoland at about 8500 ft. At that time the aural landing gear warning sounded.

The aircraft continued and was then directed by Air Traffic control towards runway 18R for an ILS approach and landing. The standard procedure for runway 18R prescribes that the aircraft is lined up at least 8 NM from the runway threshold at an altitude of 2000 feet. The glidepath is then approached and intercepted from below. Lining up at a distance between 5 and 8 NM is allowed when permitted by ATC.

Flight 1951 was vectored for a line up at approximately 6 NM at an altitude of 2000 feet. The glide slope was now approached from above.

The crew performed the approach with one of the two autopilot and autothrottle engaged.

The landing gear was selected down and flaps 15 were set. While descending through 1950 feet, the radio altimeter value suddenly changed to -8 feet. And again the aural landing gear warning sounded.

This could be seen on the captain's (left-hand) primary flight display. The first officer's (right-hand) primary flight display, by contrast, indicated the correct height, as provided by the right-hand system. The left hand radio altimeter system, however, categorised the erroneous altitude reading as a correct one, and did not record any error. In turn, this meant that it was the erroneous altitude reading that was used by various aircraft systems, including the autothrottle. The crew were unaware of this, and could not have known about it. The manuals for use during the flight did not contain any procedures for errors in the radio altimeter system. In addition, the training that the pilots had undergone did not include any detailed system information that would have allowed them to understand the significance of the problem.

When the aircraft started to follow the glidepath because of the incorrect altitude reading, the autothrottle moved into the 'retard flare' mode. This mode is normally only activated in the final phase of the landing, below 27 feet. This was possible because the other preconditions had also been met, including flaps at (minimum) position 15. The thrust from both engines was accordingly reduced to a minimum value (approach idle). This mode was shown on the primary flight displays as 'RETARD'. However, the right-hand autopilot, which was activated, was receiving the correct altitude from the right-hand radio altimeter system. Thus the autopilot attempted to keep the aircraft flying on the glide path for as long as possible. This meant that the aircraft's nose continued to rise, creating an increasing angle of attack of the wings. This was necessary in order to maintain the same lift as the airspeed reduced.

In the first instance, the pilots' only indication that the autothrottle would no longer maintain the pre-selected speed of 144 knots was the RETARD display. When the speed fell below this value at a height of 750 feet, they would have been able to see this on the airspeed indicator on the primary flight displays. When subsequently, the airspeed reached 126 knots, the frame of the airspeed indicator also changed colour and started to flash. The artificial horizon also showed that the nose attitude of the aircraft was becoming far too high. The cockpit crew did not respond to these indications and warnings. The reduction in speed and excessively high pitch attitude of the aircraft were not recognised until the approach to stall warning (stick shaker) went off at an altitude of 460 feet.

The first officer responded immediately to the stick shaker by pushing the control column forward and also pushing the throttle levers forward. The captain however, also responded to the stick shaker commencing by taking over control. Assumingly the result of this was that the first officer's selection of thrust was interrupted. The result of this was that the autothrottle, which was not yet switched off, immediately pulled the throttle levers back again to the position where the engines were not providing any significant thrust. Once the captain had taken over control, the autothrottle was disconnected, but no thrust was selected at that point. Nine seconds after the commencement of the first approach to stall warning, the throttle levers were pushed fully forward, but at that point the aircraft had already stalled and the height remaining, of about 350 feet, was insufficient for a recovery.

According to the last recorded data of the digital flight data recorder the aircraft was in a 22° ANU and 10° Left Wing Down (LWD) position at the moment of impact.

The airplane impacted farmland. The horizontal stabilizer and both main landing gear legs were separated from the aircraft and located near the initial impact point. The left and right engines had detached from the aircraft.

The aft fuselage, with vertical stabilizer, was broken circumferentially forward of the aft passenger doors and had sustained significant damage. The fuselage had ruptured at the right side forward of the wings. The forward fuselage section, which contained the cockpit and seat rows 1 to 7, had been significantly disrupted. The rear fuselage section was broken circumferentially around row 28.

Weather reported about 09:40 C (10:40 local):

EHAM 250925Z 20010KT 4500 BR ScT007 BKN008 OVc010 04/03 Q1027 TEMPO 2500=

[Wind 200 degrees, 10 kts, visibity 4500m in mist, scattered clouds 700 ft., broken clouds 800 f, overcast 1,000 ft.temperature 4 deg. C, dewpoint3 deg. C]

EHAM 250955Z 21010KT 4500 BR BKN007 OVc008 05/04 Q1027 TEMPO 2500=

[Wind 200 degrees, 10 kts, visibity 4500m in mist, broken clouds 700 f, overcast 800 ft. temperature 5 deg. C, dewpoint 4 deg. C]

Probable Cause

Source: Accident Report[147]

During the accident flight, while executing the approach by means of the instrument landing system with the right autopilot engaged, the left radio altimeter system showed an incorrect height of -8 feet on the left primary flight display. This incorrect value of -8 feet resulted in activation of the 'retard flare' mode of the autothrottle, whereby the thrust of both engines was reduced to a minimal value (approach idle) in preparation for the last phase of the landing. Due to the approach heading and altitude provided to the crew by air traffic control, the localizer signal was intercepted at 5.5 NM from the runway threshold with the result that the glide slope had to be intercepted from above. This obscured the fact that the autothrottle had entered the retard flare mode. In addition, it increased the crew's workload. When the aircraft passed 1000 feet height, the approach was not stabilised so the crew should have initiated a go around. The right autopilot (using data from the right radio altimeter) followed the glide slope signal. As the airspeed continued to drop, the aircraft's pitch attitude kept increasing. The crew failed to recognise the airspeed decay and the pitch increase until the moment the stick shaker was activated. Subsequently the approach to stall recovery procedure was not executed properly, causing the aircraft to stall and crash.

Rain Ingestion

May 24, 1988: Boeing 737, N-75356

Location	New Orleans, LA
Flight	TAI 110
Operator	TACA International Airlines
Operation	Scheduled Airline
Route	Belize City, Belize to New Orleans, LA, USA
On-Board Injuries	Minor/None <u>45</u>
Aircraft Damage	Minor

Description   Source: Aviation Safety Network[1]

TACA International Airlines Flight 110 was a regularly scheduled passenger flight between San Salvador, El Salvador, to New Orleans, Louisiana, USA with an en route stop in Belize City, Belize. The airplane, a brand new Boeing 737-300 with 38 passengers and a crew of seven on board, departed as scheduled from Belize City, and the flight was uneventful until descent into New Orleans.

During descent from FL350 for an IFR arrival to New Orleans, the flight crew noted green and yellow returns on the weather radar with some isolated red cells, left and right of the intended flight path. Before entering clouds at FL300, the captain selected continuous engine ignition and activated engine anti-ice systems. The crew selected a route between the 2 cells, displayed as red on the weather radar. Heavy rain, hail and turbulence were encountered. At about FL165, both engines flamed out. The APU was started and aircraft electrical power was restored while descending through about FL106. Attempts to wind-mill restart the engines were unsuccessful. Both engines lit-off by using starters, but neither would accelerate to idle; advancing the thrust levers increased the EGT beyond limits. The engines were shut down to avoid a catastrophic failure. An emergency landing was made on a 6060 feetx120 feet grass strip next to a levee without further damage to the aircraft.

Investigation revealed that the aircraft encountered a level 4 thunderstorm but engines flamed out, though they had met the FAA specs for water ingestion. The aircraft had minor hail damage; the #2 engine was damaged from overtemperature.

The 737 took off from the field on June 6.

Probable Cause   Source: Accident Report[148]

A double engine flameout due to water ingestion which occurred as a result of an inflight encounter with an area of very heavy rain and hail. A contributing cause of the incident was the inadequate design of the engines and the FAA water ingestion certification standards which did not reflect the waterfall rates that can be expected in moderate to higher intensity of thunderstorms.

Regulatory Outcome:[2]

When the TACA event occurred, the industry and FAA were starting to recognize engine power loss in inclement weather as a safety threat to the aircraft. An Aerospace Industries of America (AIA) task force was organized with representatives from airframe and engine manufacturers to investigate multiple engine power loss in inclement weather.

The FAA and industry worked together to study this threat and eventually revise the corresponding FAA certification standards for turbine engine rain and hail ingestion. The results of the study indicated there were a number of multiple turbine engine power loss and instability events, forced landings, and accidents in revenue service attributed to operating airplanes in extreme rain or hail conditions. Investigations revealed that ambient rain or hail concentrations can be amplified significantly through the engine core at high flight speeds and low engine power conditions. Rain or hail through the engine core may degrade compressor stability, combustor flameout margin, and fuel control run down margin. Ingestion of extreme quantities of rain or hail through the engine core may ultimately produce engine surging, power loss, and engine flameout. In 1998 a revised FAA engine rain and hail standard was issued as 14 CFR §33.78.

Comments

It has been our experience that many, if not most, certification personnel are not conversant with the limited meteorological envelopes that form the basis of much of the weather envelopes. For example, the basis for most icing certification is Appendix C which was based on the icing studies conducted in the late 1940s using reciprocating engine powered aircraft of that era. The highest altitude covered by Appendix C was 22,000 ft. We are gradually improving the range of environmental conditions that must be considered. The TACA crew and passengers were fortunate. We feel that the flight crew deserves an "Atttaboy."

Repeated Control Reversals

February 11, 1991: Airbus A310, D-AOAC

Location	Moscow, Russia
Flight	IFL D-AOA*
Operator	Interflug
Operation	Scheduled Airline
Route	Berlin, DDR to Moscow, Russia
On-Board Injuries	None Reported
Aircraft Damage	Minor

Description   Source: Source: NTSB[149]

On February 11, 1991, an Airbus A310 operated by the German airline Interflug experienced an in-flight loss of control during a missed approach to runway 25L at Sheremetyevo Airport in Moscow, Russia. None of the 9 crewmembers and 100 passengers was injured.

The flight, which had departed from Schonefeld Airport in Berlin, Germany, was uneventful until the airplane was at an altitude of 1,550 feet. At that point, ATC instructed the pilots to go around because of a blocked runway. The pilots initiated the go-around maneuver with the autopilot engaged at an altitude of 1,275 feet. Afterward, the airplane entered an extreme pitch angle, which resulted in a severe loss of speed and, at an altitude of 4,000 feet, a breakdown of airflow over the wing and a subsequent stall. The airplane descended to 1,700 feet, at which point the pilots used full engine power to make a steep climb. The airplane subsequently stalled three more times. After several minutes, the pilots stabilized the airplane at 11,000 feet. The pilots landed the airplane manually.

The probable cause of this incident was movement of the control column by the pilot while the airplane was flying in go-around mode under AFS authority. The crew was not informed about AFS behavior at this stage of the flight.159 According to FDR data (provided to the Safety Board after the flight 587 accident), the Interflug pilot made alternating rudder inputs of about one-third of the full pedal deflection. The speed of the airplane varied during this time. When the airplane reached high speed, the pedal inputs resulted in ultimate loads on the vertical stabilizer, but the same pedal inputs at low speed did not result in high loads on the vertical stabilizer.

May 12, 1997: Airbus A300B4, N-90070

Location	West Palm Beach, FL
Flight	AAL 903
Operator	American Airlines
Operation	Scheduled Airline
Route	Boston, MA to Miami, FL
On-Board Injuries	Serious 1 Minor/None 164
Aircraft Damage	Minor

Probable Cause   Source: Aviation Safety Network[1]

The flight was assigned an airspeed of 230 knots and cleared to descend from FL240 to 16,000 feet in preparation for landing at Miami. The FDR indicated that while the autopilot was engaged in the descent, the power levers moved from the mechanical autothrottle limit of 44 degrees to the manual limit of 37 degrees. As the aircraft leveled at 16,000 feet the airspeed decreased. The F/O began a right turn to enter a holding pattern and added some power, which stabilized the airspeed at 178 knots. However, the right bank and the resultant angle of attack (AOA) continued to increase, despite left aileron input by the autopilot. As the autopilot reached the maximum input of 20 degrees, bank angle increased past 50 degrees, and the AOA increased rapidly from 7 degrees to 12 degrees. At this point the stick shaker activated, the autopilot independently disconnected, the power was increased, and full left rudder was used to arrest the roll. The bank angle reached 56 degrees, and the AOA reached 13.7 degrees at 177 knots. The aircraft then pitched down, and entered a series of pitch, yaw, and roll maneuvers as the flight controls went through a period of oscillations for about 34 seconds. The maneuvers finally dampened and the crew recovered at approximately 13,000 feet. One passenger was seriously injured and one flight attendant received minor injuries during the upset. According to wind tunnel and flight test data the A300 engineering simulator should adequately represent the aircraft up to 9 degrees AOA. Unlike the accident aircraft; however, the simulator recovered to wings level promptly when the lateral control inputs recorded by the FDR were used. The roll disagreement between the simulator and accident aircraft began at 7 degrees AOA, and it appears that some effect not modeled in the simulator produced the roll discrepancy. Just prior to the upset the accident aircraft entered a cloud deck. The winds were approximately 240 degrees, 35 knots, and the ambient air temperature was approximately

---

*   Identification uses the aircraft registration *vice* the flight number.

minus 4 degrees C. An atmospheric disturbance or asymmetric ice contamination were two possible explanations considered, but unproven.

Probable Cause   Source: *Aviation Accident Report*[150]

The flightcrew's failure to maintain adequate airspeed during leveloff which led to an inadvertent stall, and their subsequent failure to use proper stall recovery techniques. A factor contributing to the accident was the flightcrew's failure to properly use the autothrottle.

Comments

According to wind tunnel and flight test data the A300 engineering simulator should adequately represent the aircraft up to 9 degrees AOA. Unlike the accident aircraft; however, the simulator recovered to wings level promptly when the lateral control inputs recorded by the FDR were used. The roll disagreement between the simulator and accident aircraft began at 7 degrees AOA, and it appears that some effect not modeled in the simulator produced the roll discrepancy.

November 12, 2001: Airbus A300B4, N-14053

Location	Belle Harbor, NY
Flight	AAL 587
Operator	American Airlines
Operation	Scheduled Airline
Route	New York (Kennedy), NY to Santo Domingo, Dominican Republic
On-Board Injuries	Fatal 260
Ground Injuries	Fatal 5
Aircraft Damage	Destroyed

Description   Source: *Aviation Safety Network*[1]

American Airlines Flight 587 was scheduled to leave New York-JFK at 08:00 for a flight to Santo Domingo, Dominican Republic. The boarding process at gate 22 took a little longer than planned due to additional security procedures that delayed boarding. The gate was closed at 08:38 and pushback from the gate was accomplished at 09:00. The crew taxied to runway 31L behind Japan Air Lines Flight 047, a Boeing 747-400 bound for Tokyo-Narita. JL047 was cleared for takeoff at 09:11:08 but started it's takeoff roll one minute later. While JL047 was still preparing for takeoff, the tower controller called AA587, cautioned the flight crew about wake turbulence and instructed them to taxi into position and hold on runway 31L: "American five eighty seven heavy Kennedy tower, caution wake turbulence runway three one left, taxi into position and hold". A little later the JAL Boeing 747 rotated and initiated a climbing left turn over Jamaica Bay (the "bridge departure").

Then, at 09:13:27 Flight 587 was cleared for takeoff: "American five eight seven heavy, wind three zero zero at niner, runway three one left, cleared for takeoff". Takeoff roll was initiated about 09:14, circa 1 minutes and 45 seconds following the 747. After leaving the ground the landing gear was retracted at 09:14:34. The tower controller then cleared the crew for the bridge departure: "American five eight seven heavy, turn left. fly the Bridge climb. contact New York departure. Good morning." Flight 587 contacted the ARTCC controller about 09:15, and stated they were climbing out of 1,300 feet for 5,000 feet. The controller responded by clearing the flight to climb to 13,000 feet, turn left, and proceed direct to WAVEY. At that moment, while in a climbing left turn, the crew heard a brief squeak and a rattling sound, possibly caused by wake turbulence. Some fifteen seconds later the Airbus began to yaw to the right. Full right and left rudder were applied and the first officer called for "max power" at 09:15:54. Again full right and left rudder were applied and sounds of a snap, a thump and a loud bang were heard when the rudder travelled full right again. The entire vertical tail fin had separated and the Airbus entered an uncontrolled descent from an altitude of about 2500 feet. During this descent both engines separated from the wings coming down within 100 feet of each other near the Newport Avenue / Beach 129th Street crossroads. The aircraft crashed into Beach 131 Street, a Queens residential area.

Probable Cause   Source: *Aviation Accident Final Report*[151]

The in-flight separation of the vertical stabilizer as a result of the loads beyond ultimate design that were created by the first officer's unnecessary and excessive rudder pedal inputs. Contributing to these rudder pedal inputs were characteristics of the A300-600 rudder system design and elements of the American Airlines Advanced Aircraft Maneuvering Program.

Regulatory Outcome

As a result of the AA587 accident, manufacturers of transport category airplanes, acting at the request of the FAA, revised the Limitations Sections of their Airplane Flight Manuals (*A300-600 AFM*) to contain a warning that rapid and large alternating control inputs, especially in combination with large changes in pitch, roll, or yaw (e.g., large sideslip angles), may result in structural failures at any speed, even below V_A.[2]

There were two proposal to amend § 25.351 to replace the single application of full rudder with repeated alternating applications. Neither has been adopted at this point.

In order to enhance FAA internal coordination regarding design related and operational issues, the FAA has established a program management position specifically tasked with tracking and follow-through on issues which span design and operational activities, with regard to safety management of the airline fleet.[2]

This was followed by the introduction of Part 60, which established qualification requirements for flight simulators used in air carrier training.[152]

January 10, 2008: Airbus A319, C-GBHZ

Location	Omak, WA
Flight	ACA 190*
Operator	Air Canada
Operation	Scheduled Airline
Route	Victoria, BC to Toronto, ON, Canada
On-Board Injuries	Serious 3 Minor/None 85
Aircraft Damage	Minor

Description   Source: *Accident Report*[153]

AC190 was en route from Victoria, BC, to Toronto, ON, with 83 passengers and 5 crew members. At 06:48 local time, a series of jolts were felt in AC190, followed by a series of rolls. The crew declared an emergency and diverted the flight to Calgary International Airport, Alberta, where it landed uneventfully at 07:28. Several passengers and crew members were injured.

Probable Cause   Source: *Accident Report*[153]

During recovery from the upset, pilot rudder and sidestick control inputs resulted in aircraft sideslip and g loadings. These contributed to the displacement of occupants and objects in the cabin, as well as placing lateral accelerations and aerodynamic loads on the vertical stabilizer structure to beyond certified limits.

During the 18-second duration of the event, vertical accelerations reached peak values of +1.57g and -0.77g. Lateral accelerations reached peak values of +0.49g (right) and 0.46g (left) during four oscillations. Some actions to rectify the upset were similar to those that contributed to damage to the vertical stabilizer attachment fittings on flight AA587 in 2001. The Airbus A300 in that event crashed after separation of the vertical stabilizer.

The TSB further notes that annual recurrent A319/A320 pilot training at Air Canada did not consistently include reference to the hazards of pilot rudder pedal reversals during upset recovery at high airspeeds. This increased the likelihood that pilots would make inappropriate rudder pedal inputs during upset recoveries.

Comments

While the accident occurred in US airspace, the NTSB, not realizing the significance of the event, delegated the investigation to the Canadian TSB. Unfortunately, the FDR data had been overwritten before it could be read.

---

* AC is the IATA code for Air Canada and is used in the source material.

376 Development of Aircraft Certification Requirements

Seats and Restraints

July 19, 1989: McDonnell Douglas DC-10-10, N-1819U

Location	Sioux City, IA
Flight	UAL 232
Operator	United Airlines
Operation	Scheduled Airline
Route	Denver, CO to Chicago, IL
On-Board Injuries	Fatal 111 Serious 47 Minor/None 138
Aircraft Damage	Destroyed

Details of this accident can be found on page 283. This discussion deals with passenger restraint.

Recommendations

There were four in-lap occupants onboard flight 232. Three of them were under 24 months, and one was 26 months old. During the preparations for the emergency landing, parents were instructed to place their infants on the floor and to hold them there when the parent assumed the protective brace position. The four in-lap occupants were held on the floor by adults who occupied seats IIF,12B, 14J and 22E.The woman in 145 stated that her son "flew up in the air" up on impact but that she was able to grab him and hold onto him. Details of what happened to the 26-month-old child at 12B during the impact sequence are not known, but he sustained minor injuries. The mother of the 11-month-old girl at 11F said that she had problems placing and keeping her daughter on the floor because she was screaming and trying to stand up. The mother of the 23-month-old at 22E was worried about her son's position. She kept asking the flight attendants for more specific instructions about the brace position and her "special situation with a child on the floor." The mothers of the infants in seats 11F and 22E were unable to hold onto their infants and were unable to find them after the airplane impacted the ground. The infant originally located at 11F was rescued from the fuselage by a passenger who heard her cries and reentered the fuselage. The infant held on the floor in front of seat 22E died of asphyxia secondary to smoke inhalation. The Safety Board addressed the infant restraint issue in Safety Recommendations A-90-78 and A-90-79 issued May 30, 1990

The NTSB issued the following recommendations to the FAA.

A-90-78   Revise 14 CFR 91,121and 135 to require that all occupants be restrained during takeoff, landing, and turbulent conditions, and that all infants and small children below the weight of40 pounds and under the height of 40 inches to be restrained in an approved child restraint system appropriate to their height and weight.

A-90-79   conduct research to determine the adequacy of aircraft seatbelts to restrain children too large to use child safety seats and to develop some suitable means of providing adequate restraint for such children.

July 2, 1994: Douglas DC-9-31, N-954VJ

Location	Charlotte, NC
Flight	USA 101
Operator	USAir (formerly Allegheny Airlines)
Operation	Scheduled Airline
Route	Columbia, NC to Charlotte, NC
On-Board Injuries	Fatal 37 Serious 16 Minor/None 4
Aircraft Damage	Destroyed

Details of this accident can be found on page 394. This discussion deals with passenger restraint.

Recommendations

The passenger cabin contained 21 rows of seats and was configured with 12 first class and 91 coach seats. There were 52 passengers on board flight1016: 27 males, 23 females, and 2 female in-lap infants (younger than 24 months) who were not listed on the passenger manifest.

A 9-month-old infant, who was unrestrained in her mother's lap in seat 21c, sustained fatal injuries. The mother was unable to hold onto her daughter during the impact sequence. Seat 21c was intact and the surrounding cabin structure sustained minor deformation. Additionally, the impact forces in this area were calculated to have been within human tolerances. According to passengers, a flash fire swept through the inside of the cabin during the impact sequence; examination found no evidence of either fire or smoke impingement in this area of the cabin.

The passenger manifest for flight 1016 listed 50 passenger names, but it did not include the names of the two "in-lap" infants. The tickets issued to the adults traveling with the in-lap infants were reviewed. One passenger's ticket had the marking "+infant" handwritten on the face of the ticket. The second passenger's ticket had no identifying information to document the carriage of the in-lap infant. 14 CFR Part 121.693 requires that children, regardless of their age or whether they are the sole occupant of a seat, must be listed by name on the passenger manifest.

The NTSB issued the following recommendations to the FAA.

A-95-50   Develop standards for forward-facing, integrated child-safety-seats for transport category aircraft.

A-95-51   Revise 14 CFR, Parts 91, 135, and 121 to required that all occupants be restrained during takeoff, landing, and turbulent conditions in a manner appropriate to their size.

The NTSB issued the following recommendations to USAir.

A-95-5   Review company procedures regarding passenger counts on manifests to ensure their accuracy and accountability of all occupants on the airplane.

June 3, 2008: Socata TBM-700N, N-849MA

Location	Iowa City, IA
Flight	N-849MA
Operator	Volunteer Part 91 "Angel" Flight
Operation	Patient Transportation
Route	Iowa City, IA to Decatur, AL
On-Board Injuries	Fatal 1 Minor/None 2
Aircraft Damage	Substantial

Description   Source: Aviation Safety Network[1]

The private pilot arrived at the accident airport as part of an Angel Flight volunteer program to provide transportation of a passenger who had undergone medical treatment at a local hospital. About 0937, the airplane landed on runway 30 (3,900 feet by 150 feet) with winds from 073-080 degrees and 5-6 knots, which continued to increase due to an atmospheric pressure gradient. The pilot met the passengers and departed the terminal about 1003, with winds at 101-103 degrees and 23-36 knots. About 1005 the airplane was near the approach end of runway 30 with wind from 089-096 degrees and 21-31 knots. The pilot stated that he began rotating the airplane about 3,000 feet down the runway. About 1006, the airplane was approximately 3,553 feet down the runway while flying about 30 feet above the runway. The airplane experienced an aerodynamic stall, and the left wing dropped before it impacted the ground. No mechanical anomalies that would have precluded normal operation of the airplane were noted during the investigation. The fatally injured passenger, who had received medical treatment, was 2 years and 10 months of age at the time of the accident. She was held by her mother during the flight, as she had been on previous Angel Flights, but was otherwise unrestrained. According to 14 CFR 91.107(3), each person on board a U.S.-registered civil aircraft must occupy an approved seat with a safety belt properly secured during takeoff, and only unrestrained children who are under the age of 2 may be held by a restrained adult. Although the accident was survivable (both the pilot and the adult passenger survived with non-life-threatening injuries), an autopsy performed on the child revealed that the cause of death was blunt force trauma of the head.

Probable Cause   Source: Accident Report[154]

The pilot's improper decision to depart with a preexisting tailwind and failure to abort takeoff. Contributing to the severity of the injuries was the failure to properly restrain (FAA-required) the child passenger.

Comment

It seems likely that the child's mother had become habituated to carrying her daughter, having taken many such "Angel Flights" with her daughter.

March 22, 2009: Pilatus PC-12/45, N-128CM

Location	Near Butte, MT
Flight	N-128CM*
Operator	Eagle Cap Leasing
Operation	Personal Transportation
Route	Orovill, CA to Bozeman, MT
On-Board Injuries	Fatal 14
Aircraft Damage	Destroyed

Details of this accident can be found on page 339. This discussion deals with passenger restraint.

Recommendations

The NTSB issued the following recommendations to the FAA.

A-10-121	Amend 14 *code of Federal Regulations* Part 91 to require separate seats and restraints for every occupant. (A-10-121)
A-10-122	Amend 14 *code of Federal Regulations* Part 91 to require each person who is less than 2 years of age to be restrained in a separate seat position by an appropriate child restraint system during takeoff, landing, and turbulence. (A-10-122)

December 22, 2012: Swearingen SA-227, C-GFWX

Location	Sanikiluaq Airport, NU, Canada
Flight	PAG 993
Operator	Perimeter Aviation
Operation	Scheduled Airline
Route	Winnipeg, MB, Canada to Sanikiluaq, NU, Canada
On-Board Injuries	Fatal 1 Serious 3 Minor/None 5
Aircraft Damage	Destroyed

Description   Source: Aviation Safety Network[1]

A Swearingen SA227-AC Metro III passenger plane was involved in an accident at Sanikiluaq Airport, NU (YSK), Canada. The flight crew sustained serious injuries, 6 passengers sustained minor injuries and one passenger, an infant, was fatally injured.

The route from Winnipeg to Sanikiluaq is normally operated by Keewatin Air, a Perimeter Aviation sister company. The normally scheduled flight on the previous day (Friday, 21 December 2012) had been cancelled due to poor weather in Sanikiluaq. With extra cargo and passengers needing to travel to Sanikiluaq before Christmas, Keewatin Air completed a flight on the morning of 22 December 2012 and had chartered Perimeter to complete an additional flight. The Perimeter flight crew had been notified in the early evening of 21 December 2012 that they would be operating flight PAG993 the next morning.

Probable Cause   Source: Accident Report[155]

Findings as to causes and contributing factors:

1. The lack of required flight documents, such as instrument approach charts, compromised thoroughness and placed pressure on the captain to find a work-around solution during flight planning. It also negatively affected the crew's situational awareness during the approaches at CYSK (Sanikiluaq).
2. Weather conditions below published landing minima for the approach at the alternate airport CYGW (Kuujjuarapik) and insufficient fuel to make CYGL (La Grande Rivière) eliminated any favourable diversion options. The possibility of a successful landing at CYGW was considered unlikely and put pressure on the crew to land at CYSK (Sanikiluaq).
3. Frustration, fatigue, and an increase in workload and stress during the instrument approaches resulted in crew attentional narrowing and a shift away from well-learned, highly practised procedures.
4. Due to the lack of an instrument approach for the into-wind runway and the unsuccessful attempts at circling, the crew chose the option of landing with a tailwind, resulting in a steep, unstable approach.
5. The final descent was initiated beyond the missed approach point and, combined with the 14-knot tailwind, resulted in the aircraft remaining above the desired 3-degree descent path.
6. Neither pilot heard the ground proximity warning system warnings; both were focused on landing the aircraft to the exclusion of other indicators that warranted alternative action.

---

* Aircraft registration used *vice* flight number.

7. During the final approach, the aircraft was unstable in several parameters. This instability contributed to the aircraft being half-way down the runway with excessive speed and altitude.

8. The aircraft was not in a position to land and stop within the confines of the runway, and a go-around was initiated from a low-energy landing regime.

9. The captain possibly eased off on the control column in the climb due to the low airspeed. This, in combination with the configuration change at a critical phase of flight, as called for in the company procedures, may have contributed to the aircraft's poor climb performance.

10. A rate of climb sufficient to ensure clearance from obstacles was not established, and the aircraft collided with terrain.

11. The infant passenger was not restrained in a child restraint system, nor was one required by regulations. The infant was ejected from the mother's arms during the impact sequence, and contact with the interior surfaces of the aircraft contributed to the fatal injuries.

Following an attempted visual approach to runway 09, a non-precision non-directional beacon (NDB) runway 27 approach was conducted. Visual contact with the runway environment was made and a circling for runway 09 initiated. Visual contact with the runway 09 environment was lost and a return to the Sanikiluaq NDB was executed. A second NDB runway 27 approach was conducted with the intent to land on runway 27. Visual contact with the runway environment was made after passing the missed approach point. Following a steep descent, a rejected landing was initiated at 20 to 50 feet above the runway; the aircraft struck the ground approximately 525 feet beyond the departure end of runway 27.

Recommendations

TSB made the following recommendations to the Canadian Department of Transport :

A15-01   The Department of Transport require commercial air carriers to collect and report, on a routine basis, the number of infants (under 2 years old), including lap-held, and young children (2 to 12 years old) traveling.

A15-02   The Department of Transport work with industry to develop age- and size-appropriate child restraint systems for infants and young children traveling on commercial aircraft, and mandate their use to provide an equivalent level of safety compared to adults.

February 17, 2014: Boeing 737-700, N-23708

Location	Near Kaycee, WY
Flight	UAL 1676
Operator	United Airlines
Operation	Scheduled Airline
Route	Denver, CO to Billings, MT
On-Board Injuries	Fatal _0_ Serious _2_ Minor/None _118_
Aircraft Damage	None

Description   Source: Accident Report[156]

On February 17, 2014, about 1254 mountain standard time (MST), United Airlines flight 1676, a Boeing 737-724, N23708, encountered severe turbulence about flight level 340 near Kaycee, Wyoming. Of the 120 passengers and crew onboard, two flight attendants were seriously injured, and one FA and nine passengers received minor injuries. The airplane was not damaged. The flight was operating under the provisions of 14 code of Federal Regulations Part 121 as a regularly scheduled passenger flight from Denver International Airport (DEN), Denver, Colorado, to Billings Logan International Airport (BIL), Billings, Montana.

The flight departed DEN normally and climbed to a cruising altitude of flight level (FL) 380. The flight encountered light turbulence and the crew requested, and was granted, a descent to FL360 followed by a descent to FL340. The flight crew extinguished the seat belt sign when the ride smoothed out. Flight data recorder (FDR) data indicated the airplane was cruising about FL340 when it encountered an area of severe turbulence for about 55 seconds. The most severe turbulence occurred for about 10 seconds when the vertical acceleration fluctuated between about -1.23 g's and +2.05 g's, the lateral acceleration fluctuated between about -0.27 g's and +0.23 g's, the pitch fluctuated between about +5.8° (nose up) and -1.8° (nose down), and the roll fluctuated between about 14° left and 20° right.

At the time of the turbulence encounter, the two flight attendants (FA) in the forward galley and the one FA in the aft galley were injured. during the turbulence encounter and suffered. The aft FA was unconscious after the event and tended to by medical personnel passengers until landing. After being notified of the serious injuries, the flight crew declared an emergency with air traffic control and arranged for medical personnel to meet the aircraft at BIL. The three FAs were transported to the hospital. One FA was diagnosed with a fractured fibula; one was diagnosed with fractures of her scapula, a rib, cervical vertebrae, and thoracic vertebrae; and one was treated and released with minor injuries. A lap-held infant flew out of its mother's arms to land in an empty seat 2 rows away; the infant was not injured.

According to a newspaper report, "A parent holding her baby in her arms could not hold the baby which lifted off too, but did not seem to have received injuries."[157] In the discussion in the accident report to Perimeter Flight 993, TSB reported that the child landed two rows away.[155]

Probable Cause   Source: Accident Report[156]

The NTSB determines the probable cause of this accident to be an encounter with severe mountain wave turbulence during cruise flight.

Grouped Recommendations

The NTSB issued several recommendations to the FAA.

> *Following the DC-10 accident in Sioux City, IA:* [158]
>
> *Following the DC-9 accident in Charlotte, NC:* [175]
>
> *Following the PC-12 accident in Butte, MT:* [113]
>
> *Following the SA227accident in* Sanikiluaq, NU:
> cHEcT Responses to these recommendations

Simulator Issues

November 12, 2001: Airbus A300B4, N-14053

Location   Belle Harbor, NY
Flight   AAL 587
Operator   American Airlines
Operation   Scheduled Airline
Route   New York (Kennedy), NY to Santo Domingo, Dominican Republic
On-Board Injuries   Fatal 260
Ground Injuries   Fatal 5
Aircraft Damage   Destroyed

Details of this accident can be found on page 374

Regulatory Outcome

This accident led to the introduction of Part 60, which established qualification requirements for flight simulators used in air carrier training.

December 11, 2018: Boeing 777, C-FITW

Location   Hong Kong Chek Lap Kok Airport, Hong Kong
Flight   ACA 15
Operator   Air Canada
Operation   Scheduled Airline
Route   Toronto, Canada to Hong Kong
On-Board Injuries   Minor/None 393
Aircraft Damage   Substantial

Description   Source: Aviation Safety Network[1]

Air Canada flight AC15, a Boeing 777-300ER, suffered a tailstrike while landing on runway 07R at Hong Kong-Chek Lap Kok International Airport.

The captain was the Pilot Monitoring (PM) and the Initial Operating Experience Training Captain for the first officer.

The crew anticipated an arrival and landing on runway 07L, however, there was a runway change to 07R. The weather was as expected with a wind velocity from 350 degrees at 12 knots.

The aircraft intercepted the ILS and was stabilised on the approach to runway 07R, on the correct descent profile with the autopilot engaged through 1000 ft. The FO disengaged the autopilot after descending through 500 ft. Following the reversion to manual flight, the approach profile became approximately half a dot above the glideslope.

At approximately 200 ft the aircraft entered into series of minor lateral roll deviations followed by a pronounced roll, first to the left and then to the right in response to the pilot's control inputs.

In response to the increasing unstable oscillations neither pilot called for or initiated a go around, nor did the other two crew members in the cockpit. At the runway contact point, the aircraft was rolling left and then right with a high rate of descent and a nose high pitch attitude. This resulted in a hard landing with the right main landing gear contacting the runway followed by the left main gear while the aft lower fuselage contacted the runway surface.

The aircraft bounced with the right-hand main landing gear contacting the runway first. The aircraft bounced again, landing on the nose gear followed by both main gears.

After the runway contact and initial bounce there was no call for a go around and after touch down from the subsequent bounce the PM removed the PF's hand from the thrust levers and selected reverse thrust. There was no formal transfer of control and there was a further distraction when a beverage container was dislodged from the PF's holder and dropped on to the floor and the PF bent forward to retrieve it.

The aircraft then completed the landing roll and continued to the parking stand.

The flight crew had recorded the descent and landing using a GoPro or similar action cam mounted on the left hand cockpit side window (without authorization from the airline). The crew voluntarily provided the footage for the investigation.

Probable Cause   Source: Accident Report[159]

An unstable approach developed due to pilot induced lateral rolling oscillations which coupled with a high rate of descent resulted in an abnormal runway contact.

Contributing Factors:
1) Stabilised Approach Criteria
   The late recognition by the PM that the stabilised approach criteria after the second (500 ft) arrival gate was outside the required tolerances.
2) Pilot Flying PIO Onset Recognition
   The over controlling (high gain) by the PF resulted in PIO. There is no requirement for PIO onset recognition or recovery actions in the operator's training procedures.
3) Go Around Decision
   The late recognition by the PM that the aircraft was in an unstable flight condition that should have resulted in an "unstabilised" or a "go around" call from the PM and required an immediate go around. The PF did not initiate a go around when the aircraft was in a PIO condition.
4) Pilot Flying Loss of Situational Awareness
   Task saturation with the lateral oscillation and high gain corrections resulted in the high descent rate up to the runway contact point.
5) Pilot Monitoring Loss of Situational Awareness
   Any decision to go around during the bounce was impeded due to the "startle effect", which delayed any response or action.

Comments

This was the First Officer's first flight in a Boeing 777! It was also his first landing in a B-777 and his first landing into Hong Kong as a crew member! He had received a type conversion course from the narrow-body Embraer E190. In our opinion, this is taking zero-time flight training too far.

The Captain was also engaged in extraneous conversations with the two relief pilots during the approach.

June 15, 2019: Boeing 757, N-26123

Location	Newark, NJ
Flight	UAL 627
Operator	United Airlines
Operation	Scheduled Airline
Route	Denver, CO to Newark, NJ
On-Board Injuries	Minor/None 272
Aircraft Damage	Substantial

Description   Source: NTSB Report[160]

On June 15, 2019, about 12:55 eastern daylight time (EDT), United Airlines flight 627, a Boeing 757-224, N26123, experienced a hard landing at Newark International Airport (KEWR), Newark, New Jersey. There were no injuries to the 172 passengers and crew onboard, and the airplane received substantial damage The flight was operating under Title 14 code of Federal Regulations Part 121 as a regularly scheduled domestic passenger flight from Denver International Airport (KDEN), Denver, Colorado.

The first officer (FO) was the pilot flying, and the captain was pilot monitoring. The takeoff, cruise, and decent were normal. The captain stated that it was an Initial Operating Experience (IOE) flight for the first officer. The auto speed brake system on the airplane was deferred in accordance with the Minimum Equipment List, and had been briefed, and all checklists had been completed. The captain stated that the FO flew a "solid" approach profile with few minor airspeed deviations, all of which were corrected immediately. The crew stated that during the descent to runway 22L the wind became gusty (220 degrees at 14 kts, gusting to 22 kts), and they increased the VREF speed in accordance with company procedures. At 500 feet, the airplane was on profile, on speed and stable. According to the crew the initial touchdown was smooth, on centerline, and in the touchdown zone. Upon touchdown on the main wheels, the captain manually deployed the speed brakes and the nose pitched up. To avoid a tailstrike, the captain said she physically blocked the yoke from moving back and instructed the FO to pitch forward. The airplane then bounced on the runway.

The airplane was equipped with a L-3/Fairchild FA2100 Flight Data Recorder (FDR). FDR data showed that at 12:54:54 EDT, the airplane was on final approach to runway 22L with the autopilot off, glideslope mode engaged, and both flight directors on. At 12:55:14 EDT, with a pitch angle of 2.8 degrees, and the speed brake handle in the down (unarmed) position, the main landing gear (MLG) weight-on-wheels (WOW) parameter changed to GROUND. At 12:55:16 variations in the vertical acceleration, pitch, elevator, and control column parameters increased in magnitude. At 12:55:17, MLG WOW parameter changed back to AIR for one second, then back to GROUND for one second, then back to AIR for one second, before finally cycling back to GROUND. At 12:55:20 the nose gear WOW indicated GROUND for the first time, along with the largest magnitude vertical acceleration of 1.6042 g's. Both left and right engine thrust reversers indicated deployed six seconds after initial MLG touchdown.

Postaccident inspection of the airplane revealed extensive structural damage to the right and left forward area (41/43 station) of the fuselage. There was extensive damage to twelve skin panels, eleven severed RH stringers and twelve buckled LH stringer sections, multiple underlying damaged structural components, and damage to the nose landing gear and support structure.

Probable Cause

Source: NTSB Report[160]

The National Transportation Safety Board determines the probable cause(s) of this accident to be:an improper landing flare, which resulted in a bounced landing and substantial damage.

Comments

It seems unconscionable to us that the handling pilot on a scheduled airline might fly the airplane for the first time with paying passengers aboard. In our opinion, we have taken "zero time flight training" too far.

Spatial Disorientation

February 3, 1959: Beech BE-35, *Bonanza*, N-3851N

Location	5 mi N of Mason City Airport, IA
Flight	N-3851N*
Operator	Dwyer Flying Service
Operation	Non-Scheduled Passenger Charter
Route	Mason City, IA to Fargo, ND
On-Board Injuries	Fatal 4
Aircraft Damage	Destroyed

Description   Source: FAA Lessons Learned[161]

The accident flight was a charter flight from Mason City, Iowa to Fargo, North Dakota. Prior to the flight, and beginning at approximately 1730 on February 2, the pilot began checking weather information for the route of flight. Weather briefings included the local Mason City area, Minneapolis, Redwood Falls, and Alexandria, Minnesota, and the terminal forecast for Fargo. During this initial check, the pilot was advised that all stations were reporting ceilings of at least 5,000 feet and visibility of ten miles or more. The Fargo terminal forecast indicated the possibility of light snow showers after 0200 February 3 and passage of a cold front at about 0400. The pilot was also made aware that a later terminal forecast would be available at 2300. At 2200, and again at 2330, the pilot repeated his weather checks. During the 2330 check, he was advised that the stations en route were reporting ceilings of at least 4,200 feet with visibility still ten miles or greater. Light snow was reported at Minneapolis. The cold front previously forecast to pass Fargo at 0400 was now forecast to pass there at 0200. The Mason City weather was a measured ceiling of 6,000 feet, overcast skies, and visibility of greater than 15 miles. Temperature was reported as 15 degrees F with a dew point of 8 degrees F. Wind was from the south at 25 to 32 knots. Barometric pressure was 29.96 inches of mercury.

Description (2)   From the Accident Report[162]

When his instrument training was taken, several aircraft were used and these were all equipped with the conventional type artificial horizon and none with the Sperry Attitude Gyro such as was installed in Bonanza N 3794N. These two instruments differ greatly in their pictorial display.

The conventional artificial horizon provides a direct reading indication of the bank and pitch attitude of the aircraft which is accurately indicated by a miniature aircraft pictorially displayed against a horizon bar and as if observed from the rear. The Sperry F3 gyro also provides a direct reading indication of the bank and pitch attitude of the aircraft, but its pictorial presentation is achieved by using a stabilized sphere whose free-floating movements behind a miniature aircraft presents pitch information with a sensing exactly opposite from that depicted by the conventional artificial horizon.

Probable Cause   Source: Accident Report[162]

The Board determines that the probable cause of this accident was the pilot's unwise decision to embark on a flight which would necessitate flying solely by instruments when he was not properly certificated or qualifies to do so. Contributing factors were serious deficiencies in the weather briefing, and the pilot's unfamiliarity with the instrument which determines the attitude of the aircraft.

Regulatory Outcome

The FAA's Lessons Learned from civil Aviation Accidents page lists the following safety initiatives:

The General Aviation Joint Steering committee (Loss of control (LOC) working group has identified safety enhancements that will result in lowering of the LOC accident rate. At this time they include strategies focused on implementing safety technologies including:

- Angle of attack sensors and displays
- Simple autopilots and cockpit automation
- Novel displays enhancing pilot situational awareness to prevent LOC inflight (LOC-I)
- Electronic parachute (emergency automatic landing)
- Flight trajectory management (advanced flight controls and fly-by-wire flight control systems for Part 23 airplanes)
- Improved flight envelope protection
- Airplane energy state awareness

---

* Aircraft registration used *vice* flight number.

In 1991, the FAA published Advisory Circular (AC) 60-22, Aeronautical Decision Making, which discusses risk and stress management tools, and how personal attitudes can affect and influence decision making relative to operational safety during flight and flight planning. The AC was published as a tool to help pilots recognize and alleviate unnecessary risk associated with flying activities.).

Additionally, in 2003 and 2011, the FAA issued advisory material regarding attitude indicators and electronic instrument displays in airplanes certificated under 14 CFR Part 23.

Advisory Circular 91.75, Attitude Indicators, provides guidance and justification relative to replacing the rate of turn indicator (turn and bank indicator) with a standby attitude indicator.

Advisory Circular 23.1311-1c, Installation of Electronic Display in Part 23 Airplanes, provides guidance for the design, installation, integration, and approval of electronic flight displays in Part 23 airplanes.

Comments:

The attitude display in the accident airplane was a non-standard type – apparently mixed dynamics: inside out in bank; outside-in in pitch.

It seems strange to us to use this accident to justify changes in policy forty-some years after the fact. It was the only general aviation accident in the FAA's Lessons Learned for all general aviation at the time we first accessed the site. (The list of transport airplane accidents totaled 81 through 2010.)

The three passengers were Buddy Holly, Ritchie Valens and The Big Bopper.

July 16, 1999: Piper PA-32, N-9253N

Location	7-1/2 miles SW Martha's Vineyard MA
Flight	N-9253N*
Operator	Private Individual
Operation	Personal Transportation
Route	Caldwell NJ to Martha's Vineyard MA
On-Board Injuries	Fatal 3
Aircraft Damage	Destroyed

Description   Source: NTSB Accident File[163]

On July 16, 1999, about 2141 eastern daylight time, a Piper PA-32R-301, Saratoga II, N9253N, was destroyed when it crashed into the Atlantic Ocean approximately 7 1/2 miles southwest of Gay Head, Martha's Vineyard, Massachusetts. The certificated private pilot and two passengers received fatal injuries. Night visual meteorological conditions (VMC) prevailed, and no flight plan had been filed for the personal flight conducted under the provisions of 14 code of Federal Regulations (CFR) Part 91. The flight originated from Essex County Airport (CDW), Caldwell, New Jersey, and was destined for Barnstable Municipal-Boardman/Polando Field (HYA), Hyannis, Massachusetts, with a scheduled stop at Martha's Vineyard Airport (MVY), Vineyard Haven, Massachusetts.

During interviews, witnesses stated that the purpose of the flight was to fly to Martha's Vineyard to drop off one passenger and then continue to HYA. An employee of a fixed-base operator (FBO) at cDW stated that he had called the pilot about 1300 on the day of the accident to verify that the pilot intended to fly the airplane, N9253N, over the weekend. The pilot informed the employee that he did plan to fly the airplane and that he would arrive at the airport between 1730 and 1800. The employee informed the pilot that he would have the airplane parked outside of the hangar.

Witnesses who were at cDW on the night of the accident stated that they saw the pilot and a female near the accident airplane. The witnesses also reported that they saw the pilot using crutches and loading luggage into the airplane. One witness stated that he watched the pilot perform an engine run-up and then take off about 2040. The witness further stated that "takeoff and right downwind departure seem[ed] normal."

According to air traffic control (ATC) transcripts from cDW's tower, about 2034, the pilot of N9253N contacted the ground controller and stated, "...Saratoga niner two five three november ready to taxi with mike...right turnout northeast bound." The ground controller instructed the pilot to taxi to runway 22, which the pilot acknowledged. At 2038:32, the pilot of N9253N contacted the tower controller and advised that he was ready to take off from runway 22. At 2038:39, the tower controller cleared N9253N for takeoff; at 2038:43, the pilot acknowledged the clearance. A few seconds later, the tower controller asked the pilot if he was heading towards Teterboro, New Jersey. The pilot replied, "No sir, I'm uh actually I'm heading a little uh north of it, uh eastbound." The tower controller then instructed the pilot to "make it a right downwind departure then." At 2038:56, the pilot acknowledged the instruction stating, "right downwind departure two two." No records of any further communications between the pilot and ATC exist.

According to radar data, at 2040:59, a target transmitting a visual flight rules (VFR) code was observed about 1 mile southwest of cDW at an altitude of 1,300 feet. The target proceeded to the northeast, on a course of about 55 degrees, remaining below 2,000 feet. The target was at 1,400 feet when it reached the Hudson River. When the target was about 8 miles northwest of the Westchester county Airport (HPN), White Plains, New York, it turned north over the river and began to climb. After proceeding north about 6 miles, the target turned eastward to a course of about 100 degrees. The target continued to climb and reached 5,500 feet about 6 miles northeast of HPN. When the target's course was plotted on a New York VFR navigational map, the extended course line crossed the island of Martha's Vineyard.

---

*   Aircraft registration used *vice* flight number.

The target continued eastward at 5,500 feet, passing just north of Bridgeport, Connecticut, and crossed the shoreline between Bridgeport and New Haven, Connecticut. The target ground track continued on the 100-degree course, just south and parallel to the Connecticut and Rhode Island coastlines. After passing Point Judith, Rhode Island, the target continued over the Rhode Island Sound.

A performance study of the radar data revealed that the target began a descent from 5,500 feet about 34 miles west of MVY. The speed during the descent was calculated to be about 160 knots indicated airspeed (KIAS), and the rate of descent was calculated to have varied between 400 and 800 feet per minute (fpm). About 2138, the target began a right turn in a southerly direction. About 30 seconds later, the target stopped its descent at 2,200 feet and began a climb that lasted another 30 seconds. During this period of time, the target stopped the turn, and the airspeed decreased to about 153 KIAS. About 2139, the target leveled off at 2,500 feet and flew in a southeasterly direction. About 50 seconds later, the target entered a left turn and climbed to 2,600 feet. As the target continued in the left turn, it began a descent that reached a rate of about 900 fpm. When the target reached an easterly direction, it stopped turning; its rate of descent remained about 900 fpm. At 2140:15, while still in the descent, the target entered a right turn. As the target's turn rate increased, its descent rate and airspeed also increased. The target's descent rate eventually exceeded 4,700 fpm. The target's last radar position was recorded at 2140:34 at an altitude of 1,100 feet. (For a more detailed description of the target's [accident airplane's] performance, see Section, "Tests and Research," Subsection, "Aircraft Performance Study.")

On July 20, 1999, about 2240, the airplane's wreckage was located in 120 feet of water, about 1/4 mile north of the target's last recorded radar position.

The accident occurred during the hours of darkness. In the area of and on the night of the accident, sunset occurred about 2014. civil twilight ended about 2047, and nautical twilight ended about 2128. About 2140, the moon was about 11.5 degrees above the horizon at a bearing of 270.5 degrees and provided about 19 percent illumination. The location of the accident wreckage was about 41 degrees, 17 minutes, 37.2 seconds north latitude; 70 degrees, 58 minutes, 39.2 seconds west longitude.

Probable Cause   Source: NTSB Accident File [163]

The pilot's failure to maintain control of the airplane during a descent over water at night, which was a result of spatial disorientation. Factors in the accident were haze, and the dark night.

Comments

This accident is typical of the many similar spatial disorientation accidents to non-instrument-rated pilots. We included it because of the similarities with the following accident to ZHA-604.

January 3, 2004: Boeing 737-300, SU-ZCF

Location	Sharm el Sheikh, Egypt
Flight	ZHA 604
Operator	Flash Airlines
Operation	Nonscheduled Passenger
Route	Sharm el Sheikh, Egypt to Cairo, Egypt
On-Board Injuries	Fatal 148
Aircraft Damage	Destroyed

Probable Cause   Source: Aviation Safety Network [1]

Weather was perfect (excellent visibility, 17 degrees C and a light breeze) when Flash Air flight 604 departed the Red Sea resort of Sharm el Sheikh for a flight to Paris-CDG with an intermediate stop at Cairo. On board were 135, mostly French, holidaymakers who were heading home. At 04:38 the flight was cleared to taxi to runway 22R for departure. After takeoff, at 04:42, the plane climbed and maneuvered for a procedural left turn to intercept the 306 radial from the Sharm el-Sheikh VOR station. When the autopilot was engaged the captain made an exclamation and the autopilot was immediately switched off again. The captain then requested Heading Select to be engaged. The plane then began to bank to the right. The copilot then warned the captain a few times about the fact that the bank angle was increasing. At a bank angle of 40 degrees to the right the captain stated "OK come out". The ailerons returned briefly to neutral before additional aileron movements commanded an increase in the right bank.

The aircraft had reached a maximum altitude of 5460 feet with a 50 degrees bank when the copilot stated: "Overbank". Repeating himself as the bank angle kept increasing. The maximum bank angle recorded was 111 degrees right. Pitch attitude at that time was 43 degrees nose down and altitude was 3470 feet.

The observer on the flight deck, a trainee copilot, called "Retard power, retard power, retard power". Both throttles were moved to idle and the airplane gently seemed to recover from the nose-down, right bank attitude. Speed however increased, causing an overspeed warning. At 04:45 the airplane struck the surface of the water in a 24 degrees right bank, 24 degrees nose-down, at a speed of 416 kts and with a 3,9 G load.

The wreckage sank to a depth of approx. 900 metres.

Probable Cause
Source: Accident Report[164]

No conclusive evidence could be found from the findings gathered through this investigation to determin the probable cause. However, based on the work done, it could be concluded that any combination of these findings could have caused or contributed to the accident.

Although the crew at the last stage of this accident attempted to correctly recover, the gravity upset condition with regards to attitude, altitude and speed made this attempt insufficient to achieve a successful recovery.

Additional comment

One of the comments on the draft of the Final Report made by the US as a participant in the Investigation was that

"The only scenario identified by the investigative team that explained the accident sequence of events and was supported by the available evidence was a scenario indicating that the captain experienced spatial disorientation, which resulted in his making inadvertent actions that caused the accident. The remaining scenarios and possible causes were not consistent with the evidence and did not explain the sequence of events identified by the investigative team.

"To which the response of the MCA, as the body with responsibility for the conduct and findings of the Investigation was that:

"With regard to the statement that there was supporting evidence that the captain experienced spatial disorientation is inaccurate to say the least. The investigation team studied this scenario extensively, numerous conflicting evidences appeared leading to the MCA adopting the position that no conclusive evidence could be found to explain this accident." [165]

Comments

One of your authors (RLN) participated in the accident investigation. When he briefed is manager at the FAA, surprise was expressed because "Airline pilots don't get disoriented." The previous accident to John Kennedy, Jr was included because of the similarities to this accident.

It was clear that the pilot, an experienced Egyptian Air Force pilot was overcome by spatial disorientation. He had over 7000 flying hours, mostly in C-130's, but limited recent experience. It is interesting to note that early in his career, he had flown a Russian MiG fighter with a moving airplane attitude indicator which moves in the opposite sense to the more common moving horizon indicator. It would seem that he reverted to his early experience.

The moving airplane attitude indicator, referred to as the *outside-in display* was proposed in the 1930s by two US Army pilots, Ocker and Crane.[166] It was not adopted in the US, but was found in many Russian airplanes, including MiG 21. In the early 1970s, we had the opportunity to fly such display with Carl Crane, in a Cessna 172.[167] The *outside-in* format has been thought to be better for naive subjects. However, when introducing such a display, the large number of pilots who have been over-trained on a different format cannot be ignored.

The attitude indicator in the N-3794N accident was a mixed inside-out and outside-in design – possibly the worst of both formats.

January 26, 2020: Sikorsky S-76B, N-72EX

Location	Calabasas, CA
Flight	ISHA* N-72EX†
Operator	Island Express Helicopters
Operation	Nonscheduled Passenger
Route	Santa Ana, CA to Camarillo, CA
On-Board Injuries	Fatal 9
Aircraft Damage	Destroyed

Probable Cause
Source: Aviation Safety Network[1]

The on-demand flight was operated by Island Express Helicopters Inc. under visual flight rules. The flight departed from John Wayne Airport-Orange county (SNA), Santa Ana, California, about 09:07 and was destined for Camarillo Airport (CMA), Camarillo, California, about 24 miles west of the accident site.

After the helicopter departed from SNA, it flew at altitudes that remained below 1,700 ft mean sea level (msl) and generally between 400 to 600 ft above ground level (agl), and the flight's progress through controlled airspace en route to CMA was uneventful. Weather conditions reported to the pilot by air traffic controllers during the flight included an overcast ceiling at 1,100 ft agl, visibility of 2.5 miles with haze, and cloud tops at 2,400 ft msl.

At 09:44:34 (about 2 minutes before the accident), while the helicopter was flying west at an altitude of about 1,370 ft msl (450 ft agl) over US Route 101 and rising terrain, the pilot announced to an air traffic control facility that he was initiating a climb to get the helicopter "above the [cloud] layers," and the helicopter immediately began climbing at a rate of about 1,500 ft per minute. About the same time, the helicopter began a gradual left turn, and its flight path generally continued to follow US 101 below. About 36 seconds later and while still climbing, the helicopter began to turn more tightly to the left, and its flight path diverged from its overflight of US 101.

*   ISHA is the FAA operator code for Island Express Helicopters
†   Aircraft registration used *vice* flight number.

The helicopter reached an altitude of about 2,370 ft msl (about 1,600 ft agl) at 09:45:15, then it began to descend rapidly in a left turn to the ground. At 09:45:17 (while the helicopter was descending), the air traffic controller asked the pilot to "say intentions," and the pilot replied that the flight was climbing to 4,000 ft msl. A witness near the accident site first heard the helicopter then saw it emerge from the bottom of the cloud layer in a left-banked descent about 1 or 2 seconds before impact.

Probable Cause   Source: Accident Report[168]

The National Transportation Safety Board determines that the probable cause of this accident was the pilot's decision to continue flight under visual flight rules into instrument meteorological conditions, which resulted in the pilot's spatial disorientation and loss of control. Contributing to the accident was the pilot's likely self-induced pressure and the pilot's plan continuation bias, which adversely affected his decision-making, and Island Express Helicopters Inc.'s inadequate review and oversight of its safety management processes.

Safety Recommendations

The NTSB issued two recommendations to the FAA regarding training for spatial disorientation prevention and recovery.

Require the use of appropriate simulation devices during initial and recurrent pilot training for Title 14 code of Federal Regulations Part 135 helicopter operations to provide scenario-based training that addresses the decision-making, skills, and procedures needed to recognize and respond to changing weather conditions in flight, identify and apply mitigation strategies for avoiding adverse weather, practice the transition to the use of flight instruments to reduce the risk of spatial disorientation, and maintain awareness of a variety of influences that can adversely affect pilot decision-making. (A-21-5)

Convene a multidisciplinary panel of aircraft performance, human factors, and aircraft operations specialists to evaluate spatial disorientation simulation technologies to determine which applications are most effective for training pilots to recognize the onset of spatial disorientation and successfully mitigate it, and make public a report on the committee's findings. (A-21-6)

Volcanic Ash Ingestion

June 24, 1982: Boeing 747, G-BDXH

Location	150 SE Jakarta, Indonesia
Flight	BAW 009
Operator	British Airways
Operation	Scheduled Airline
Route	Kuala Lumpur, Malaysia to Perth, Australia
On-Board Injuries	Minor/None 262
Aircraft Damage	Minor

Probable Cause   Source: Aviation Safety Network[1]

Cruising at FL370 the aircraft entered a cloud of volcanic dust (Mount Galunggung had erupted). The crew noticed St. Elmo's fire and smoke and dust were present in the cabin. All four engines failed and the aircraft started to descend. The crew managed to restart engine no. 4 at FL130 and restarted the other engines in succession. Because engine no. 2 continually surged, a 3-engined emergency landing was carried out at Jakarta.

Probable Cause:*[169]

British Airways Flight 009, sometimes referred to by its callsign Speedbird 9 or as the Jakarta incident,[1] was a scheduled British Airways flight from London Heathrow to Aukland, with stops in Bombay, Kuala Lumpur, Perth and Melbourne.

On 24 June 1982, the route was flown by ... a Boeing 747-200 registered as G-BDXH. The aircraft flew into a cloud of volcanic ash thrown up by the eruption of Mount Galunggung around 110 miles (180 km) south-east of Jakarta, Indonesia, resulting in the failure of all four engines. The reason for the failure was not immediately apparent to the crew or air traffic control. The aircraft was diverted to Jakarta in the hope that enough engines could be restarted to allow it to land there. The aircraft glided out of the ash cloud, and all engines were restarted (although one failed again soon after), allowing the aircraft to land safely at the Halim Perdanakusuma Airport in Jakarta.

The crew members of the accident segment had boarded the aircraft in Kuala Lumpur, while many of the passengers had been aboard since the flight began in London.

December 15, 1989: Boeing 747, PH-BFC

Location	Near Anchorage, AK
Flight	KLM 867†
Operator	KLM Royal Dutch Airlines
Operation	Scheduled Airline
Route	Amsterdam, Netherlands to Anchorage, AK, USA
On-Board Injuries	Minor/None 245
Aircraft Damage	Substantial

Probable Cause   Source: Aviation Safety Network[1]

At FL250 the aircraft flew into a normal looking cloud, which turned out to be a volcanic ash cloud (the result of an eruption of Mount Redoubt). Power was added to climb out of the cloud. About 10-15 seconds later all 4 engines failed and the standby electrical system failed. The crew were able to restart the engines after numerous attempts. The no. 1 and 2 engines were relit while descending through FL130 and the remaining 2 engines were relit at FL110. The aircraft landed safely at Anchorage, substantially damaged by the in-flight blasting by volcanic ash. The windshields were damaged, as were internal aircraft systems, avionics and electronics.

Probable Cause   Source: Accident Report[170]

Inadvertent encounter with volcanic ash cloud which resulted in damage from foreign material (FOREIGN OBJECT) and subsequent compressor stalling of all engines. A factor related to the accident was the lack of available information about the ash cloud to all personnel involved.

---

*   We could not locate any report from any of the likely agencies. The summary was obtained from the internet.

†   The source material uses the IATA code, KL.

Uninstalled & Unapproved Equipment

December 8, 2005: Boeing 737, N-471WN

Location	Chicago-Midway Airport, IL
Flight	SWA 1248
Operator	Southwest Airlines
Operation	Scheduled Airline
Route	Baltimore MD to Chicago IL
On-Board Injuries	Minor/None <u>103</u>
Ground Injuries	Fatal <u>1</u>
Aircraft Damage	Substantial

Probable Cause   Source: Aviation Safety Network[1]

Southwest Flight 1248 was scheduled for a 14:55 departure out of Baltimore/Washington International Airport, but the aircraft's departure was delayed until 16:50 because of weather. After circling for over 30 minutes the crew were cleared for landing at Midway's runway 31c (6522 feet long). The crew choose to use the auto-braking system at "max". The approach speed was 120 kts and 132 kts by the time it landed. The aircraft touched down with about 4,500 feet of remaining runway. The flying pilot (captain) stated that he could not get the reverse thrust levers out of the stowed position. The first officer, after several seconds, noticed that the thrust reversers were not deployed and activated the reversers without a problem. The reversers were activated 18 seconds after touchdown. [Noticing] that the airplane was not decelerating normally, both crew members applied maximum braking manually. The airplane departed the end of the runway, rolled through a blast fence, a perimeter fence, and onto a roadway. The 737 collided with some vehicles and came to rest on the corner of S central Avenue and W 55th Street. A six-year-old boy in one of the vehicles has died as a result of injuries sustained in the accident.

Probable Cause   Source: Accident Report[171]

The National Transportation Safety Board determined that the probable cause of this accident was the pilots' failure to use available reverse thrust in a timely manner to safely slow or stop the airplane after landing, which resulted in a runway overrun. This failure occurred because the pilots' first experience and lack of familiarity with the airplane's autobrake system distracted them from thrust reverser usage during the challenging landing.

Contributing to the accident were Southwest Airlines' 1) failure to provide its pilots with clear and consistent guidance and training regarding company policies and procedures related to arrival landing distance calculations; 2) programming and design of its on board performance computer, which did not present critical assumption information despite inconsistent tailwind and reverse thrust assessment methods; 3) plan to implement new autobrake procedures without a familiarization period; and 4) failure to include a margin of safety in the arrival assessment to account for operational uncertainties. Contributing to the severity of the accident was the absence of an engineering materials arresting system, which was needed because of the limited runway safety area beyond the departure end of runway 31c.

Regulatory Outcome

The FAA published an "Announcement of Policy for Landing Performance After Departure for All Turbojet Operators". However there was considerable industry opposition and the FAA decided not to issue the mandatory changes to operations specifications. Instead, the FAA decided to pursue formal rulemaking. In the interim, the FAA issued a Safety Alert for Operators (SAFO).[172]

The FAA issued Order 5200.8 to establish a program to bring airport runway safety areas up to current standards.[22] However, a search of the RGL website did not show any Order 5200.8.

All of the regulatory changes to date involved publishing advisory material.

Comments

This accident points up the "unintended consequences" of uncertified software. There was no check to ensure that the software performed its intended function. As it happened, with a tailwind in excess of the maximum calculated the runway requirements based on the maximum not the actual tailwind! We also note that in our days as an airline pilot, our company's maximum tailwind on an icy runway was zero.

A final observation: this was the crew's first ever using autobrakes in an actual airplane. We agree with the NTSB that the crew delayed applying reverse thrust, but feel that was because they wanted to make sure the system functioned. A pilot should never have to use a system for the first time in a critical situation. The first real autobrake use should be on a dry long runway.

Windshear

August 2, 1985: Lockheed L-1011, N-726DA

Location	Dallas-Fort Worth, TX
Flight	DAL 191
Operator	Delta Air Lines
Operation	Scheduled Airline
Route	Fort Lauderdale, FL to Dallas-Fort Worth, TX
On-Board Injuries	Fatal <u>134</u> Serious<u>27</u> Minor/None <u>2</u>
Ground On-Board Injuries	Fatal <u>1</u> Serious <u>1</u>
Aircraft Damage	Destroyed

Probable Cause

Source: Aviation Safety Network[1]

Delta Air Lines flight 191 was a regularly scheduled passenger flight between Fort Lauderdale, FL (FLL), and Los Angeles, CA (LAX), with an en route stop at the Dallas/Fort Worth International Airport, TX (DFW). Flight 191, a Lockheed L-1011 TriStar airplane, departed Fort Lauderdale on an IFR flight plan with 152 passengers and a crew of 11 on board at 15:10 EDT. The DFW Airport terminal weather forecast contained in the flightcrew's dispatch document package stated, in part, that there was a possibility of widely scattered rain showers and thunderstorms, becoming isolated after 20:00 CDT.

The flight was uneventful until passing New Orleans, Louisiana. A line of weather along the Texas-Louisiana gulf coast had intensified. The flightcrew elected to change their route of flight to the more northerly Blue Ridge arrival route to avoid the developing weather to the south. This change necessitated a 10 to 15-minute hold at the Texarkana, Arkansas, WORTAC for arrival sequencing at the DFW Airport. At 17:35, the flightcrew received the following ATIS broadcast: "DFW arrival information romeo, two one four seven Greenwich, weather six thousand scattered, two one thousand scattered, visibility one zero, temperature one zero one, dew point six seven, wind calm, altimeter two niner niner two, runway one eight right one seven left, visual approaches in progress, advise approach control that you have romeo".

Fort Worth Air Route Traffic control center (ARTCC) then cleared flight 191 to the Blue Ridge, Texas, WORTAC for the Blue Ridge Nine arrival, and to begin its descent. At 17:43:45, Fort Worth ARTCC cleared flight 191 to descend to 10,000 feet, gave it a 29.92 in Hg altimeter setting, and suggested that the flight turn to a heading-of 250 degrees "to join the Blue Ridge zero one zero radial inbound and we have a good area there to go through.!' The captain replied that he was looking at a "pretty good size" weather cell, "at a heading of two five five ... and I'd rather not go through it, I'd rather go around it one way or the other." Fort Worth ARTCC then gave the flight another heading and stated "when I can I'll turn you into Blue Ridge, it'll be about the zero one zero radial."

At 17:46, the center cleared flight 191 direct to Blue Ridge and to descend to 9,000 feet, and flight 191 acknowledged receipt of the clearance. At 17:48, the captain told the first officer, "You're in good shape. I'm glad we didn't have to go through that mess. I thought sure he was going to send us through it." Three minutes later, the flight engineer said, "Looks like it's raining over Fort Worth."

At 17:51, Forth Worth ARTCC instructed flight 191 to contact DFW Airport Approach control. At 17:56:28, Regional Approach control's Feeder East controller transmitted an all aircraft message which was received by flight 191. The message stated in part, "Attention, all aircraft listening... there's a little rainshower just north of the airport and they're starting to make ILS approaches ... tune up one oh nine one for one seven left." At 17:59, the first officer stated, "We're gonna get our airplane washed," and the captain switched to Regional Approach control's Arrival Radar-1 (AR-1) frequency and told the controller that they were at 5,000 feet. At 18:00, the approach controller asked American Air Lines flight 351 if it was able to see the airport. (Flight 351 was two airplanes ahead of flight 191 in the landing sequence for runway 17L.) Flight 351 replied, "As soon as we break out of this rainshower we will." The controller then told flight 351 that it was 4 miles from the outer marker, and to join the localizer at 2,300 feet; the controller then cleared the flight for the ILS approach to runway 17L.

At 18:00, the approach controller asked flight 191 to reduce its airspeed to 170 knots, and to turn left to 270 degrees; flight 191 then acknowledged receipt of the clearance. Flight 191 had been sequenced behind a Learjet 25 for landing on runway 17L. At 18:02, the approach controller told flight 191 that it was 6 miles from the outer marker, requested that it turn to 180 degrees to join the localizer at or above 2,300 feet, and stated, "cleared for ILS one seven left approach." The flight acknowledged receipt of the transmission.

At 18:03:03, the approach controller requested flight 191 "to reduce your speed to one six zero please," and the captain replied, "Be glad to." Thereafter, at 18:03:30, he broadcast, "And we're getting some variable winds out there due to a shower... out there north end of DFW." This transmission was received by flight 191.

At 18:03:46, the approach controller requested flight 191 to slow to 150 KIAS, and to contact the DFW Airport tower. At 18:03:58, the captain, after switching to the tower's radio frequency, stated, "Tower, Delta one ninety one heavy, out here in the rain, feels good." The tower cleared the flight to land and informed it, "wind zero nine zero at five, gusts to one five." At 18:04:07, the first officer called for the before-landing check. The flightcrew confirmed that the landing gear was down and that the flaps were extended to 33 degrees, the landing flap setting. At 18:04:18, the first officer said, "Lightning coming out of that one." The captain asked, "What?" and the first officer repeated "Lightning coming out of that one." The captain asked, and at 18:04:23, the first officer replied, "Right ahead of us." Flight 191 continued descending along the final approach course. At 18:05:05 the captain called out "1,000 feet." At 18:05:19, the captain cautioned the first officer to watch his indicated airspeed and a sound identified as rain began. The captain then warned the first officer, "You're gonna lose it all of a sudden, there it is." The captain stated, "Push it up, push it way up." At 18:05:29, the sound of engines at high rpm was heard on the CVR, and the captain said "That's it." At 18:05:44, the Ground Proximity Warning System's (GPWS) "Whoop whoop pull up" alert sounded and the captain commanded "TOGA". The CVR recording ended at 18:05:58. Witnesses on or near State Highway 114 north of the airport saw flight 191 emerge from the rain about 1.25 miles from the end of runway 17L and then strike an automobile in the westbound lane of State Highway 114. Subsequent investigation showed that the airplane had touched down earlier and became airborne again before striking the automobile. After the plane struck the car and a light pole on the highway, other witnesses saw fire on the left side of the airplane in the vicinity of the wing root. The witnesses generally agreed that the airplane struck the ground in a left-wing-low attitude, and that the fuselage rotated counterclockwise after the left wing and cockpit area struck a water tank on the airport. A large explosion obscured the witnesses' view momentarily, and then the tail section emerged from the fireball, skidding backwards. The tail section finally came to rest on its left side with the empennage pointing south and was subsequently blown to an upright position by wind gusts. One hundred and thirty four persons on board the airplane and the driver of the automobile which was struck by the airplane were killed in the accident; 27 persons on board the airplane and 1 rescue worker at the accident site were injured, 2 passengers on the airplane were uninjured.

Probable Cause   Source: Accident Report[173]

The National Transportation Safety Board determines that the probable causes of the accident were the flightcrew's decision to initiate and continue the approach into a cumulonimbus cloud which they observed to contain visible lightning; the lack of specific guidelines, procedures, and training for avoiding and escaping from low-altitude windshear; and the lack of definitive, real-time windshear hazard information. This resulted in the aircraft's encounter a t low altitude with a microburst-induced, severe windshear from a rapidly developing thunderstorm located on the final approach course.

Safety Recommendations[173]

Pilot training which stresses avoidance of windshear and discusses the meteorological conditions conducive to the development of windshears, particularly convective windshears.

Pilot training programs which

(1) discuss the aerodynamic performance problems associated with windshear penetrations as well as simulations of windshear encounters during all low-altitude phases of flight,

(2) stress the need for rapid recognition and response by using all of the airplane's performance capability, and

(3) address the effect of an out-of-trim speed condition on the control forces needed to use the airplane's performance.

Development, certification, and installation of airborne equipment which can provide the pilot early warning of windshear encounters and optimize the logic of command guidance instruments to enhance the pilot's response to the encounter.

August 16, 1987: McDonnell Douglas MD-82, N-312RC

Location	Detroit, MI
Flight	NWA 255
Operator	Northwest Airlines
Operation	Scheduled Airline
Route	Detroit, MI, FL to Phoenix, AZ
On-Board Injuries	Fatal 154 Serious 1
Ground Injuries	Fatal 2
Aircraft Damage	Destroyed

Description   Source: Aviation Safety Network[1]

A McDonnell Douglas DC-9-82 operating Northwest Airlines flight 255 was destroyed when it crashed onto a road during takeoff from Detroit-Metropolitan Wayne County Airport, Michigan, USA. Just one of the 155 occupants survived the accident. Additionally, Two persons on the ground were killed.

Flight NW255 was a regularly scheduled passenger flight between Saginaw, Michigan and Santa Ana, California, with en route stops at Detroit and Phoenix, Arizona. About 18:53, flight 255 departed Saginaw and about 19:42 arrived at its gate at Detroit.

About 20:32, flight 255 departed the gate with 149 passengers and 6 crewmembers on board. During the pushback, the flightcrew accomplished the BEFORE (engine) START portion of the airplane checklist, and, at 20:33, they began starting the engines. The flight was then cleared to "taxi via the ramp, hold short of (taxiway) delta and expect runway three center [3C] (for takeoff)..." The ground controller amended the clearance, stating that the flight had to exit the ramp at taxiway Charlie. The crew was requested to change radio frequencies. The first officer repeated the taxi clearance, but he did not repeat the new radio frequency nor did he tune the radio to the new frequency.

At 20:37, the captain asked the first officer if they could use runway 3C for takeoff as they had initially expected 21L or 21R. After consulting the Runway Takeoff Weight Chart Manual, the first officer told the captain runway 3C could be used for takeoff.

During the taxi out, the captain missed the turnoff at taxiway C. When the first officer contacted ground control, the ground controller redirected them to taxi to runway 3C and again requested that they change radio frequencies. The first officer repeated the new frequency, changed over, and contacted the east ground controller. The east ground controller gave the flight a new taxi route to runway 3C, told them that windshear alerts were in effect, and that the altimeter setting was 29.85 inHg. The flightcrew acknowledged receipt of the information.

At 20:42, the local controller cleared flight 255 to taxi into position on runway 3C and to hold. He told the flight there would be a 3-minute delay in order to get the required "in-trail separation behind traffic just departing." At 20:44:04, flight 255 was cleared for takeoff.

Engine power began increasing at 20:44:21. The flightcrew could not engage the autothrottle system at first, but, at 20:44:38, they did engage the system, and the first officer called 100 knots at 20:44:45. At 20:44:57, the first officer called "Rotate." Eight seconds later, the stall warning stick shaker activated, accompanied by voice warnings of the supplemental stall recognition system (SSRS). The takeoff warning system indicating that the airplane was not configured properly for takeoff, did not sound at any time prior or during takeoff.

After flight 255 became airborne it began rolling to the left and right before the left wing hit a light pole in a rental car lot.

After impacting the light pole, flight 255 continued to roll to the left, continued across the car lot, struck a light pole in a second rental car lot, and struck the side wall of the roof of the auto rental facility in the second rental car lot. The airplane continued rolling to the left when it impacted the ground on a road outside the airport boundary. The airplane continued to slide along the road, struck a railroad embankment, and disintegrated as it slid along the ground.

Fires erupted in airplane components scattered along the wreckage path. Three occupied vehicles on the road and numerous vacant vehicles in the auto rental parking lot along the airplane's path were destroyed by impact forces and or fire. One passenger, a 4-year-old child was injured seriously.

Probable Cause   Source: Accident Report[174]

The National Transporntion Safety Board determines that the probable cause of the accident was the flightcrew's failure to use the taxi checklist to ensure that the flaps and slats were extended for takeoff. Contributing to the accident was the absence of electrical power to the airplane takeoff warning system which thus did not warn the flightcrew that the airplane was not configured properly for takeoff. The reason for the absence of electrical power could not be determine

Safety Recommendations[174]

conduct a directed safety investigationto determine the reliability of circuit breakers and the mechanisms by which failures internal to the circuit breakers can disable operating systems and to identify appropriate correctiveactions as necessary. (class 11, PriorityAction) (A-88-64)

Require the modification of the DC-9-60 series airplanes to illuminate the existing central aural warning system (CAWS) fail light on the overhead annunciator panel in the event of CAWS input circuit power loss so that the airplane conforms to the original certification configuration. (class 11, PriorityAction) (A-88-65)

Develop and disseminate guidelines for the design of central aural warning systems to include a determination of the warning to be provided, the criticality of the provided warning, and the degree of system self- monitoring. (class 11, PriorityAction) (A-88-66)

Require that all Parts 121 and 135 operators and principal operations inspectors emphasize the importance of disciplined application of standard operating procedures and, in particular, emphasize rigorous adherence to prescribed checklist procedures. (class 11, Priority Action) (A-88-67)

Convene a human performance research group of personnel from the National Aeronautics and Space Administration, industry, and pilot groups to determine if there is any type or method of presenting a checklist which produces better performance on the part of user personnel. (class 11, Priority Action) (A-88-68)

Expedite the issuance of guidance materials for use by Parts 121 and 135 operators in the implementation of team-oriented flightcrew training techniques, such as cockpit resources management, line-oriented flight training, or other techniques which emphasize crew coordination and management principles. (class I!, Priority Action) (A-88-69)

Comments

The recommendation to convene a human performance research group led to some very important human factors studies of airline flight crews with very positive results.

July 2, 1994: Douglas DC-9, N-954VJ

Location	Charlotte, NC
Flight	USA 101
Operator	USAir (formerly Allegheny Airlines)
Operation	Scheduled Airline
Route	Columbia, NC to Charlotte, NC
On-Board Injuries	Fatal <u>37</u> Serious <u>16</u> Minor/None <u>4</u>
Aircraft Damage	Destroyed

Description   Source: Aviation Safety Network[1]

USAir Flight 1016 was a domestic flight from Columbia (CAE) to Charlotte (CLT). The DC-9 departed the gate on schedule at 18:10. The first officer was performing the duties of the flying pilot.

The weather information provided to the flightcrew from USAir dispatch indicated that the conditions at Charlotte were similar to those encountered when the crew had departed there approximately one hour earlier. The only noted exception was the report of scattered thunderstorms in the area.

Flight 1016 was airborne at 18:23 for the planned 35 minute flight. At 18:27, the captain of flight 1016 made initial contact with the Charlotte Terminal Radar Approach control (TRACON) controller and advised that the flight was at 12,000 feet mean sea level (msl). The controller replied "USAir ten sixteen ... expect runway one eight right." Shortly afterward the controller issued a clearance to the flightcrew to descend to 10,000 feet. At 18:29, the first officer commented "there's more rain than I thought there was ... it's startin ...pretty good a minute ago ... now it's held up." On their airborne weather radar the crew observed two cells, one located south and the second located east of the airport. The captain said "looks like that's [rain] setting just off the edge of the airport." One minute later, the captain contacted the controller and said "We're showing uh little buildup here it uh looks like it's sitting on the radial, we'd like to go about five degrees to the left to the ..." The controller replied "How far ahead are you looking ten sixteen?" The captain responded "About fifteen miles." The controller then replied "I'm going to turn you before you get there I'm going to turn you at about five miles northbound." The captain acknowledged the transmission, and, at 18:33, the controller directed the crew to turn the aircraft to a heading of three six zero. One minute later the flightcrew was issued a clearance to descend to 6,000 feet, and shortly thereafter contacted the Final Radar West controller.

At 18:35 the Final Radar West controller transmitted "USAir ten sixteen ... maintain four thousand runway one eight right." The captain acknowledged the radio transmission and then stated to the first officer "approach brief." The first officer responded "visual back up ILS." Following the first officer's response, the controller issued a clearance to flight 1016 to "...turn ten degrees right descend and maintain two thousand three hundred vectors visual approach runway one eight right."

At 18:36, the Final Radar West controller radioed flight 1016 and said "I'll tell you what USAir ten sixteen they got some rain just south of the field might be a little bit coming off north just expect the ILS now amend your altitude maintain three thousand." At 18:37, the controller instructed flight 1016 to "turn right heading zero niner zero." At 18:38, the controller said "USAir ten sixteen turn right heading one seven zero four from SOPHE [the outer marker for runway 18R ILS] ... cross SOPHE at or above three thousand cleared ILS one eight right approach." As they were maneuvering the airplane from the base leg of the visual approach to final, both crew members had visual contact with the airport. The captain then contacted Charlotte Tower. The controller said "USAir ten sixteen ... runway one eight right cleared to land following an F-K one hundred short final, previous arrival reported a smooth ride all the way down the final." The pilot of the Fokker 100 in front also reported a "smooth ride". About 18:36, a special weather observation was recorded, which included: ... measured [cloud] ceiling 4,500 feet broken, visibility 6 miles, thunderstorm, light rainshower, haze, the temperature was 88 degrees Fahrenheit, the dewpoint was 67 degrees Fahrenheit, the wind was from 110 degrees at 16 knots .... This information was not broadcast until 1843; thus, the crew of flight 1016 did not receive the new ATIS.

At 18:40, the Tower controller said "USAir ten sixteen the wind is showing one zero zero at one nine." This was followed a short time later by the controller saying "USAir ten sixteen wind now one one zero at two one." Then the Tower controller radioed a wind shear warning "windshear alert northeast boundary wind one nine zero at one three."

On finals the DC-9 entered an area of rainfall and at 18:41:58, the first officer commented "there's, ooh, ten knots right there." This was followed by the captain saying "OK, you're plus twenty [knots] ... take it around, go to the right." A go around was initiated. The Tower controller noticed Flight 1016 going around "USAir ten sixteen understand you're on the go sir, fly runway heading, climb and maintain three thousand."

The first officer initially rotated the airplane to the proper 15 degrees nose-up attitude during the missed approach. However, the thrust was set below the standard go-around EPR limit of 1.93, and the pitch attitude was reduced to 5 degrees nose down before the flightcrew recognized the dangerous situation. When the flaps were in transition from 40 to 15 degrees (about a 12-second cycle), the airplane encountered windshear. Although the DC-9 was equipped with an on-board windshear warning system, it did not activate for unknown reasons. The airplane stalled and impacted the ground at 18:42:35.

Investigation revealed that the headwind encountered by flight 1016 during the approach between 18:40:40 and 18:42:00 was between 10 and 20 knots. The initial wind component, a headwind, increased from approximately 30 knots at 18:42:00 to 35 knots at 18:42:15. The maximum calculated headwind occurred at 18:42:17, and was calculated at about 39 knots. The airplane struck the ground after transitioning from a headwind of approximately 35 knots, at 18:42:21, to a tailwind of 26 knots (a change of 61 knots), over a 14 second period.

The passenger cabin contained 21 rows of seats and was configured with 12 first class and 91 coach seats. There were 52 passengers on board flight1016: 27 males, 23 females, and 2 female in-lap infants (younger than 24 months) who were not listed on the passenger manifest.

A 9-month-old infant, who was unrestrained in her mother's lap in seat 21c, sustained fatal injuries. The mother was unable to hold onto her daughter during the impact sequence. Seat 21c was intact and the surrounding cabin structure sustained minor deformation. Additionally, the impact forces in this area were calculated to have been within human tolerances. According to passengers, a flash fire swept through the inside of the cabin during the impact sequence; examination found no evidence of either fire or smoke impingement in this area of the cabin.

The passenger manifest for flight 1016 listed 50 passenger names, but it did not include the names of the two "in-lap" infants. The tickets issued to the adults traveling with the in-lap infants were reviewed. One passenger's ticket had the marking "+infant" handwritten on the face of the ticket. The second passenger's ticket had no identifying information to document the carriage of the in-lap infant. 14 CFR Part 121.693 requires that children, regardless of their age or whether they are the sole occupant of a seat, must be listed by name on the passenger manifest.

Probable Cause   ## Source: Accident Report[175]

1) the flight crew's decision to continue an approach into severe convective activity that was conducive to a microburst; 2) the flight crew's failure to recognize a windshear situation in a timely manner; 3) the flight crew's failure to establish and maintain the proper airplane attitude and thrust setting necessary to escape the windshear; and 4) the lack of real-time adverse weather and windshear hazard information dissemination from air traffic control, all of which led to an encounter with and failure to escape from a microburst-induced windshear that was produced by a rapidly developing thunderstorm located at the approach end of runway 18R.

Contributing to the accident were: 1) the lack of air traffic control procedures that would have required the controller to display and issue ASR-9 radar weather information to the pilots of flight 1016; 2) the Charlotte tower supervisor's failure to properly advise and ensure that all controllers were aware of and reporting the reduction in visibility and the RVR value information, and the low level windshear alerts that had occurred in multiple quadrants; 3) the inadequate remedial actions by USAir to ensure adherence to standard operating procedures; and 4) the inadequate software logic in the airplane's windshear warning system that did not provide an alert upon entry into the windshear.

Safety Recommendations

The NTSB issued several recommendations to the FAA.

Regarding Occupant Restraints:

A-95-50: Develop standards for forward-facing, integrated child-safety-seats for transport category aircraft.

A-95-51: Revise 14 CFR, Parts 91, 135, and 121 to required that all occupants be restrained during takeoff, landing, and turbulent conditions in a manner appropriate to their size.[175]

Wiring Accidents

July 17, 1996: Boeing 747, N-93119

Location	East Moriches, NY
Flight	TWA 800
Operator	Trans World Airlines
Operation	Scheduled Airline
Route	New York, NY to Paris, France
On-Board Injuries	Fatal <u>230</u>
Aircraft Damage	Destroyed

Description   Source: Accident Report[176]

On July 17, 1996, about 2031 eastern daylight time, Trans World Airlines, Inc. (TWA) flight 800, a Boeing 747-131, N93119, crashed in the Atlantic Ocean near East Moriches, New York. TWA flight 800 was operating under the provisions of Part 121 as a scheduled international passenger flight from John F. Kennedy International Airport (JFK), New York, New York, to Charles De Gaulle International Airport, Paris, France. The flight departed JFK about 2019, with 2 pilots, 2 flight engineers, 14 flight attendants, and 212 passengers on board. All 230 people on board were killed, and the airplane was destroyed. Visual meteorological conditions prevailed for the flight, which operated on an instrument flight rules flight plan. The investigation revealed that the crash occurred as the result of a fuel/air explosion in the airplane's center wing fuel tank (CWT) and the subsequent in-flight breakup of the airplane. The investigation further revealed that the ignition energy for the CWT explosion most likely entered the CWT through the fuel quantity indication system wiring; neither the ignition energy release mechanism nor the location of the ignition inside the CWT could be determined from the available evidence. There was no evidence of a missile or bomb detonation.

Probable Cause   Source: Accident Report[176]

An explosion of the center wing fuel tank (CWT), resulting from ignition of the flammable fuel/air mixture in the tank. The source of ignition energy for the explosion could not be determined with certainty, but, of the sources evaluated by the investigation, the most likely was a short circuit outside of the CWT that allowed excessive voltage to enter it through electrical wiring associated with the fuel quantity indication system. Contributing factors to the accident were the design and certification concept that fuel tank explosions could be prevented solely by precluding all ignition sources and the design and certification of the Boeing 747 with heat sources located beneath the CWT with no means to reduce the heat transferred into the CWT or to render the fuel vapor in the tank nonflammable.

September 2, 1998: McDonnell Douglas MD-11, HB-IWF

Location	Near Peggy's Cove, NS, Canada
Flight	SWR 111
Operator	Swissair
Operation	Scheduled Airline
Route	New York, NY to Geneva, Switzerland
On-Board Injuries	Fatal <u>229</u>
Aircraft Damage	Destroyed

Probable Cause   Source: Aviation Safety Network[1]

The accident to Swissair flight SWR 111 was described in previously on page 291. That discussion dealt primarily with issues dealing with an extensive fire while enroute. The following discussion concentrates on the cause of the fire.

Probable Cause   Source: Accident Report[52]

Findings as to causes and contributing Factors:

- Aircraft certification standards for material flammability were inadequate in that they allowed the use of materials that could be ignited and sustain or propagate fire. consequently, flammable material propagated a fire that started above the ceiling on the right side of the cockpit near the cockpit rear wall. The fire spread and intensified rapidly to the extent that it degraded aircraft systems and the cockpit environment, and ultimately led to the loss of control of the aircraft.
- Metallized polyethylene terephthalate (MPET)-type cover material on the thermal acoustic insulation blankets used in the aircraft was flammable. The cover material was most likely the first material to ignite, and constituted the largest portion of the combustible materials that contributed to the propagation and intensity of the fire.

- Once ignited, other types of thermal acoustic insulation cover materials exhibit flame propagation characteristics similar to MPET-covered insulation blankets and do not meet the proposed revised flammability test criteria. Metallized polyvinyl fluoride–type cover material was installed in HB-IWF and was involved in the in-flight fire.
- Silicone elastomeric end caps, hook-and-loop fasteners, foams, adhesives, and thermal acoustic insulation splicing tapes contributed to the propagation and intensity of the fire.
- The type of circuit breakers (CB) used in the aircraft were similar to those in general aircraft use, and were not capable of protecting against all types of wire arcing events. The fire most likely started from a wire arcing event.
- A segment of in-flight entertainment network (IFEN) power supply unit cable (1-3791) exhibited a region of re-solidified copper on one wire that was caused by an arcing event. This resolidified copper was determined to be located near manufacturing station 383, in the area where the fire most likely originated. This arc was likely associated with the fire initiation event; however, it could not be determined whether this arced wire was the lead event.
- There were no built-in smoke and fire detection and suppression devices in the area where the fire started and propagated, nor were they required by regulation. The lack of such devices delayed the identification of the existence of the fire, and allowed the fire to propagate unchecked until it became uncontrollable.
- There was a reliance on sight and smell to detect and differentiate between odour or smoke from different potential sources. This reliance resulted in the misidentification of the initial odour and smoke as originating from an air conditioning source.
- There was no integrated in-flight firefighting plan in place for the accident aircraft, nor was such a plan required by regulation. Therefore, the aircraft crew did not have procedures or training directing them to aggressively attempt to locate and eliminate the source of the smoke, and to expedite their preparations for a possible emergency landing. In the absence of such a firefighting plan, they concentrated on preparing the aircraft for the diversion and landing.
- There is no requirement that a fire-induced failure be considered when completing the system safety analysis required for certification. The fire-related failure of silicone elastomeric end caps installed on air conditioning ducts resulted in the addition of a continuous supply of conditioned air that contributed to the propagation and intensity of the fire.
- The loss of primary flight displays and lack of outside visual references forced `the pilots to be reliant on the standby instruments for at least some portion of the last minutes of the flight. In the deteriorating cockpit environment, the positioning and small size of these instruments would have made it difficult for the pilots to transition to their use, and to continue to maintain the proper spatial orientation of the aircraft.

Certification Issues
The TSB found several certification irregularities:
- The inflight entertainment system STC project management structure did not ensure that the required elements were in place to design, install, and certify a system that included emergency electrical load-shedding procedures compatible with the MD-11 type certificate. No link was established between the manner in which the system was integrated with aircraft power and the initiation or propagation of the fire.
- The FAA STC approval process for the inflight entertainment system did not ensure that the DAS employed personnel with sufficient aircraft-specific knowledge to appropriately assess the integration of the system's power supply with aircraft power before granting certification.
- The FAA allowed a de facto delegation of a portion of their AEG function to the DAS even though no provision existed within the FAA's STC processes to allow for such a delegation.

Regulatory Outcome
Following the TWA 800 Accident, The FAA instituted a program to ensure the airworthiness of aging aircraft fuel systems and wiring.[177] This was the Enhanced Airworthiness Program for Airplane Systems (EAPAS). In May 2001, the FAA issued SFAR 88 to detect and correct potential system failures that can cause ignition.[178] SFAR 88 places requirements on some turbine-powered transport airplanes for a safety review of the fuel tank system to determine if the design meets the requirements of §§25.901, 25.981(a), and 25.981(b) and to develop maintenance and inspection instructions necessary to preclude the existence or development of ignition sources within the fuel tank system throughout the operational life of the airplane.

Following the TWA 800 and the Swissair accident the FAA also issued a new set of rules, Part 26. This rule amended FAA regulations for certification and operations of transport category airplanes. According to the rule preamble,

"These changes are necessary to help ensure continued safety of commercial airplanes. They improve the design, installation, and maintenance of airplane electrical wiring systems and align those requirements as closely as possible with the requirements for fuel tank system safety. This final rule organizes and clarifies design requirements for wire systems by moving existing regulatory references to wiring into a single section of the regulations specifically for wiring and by adding new certification rules. It requires holders of type certificates for certain transport category airplanes to conduct analyses of their airplanes and make necessary changes to existing Instructions for continued Airworthiness (ICA) to improve maintenance procedures for wire systems. It requires operators to incorporate ICA for wiring into their maintenance or inspection programs. And finally, this final rule clarifies requirements of certain existing rules for operators to incorporate ICA for fuel tank systems into their maintenance or inspection programs.

References

1   http:// aviation-safety.net/database/
2   http://lessonslearned.faa.gov/transport.cfm
3   L M Nicolai, *Lessons Learned: A Guide to Improved Aircraft Design*, Reston, VA: AIAA, 2018, p. 28
4   *Aircraft Accident Report. Braniff Airways, Inc., Lockheed Electra, N 9705C, Buffalo, Texas, September 29, 1959*, CAB AAR File No. 1-0060, May 1961
5   *Aircraft Accident Report. Northwest Airlines Lockheed Electra, N 121US, Near Cannelton, Indiana, March 17, 1960*, CAB AAR File No. 1-0003, April 1961
6   *Aircraft Accident Report. American Airlines, Inc. McDonnell Douglas DC-10-10, N103AA, Near Windsor, Ontario, Canada, June 12, 1972*, NTSB-AAR-73-2, February 1973
7   C Bartlett, *Air Crashes and Miracle Landings*, Gillingham, Dorset, UK: OpenHatch Books, Second Edition, 2018
8   *Rapport final de la commission d'Enquête sur l'accident de l'avion DC 10 TC-JAV des Turkish Airlines survenu à Ermenonville le 3 mars 1974*. Journal Officiel du 12 mai 1976; BEA report tc-v740303
9   R Witkin, "Change on DC-10 Called Optional,' *New York Times*, March 27, 1974, p 9; accessed from http://www.nytimes.com/1974/03/27/archives/change-on-dc10-called-optional-faa-aide-says-an-accord-with-builder.html, accessed December 18, 2021.
10 *Eastern Air Lines, Inc., Lockheed Electra L-188, N-5533, Logan International Airport, Boston, Massachusetts, October 4, 1960*, CAB Accident Report, File 1-0043, July 1962
11 *Aircraft Accident Report, Overseas National Airways, Inc., Douglas DC-10-30, N1032F, John F Kennedy International Airport, Jamaica, New York, November 12, 1975*, NTSB AAR-76-19, December 1976
12 *Turbine-engine Foreign Object Ingestion and Rotor Blade Containment Type Certification Procedures*, FAA Advisory Circular 33-1A, June 1968
13 "Foreign Object Ingestion," *Airworthiness Standards: Aircraft Engines*, §33.77, amendment 33-6, effective October 31, 1974
14 "Bird Ingestion," *Airworthiness Standards: Aircraft Engines*, §33.76, amendment 33-20, effective December 13, 2000
15 *Aviation Accident Final Report*, NTSB Accident File CEN09MA117, November 2010
16 *Bird Strike Avoidance*, Helicopter Safety Advisory Conference Recommended Practice, No. 2010-3, Rev 1, May 2010
17 "Statement of Probable Cause Concerning an Aircraft Accident Which Occurred to Plane of Transcontinental and Western Air, Inc. on May 6, 1935, near Atlanta, Macon County, Mo.," *Air Commerce Bulletin*: 7, July 15, 1935, pp. 16-18; cited by Komons[18]
18 N A Komons, *The Cutting Air Crash. A Case Study in Early Federal Aviation Policy*, Washington: Government Printing Office, 1973
19 *Aircraft Accident Report: Alaska Airlines, Inc., Boeing 727, N2969G, Near Juneau, Alaska, September 4, 1971*, NTSB AAR-72-28, October 1972
20 *Accident Report. Lockheed L188C, CF-PAB, 30 Oct 1975, Rea Point, Melville Is NWT*, Ministry of Canada Report, undated
21 *Aircraft Accident Report: Trans World Airlines, Inc., Boeing 727-231, N54328, Berryville, Virginia, December 1, 1974*, NTSB AAR-75-16, November 1975
22 Lessons Learned from Civil Aviation Accident Home Page, http://lessonslearned.faa.gov/ll_main.cfm?TabID=1&LLID=63&LLTypeID=10
23 R O Phillips, *Investigation of Controlled Flight into Terrain: Descriptions of Flight Paths for Selected Controlled Flight into Terrain (CFIT) Aircraft Accidents, 1985-1997*, Volpe National Transportation Systems Center Report DOT-TSC-FA9D1-99-01, March 1999
24 R G Rodriguez, "Information Involving Operation of EAL 980, La Paz, Bolivia, January 1, 1985," Memorandum included in Aviation Accident Report (AAR) DCA85RA007, NTSB, May 6, 1985, pp. 1-7
25 "Team Fails to Find Recorders in Eastern Crash," *Aviation Week and Space Technology*, October 28, 1985, p. 29

26   T G Farr *et al.*, "The Shuttle Radar Topography Mission", *Reviews of Geophysics*: 45, June 2007, RG2004, doi:10.1029/2005RG000183; http://agupubs.onlinelibrary.wiley.com/doi/epdf/10.1029/2005RG000183 , accessed November 5, 2020

27   "Terrain Awareness and Warning System," *Operating Requirements: Domestic, Flag, and Supplemental Operations*, §121.354, amendment 121-273, effective March 29, 2001

28   *Rapport de la commission d'enquête sur l'accident survenu le 20 janvier 1992 près du Mont Sainte-Odile (Bas Rhin) à l'Airbus A 320 immatriculé F-GGED exploité par la compagnie Air Inter*, BEA Report F-ED920120, November 1993

29   *Aircraft Accident Report. Controlled Flight Into Terrain, American Airlines Flight 965, Boeing 757-223, N103AA, Near Cali, Columbia, December 20, 1995*, Aeronautica Civil of the Republic of Columbia Final Report

30   Formal Report Accident. Sikorsky S-92A, EI-ICR, Black Rock, Co. Mayo, Ireland, 14 March 2017, AIAA Report 2021-008, Published November 2021

31   Safety Recommendaton A-03-55 and -56, Dated December 2, 2003

32   *Instrument system event - Boeing 747-400 aircraft, 9V-SPP, 19 km SE Nyngan Airport, 6 November 2001*, ATSB Investigation Number 200105338. http://www.atsb.gov.au/publications/investigation_reports/2001/aair/aair200105338/, accessed April 16, 2022

33   NTSB Safety Recommendaton A-03-55 and -56, Dated December 2, 2003

34   *Accident Investigation Report, Pan American World Airways, Inc., Boeing 377, N90943 in the Pacific Ocean, Between Honolulu and San Francisco, October 16, 1956*, CAB File No. 1-0121, July 1957

35   "This ia it!", published in Life Magazine, October 29, 1956, http://www.vintagewings.ca/stories/this-is-it, Accessed April 7, 2022

36   NTSB Docket BFO85LA031

37   *Pilot's Operating Handbook, Cessna Stationair, Cessna Model U206F*, Cessna Aircraft Co, n. d.

38   *Egress Through Cargo Doors During Emergency*, FAA Airworthiness Concern Sheet, February 27, 2020

39   *Cessna 206, T206, TU206, and U206 Aircraft Series – Emergency Egress Difficulty*, CAA New Zealand, Continuing Airworthiness Notice, 25-003, April 20, 2020

40   *Collision With Water Involving Textron Aviation (Cessna) 206, VH-AEE, Near Happy Valley, Fraser Island, Queensland on January 2020*, ATSB AO-2020-010, dated July 2021

41   *Aviation Incident Final Report*, NTSB Accident File CHI08FI152, May 2009

42   C L Scovel, "FAA's Certification of the Eclipse EA-500 Very Light Jet," *Statement for Committee on Transportation, US House of Representatives*, September 17, 2008

43   H U Hamid, R J Margason, and G Hardy, *Investigation of Wing Upper Surface Flow-Field Disturbance Due to NASA DC-8-72 In-flight Inboard Thrust Reverser Deployment*, NASA TM 110351, June 1995

44   *Aircraft Accident Report. United Airlines Flight 232, McDonnell Douglas DC-10-10, Sioux Gateway Airport, Sioux City, Iowa, July 19, 1989*, NTSB AAR-90-06, November 1990

45   *In-flight Uncontained Engine Failure, Airbus A380-842, VH-OQA, Overhead Batam Island, Indonesia, 4 November 2010*, ATSB AO-2010-089, 27 June 2013

46   R De Crespigny, *QF32*, Sydney: Pan Macmillan Australia, 20q2

47   *Rapport Final. de la Commission d'Enquêtes sure l'accident survenu au Boeing 707 PP-VJZ de la Compagnie Varig à Saulx-les-Chartreux, le 11 juillet 1973*, BEA pp-z730711, April 1976

48   M E F Plave,, "United States v. Varig Airlines. The Supreme Court Narrows the Scope of Government Liability Under the Federal Tort Claims Act," *Journal of Air Law and Commerce*: 51, 1985, 197-239

49   *Aircraft Accident Report. Air Canada Flight 797, McDonnell Douglas DC-9-32, C-FTLL, Greater Cincinnati International Airport, Covington, Kentucky, June 2, 1983*, NTSB AAR-86-02, January 1986; Supersedes NTSB-AAR-84-09

50   *Report of the Board of Inquiry into the Loss of South African Airways Boeing 747-244B Combi Aircraft "Helderberg" in the Indian Ocean on November 28th 1987*, Government Printer, Republic of South Africa, ISBN 0-621-13030-2, May 1990

51 *Aircraft Accident Report, In-Flight Fire and Impact with Terrain, Valuejet Airlines Flight 592, DC-9-32, N904VJ, Everglades, Near Miami, Florida, May 11, 1996*, NTSB AAR-97-06, corrected copy, August 1997

52 *Aviation Investigation Report, In-Flight Fire Leading to Collision with Water, Swissair Transport Limited, McDonnell Douglas MD-11 HB-IWF, Peggy's Cove, Nova Scotia 5 NM SW, September 2, 1998*, TSB Report Number A98H0003

53 *Final Air Accident Investigation Report. Uncontained Cargo Fire Leading to Loss of Control Inflight and Uncontrolled Descent into Terrain. Boeing 747-44AF, N571UP, Dubain, United Arab Emirates, 03 September 2010*, General Civil Aviation Authority (UAE) Report 13/2010, August 2013

54 *Boeing and McDonnell Douglas Models 707, 727, 737, 747, and 757 and McDonnell Douglas Models DC-8, DC-9, and DC-10 Series Airplanes*, Airworthiness, Directive 93-07-15, amendment 39-8547, effective May 2, 1993

55 https://boeing.mediaroom.com/2010-11-11-Updated-Boeing-Statement-on-787-Dreamliner-ZA002-Incident

56 D Gates, "Electrical Fire Forces Emergency Landing Of 787 Test Plane," *Seattle Times*, November 9, 2010

57 "Immediate Notification," *Notification and Reporting of Aircraft Accidents or Incidents and Overdur Aircraft, and Preservation of Aircraft Wreckage, Mail, Cargo, and Records*, 49CFR Part 830, §830.5, as amended August 7, 1995

58 *Emergency Evacuation Using Slides, All Nippon Airways Co., Ltd., Boeing 787-8, JA804A, Takamatsu Airport at 08:49 JST, January 16, 2013*, Japan Transportation Safety Board Aircraft Serious Incident Investigation Report A12014-4, September 2014

59 *Aircraft Incident Report. Auxiliary Power Unit Battery Fire, Japan Airlines Boeing 787-8, JA829J, Boston, Massachusetts, January 7, 2013*. NTSB-AIR-14-01, November 2014.

60 *Report on the Serious Incident to Boeing B787-8, ET-AOP, London Heathrow Airport on 12 July 2013*, Air Accident Investigation Branch AAR-2/2015, August 1997, published 2015

61 *Uncontrolled Descent and Collision with Terrain, United Airlines Flight 585, Boeing 737-200, N999UA, 4 Miles South of Colorado Springs Municipal Airport, Colorado Springs, Colorado, March 3, 1991*, NTSB AAR-01-01, Amended March 27, 2001

62 *Aircraft Accident Report. Uncontrolled Descent and Collision with Terrain, USAir Flight 427, Boeing 737-300, N513AU Near Aliquippa, Pennsylvania, September 8, 1994*, NTSB AAR-99-01, March 24, 1999, corrected November 4, 1999 and February 16, 2000

63 *Aviation Incident Final Report*, NTSB File DCA96IA061, November 30, 2007

64 *Aviation Investigation Report. Spin—Loss of Directional Control, Laurentide Aviation, Cessna 152, C-GZLZ, Lac Saint-François, Quebec, 18 July 1998*, TSB Report A98Q0114, July 2000

65 *Investigation of Locked Rudder*, TSB Report cited in Accident Report a98Q011464

66 *Rudder Stop Modification*, Cessna Single Engine Service Bulletin SEB01-1, January 2001

67 *Cessna 150/152 – Rudder Stop*, Transport Canada Airworthiness Directive, CF-2000-20R1; cancelled by CF-2000-20R2, dated September 2003

68 *Aviation Accident Final Report*, NTSB Accident File NYC05FA069, December 2005

69 NTSB Safety Recommendation A-07-33, March 21, 2007

70 *Aircraft Accident Report, Loss of Control and Impact with Pacific Ocean, Alaska Airlines Flight 261, McDonnell Douglas MD-83, N863AS, About 2.7 Miles North of Anacapa Island, California, January 31, 2000*, NTSB AAR-02-01, December 2002

71 D L Cheney *et al.*, *Commercial Airplane Certification Process Study. An Evaluation of Selected Aircraft Certification Operations, and Maintenance Processes*, FAA Study Report, March 2002, available at http://lessonslearned.faa.gov/Alaska261/CPS_Report_2002_CD_3rd.pdf

72 *Aircraft Accident Report, Loss of Pitch Control on Takeoff, Emery Worldwide Airlines, Flight 17, McDonnell Douglas DC-8-71F, N8079U, Rancho Cordova, California, February 16, 2000*, NTSB AAR-03-02, August 2003

73 *Aircraft Accident Report, Northwest Airlines, Inc., Lockheed Electra, L-188C, N-137US, O'Hare International Airport, Chicago, Illinois, September 17, 1961*, CAB Report File 1-0018, Report dated December 1962

74 *Lufthansa A320, Incorrectly-wired Sidestick, Frankfurt Am Main, March 2001*, BFU Report 5X004-0/01, Report date April 2003

75   *Aviation Accident Final Report*. NTSB File CHI01FA104

76   *Roll Oscillations on Landing, Airbus 321-211, C-GJVX and C-GGIUF, Toronto/Lester B Pearson international Airport, Ontario, 07 December 2002*, TSB Report A02O0406, Report Date July 4, 2005

77   *Aircraft Accident Report. Air South, Inc., Beechcraft B-99, N844NS Near Monroe, Georgia, July 6, 1969*, NTSB AAR-70-18, August 1970

78   *Airbus A330 C-GGWD and Airbus A340 TC-JDN, 2 October 2000*, AAIB Bulletin No. 6/2001

79   *Software Approval Guidelines*, FAA Order 8110.49A, effective 3/29/2018

80   *Software Considerations in Airborne Systems and Equipment*, RTCA DO-178B, December 1992

81   *Accident of aircraft Airbus A-320-214, registration EC-HKJ, at Bilbao Airport on 7 February 2001*, CIAIAC Report A-006/201

82   *In-Flight Upset, 154 km West of Learmonth, WA, 7 October 2008, VH-QPA, Airbus A330-303*, ATSB Aviation Occurrence Investigation AO-2008-070, 2011, October 2019

83   *Aircraft Accident Investigation Report. PT. Lion Mentari Airlines Boeing 737-8(MAX); PK-LQP, Tanjung Karawang, West Java, Republic of Indonesia, October 29, 2018*, KNKT Final Report 19.10.35.04

84   *Interim Investigation Report on Accident to the B-737-8 (MAX) Registered ET-AVJ Operated by Ethiopian Airlines on 10 March 2019*, Aircraft Accident Investigation Bureau Report AI-01/19, March 2020

85   *Collision with Terrain, Newcal Aviation, Inc, Modified DHC-4A (prototype connversion L400NC, Gimli Industrial Park,Manitoba, 27 August 1992*, TSB Air Transportation Safety Investigation Report A92C0154, Report undated

86   J Donnelly, "Remarks," *2001 Australasian Air Safety Seminar*, June 2001

87   *Aircraft Accident Report: Lockheed L382E, N-130X, Dobbins AFB, Marietta, GA, 3 February 1993*, NTSB Aviation Accident Final Report, File ATL93MA055, March 1994

88   "A330 Crashed On CAT 3 Test Flight," *Aviation Week*, April 3, 1995, pp. 72-73; Apr 10, p. 60; Apr 17, p. 44-45; May 15, p. 58-59; May 22, p. 54-56; May 29, p. 69-70; reprint of final BEA report (which is unavailable)

89   *Accident on 27 November 2008 off the coast of Canet-Plage (66) to the Airbus A320-232 registered D-AXLA operated by XL Airways Germany*, BEA d-la081127, report undated

90   *Aircraft Accident Report, Crash During Experimental Test Flight, Gulfstream Aerospace Corporation GVI (G650), N652GD, Roswell, New Mexico, April 2, 2011*, NTSB AAR-12-02, October 19, 2012

91   *Crashed During Approach, Boeing 737-800, Near Amsterdam Schiphol Airport, 25 February 2009*, Dutch Safety Board, Report Dated May 2010

92   *Accident on 25 July 2000, at La Patte d'Oie in Gonesse (95) to the Concorde registered F-BTSC operated by Air France*, BEA Report f-sc000725

93   *Aviation Accident Final Report*, NTSB File ATL90FA099, March 1993

94   *Model Skyship 600 (Normal Category)*, FAA Type Certificate AS1EU, Approved May 9, 1989

95   *Aviation Accident Final Report. Aberdeen, SD, Learjet 35, N47BA, 10/25/1999*. NTSB Accident DCA00MA005, November 2000

96   *Accident of the Aircraft 5B-DBY of Helios Airways, Flight HCY522 on August 14, 2005, in the Area of Grammtiko, Attikis, 35 km Northwest of Athens International Airport*, Ministry of Transport & Communications AAR-11/2006, November 2006

97   amendment 25-131

98   *Aviation Accident Final Report. Caribbean Sea, Socata TBM 700, N900KN, 9/5/2014*. NTSB Accident File ERA14LA424, November 2017

99   *Review of High Altitude Operations*, Daher-Socata Service Information 2014-008, September 2014

100   *Aircraft Accident Report, In-Flight Icing Encounter and Loss of Control, Simmons Airlines, d.b.a. American Eagle Flight 4184, Avions de Transport Regional (ATR) Model 72-212, N401AM, Roselawn, Indiana, October 31, 1994*, NTSB AAR-96-01 Volume 1: Safety Board Report, July 1996

101   *Aircraft Accident Report, In-Flight Icing Encounter and Loss of Control, Simmons Airlines, d.b.a. American Eagle Flight 4184, Avions de Transport Regional (ATR) Model 72-212, N401AM, Roselawn, Indiana, October 31, 1994*, NTSB AAR-96-01 Volume 2: Response of Bureau Enquetes-Accidents to Safety Board's Draft Report, July 1996

102   W Lewis, *A Flight Investigation of the Meteorological Conditions Conducive to the Formation of Ice on Airplanes*, NACA TN-1393, 1947

103 A Jones and W Lewis, *Recommended Values of Meteorological Factors to be Considered in the Design of Aircraft Ice Protection Equipment*, NACA TN-1855, 1949

104 *Effect of Icing on Aircraft Control and Airplane Deice and Anti-ice System*, FAA AC-91-51A, July 17, 1996

105 *Aircraft Accident Report, In-Flight Icing Encounter and Uncontrolled Collision with Terrain, Comair Flight 3272, Embraer EMB-120RT, N265CA, Monroe, Michigan, January 9, 1997*, NTSB AAR-98-04, November 1998

106 *Effect of Icing on Aircraft Control and Airplane Deice and Anti-ice System*, FAA AC-91-51A, July 17, 1996

107 *Pilot Guide: Flight In Icing Conditions*, FAA AC-91-74A, December 31, 2007

108 *Loss of Control on Approach, Colgan Air, Inc. Operating as Continental Connection Flight 3407, Bombardier DHC-8-400, N200WQ, Clarence Center, New York, February 12, 2009*, NTSB AAR-10-01, February 2010

109 *Panarctic Explorations, Lockheed L-188, CF-PAB, Accident at Rea Point, Melville Island, Canada, on 30 October 1974*, ICAO Circular 132-AN/93, Accident Report not dated, released by the Ministry of Transport, Canada

110 AAIB Report 1/2010, EW/C2008/01/01

111 "Airworthiness Directive: Rolls-Royce plc RB211-Trent 500, 700, and 800 Series Turbofan Engines," Docket No. FAA-2009-0674; Directorate Identifier 2009-NE-25-AD; amendment 39-16244; AD 2010-07-01

112 *Aviation Incident Final Report*, NTSB File DCA09IA014, Undated

113 *Aircraft Accident Report. Loss of Control While Maneuvering, Pilatus PC-12/45, N128CM, Butte, Montana, March 22, 2009*, NTSB AAR-11-05, Corrected Copy, June 5, 2012

114 Incident Report, February 15, 1975

115 *In-flight upset event 240 km north-west of Perth, WA Boeing Company 777-200, 9M-MRG*, ATSB AO-2005-3722, dated March 2007

116 *Report on the Accident to Aerospatial AS33L, Super Puma, G-TIGK, in North Sea 6nm South-West of Brae Alpha Oil Production Platform on January 19, 1995*, Air Accident Investigation Branch AAR-2/97, August 1997

117 *Protection of Airplane Fuel Systems Against Fuel Vapor Ignition Due to Lightning*, FAA AC-20-53A, dated April 12, 1986; cancelled by AC-20-53B, effective June 5, 2006; and by AC-20-53C. effective September 24, 2018

118 *Aircraft Accident Report. American Airlines, Inc. DC-10-10, N110AA, Chicago-O'Hare International Airport, Chicago, Illinois, May 25, 1979*, NTSB-AAR-79-17, December 1979

119 *Aircraft Accident Report. American Airlines, Inc. DC-10-10, N110AA, Chicago-O'Hare International Airport, Chicago, Illinois, May 25, 1979*, NTSB-AAR-79-17, December 1979

120 *Civil Aircraft Accident. Report of the Court of Inquiry into the Accidents to Comet G-ALYP on 10th January 1954 and Comet G-ALYY on 8th April 1954*, London: Her Majesty's Stationery Office, 1955

121 *Damage Tolerance and Fatigue Evaluation of Structure*, FAA AC-25.571-1D, January 13, 2011, Superseding AC-25..571-1C, dated April 29, 1998

122 *Aircraft Accident Report, Aloha Airlines Flight 243, Boeing 737-200, N73711, Near Maui, Hawaii, April 28, 1988*, NTSB AAR-89-03, June 1989

123 *Aircraft Accident Report. El Al Flight 1862. Boeing 747-258F, 4X-AXG, Bijmermeer, Amsterdam, October 4, 1992*, Nederlands Aviation Safety Board, AAR 1862, February 1994

124 *Accident Investigation Report, Trans World Airlines, Inc., Lockheed 1049A, N 6902C, and United Airlines, Inc., Douglas DC-7, N 6324C, Grand Canyon, Arizona, June 30, 1956*, CAB AAR File No. 1-0090, April 1957

125 *Aeronautical Accident: PR-GTD and N600XL, Models: B-737 8EH and EMB-135 BJ Legacy, Date: 29 September 2006*, CENIPA Final Report A-00X/CENIPA/2008

126 "Instrument systems," *Airworthiness Standards: Transport Category Airplanes*, 14 CFR Part25, §25.1333, amendment 25-23, effective May 8, 1970

127 "Flight and navigational equipment," *Operating Requirements: Domestic, Flag, and Supplemental Operations*, 14 CFR Part121, §121.305, amendment 121-60, effective March 1, 1997

128 *Incident to Airbus A432-200, I-BIKE, 25 June 2005, On Approach to Runway 09L at London Heathrow Airport*, AAIB Bulletin 6/2006, 2006

129 Aircraft Accident Report. Pan-Alask Airways, Ltd., Cessna 310C, N1812C, Missing Between Anchorage and Juneau, Alaska, October 16, 1972, NTSB AAR-73-01, January, 1973

130 http://www.aopa.org/advocacy/aircraft/aircraft-operations/emergency-locator-transmitters   ,   Accessed May 27, 2022

131 http://www.aopa.org/advocacy/aircraft/aircraft-operations/emergency-locator-transmitters   ,   Accessed May 27, 2022

132 D. Jonz, "Ice Without Fear," *Flying*, October 1972, pp. 67-68, 123

133 *Safety Investigation Report. Malaysia Airlines Boeing B777-200ER (9M-MRO) 08 March 2014*, Malaysian ICAO Annex 13 Safety Investigation Team for MH370 Report MH370/01/2018, Issued July 2018

134 *MH 370 Preliminary Report*, Ministry of Transport, Malaysia, 03/2014, Dated 9 April 2014

135 *Aircraft Accident Investigation Report. China Airlines, Airbus Industrie A300B4-622R, B1816, Nagoya Airport, April 26, 1994*, Aircraft Accident Investigation Commission (Japan) AAIR-96-5, July 1996

136 *Aviation Accident Final Report*, NTSB File NYC96LA148, March 1999

137 K Abbott *et al.*, *The Interfaces Between Flightcrews and Modern Flight Deck Systems*, Report of the FAA Human Factors Team, June 18, 1996

138 *Aircraft Accident Report, Northwest Airlines, Incorporated, Boeing 727-25, N264US, Near Theills, New York, December 1, 1974*, NTSB AAR-75-13, August 1975

139 R L Newman, *Pitot-Static Accidents: A Tale of Two Airplanes*, Crew Systems Report TN-16-10, July 2016; available from Crew Systems, PO Box 25054, Seattle WA 98165

140 *Reporte Final Accidente Aereo Birgenair, Vuelo ALW-301, Febrero 06, 1996*, DGAC Report

141 "Erroneous Airspeed Indications Cited in Boeing 757 Loss," *Flight Safety Foundation-Accident Prevention*: 56 [10], October 1999, pp. 1-8

142 *G-STRZ Serious Incident*, AAIB Bulletin 12/2009, pp. 17-20

143 *Final Report on the Accident on 1st June 2009 to the Airbus A330-203 registered F-GZCP, Operated by Air France, Flight AF 447 Rio de Janeiro - Paris.* BEA Report f-cp090601en, Published July 2012

144 *Serious Incident Investigation*, GCAA Report AIFN/0005/2013, Dated September 2015

145 *Aircraft Accident Report, Los Angeles Airways, Inc., Sikorsky S-61L, N303Y, Paramount, California, May 22, 1968*, NTSB AAR-70-1, Dated December 1969

146 *Aircraft Accident Report, Atlantic Southeast Airlines, Inc., Flight 2311, Uncontrolled Collision with Terrain, An Embraer EMB-120RT, N270AS, Brunswick, Georgia, April 5, 1991*, NTSB AAR-92-03, April 1991

147 *Crashed During Approach, Boeing 737-800, Near Amsterdam Schiphol Airport, 25 February 2009*, Dutch Safety Board, Report Dated May 2010

148 Aviation Incident Final Report, NTSB File FTW88IA109, Dated March 25, 1991

149 *Aircraft Accident Report. In-Flight Separation of Vertical Stabilizer, American Airlines Flight 587, Airbus Industrie A300-606R, N14053 Belle Harbor, New York, November 12, 2001*, NTSB AAR-04-04, October 26, 2004

150 *Aircraft Accident Report. In-Flight Separation of Vertical Stabilizer, American Airlines Flight 587, Airbus Industrie A300-606R, N14053 Belle Harbor, New York, November 12, 2001*, NTSB AAR-04-04, October 26, 2004

151 *Aircraft Accident Report. In-Flight Separation of Vertical Stabilizer, American Airlines Flight 587, Airbus Industrie A300-606R, N14053 Belle Harbor, New York, November 12, 2001*, NTSB AAR-04-04, October 26, 2004

152 Docket No. FAA-2002-12461; amendment Nos. 1-54, 11-52, 60-1, 121-327

153 *Aviation Investigation Report Encounter with Wake Turbulence, Air Canada, Airbus A319-114, C-GBHZ, Washington State, United States, 10 January 2008*, TSB t-A08W0007, Published April 201-

154 *NTSB File Aviation Accident Final Report*, NRSB FileCHI08FA150

155 *Low-Energy Rejected Landing and Collision with Terrain. Perimeter Aviation LP, Fairchild SA227-AC, Metro III, C-GFWX, Sanikiluaq, Nunavut, 22 December 2012*, Transportation Safety Board of Canada Accident Investigation Report TSB-AIR-A12Q0216

156 *NTSB File Aviation Accident Final Report*, NTSB File DCA14LA060

157 "United Boeing 737-700 near Billings on Feb 17th 2014, Turbulence Injures 12,"

158 *Aircraft Accident Report, United Airlines Flight 232, McDonnell Douglas DC-10-10, Sioux Gateway Airport, Sioux City, Iowa, July 19, 1989*, NTSB AAR-90-06, November 1990

159 *Abnormal Runway Contact (ARC), Boeing 777-333ER, C-FTIW, Hong Kong International Airport, Hong Kong, 11 December 2018*, Air Accident Investigation Authority Report AAIA-05-2021

160 *Aviation Accident Final Report*, NTSB File DCA19CA167, Dated February 24, 2022

161 Lessons Learned from Civil Aviation Accident Home Page, http://lessonslearned.faa.gov/ll_main_ga.cfm?TabID=1&LLID_GA=1&LLTypeID=2

162 *Aircraft Accident Report. Beech Bonanza, N 3794N, Mason City, Iowa, February 3, 1959, CAB Report, File 2-0001, September 1959*

163 *NTSB File Aviation Accident Final Report*, NRSB File NYC99MA178, report date July, 6, 2000

164 *Final Report of the Accident Investigation, Flash Airlines Flight 604, January 3, 2004, Boeing 737-300 SU-ZCG, Red Sea off Sharm El-Sheikh, Egypt*, Egyptian Ministry of Civil Aviation Report su=f040103a, February 2021

165 http://skybrary.aero/accidents-and-incidents/b733-vicinity-sharm-el-sheikh-egypt-2004, accessed December 24, 2021

166 W C Ocker and C J Crane, *Blind Flight in Theory and Practice*, San Antonio: Naylor Publishing, 1932

167 R L Newman, " Profile of a Display," *Professional Pilot*, May 1975, pp. 32-34

168 *Aircraft Accident. Report, Rapid Descent into Terrain, Island Express Helicopters Inc., Sikorsky S-76B, N72EX, Calabasas, California, January 26, 2020*, NTSB AAR-21-01, February 2021

169 http://en.wikipedia.org/wiki/British_Airways_Flight_009, Accessed April 25, 2022

170 Aircraft Accident Final Report, NTSB File ANC90FA020, Dated June, 30, 1992

171 *Aircraft Accident Report, Runway Overrun and Collision, Southwest Airlines Flight 1248, Boeing 737-7H4, N471WN, Chicago Midway International Airport, Chicago, Illinois, December 8, 2005*, NTSB AAR-07-06, October 2007

172 *Landing Performance Assessment at Time of Arrival (Turbojets)*, FAA SAFO-06012, August 31, 2006

173 *Aircraft Accident Report, Delta Air Lines, Inc., Lockheed L-1011-385-1, N726DA Dallas/Fort Worth International Airport, Texas, August 2, 1985*, NTSB AAR-86-05, August 15, 1986

174 *Aircraft Accident Report, Northwest Airlines, Inc., McDonneell Douglas DC-9-83, N312RC Detroit Metropolitan Wayne County Airport, Romulus, Michigan, August 16, 1987*, NTSB AAR-88-06, May 10, 1988

175 *Aircraft Accident Report, Flight Into Terrain During Missed Approach, USAir Flight 1016, DC-9-31, DC-9-31, Charlotte/Douglas International Airport, Charlotte, North Carolina*, July 2, 1994, NTSB AAR-95-03, April 1995

176 *Aircraft Accident Report, In-Flight Breakup Over the Atlantic Ocean, Trans World Airlines Flight 800, Boeing 747-1305-20211, N93119, Near East Moriches, New York, July 17, 1996*, NTSB AAR-00-03, August 2000

177 *Review of Federal Program for Wire System Safety*, National Science and Technology Council, Final Report. November 2000

178 *Fuel Tank System Fault Tolerance Evaluation Requirements*, SFAR 88, effective December 9, 2002

Index